Analysis of Organic and Biological Surfaces

CHEMICAL ANALYSIS

A SERIES OF MONOGRAPHS ON ANALYTICAL CHEMISTRY AND ITS APPLICATIONS

VOLUME 71

A WILEY-INTERSCIENCE PUBLICATION

JOHN WILEY & SONS

New York / Chichester / Brisbane / Toronto / Singapore

Analysis of Organic and Biological Surfaces

PATRICK ECHLIN

University Lecturer
and
Fellow, Clare Hall
University of Cambridge

A WILEY-INTERSCIENCE PUBLICATION

JOHN WILEY & SONS

New York / **Chichester** / **Brisbane** / **Toronto** / **Singapore**

Library of Congress Cataloging in Publication Data:

Main entry under title:

Analysis of organic and biological surfaces.

(Chemical analysis, ISSN 0069-2883 ; v. 71)
"A Wiley-Interscience publication."
Includes index.

1. Surfaces (Technology)—Analysis.   2. Spectrum
analysis.   3. Microscope and microscopy.   4. Chemistry,
Organic.   5. Biological chemistry.   I. Echlin, Patrick.
II. Series.

QD506.A55 1984   547.1'3453   83-23585
ISBN 0-471-86903-1

Printed in the United States of America

10 9 8 7 6 5 4 3 2 1

CONTRIBUTORS

P. C. ROBINSON
Department of Ceramic
   Technology
North Staffordshire Polytechnic
College Road
Stoke-on-Trent
England

R. R. HOLM
Department of Applied Physics
Bayer AG, Leverkusen
Germany

DR. PETER FRITZ SCHMIDT
Institute for Medical Physics
University of Munster
Munster, Germany

II. K. WICKRAMASINGHE
Department of Electrical
   Engineering
University College London
London, England

I. PASQUALI RONCHETTI
C. FORNIERI
G. MORI
Institute of General Pathology
University of Modena
Modena, Italy

A. P. JANSSEN
Hamlyns House
Exwick, Exeter
Devonshire, England

J. A. PANITZ
Sandia National Laboratories
Surface Physics Division
Albuquerque, New Mexico

V. N. E. ROBINSON
Faculty of Applied Science
University of New South Wales
Kensington, New South Wales
Australia

H. F. DYLLA
Plasma Physics Laboratory
Princeton University
Princeton, New Jersey

J. H. ABRAMS
Department of Surgery
University of Minnesota
Minneapolis, Minnesota

BARRY G. COHEN
Research Devices, Inc.
Berkeley Heights, New Jersey

MARGARET S. BURNS
Departments of Opthalmology and
   Biochemistry
Albert Einstein College of
   Medicine
Bronx, New York

G. J. BRAKENHOFF
Department of Electron
   Microscopy and Molecular
   Cytology
University of Amsterdam
Amsterdam, The Netherlands

D. B. HOLT
Department of Metallurgy and
  Materials Science
Imperial College of Science &
  Technology
London, England

PETER G. T. HOWELL
ALAN BOYDE
Department of Anatomy and
  Embryology
University College London
London, England

NINA FAVARD
PIERRE FAVARD
Centre de Cytologie
  Experimentale
Ivry sur Seine, France

C. E. FIORI
C. R. SWYT
Division of Research Services
Biomedical Engineering and
  Instrumentation Branch
National Institutes of Health
Bethesda, Maryland

ANDREW W. WAYNE
Department of Haematology
King's College Hospital
London, England

O. HAYES GRIFFITH
Institute of Molecular Biology and
  Department of Chemistry
University of Oregon
Eugene, Oregon

HANS JØRGEN G. GUNDERSEN
Second University Clinic of
  Internal Medicine
Electronmicroscopic Laboratory
  for Diabetes Research and
  Institute of Experimental
  Clinical Research
University of Aarhus
Denmark

RICHARD P. C. JOHNSON
Department of Botany
Aberdeen University
Aberdeen, Scotland

DOUGLAS A. ROSS
The Department of Electrical and
  Electronic Engineering
Queen Mary College
London, England

PATRICK ECHLIN
The Botany School
University of Cambridge
Cambridge, England

DALE E. JOHNSON
Center for Bioengineering
University of Washington
Seattle, Washington

MICHAEL L. KNOTEK
Sandia National Laboratories
Albuquerque, New Mexico

J. S. PLOEM
Department of Histochemistry
  and Cytochemistry
University of London
London, England

PREFACE

A wide range of analytical techniques are now available for the characterization of inorganic, organic, and biological materials. Not all these methods are applicable to organic samples, either because of limitations due to specimen preparation and examination and/or the damaging effect some of the techniques can have on organic samples. This book collects together 24 analytical techniques and methods that can be used to give a total characterization of the surfaces of organic and biological samples. Each chapter is complete in itself in that it explains the basis of the analytical method, how it can be applied to organic material, and the precautions that must be observed during the experimental process and in the analysis of the results. No single method will give all the answers; a judicious combination of techniques will allow quantitative information to be obtained from organic samples at the atomic, molecular, and morphological level.

At first glance it would appear that some methods have been omitted from this book. This is quite deliberate, for some of the methods are now commonplace and already adequately described in the scientific literature. Other methods have not been included either because they are too destructive or simply because they are at present not applicable to organic samples. The 24 chapters in this book, each written by an acknowledged expert in the field, contain methods and techniques that allow a proper analysis of organic and biological surfaces. It is recommended that the introductory chapter is read first because it gives an overview of the whole subject. In addition, it mentions others methods which, although not at present being used to analyze organic surfaces, offer considerable potential for studying this type of specimen.

PATRICK ECHLIN

Cambridge, England
June 1984

vii

CONTENTS

ix

INTRODUCTION

Although we live in a complex, three-dimensional world lit from above, our cognition of this environment takes place by a process of visual perception of interlocking two-dimensional surfaces. Through experience, we can recognize patterns and shapes, and because we possess the remarkable facility of binocular vision, we are able to synthesize a series of planar surfaces into familiar three-dimensional objects. In many instances, and in particular in microscopy, we are able to codify and recognize new images that we have never seen before. With a few exceptions we are unable to see below the surface of natural objects, yet our experience tells us that many of the subsurface features of an object are important in determining the surface characteristics that are presented to us. Organic surfaces, and in particular biological surfaces, are heterogeneous and generally more complex than inorganic surfaces. The surfaces of a block of metal, a crystal of an inorganic salt, or a piece of concrete appear the same from whichever direction they are viewed, although important submicroscopic and crystallographic differences do exist. If we section or fracture these homogeneous materials, the internal surfaces we reveal are, at a macroscopic level, identical to the external surface. This is not true of many organic materials, and much of their functional significance can be related to the internal surfaces that are normally hidden from our view.

An analysis of organic surfaces must necessarily look below the surface and it is important to consider some of the ways we can expose internal surfaces without introducing artifacts during the preparative procedures. These manipulations are particularly important in the examination of biological surfaces, and great care must be taken in converting or replicating the specimens into a form that is conducive to the alien environment of the system used to image and abstract the information.

It is convenient to consider the properties of surfaces from the point of view of their *morphology* and their *composition*. The morphological features of a surface may be divided into an examination and measurement of the microstructural features (topography) and a quantitation of their interrelationships (topology). This morphological information is essentially two-dimensional and is frequently represented as a series of picture points along the X and Y coordinates. The real surface, as seen by our eyes, and

1

images of the surface (photographs, drawings, maps, etc.) only become three-dimensional because our previous experience in pattern recognition tells us that the various surface features are separated along an axis normal to the mean depth of the surface. This recognition is usually quite easy in the real world but may become problematic when we examine images of unfamiliar objects, for here the third dimension may not be directly observed but only perceived. In these circumstances it is essential to subject the planar surface projections to stereopair analysis, which makes it easier to perceive the distances between points on a nonplanar surface. Stereogrammetric techniques are the only way in which accurate measurements may be made on nonplanar surfaces. The chapter by Howell and Boyde considers these matters in some detail.

The compositional analysis of a surface must necessarily delve below the surface, although it could well be argued that the wavelength contrast (color), albedo (solar reflectivity), and specular features (shininess) can give us compositional information about an object (the shiny yellow color of gold is a good example). The depth analysis may be confined to the first few atomic layers of the surface or penetrate several micrometers below the surface. The information gained can be related to the elemental, molecular, and even macromolecular composition of the surface.

Although it is convenient to consider morphology and composition separately, they are in reality very closely linked. This link sometimes puts very severe restraints on the effectiveness of the imaging systems, for while one particular technique may give accurate compositional information the very act of quantitation may destroy, and in most instances the process of preserving and subsequently observing may only be achieved at the expense of the natural composition of the object. This paradox, "the very act of observing introduces artifacts," is alas one of the problems we have to accept in the microscopy and analysis of organic materials.

The 24 chapters in this book are concerned with the in situ analysis of organic samples, as they usually present greater problems in preparation, examination, and analysis than do their inorganic counterparts. Biological specimens present the most severe challenge to the microscopist and analyst, because they are nearly always highly hydrated. The chapter by Robinson considers the procedures whereby hydrated specimens can be examined directly by the microscope. The presence of water is less of a problem in the other types of organic samples, such as natural and artificial polymers, and elastomers, and the low water content and structural characteristics of some of the more robust plant and animal products (chitin, wood, bone, teeth, etc.) are such that they require little preparation prior to examination in the microscope.

ANALYSIS OF SURFACE MORPHOLOGY

The structural features of the specimen surface may be readily localized using a beam of photons or electrons. Other imaging systems exist e.g., those based on sound waves, x rays, and ions, that give a different view of surfaces. *Ultrasoft* x-ray microscopy, although not a new technique, has recently become a subject of renewed interest (Parsons, [1]). The x-ray imaging depends on the differential absorption of x rays in the wavelength range 1–10 nm, by the various components in the specimen. Recent work by Rudolph et al. [2] has shown that it is possible to obtain high resolution (70 nm) images of organic material using zone plate optics in conjunction with a high x-ray flux. Alternatively, the image need not be viewed directly but a replica is formed on a photoresist by the x rays that pass through the specimen. In areas of high mass density the x rays are absorbed and these regions appear in high relief after the photoresist is chemically developed and examined in a scanning electron microscope. The spatial resolution varies but can be as good as 5 nm. The advantages of the system are the reduced specimen damage and the ability to examine hydrated samples. This procedure is quite distinct from *direct x-ray microscopy,* which although giving information from wet specimens up to 5 μm thick can only produce images up to a magnification of $\times 200$.

The advantages of *acoustic microscopy* are discussed in the chapter by Wickramasinghe. Acoustic waves are reflected or deflected by variations in specimen density on unstained specimens immersed in water. The resolution is consequently no better than 2.0 μm. By using cryogenic liquids instead of water it has been possible to obtain a spatial resolution of about 100 nm.

The limited resolution of *light microscopes* is the only restraint that can be put on this form of imaging system. A wide range of light microscope systems is discussed in separate chapters by Robinson, Ploem, Wayne, Cohen, Brackenoff, and Johnson. If we use light in the visible range, we are able to obtain information about the surfaces of all organic materials (except those that are photosensitive) and of course living, physiologically intact specimens. The information is presented to the observer in both wavelength and amplitude contrast, albeit at somewhat limited spatial resolution (~ 100–200 nm). The process of information transfer does not appear to damage or unduly perturb the organic specimen and we assume we are observing the specimen close to its natural state.

To increase the spatial resolution of the imaging system, we must use an illuminating system of much reduced wavelength. High energy electrons are a convenient illuminating and imaging system and although the spatial resolution is decreased to a few tens of nanometers, the use of electron beams creates serious problems for many organic specimens. The short mean free path of high energy electrons in air necessitates the use of a high vacuum inside the microscopes. Such low pressures put severe restraints on many organic samples, particularly those that contain volatile material and an appreciable amount of unbound water.

The beam of electrons may also cause thermal and radiation damage to specimens, and organic samples are more susceptible to these deleterious effects than are inorganic samples. A possible way of avoiding these radiation effects would be to use a *mirror electron microscope* (Gvosdover and Zel'dovich, [3]). The primary electron beam reverses its direction and is reflected by an equipotential very close to the specimen, which is at a slightly negative bias voltage. The spatial resolution is only 100–200 nm but it does permit surface inhomogeneities to be imaged. There is some evidence that ion currents may develop at the surface of the sample, which could damage organic specimens.

As microscopists, we are familiar with the basic instrumentation, optical pathways, and mechanisms of image formation of microscope systems such as TEM, HVEM, SEM, and STEM. Suffice it to say that to obtain morphological information about surfaces it is necessary either to collect a signal reflected from a surface or one transmitted through a thin section of the surface. The familiar *scanning electron microscope* and its more recent derivative, the *low-loss electron microscope,* can provide a wealth of morphological detail at a resolution between 2 and 8 nm by collecting electrons reflected from the surface. The low-loss image is formed by collecting electrons that have lost only a small amount of energy (less than 0.1%) while being backscattered from a solid specimen. The primary electrons strike the specimen at oblique incidence and an energy filter is used to collect the electrons that have undergone the smallest deflection and the least energy loss. The microscope described by Wells et al. [4] and Broers [5] collects "low-loss" electrons that emerge from the bottom of the condenser-objective after they have been scattered from the surface of solid specimens. The specimens are placed approximately at the center of the lens gap, which means the lens operates at a focal length of 1 mm. The lens aperture is 30 μm in diameter, which yields a beam half-angle of 1.5×10^{-2} rad. The beam current is 20 pA and the beam diameter is 1 nm. The microscope can give very high resolution (~3 nm) images of solid surfaces. (A detailed chapter on scanning electron microscopy has deliberately not been included in this book because a number of good texts are available on the subject; see, for example, Goldstein et al., [6].) A related technique is *photoemission microscopy* discussed in the chapter by Griffith, where the sample itself is the source of electrons that are generated by bombarding the sample with UV photons. Small changes of topographic detail can be readily discerned at a spatial resolution of between 5 and 20 nm. This technique can also give chemical information about surfaces.

The transmission electron microscope (TEM) and the scanning transmission electron microscope (STEM) reveal information contained within thin sections of surfaces, or in replicas derived from surfaces, at a spatial resolution of 0.2–3 nm. As the chapter by Pasquali-Ronchetti shows, with a conventional TEM operating in the 80–120 keV range it is necessary to limit section thickness to between 50 and 80 nm, and thinner still if high resolution is required. With a high voltage electron microscope, Carasso shows in her chapter on this technique that the thickness can be increased up to 1–3 μm but at the expense of resolution.

There is some doubt regarding the *amount* as distinct from the *quality* of information that can be obtained from a very thin section cut across an organic surface. A single section, although informative, is generally unrepresentative and it is necessary to reconstitute several hundred planar images into a three-dimensional model before true details of the surface become apparent. Complementary procedures may be found in morphometry and stereology in which the profiles and topological parameters of thin sections are reduced to mathematical terms that may be processed to give dimensions that may be related to the bulk material from which the section was cut. Quantitative relationships exist between the average dimensions of inclusions within a bulk sample and those of their profiles in sections. When certain conditions are met, the sum of profiles in a unit area of section may be quantitatively related to the sum of the inclusions contained in a unit volume of the solid. The chapter by Gundersen should be consulted for details of these procedures and for some of the ambiguities that may arise from three-dimensional interpretation of projections of thin sections. The process of image formation using a reflected signal and/or a transmitted signal is unaffected by the nature of the specimen, although the quality of the image, that is, the signal-to-noise ratio, can show considerable variation.

The fidelity of detail and quality of information we can obtain from organic surfaces are dependent on the way the specimen has been prepared prior to examination. Organic samples are usually composed of low atomic number polymers that are thermolabile and radiation sensitive, frequently contain substances with a high vapor pressure, and, in the case of biological material, contain a substantial amount of water. In addition, one of the many special features of living material is that it has extensive internal surfaces that are quite different in structure and function from the external surface that forms an interface with the environment. Much effort has been put into discovering ways to preserve these internal surfaces so that they may be examined and analyzed by different forms of electron beam instrumentation. A multitude of papers exists giving details of specimen preparation for both biological and organic samples. A comprehensive discussion of these techniques may be found in the recent book by Goldstein et al. [6]. Briefly, the preservation involves chemical stabilization of the macromolecular architecture, careful removal or immobilization of any water and, under some circumstances, the introduction of high atomic weight elements to specific sites in the sample. The perfect preparative technique should aim to transform the thermodynamically unstable organic matrix into a heat- and ionizing-radiation-resistant inorganic replica. For scanning microscopy one should aim to preserve the spatial relationship of the organic matrix and then proceed to cover the surface with a thin layer of metal in order to increase the electron emissivity of the sample.

Surprisingly enough, these purely chemical procedures result in the preservation of organic structures of consummate functional excellence and beauty. However, in the case of living material, this preservation is only achieved at the expense of up to 99% of the contents of the cells being lost during the process. This loss is of great concern where the chemical (elemental and molecular) composition is the main objective of the analytical procedures. Following preservation, the internal surfaces of the organic material may be exposed by a number of means, but only after the mechanical strength and stability of the sample has been enhanced. This may be achieved either by replacing any fluids by liquid resins that are then polymerized to a solid or by converting the low viscosity phases of the sample to a high viscosity form (solid) by quench cooling. The strengthened material may now be sectioned, fractured, or replicated to reveal the internal surfaces.

It should be apparent that the validity and usefulness of information obtained from organic material examined and analyzed by any form of high energy beam instrumentation is vitally dependent on the preparative procedures. Of all the methods that have been devised, low temperature techniques show the greatest promise, particularly where an accurate chemical characterization is to be made of the sample. Cryotechniques, if properly and conscientiously applied, result in minimal disruption of organic and biological material, leaving the soluble constituents of the sample more or less in their natural location. Low temperature methods are an easy and convenient way of increasing the mechanical strength of many plastics and elastomers. Rubber, for example, can be easily thin-sectioned at 133 K and polyethylene all too readily fractures when cooled to 77 K. For the biologist, low temperature methods have been used to produce high resolution images of macromolecular assemblages within and on cell membranes, by making a replica of the rapidly frozen sample. This freeze-fracture technique only allows an examination of a surface replica, whereas it is now possible to examine *and analyze* the frozen surface directly by maintaining it at a low temperature inside the microscope. The chapter by Echlin discusses these procedures in some detail.

It is convenient to record all the morphological information we see in microscopes in the form of photographic images. We do this as a consequence of our fallibility as we can more readily convince our colleagues about the existence of a new surface feature if we can show them a picture, photograph, or map.

It is opportune to consider briefly here the effectiveness of the ways by which the observer conveys information seen in the microscope to other observers. There are two methods by which this information transfer can take place: the verbal or literal and the visual. With little conscious effort, we are able to recognize patterns and an immense amount of scientific information has been gathered by the direct visual recognition of similarities or analogues between pictorial patterns. This process of intersubjective pattern recognition is, as Polanyi [7] points out, a fundamental element in the creation of all scientific knowledge. But how true is the maxim "seeing is believing." This initial part of the information transfer process must be determined to a large extent by the primary observer, who is responsible for distinguishing fact from preparation-induced artifact and datum points from instrumental malfunction. Having, to the best of his or her ability, validated the acuity of the visual observation, the microscopist frequently takes a representative photograph of the observed surface. A micrograph is an important item of scientific observation, as it accurately locates observed details, as well as accentuating the consensual* elements in the image. But the micrograph, although undeniably an accurate representation of the surface image, is of limited usefulness. It may be limited by transient accidents of illumination, by absorbance and reflectivity, and by spatial effects of color. Although representative of a small area of the surface at a particular instant in time, it may well not be representative of the whole surface. This is a particularly acute problem with biological material, which is immensely variable, and a better graphic representation of a surface would be achieved by means of a drawing or a map. A drawing is able to condense information that may be read at normal optical resolution and is an expression of structural understanding. A drawing arises from a unique combination of

---

* Ziman [8] distinguishes between *consensible* messages, which are unambiguous statements describing an event and which may contribute to the general body of scientific knowledge, and *consensual* statements that have been fully tested and are universally agreed.

the precision of the hand, eye, and mind and unlike the photograph is timeless. In this way it should be possible to summarize the information gained by examining the images of many surfaces of the same object and give a representative picture. Alternatively, the information in the photographs could be codified either as a series of gray levels delineating areas of interest or could be transformed into a wider range of wavelength contrast levels. Alternatively, a series of measurements may be taken from stereopair micrographs to emphasize spatial characteristics such as shape connection and intersections. This information could be stored and added to, in this way giving some measure of the variability one might expect to see in a single micrograph.

The use of computers in microscopy is now becoming more popular particularly where the image is produced as a result of digital scanning methods. Simple analogue processing techniques such as black level subtraction, Y modulation, signal differentiation, artificial color coding, and contrast variation enable the operator to enhance the information content of the image but are limited when compared to digital processing. Provided one has a sufficiently high storage capacity ($\sim$1 megabyte per micrograph [Smith [9]), the photographic process can be eliminated, and the image stored on a magnetic disk. This stored image may be modified in light of information obtained from subsequent images and a hard copy representative image produced. This image could be in the form of a graph, a map, instructions for a three-dimensional model, or a photograph.

Although our consciousness is dominated by visual images, we tend to communicate scientific information by linguistic means. This requires a very precise use of language but can in some circumstances be a more accurate way of communicating information than a purely visual image.

A single photograph can do no more than convey basic information from observers to their colleagues. Hanson [10] states that "a picture is worth a thousand words." Most people forget the rest of the quotation, which continues "a picture is a thousand times less specific than a short sharp statement. A statement can supply a focus for the attention that is different in type from anything generated by confrontation with a picture." We need to combine the visual and the literal. How might we best convey the images we see in microscopes? The photograph is a convenient, high density information storage medium and relatively free of bias but with the limitation of only recording an unrepresentative two-dimensional planar projection of a surface that invariably has a third dimension. The photograph is an intermediary in the process of information transfer; it is not the end product. We should therefore use micrographs, together with the precise verbal information we record when examining the image of the real object to construct the permanent baseline information either as a model or a map, and use this to convey new information to our colleagues or to store in the scientific archives.

ANALYSIS OF SURFACE COMPOSITION

The total characterization of a surface involves more than just obtaining high resolution topographical and topological information along the X and Y coordinates and the chemical identity of the surface layer of atoms. It should also include a coordinated in-depth analysis along the Z axis at as high a spatial resolution as may be obtained with a particular form of instrumentation. A surprisingly large number of techniques, based on exciting the sample with photons, electrons, or ions, are available for the analysis of organic surfaces.

TRANSMISSION ELECTRON MICROSCOPY

As Pasquali-Ronchetti shows, the TEM may be used to give analytical information by virtue of the increased electron scattering from high atomic number inclusions within the low atomic number matrix that characterizes most organic material. Unlike visible photon radiation, which is sensitive only to chemical bond structures, electron radiation is sensitive only to elemental atomic number. This seriously limits the ability of TEM to localize directly the molecular and macromolecular species that are the principal constituents of organic material and also makes it virtually impossible to localize directly the elements (C, H, O, N, S, P) that form the bulk of organic compounds. The direct localization of molecular species and chemical bonds can be carried out albeit at reduced resolution using the light microscope by virtue of natural pigments, that is, chlorophyll in leaves, vital and nonvital dyes, and the natural fluorescence of some molecules when irradiated with UV light. The chapters by Ploem, Wayne, and Cohen provide details of such methods.

Occasionally, high atomic number material, usually as heavy metals, may be present as a natural deposit in organic samples. Under these circumstances the TEM can only indicate that a high atomic weight particle is present; it cannot give information on the precise nature of the inclusion. If the inclusion is in a crystalline form, it may be identified by means of its characteristic *electron diffraction pattern* using the selected area diffraction device fitted to most modern transmission microscopes.

It is more usual to introduce heavy metals into a specimen artificially by manipulating known chemical or enzymatic reactions so that the end product is a heavy metal deposit. Thus in biological systems alkaline and acid phosphatases may be localized by means of a lead deposit, acid mucopolysaccharides by means of iron, and nucleic acids as uranyl groups. Alternatively, it is possible to precipitate low atomic number elements as insoluble heavy metal salts. Silver salts can be used to precipitate chloride, and antimony compounds can be used to localize monovalent and divalent cations, such as Na^+, K^+, Ca^{2+}, and Mg^{2+}. In addition, sites of specific molecular activity may be localized in living tissue by the techniques of autoradiography and immunochemistry.

In the former technique radioactive tracers are incorporated into biological material via a known metabolic pathway. The sites of radioactivity are subsequently revealed by exposing the tissues to a photographic emulsion, and the developed silver grains may be related to specific morphological features of the cell. The immunochemical methods take advantage of the affinity of antibodies, tagged with a heavy metal compound such as ferritin, for specific antigenic sites on the cell surface. These five approaches of positive staining, enzyme cytochemistry, precipitation, autoradiography, and immunochemistry can only be used qualitatively because of the difficulty of objectively assessing differences in electron density in an organic sample.

THE SCANNING ELECTRON MICROSCOPE

The SEM can give much more chemical information about a specimen than TEM if only because of the multiplicity of signals that may be obtained from the specimen during its interaction with the primary electron beam. The book by Goldstein et al. [6] reviews the available methods.

The most familiar of the specimen beam interactions are the *secondary electrons,* which have an energy of between 0 and 50 eV. They arise from within the first 10 nm of the surface and are the principal source of the signal used to provide morphological information about surfaces using the topographic contrast mechanisms. Under special circumstances the secondary electrons may also be used to obtain chemical information about the specimen. On a highly polished flat surface (where there is minimal topographic contrast) there is sufficient difference in atomic number contrast between elements that are widely spaced in the periodic table to separate them one from another. For example, it has been possible to localize copper inclusions (Cu = 29) in a highly polished aluminum matrix

(Al = 13) by virtue of differences in secondary electron emission at 5 and 20 keV. This approach has not been applied to organic samples as it is difficult to obtain a smooth, highly polished surface.

In a rather roundabout way, the topographic contrast mechanism has been used to obtain chemical information about biological surfaces. Sites of differing antigenic specificity may be localized on a biological surface using standard immunologic techniques. Instead of using a fluorescent marker, a morphologically recognizable marker is attached to the antibody and the specific antigenic sites may be localized by recognizing the structural marker. By using the bacteriophage T_2, it has been possible to map out specific antigenic sites on the surface of red blood cells. Silica spheres and polystyrene latex spheres of a known and consistent diameter have been used for the same purpose. An alternative approach is to use an organic marker of known molecular dimensions.

Because electron radiation is sensitive to elemental atomic number it is more convenient to use the *backscattered primary electrons* for simple qualitative analysis. The backscattered electrons only suffer a small energy loss during interaction with the specimen, but the amount of scattering is, to a first approximation, a function of atomic number. The backscattering coefficient and atomic number separation are also functions of surface topography, incident beam energy, and specimen tilt angle. In a flat specimen the backscattering coefficient increases with atomic number and the best atomic number separation that has been achieved is aluminum ($Z = 13$) in a plastic matrix (mainly carbon $Z = 6$), two elements only seven atomic numbers apart (Goldstein et al., [6]). A more usual application in organic samples is to distinguish iron ($Z = 26$) and silver ($Z = 47$) in an organic matrix that has an average atomic number of seven. By combining this method with the wet-chemical preparative procedures that can localize specific chemical groups by linking them to high atomic number materials, it is possible to provide useful quantitative analytical information about biological surfaces.

An alternative to using the backscattered signal is to make use of the *specimen current* signal. This particular mechanism will have a limited application with organic samples because of their inherently low conductivity.

The scanning electron microscope can also give information about the molecular and macromolecular composition of an organic surface by measuring the *cathodoluminescence* that arises from the interaction of many materials with a high energy beam of electrons. For efficient observation of the signal the sample should be transparent in the visible light region— a property of thin sections of most organic materials. The emitted signal may be categorized either as autoluminescence, which occurs naturally in some materials, or staining and cathodoluminescence, which occurs when appropriate fluorescent probes are added to the specimen.

The wavelength of the emitted photons is characteristic of the particular molecular species, and provided care is taken in measuring the wavelength and energy of light, the method can be used semiquantitatively.

It can be used to localize artificially introduced fluorescent dyes that have an affinity for particular chemical groups. Quinacrine dihydrochloride has been used as a probe in the examination of cells in cervical smears, and acridine orange has been used to localize nucleic acids.

As the chapter by Holt shows, cathodoluminescence would appear to be an ideal method to examine organic surfaces, particularly when near infrared detection systems are developed. The difficulties appear in the application of suitable fluorescent probes that are not quenched by the electron beam, and the fact that the cathodoluminescent yield is usually rather low. The latter problem may be overcome by using large probe sizes and increased probe current, but only at the expense of specimen damage and diminished spatial resolution due to the increased size of the interaction volume. Further improvements in cathodoluminescence can be achieved by cooling the photomultiplier to 250 K, and the beam damage effects are lessened and light emission increased by holding the specimens at 4.5 or 77 K. Holt considers that the development of a liquid helium stage for cathodoluminescent studies is of prime importance. The very low temperatures will increase the luminescent efficiency, decrease specimen damage, and allow the spectra to be revolved as sharp lines rather than broad curves. Electrons scanned over a sample may also be used to study surfaces by operation in a scanning electron stimulated desorption mode. The chapter by Dylla considers this mode of operation in some detail and the chapter by Knotek discusses the application of electron- and photon-stimulated desorption to the study of organic surfaces.

MICROANALYSIS

It is only when we turn to *x-ray microanalysis* that we begin to realize the analytical potential of electron beam instruments. For general discussion see Goldstein et al. [6]. X rays are produced when an electron beam of sufficient energy strikes any material. The x rays that are produced have a characteristic energy and wavelength for each and every element, and this feature is the basis of the x-ray analytical technique. The x rays may be collected and measured using a wavelength diffracting spectrometer

(WDS) and/or an energy dispersive spectrometer (EDS). The former method has a high analytical resolution and can readily separate closely related elements, and electron probes are usually equipped with several wavelength spectrometers each adjusted to simultaneously analyze one of the elements of interest. If only one spectrometer is available, then each of the elements of interest have to be analyzed in sequence. The energy dispersive spectrometer has a lower analytical resolution, but is able to analyze and display all the elements at one time, which is particularly useful if one wishes to obtain information relating to the total elemental (above $Z = 8$) composition of the specimen. X-ray microanalysis gives information about the distribution and concentration of elements, and with a few minor exceptions makes no distinction between elements that are covalently bound within the organic matrix, in an ionized state, or in the form of a crystalline inclusion. It must also be remembered that there is frequently a disparity between the number and type of x rays generated within the specimen, emitted from the specimen, and measured by the detecting systems. This disparity is due to problems of secondary fluorescence, absorption of soft x rays, and production of spurious signals from the immediate environment surrounding the site of analysis. Much effort is being made to refine the quantitative procedures and these are discussed in great detail in the chapter by Fiori. We are now at the point where it is possible under some circumstances to detect 0.1 attograms (10^{-19}) of an element with an analytical spatial resolution of 25 nm. The spatial resolution is greatest in thin sections and quickly decreases if the analysis is carried out on bulk samples. Most elements of biological interest can be analyzed using x-ray microanalysis but it becomes progressively more difficult to carry out quantitative analysis on light elements, and sodium ($Z = 11$) is the practical lower limit.

X-ray microanalysis may be carried out using a wide range of electron beam instruments, and it is usual to find one or both types of x-ray detector attached to a scanning or a transmission electron microscope, rather than to find the analysis being carried out using a purpose built and dedicated electron microprobe. X-ray microanalysis may usually be carried out at different levels of sophistication. In its simplest form it may be used in conjunction with a scanning electron microscope to map the distribution of elements within a section of a sample or on its surface. It can be used to measure relative concentrations of elements in different parts of the specimen and, in conjunction with the appropriate standards, can be used to measure absolute concentrations.

The final quality and significance of x-ray microanalysis is dependent on two factors. It is entirely dependent on the specimen preparation procedures whose prime aim must be to retain the elements in their natural position and natural physiological concentration in the tissues. Low temperature preparation techniques offer the best prospect of achieving this goal, and the chapter by Echlin reviews many of the methods currently being used to prepare specimens for low temperature microscopy and analysis. Unfortunately, compromises have to be made, as our preparative procedures sadly lack the sophistication and potential accuracy of the analytical instruments. The quality of the analysis also relies on the way the raw x-ray data are handled and it is here that considerable advances have been made. Thanks to microprocessors and sophisticated computer programs it is now possible to deconvolute the complex spectra that emerge from an x-ray microanalyzer and express the data as elemental concentrations, free from the influence of overlapping spectra, background signal, and instrumental variations. The chapter of Fiori gives further details of such methods.

One of the limitations of x-ray microanalysis is the difficulty associated with the analysis of light elements. These limitations are to some extent compensated for by using *electron energy loss spectroscopy* (EELS) described in the chapter by Johnson. In this technique one detects and measures the changes in the angular and energy distribution of the electrons that have passed through the specimen. From the point of view of microanalysis the most interesting portions of the energy loss spectra are in the 0–10 eV range, and at inner-shell excitations occurring above 200 eV loss as these regions provide the most information about the elemental composition of the specimen. By placing an energy loss spectrometer below the specimen one may measure the angle through which an electron is scattered and its energy relative to its incident energy. It is possible to collect between 20 and 50% of the inelastically scattered electrons, which compares favorably with the 1% collection efficiency of an EDS x-ray system. The collected spectra are characteristic of the material being irradiated, and provided the probe diameter of the incident beam is sufficiently small ($\sim$3–10 nm) and the specimen is sufficiently thin, high spatial resolution analytical information may be obtained from organic material, particularly for the light elements ($Z = 2$–20). Under the appropriate conditions, unlike x-ray microanalysis, EELS can also provide information about the various bonding states as well as measuring the amount of an element in the sample. This aspect is of particular importance in biological analysis where one frequently wishes to measure the ratio between bound and free ions. Electron spectroscopy can be used to greater effect in the chemical analysis of organic samples as shown in the chapter by Holm.

The third major technique involving the use of electrons that may be used for analyzing surfaces is *Auger electron spectroscopy* (AES), discussed in the chapter by Janssen. When a beam of electrons of sufficient energy strikes a specimen, electrons may be ejected from atoms, leaving orbital vacancies behind. Auger electrons are the result of deexcitation processes involving these vacancies and electrons from other shells and a reemission of an electron to carry away excess energy. Auger electrons from low atomic number elements have energies in the range from 30–50 eV up to 2 keV and the spectra have much narrower peaks (5–20 eV) compared with edges in electron energy loss spectra.

The production efficiency of AES is complementary to the x-ray signal, but the AES signal tends to be smaller because of the shallow escape depth. Herein lies the analytical effectiveness of AES because the electrons that are emitted have a very short mean free path and are all derived from the first few atomic layers of the surface. The spatial resolution of AES is about the same as x-ray microanalysis in thin sections but is poorer than that found in EELS. Whereas the depth penetration of EELS and x-ray microanalysis is limited by the section thickness, in AES the analysis depth is of the order of 0.5–2.00 nm and the analyzed volume is 10^3 smaller than that normally seen in x-ray microanalysis.

The analytical potential of Auger spectroscopy in the investigation of organic surfaces is limited by the nature of the sample, and the lack of adequate specimen preparation still remains the most severe constraint to the full use of this procedure. By maintaining the specimen at low temperatures, the deleterious effects of contamination and beam damage will be diminished, but the progressive loss of low atomic number species and volatile components can only be prevented by chemical stabilization or by working at very low temperatures.

The *photoemission microscope* (PEM), or photoelectron microscope as it is sometimes called, has been in use for some time in the examination and analysis of inorganic materials and metals. As the chapter by Griffith shows, the method is particularly useful for the analysis of organic surfaces as the image contrast is due to differences in ionization potential of molecules, unlike CTEM, where the contrast is a function of differences in scattering cross section. In the CTEM there is little difference in the contrast between molecules of similar elemental composition; in the PEM very large contrast differences can be observed.

In the photoemission microscope the sample is maintained at a high negative potential and irradiated with an intense beam of UV light. This causes the surface to emit electrons that are accelerated to high velocity, passed through an electron lens system, and imaged on a fluorescent screen. In many respects PEM is similar to fluorescence microscopy and has been used to advantage in analyzing the photoelectric properties of organic compounds. The photoelectron quantum yields are low and extreme topographical detail can produce image distortions. The lateral resolution is between 10 and 20 nm but the excellent depth resolution of between 1 and 5 nm means that it is possible to carry out in situ analysis of the molecular chemistry at the surface of a specimen. Sample preparation is limited to simple macromolecular stabilization and dehydration and it is not necessary to increase the mass density of the organic components by the addition of heavy metals. Clean samples are essential as the image quality is sensitive to contamination. A useful extension of PEM would be to carry out the investigation of freshly fractured surfaces of frozen hydrated samples maintained at low temperatures inside the microscope.

The very high sensitivity to surface relief even on flat specimens due to the low kinetic energy of the emitted electrons means that the PEM can be used to provide morphological details of surfaces. However, the real potential of the technique would seem to lie in its ability accurately to localize specific molecules on an organic surface.

PEM has also been used to great effect in cytochemical studies of intact biological tissue that has been infiltrated with vital stains. The material is subsequently processed for conventional TEM, but without using heavy metal fixatives and stains. An additional advantage of this technique is that it is possible to use large (10 mm^2), semithin (100 nm) sections, which enable the observer to follow transport pathways over large areas of tissue. Advantage can also be taken of the natural emissive properties of biological products under UV irradiation.

In addition to the analytical techniques based on electron irradiation and electron emission, there are methods that depend on other forms of high-energy beams. One of the disadvantages of x-ray microanalysis is the large background of low energy x rays, which can limit the minimum concentration of trace elements that may be detected to between 100 and 200 ppm, even using wavelength spectrometers. This advantage can be overcome by using a beam of protons rather than electrons to irradiate the specimen. Protons are approximately 2000 times heavier than electrons and produce a much smaller x-ray background. Martin [11] has recently reviewed the instrumentation and application of *proton microprobes* to a wide range of inorganic and organic specimens. A 2–10 μm beam of 1–4 MeV protons bombards a specimen kept either in vacuum or in air, and the emitted x rays are collected on SiLi detectors. The excellent peak-to-background ratio of the x-ray spectra allows identification of trace elements down to the 1 ppm level at a spatial resolution promising results on the distribution of elements in organic samples, in particular in biological specimens.

A derivative of the proton microprobe is the *photon microprobe* (Sparks, [12]), which used synchrotron radiation as the excitation source for x-ray fluorescence. For a given beam energy the ionization cross section for x-rays is much larger than those for either protons or electrons with a consequent improvement in the signal-to-noise ratio and a minimum detection of 50 ppb. This is very much a developing technique but promises to be a valuable technique for microanalysis. The high-energy particle beams can also be used to induce nuclear reactions in the target material which can be identified by the gamma emission. Elastic scattering can be used to identify elemental distributions across the surface to a depth of 10 nm. Nuclear reactions induced by He^3 ions are suited to light elemental analysis, and it has for example been possible to identify the location of a protein in a sample on the basis of the nitrogen distribution (Pierce, [13]). Because nuclear reactions give a measure of the interaction of the incident ion beam with a specific isotope, the method can also be used as a means of isotope analysis. One unique advantage of the particle probe methods is that the high beam energy allows organic material to be examined *al fresco* and not within the high vacuum of the instrument column.

The stability of organic specimens irradiated by high energy particles is still an open question, but would appear to be no worse than the problems experienced in x-ray microanalysis although the additional radiation damage may easily cause volatile components to be lost from the specimen. One disadvantage to using high energy particles for the in situ analysis of organic surfaces is the depth of penetration below the surface. This is not a problem with sectioned material, but would give misleading results on the surfaces of bulk samples.

Martin [11] has calculated that the power dissipation of a 2 MeV proton beam is approximately 1.5 kW/mm³ and this energy must be removed in order to prevent a bulk sample from vaporizing. 2 MeV protons will pass through about 50 μm of a sample without significant scattering. While this is not a problem with sections of material it would mean that the emitted x-ray signal would represent an average of the elemental concentration over an appreciable depth. On balance it would appear that the proton probe analysis is a very sensitive technique for analyzing low concentrations of elements in thin ($\sim$2–5 μm) surfaces or in sections of bulk material, but it is a less useful method for the *in situ* analysis of the surface of organic material.

A related technique discussed in the chapter by Burns-Bellhorn is *secondary ion mass spectrometry* (SIMS) in which a low energy beam of primary ions such as Ar^+, O_2^+, or Cs^+ is used to sputter away surface species from unfixed and unstained solid samples. The erosion rate may be accurately controlled—atomic layer by layer, a feature of considerable value in depth profiling a specimen. The sputtered material is analyzed using mass spectrometry and provides a uniquely sensitive technique for both surface and depth profile analysis with no significant background signal. The lateral resolution is between 0.5 and 2.0 μm, the depth resolution is between 5 and 50 nm, depending on the amount of surface sputtering, and the elemental detection can be as good as 10^{-20} g, which is equivalent to about 300 atoms.

There are two ways in which SIMS may be carried out, by using either a scanning ion probe or a direct focusing system. The *ion microprobe* rasters a focused ion beam over the surface of the specimen and the resultant secondary ion intensity is displayed on a TV screen that is being rastered in synchrony with the primary beam. It is an optical system analogue to an SEM. The *ion microscope* produces a magnified image of the surface using secondary optics to transmit the secondary ion emission of interest to an imaging system. In both instances the image or ion micrograph is a map of the specimen surface showing the distribution of a particular element. By localizing a static probe or a reduced raster on the surface, it is possible to make a qualitative elemental analysis of a given feature with the information being expressed as an ion probe mass spectrum. An additional advantage of SIMS over other surface analytical methods is the ability to resolve many of the natural isotopes, which would permit an *in situ* study of isotopically labeled compounds. It is especially suitable for the analysis of light elements such as Li, Be, and B and of the electrolytes important in cell metabolism.

Another form of ion microscopy is *field ion* microscopy (FIM), which has been a most useful surface analytical technique for metallurgists and material scientists. In the FIM technique described here in the chapter by Panitz the specimen holder is a very small tip of tungsten held at low ($\sim$20–80 K) temperature and positively charged in a high vacuum. Substances to be analyzed may be evaporated onto this tip and a gas such as helium is then introduced into the system. A polarized helium atom is attracted to the metal tip and becomes ionized and the helium atom accelerates toward the negatively charged fluorescent screen. The field ion microscope is a point projection microscope and the magnification is simply obtained as a result of the very small size of the specimen tip and the large size of the fluorescent screen some distance away from the sample. It is possible to see atoms with a FIM as the instrument has a practical resolution of 0.2–0.5 nm. As one is only looking at a surface layer of atoms, the system is very sensitive to contamination and it is probably for this reason that, up to now, very little work has been done on organic material.

An alternative procedure that may help to overcome some of the limitations of both electron probe microanalysis and ion beam microanalysis is the new technique of *laser microprobe mass analysis* (LAMMA). Details of the technique together with a wide range of applications may be found in the chapter by Schmidt. Pulses of Q-switched lasers with a pulse width of about 10 ns are used to evaporate and partially ionize microvolumes of organic materials. The ions are extracted and analyzed using a time of flight mass spectrometer. Although the analysis is usually carried out on thin (0.5 μm) sections or single cells up to 1.0 μm thick, recent studies have shown that the surfaces of bulk specimens may also be analyzed. Because visible photons are used to image the morphological details the lateral spatial resolution on sectioned material is between 0.5 and 1.0 μm. With bulk samples the lateral resolution is about 3.0 μm with a depth of analysis between 0.1 and 1.0 μm. The absolute sensitivity of LAMMA is down to 10^{-20} g, and it is possible to detect all elements in the periodic table in addition to isotope discrimination and the analysis of molecules up to $M^w = 1000$. The analysis of organic molecules is complex and depends on fingerprinting methods, while quantitative studies depend on an assured equivalence between standard and sample. Although the visible photon imaging system limits the spatial resolution, it has the advantage of permitting unstained and unfixed organic material to be examined using dark field illumination or phase contrast. The laser microprobe analyzer is a most versatile instrument for the analysis of organic surfaces, and once the problems of quantitation have been resolved it will provide as much analytical information as may be obtained by the combined effort of the energy loss electron spectroscopy, x-ray microanalysis, and secondary ion mass spectrometry—albeit at a slightly inferior spatial resolution.

Another form of laser mediated analysis is *laser Raman microanalysis* (Etz, [14]). This is capable of molecular analysis at a spatial resolution of ~10 μm. Samples may be analyzed under ambient conditions, and the technique might be useful in analyzing membranes and single living cells.

CONCLUSION

It is clear that as far as the analysis of organic surfaces is concerned, the specimen preparation techniques are to a large extent the limiting factor. A further hazard centers around the damage that may be caused by the interaction of the high energy beam with the specimen. This damage may range from volatilization of thermolabile substances, radiation damage, mass loss from the specimen, selective loss of elements from the specimen, and contamination of the sample. These are the problems common to the microscopy and analysis of all materials but are particularly serious for organic samples and biological materials. It will also be realized that the different analytical techniques cause varying degrees of specimen damage. Laser probe microanalysis completely destroys that part of the specimen being analyzed. Secondary ion mass spectrometry and photo-emission microscopy progressively erode away the sample surface. The deleterious effects of x-ray microanalysis and electron beam energy loss spectrometry are more insidious and care must be taken to monitor the results of analysis during the experimental procedures. Many different analytical methods will be discussed in this book and it is important to distinguish between those techniques that allow an *in situ* analysis of the surface and those that only permit analysis on pieces or patterns of the surface away from the original material. The former approach is more likely to give relevant information.

Many of the problems of specimen damage and specimen preparation can be overcome by holding the specimens at low temperature during preparation, examination, and analysis. It is interesting to observe that nearly all the analytical techniques for organic specimens are using or propose to incorporate cold stages in the instrumentation.

The organic surfaces analysts have many sophisticated techniques at their disposal, but none of them are perfect. The analytical capabilities of some of the techniques must be offset against their low spatial resolution and vice versa. If you can see it with high resolution you cannot analyze it accurately, and a detailed chemical analysis is invariably accompanied by a diminished spatial resolution. We are left with a series of compromises.

The problem still remains as to how we may best convey information obtained from surfaces to other interested observers. With chemical analysis it is convenient (and reassuring) to express our results as statistically massaged numerical data. Because we have a touching faith in believing what we can see, we will probably continue to display the detailed morphological features of surfaces by means of data-rich but information-poor photographs. It would be far more efficient, although more time-consuming and cumbersome, to express these findings as well-annotated drawings, three-dimensional models, or holograms.

REFERENCES

1. Parsons, D. F., Ed., *Ultrasoft X-ray Microscopy: Its application to Biological and Physical Sciences,* New York Academy of Sciences, New York, 1980. A general reference.
2. Rudolph, D., and G. Schmahl, Ann. N.Y. Acad. Sci. *342,* 94–104 (1980).
3. Gvosdover, R. S. and B. Y. Zel'dovich, Mirror electron microscopy, in *Prin-*

ciples and Techniques of Electron Microscopy, M. A. Hayat, Ed., Van Nostrand, New York, 1977, chap. 7.

4. Wells, O. C., A. N. Broers, and C. G. Bremer, Appl. Phys. Lett. *22,* 353–355 (1973).

5. Broers, A. N., Appl. Phys. Lett. *22,* 610–612 (1973).

6. Goldstein, J. I., D. Newbury, P. Echlin, D. C. Joy, E. Lifshin, and C. Fiori, *Scanning Electron Microscopy and X-Ray Microanalysis: A Text for Biologists, Materials Scientists and Geologists,* Plenum, New York, 1981. A general reference.

7. Polanyi, M., *Personal Knowledge,* Routledge and Kegan Paul, London, 1958, chap. 12.

8. Ziman, J., *Reliable Knowledge,* Cambridge University Press, Cambridge, England, 1978, p. 6.

9. Smith, K. C. A., Ed., Computer techniques in electron microscopy and analysis, J. Microsc. *27*(1), 1982 Special Issue.

10. Hanson, N. R. A picture theory of meaning, in *The Nature and Function of Scientific Theories,* R. G. Colodney, Ed., University of Pittsburgh Press, Pittsburgh, 1970, p. 243.

11. Martin, B. W. Scanning Electron Miscrosc. I, 419–438 (1980).

12. Sparks, C. J., Chemical analysis, in *Workshop on X-ray Instrumentation for Synchrotron Radiation Research,* S. Donlach and H. Winnick, Eds., Plenum, New York, 1979, pp. 123–138.

13. Pierce, T. B., The nuclear microprobe, actual state of instrumentation and possible biomedical implication, in *Microprobe Analysis in Biology and Medicine,* P. Echlin and R. Kaufmann, Eds., Microsc. Acta Supp. II, 318–330 (1978).

14. Etz, E. S., Scanning Electron Miscrosc. *1,* 67–82 (1979).

LIGHT OPTICAL METHODS

P. C. ROBINSON

Department of Ceramic Technology
North Staffordshire Polytechnic
Stoke-on-Trent, England

1.1 INTRODUCTION: LIGHT MICROSCOPICAL CAPABILITIES

Light microscopes offer several advantages over electron beam instruments, and there are some types of information that can be gained in no other way. Specimens are placed in an environment that is much less hostile than those of electron microscopes. Heating and some ultraviolet radiation remain as principal hazards. There are a wide choice of operating modes, in both transmission and reflection, each of which may extract a different aspect of information about the specimen. It is possible to use these methods singly and in some combinations (e.g., incident excitation fluorescence with transmitted light phase contrast). If restriction of x–y plane resolution can be tolerated, z-direction resolving power using interferometric techniques can be astonishingly good (down to 2.0 nm has been claimed) [1]. Fields of view up to 2.5 cm for "compound" microscopes and double that for stereo microscopes are possible. Such large fields, when used properly, can reduce the extrapolation hazards necessarily arising when examining very small fields of view and making deductions about bulk material or whole organisms. Skills required for using instruments of modest capabilities are not high, although the requirements for highly specialized microscopes are very high. Some molecular orientation information may be deduced using polarized light methods.

Restrictions on uses of light microscopes would include resolving power in the x–y (specimen) plane. The often quoted limit of 0.2-μm "point-to-point" resolution is realized only with well-adjusted microscopes and 100× oil immersion objectives. It is interesting and salutary to work out the resolving power of each objective in use using the expression

$$d = \frac{0.61\lambda}{\text{N.A.}}$$

(λ being the wavelength of light being used and N.A. being the numerical aperture of the objective).

Depths of fields are often uncomfortably small, being given by

$$D = \frac{0.61\lambda \cos \alpha}{n \sin^2}$$

[α being the half-angle of the cone of rays just accepted by the objective (i.e., in N.A. = $n \sin \alpha$), and n being the refractive index in the same equation]. As a rule, depth of field is of the same order of magnitude as the resolving power of the objective.

There is a relatively narrow band of radiation defined by the term *light,* and the types of information that may be available are governed by this band (i.e., absorption, reflectance, refraction, diffraction, fluorescence, polarization, and phase changes).

The emergent signal consists of photons, so electronic image processing is available only through closed-circuit television systems.

Specimens for transmitted light must be transparent, of such thickness that features are "one-high" within the section, and preferably with enough contrast produced within the specimen for comfortable bright-field observation. This last requirement is mentioned in the hope that the all too common trick of using inadequate illuminating apertures to obtain image contrast may be discouraged.

Specimens for reflected light examination should, for preference, be flat, horizontal (i.e., at right angles to the optical axis of the microscope), and specularly reflecting; they should also have high reflectance and have interior material of high absorbance. It is not easy to think of any biological specimen that has a surface that even approaches such a combination of properties; this is regrettable, since reflected light methods are a first choice for investigation of exterior surfaces.

1.2   TOPOGRAPHICAL EXAMINATION OF EXTERIOR SURFACES

Exterior surfaces may be defined as those surfaces of organic specimens that are in contact with the atmosphere or hydrosphere.

In a bright-field, reflected-light system the objective acts as its own condenser, and so is automatically focused when the objective is in focus, when the illuminating aperture matches the collecting (numerical) aperture, and the condenser corrections match those of the objective.

Considerations of the specimen surface include

1. Flatness. Besides the obvious depth of field requirement, unless any slope on the specimen is at an angle less than 0.5α to the stage, where α is the half-angle of the cone of rays collected by the objective, light from that slope will not re-enter the objective. Slopes will thus appear dark in the image. Clearly, objectives of higher numerical aperture are able to collect more of the light scattered from such surfaces, but they also collect more glare, so contrast is reduced, and they have lower depths of field.

2. Unless the surfaces under examination are perpendicular to the optical axis of the microscope, not only is there the problem of focus along one strip across the field of view, but the illumination becomes oblique. This condition, which arises only with wedge-shaped preparations in transmitted light, is more serious with low numerical aperture objectives.

3. Specularly reflecting surfaces, being those whose surface is smooth enough to give an image by reflection, are those for which reflected light examination was designed. Contrast in bright field results from light scatter from imperfections, on the surface, either natural or as artifacts. Diffusely reflecting surfaces scatter so much of the illuminating beams that image intensity and information content are low. One illuminating method that may offer some success on nonspecular surfaces is dark ground illumination, where some of the surface roughness can be expected to be at angles favorable to reflect toward the objective some of the "all-the-way-round," external oblique illuminating beams.

4. The principal difficulty in examining organic surfaces with reflected light arises from the low reflectance of the surfaces. Reflectance is defined, for specularly reflecting surfaces, as that percentage of light incident normally on the surface, which is reflected

normally. Although this ideal situation is very rare indeed, the principle is useful in understanding the difficulty of examining organic surfaces.

Reflectance for a medium of refractive index (RI) n_1 in a medium of RI n_2 is given by

$$R = \frac{(n_1 - n_2)^2}{(n_1 + n_2)^2}$$

and this simplifies in air to

$$R_{air} = \frac{(n_1 - 1)^2}{(n_1 + 1)^2}$$

If a medium has a RI of 1.500 in air, then its reflectance will be 4%; not only does this indicate that image intensities will be low, but it indicates that some 96% of the light will proceed past the surface and be available for glare when scattered by features behind the surface.

There are three obvious ways of improving this situation.

1. The reflectance of the surface may be increased by metallizing. Not only does this increase the percentage of image-forming light reflected by the surface, but it reduces the amount of light available for glare, and much of this is absorbed by the metal itself. Unfortunately, the exposure to heat and vacuum required for metallizing may well modify the surface, and sputtering is probably less harmful to many organic materials than evaporative coating.

2. It may be feasible to arrange for as much as possible of the light which enters the specimen to be absorbed, and so to reduce its contribution to glare. Certainly the surfaces of dark-colored specimens give better contrast than those of pale ones (see Figs. 1 and 2), and practically, it may be possible to remove as much of the material behind the specimen surface as possible and to place the specimen on top of a matt, black surface. Improvements obtained in these ways are seldom dramatic, but contrast in images of surfaces of poorly absorbing specimens is so low, especially when high numerical aperture objectives are used, that any improvement is worthwhile.

3. Replication offers the opportunity to remove light-scattering detail from behind the surface under investigation, plus the chance to use transmitted light methods (see Figs. 3 and 4). Replication of the

Figure 1. Interior of daffodil calyx, reflected bright field. The marginal shadow is the image of the illuminated field iris. Note how contrast improves in this region.

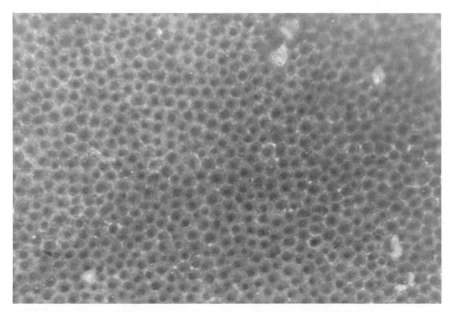

Figure 2. Back of primula petal, reflected bright field. Image contrast from the purple petal is better than that from the daffodil.

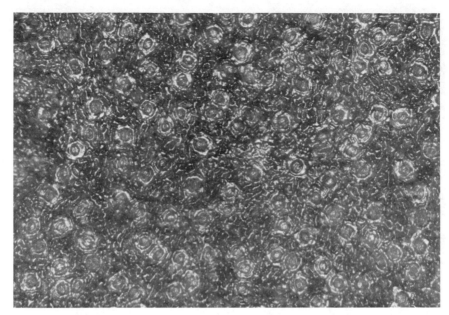

Figure 3. Underside of holly leaf, reflected light, bright field.

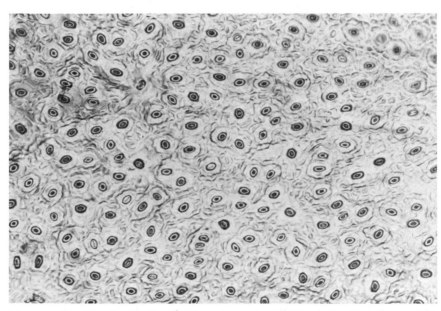

Figure 4. Replica of underside of holly leaf, transmitted light, bright field. Compare this image with that in Fig. 3.

more fragile specimens is not easy and must be regarded as a last resort.

1.2.1   Methods of Examining Surfaces Using Reflected Light

Stereomicroscopes invariably use external oblique illumination when examining surfaces of specimens (external oblique illumination is that azimuthal set of beams whose angle of approach to the specimen is larger than the numerical aperture of the objective, both being measured from the optical axis of the objective). The compromise of long working distance and depth of field versus resolving power is biased toward the former, and most stereomicroscope images become "grainy," with diffraction halos around fine detail at magnifications of about 50×.

Familiarity with such instruments should not breed contempt for them, since they offer a vital link from ordinary viewing to "proper" microscopy, and their use can avoid both time-wasting and elementary mistakes.

Bright-field imaging provides contrast from features that scatter light more effectively than the remainder of the field of view. It is used only on the smoothest of organic surfaces. The technique of closing the illuminated field iris to illuminate less than the full field of view offers improvements in contrast both inside and outside the edge of the iris image with no loss of resolving power.

Dark ground illumination and external oblique illumination are effective on less than smooth surfaces, since both provide imaging using scattered light once the direct (or zero-order diffracted) light is diverted clear of the objective.

Differential interference contrast (DIC) is effective only on smooth surfaces where gradients are not too steep for light to reenter the objective and where relief is less than perhaps 700 nm between points separated by the resolving power of the objective.

It will be recalled that DIC involves splitting each beam approaching the specimen into a pair of component beams that interact with the specimen. It is usual for the separation of the pairs of "target areas" to be two-thirds of the resolving power of the objective and that the optical path difference (OPD) between each pair is twice the height difference between the two target areas when using reflected light. The specimen OPD is converted to a polarization color after passage through the analyzer, and the system can be "tuned" by introducing a second OPD from the beam-combining prism. It is not easy to provide contrast at specimen OPD's greater than 1400 nm, as reference to a Michel-Levy chart will confirm. Image interpretation is not simple.

Reflection contrast [2–4] serves to illustrate the difficulties that must be overcome when using interferometric methods on live specimens using reflected light. Thin-film interference effects are familiar, and the principle has been adapted to investigations of thin-film effects of cells sedimented onto dish floors and stained sections. Consider a cell resting on a glass surface (see Fig. 5). Pairs of surfaces that may give rise to interference effects are the glass and nearer cell surface and the two cell surfaces. Other combinations of surfaces add to interpretation problems and loss of contrast. Since refractive index differences between the cell and water and between glass and water are small, reflectances are very small indeed. This means that interference fringes have low intensity and lower contrast.

If they are to be perceptible, then all sources of stray light must be eliminated: (1) A central stop in the illuminating aperture plane is used to suppress reflections from glass–air surfaces in the system that are perpendicular to the optical axis of the microscope. (2) Oil immersion objectives are used so that reflections from the objective front lens front surface and the "air" surface of the culture chamber are suppressed. (3) Crossed polars suppress further any reflections from glass–air surfaces within the objectives, and so that light carrying information from cell surfaces of interest can pass the analyzer, a quarter-wave plate is situated in front of the objective front lens. The essential parts of the diagram are shown in Fig. 6.

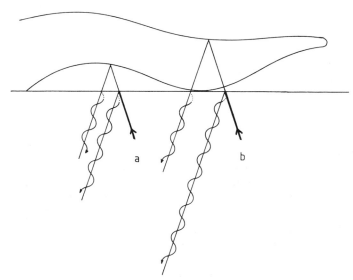

Figure 5. Cell resting on the floor of a glass dish. The pairs of waves a and b reflected from the three surfaces can produce interference fringes, constructively for a and destructively for b.

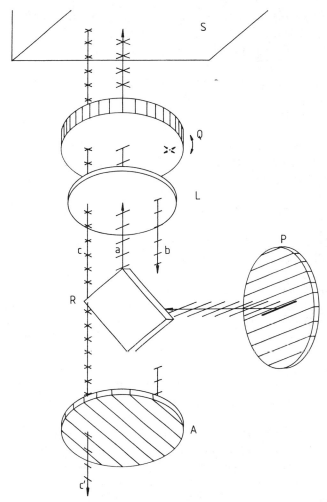

Figure 6. Principle of confocal scanning light microscope. The illuminating pinhole is imaged on the specimen, and again on the detector. The specimen is scanned mechanically through the first pinhole image, and intensity variations are recorded by the detector. The CRT is scanned synchronously with the object scan. P, polarizer; R, reflector; A, analyzer; Q, quarter wave plate; S, specimen; L, objective lens glass–air surface; a, illuminating beam; b, light reflected from lens surface—this does not leave the analyzer; c, circularly polarized light from specimen, component c′ does leave the analyzer.

31

Specimen chambers of anisotropic polymers produce such gross polarizing effects that they cannot be used, and glass chambers should have a base of about 0.17 mm (standard No. 1½ cover slip) thickness for optimal results.

It will be appreciated that information on mechanisms of cell adhesion to surfaces cannot be obtained in other ways and that the difficulties of getting this information are formidable.

More recently, reflection contrast has been used to examine weakly stained sections of histological and cytological material. The information gained is primarily from the stain employed and is used to supplement that from bright-field, phase contrast, and incident fluorescence methods.

1.2.2 Methods Using Transmitted Light

Difficulties of examining external surfaces in transmission are essentially, first, that the specimen must be sufficiently transparent, and, second, that light–matter interactions, and so information pickup, occur at both surfaces of the specimen and within the specimen. While it is possible to match the refractive indices of the outer surfaces and thus eliminate their interactions (which facilitates examination of internal features), the reverse is not true. Only when there is minimal light scatter within a specimen is it possible to elucidate from the image information about the surfaces. It is worth remembering that replicas should have no light-scattering features except those generated by the specimen and so lend themselves to transmitted light examination.

Bright-field and dark ground methods are useful only where internal features are absent, and sufficient light scatter to generate contrast is produced by surface detail. The principle that light scattering at a boundary increases in efficiency with refractive index change across the boundary may be used both to increase and to reduce contrast under appropriate circumstances. Phase contrast requires modest optical path differences from features small enough to diffract light away from the phase ring and is intolerant of light-scattering features outside the plane of focus.

Differential interference contrast provides excellent optical sectioning with good tolerance of features above and below the plane of focus. Optimal contrast, however, in the image calls for matching of specimen optical path differences with beam-combiner optical path difference; this gives the darkest, compensation "shadows." Those specimens providing optical path differences less than 50 nm give uncomfortably low-intensity images. Allen et al. [5,6] have developed techniques for video processing DIC images in the first-order white range, which have high intensity but low contrast, to yield high intensity and high contrast. Their methods have been so effective as to detect artifacts produced by the microscope lens systems. A recent development [7] has been to use a computer frame memory to store the image of an empty field of view under selected viewing conditions and then to subtract this from the working image, so removing the glassware aberrations. It is claimed that this system is capable of resolving tubules down to 25 nm diameter.

Use of interferometric methods to measure dry masses of cells is well established.

1.3   EXAMINATION OF INTERIOR SURFACES

Unless sectioning is used, examination of interior surfaces requires transmitted light illumination and may be successful only on thin specimens whose exterior surfaces are matched in refractive index with that of the mountant. When sections have been prepared, the usual criteria for generating contrast by staining or refractive index mismatch and by illumination techniques apply. Phase and modulation contrast, differential interference contrast, and interferometric methods may be appropriate to specimens whose characteristics suit the relevant method.

1.4   RECENT DEVELOPMENTS

1.4.1   Holographic Microscopy [8,9]

Since holography can be used to record on a plane photograph enough information about three-dimensional objects, it offers the chance to overcome the usual depth-of-field limitations of light microscopy. It is possible to record on one hologram information about the entire thickness of a section and then to reconstruct images where any selected plane is in focus. Disadvantages of holographic microscopy are such that instruments, by and large, are still research instruments dedicated to individual tasks. The number of steps involved in using the instruments properly and the time taken to master and perform the adjustments mean that only where the required information is not available by other means are holographic methods used. Holographic microscopy calls for a combination of a knowledge of physical principles for efficient holographic results and of biological science that may not be readily available in a team, let alone in an individual. The potential of holography is likely to remain unrealized until there is sufficient demand to warrant the efforts involved.

1.4.2   Scanning Light Microscopy [10]

The production of microscopical images using point-by-point scanning of objects with a fine spot of light and electronic image buildup offers advantages of electronic image processing and the opportunity of a variety of imaging modes. (See Chap. 12 for further details.)

Confocal scanning [11] light microscopy (CSLM) offers an improvement in x–y plane resolving power. A schematic diagram of such a system is shown in Fig. 7. Although it is necessary to achieve scanning by moving the specimen past a stationary light spot, it is claimed that small areas, perhaps 4 μm by 4μm with a scanline distance of 0.2 μm at 500 Hz takes 1/25 of a second.

Modifications of CSLM make available an interferometer system of the Mach–Zehnder type to measure refractive index variations in the specimen, stereoscopic devices, and scanning fluorescence microscopy. An advantage claimed for the last technique is that relatively thick specimens may be examined, since only the area of study is at the focus of the illuminating beams and detection of light from that area only is optimized.

1.4.3   Combined Light and Electron Microscopes

Two combined light and electron microscopes have been marketed. One has a light microscope alongside a transmission electron microscope, and a specimen acceptable to both methods is examined first with the light microscope and then moved to the stage of the TEM.

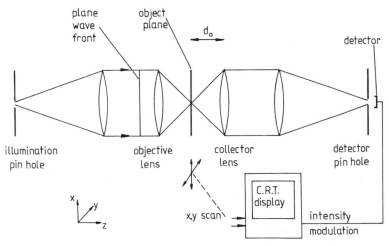

Figure 7.

More recently, [13], a light microscope has been incorporated into an SEM, so that an oil immersion objective viewed the side of the specimen remote from the electron beam. The specimen is situated on a quartz glass cover slip and is illuminated by a long working distance condenser. It is possible to combine SEM and light microscopy imaging simultaneously, for instance, secondary electron imaging with fluorescence, or transmitted light bright field. The oil immersion objective is a very efficient collector of cathodoluminescence.

REFERENCES

1. G. Nomarski (private communication).
2. A. S. G. Curties, J. Cell Biol. *20*, 199–215 (1964).
3. W. J. Patzelt, Mikrokosmos, 78–81, 1977.
4. J. S. Ploem, in *Mononuclear Phagocytes in Immunity, Infection and Pathology*, R. V. Furth, Ed., Blackwell, London, 1975.
5. R. D. Allen, J. L. Travis, N. S. Allen, and H. Yilmaz, Cell Motility, 275–289, 1981.
6. R. D. Allen, N. S. Allen, and J. L. Travis, Cell Motility *1*, 291–302, 1981.
7. R. D. Allen, J. Microsc. *129*(1), 3–18 (1983).
8. M. E. Cox, Microscope *19*, 137 (1971).
9. M. E. Cox, Microscope *22*, 361 (1974).
10. C. J. R. Sheppard and A. Choudhury, Optica *24*, 1051 (1977).
11. G. J. Brakenhoff, J. Microsc. *117*, 219 (1979).
12. G. J. Brakenhoff, J. Microsc. *117*, 213 (1979).
13. J. S. Ploem and A. Thaer, Proc. RMS *15*(2), Suppl. 9 (1980).

X-RAY PHOTOELECTRON SPECTROSCOPY

R. HOLM

Department of Applied Physics
Bayer AG
Leverkusen, Germany

2.1 INTRODUCTION

The abbreviation *ESCA*, standing for *e*lectron *s*pectroscopy for *c*hemical *a*nalysis, was created by Siegbahn [1] and is synonymous with the term *XPS* (x-ray photoelectron spectroscopy), which is often used in American literature. The process for which these terms are used is the measurement of the kinetic energy of electrons that are emitted by a substance under the influence of x rays.

The incidence of x rays on materials causes two groups of interactions: absorption and x-ray fluorescence, on the one hand, and electron emission caused by the photo and Auger effects, on the other. For experimental reasons x-ray spectroscopy has been favored, and electron spectroscopy neglected, since 1920. It was not until 1952 that Siegbahn began to construct electron spectrometers that provide more information than may be obtained by x-ray spectroscopy. The discovery of sharp lines and of chemical shifts similar to those seen in NMR spectroscopy raised the expectation that ESCA would permit the study of binding effects on all the elements involved in a bond and that it would therefore create new opportunities in the structural investigation of organic and inorganic molecules. Siegbahn and his co-workers at Uppsala extended their investigations to substances in all states of aggregation. But their hopes as to the possibility of universal structural investigation were fulfilled only partly. Instead another aspect of ESCA, its surface sensitivity in the investigation of solids, became increasingly important, but nevertheless the experience gained in the investigation of structures has laid the foundations for the detection of compounds in thin layers.

In the investigation of organic surfaces there is at present no method that—in the case of the uppermost molecular layers and in the subsurface regions, whose analysis presents considerable difficulty—permits such accurate quantitative determinations and such precise statements as to binding states as may be provided by ESCA.

2.2   DESCRIPTION OF THE METHOD

2.2.1   Physical Basis

ESCA is based on the photoeffect and Auger effect. An x-ray quantum having the energy $h\nu$ contains sufficient energy to release an electron from the K shell of an atom whose ionization energy is E_K (often referred to as the binding energy of the electron). The electron released in this way by the photoeffect then has approximately the kinetic energy

$$E_{kin} = h\nu - E_K \tag{1}$$

If the sample is stimulated by AlK_α radiation ($h\nu = 1486$ eV), photoelectrons from, for example, the C1s level (ionization energy $E_K \approx 284$ eV) may be expected to give a line at $E_{kin} \approx 1202$ eV. For accurate determination of ionization energies the following additional terms must be incorporated in Eq. (1): the electrostatic charging and the work function of the spectrometer material. For analytical applications it is generally sufficient to take account of the electrostatic charging of the sample caused by the emission of the electrons, whereas the work function of the spectrometer material can be looked on as constant and can be compensated electronically. As a relaxation mechanism the Auger effect (together with the x-ray emission) is always coupled with the photoemission. As the Auger effect too leads to the emission of electrons with characteristic kinetic energies, lines appear in the ESCA spectrum both because of the photoeffect and because of the Auger effect.

Figure 1 presents some wide-scan spectra of common polymers. The kinetic energy of the electrons is plotted on the abscissa and the number of electrons is plotted on the ordinate axis. The spectra contain lines attributable to the photoeffect and also several Auger lines (e.g., those of F and O).

The lines provided by the photoeffect directly reflect the energy levels of the detected atoms, insofar as these are ionized by the stimulating x radiation. The energy level is characterized by the capital letters K, L, M . . . , as customary in x-ray spectroscopy, or by stating the principal quantum number $n = 1, 2, 3 . . .$, the orbital angular momentum $\mathbf{1}$ (the symbol s is used for $\mathbf{1} = 0$, p for $|\mathbf{1}| = 1$, d for $|\mathbf{1}| = 2$, etc.), and the sum of the orbital angular momentum $\mathbf{1}$ and spin $\mathbf{s}$: $\mathbf{j} = \mathbf{1} + \mathbf{s}$ (half-integral)). The following terms are therefore equivalent: K shell and 1s level, L_I and 2s, L_{II} and $2p_{1/2}$, L_{III} and $2p_{3/2}$, etc. The Auger electrons are characterized by stating the levels that contribute to their emission.

Electrons lose energy as they pass through matter. The losses of energy, however, are not continuous but quantized (e.g., discrete plasmon

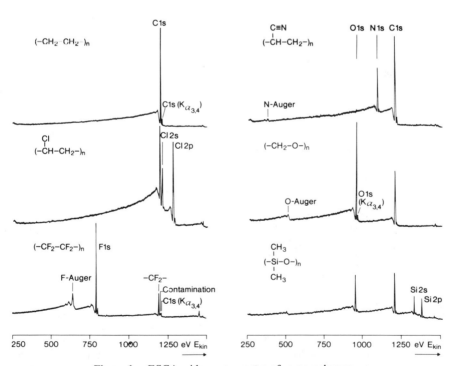

Figure 1. ESCA wide-scan spectra of some polymers.

losses). This is to be seen in the fact that a line in the electron spectrum is always accompanied by a tail on the side of lower kinetic energies; the line is separated from this tail by at least one minimum. One can therefore be certain that the electrons in the main line are the ones that have lost no energy. Naturally only electrons that originate from the uppermost atomic layers can contribute to this line. This fact makes ESCA suitable for the investigation of surfaces.

The investigated layer thickness is given by the escape depth of the photoelectrons and Auger electrons. Their determination and the investigation of their dependence on kinetic energy and matrix have formed the subjects of numerous studies. A compilation of data is given in Fig. 2 [2]. The data show that at kinetic energies of 300 and 1500 eV the mean escape depths are approximately 0.8 nm and 1.5 nm, respectively. The thickness of the layer actually measured must then be assumed to be two to three times greater. Figure 2 applies mainly to metals. Somewhat greater escape depths must be assumed for oxides and organic substances. It follows that the upper limit of the measured layer thickness in the high energy part of the spectrum is 10 nm.

The fact that ESCA gives not just an analysis of the uppermost monolayer but also signal contributions from a layer with a thickness of about 6–10 nm has an important advantage in routine investigations. The uppermost layers of a sample produced under ordinary industrial or laboratory conditions consist almost exclusively of chance impurities, known as con-

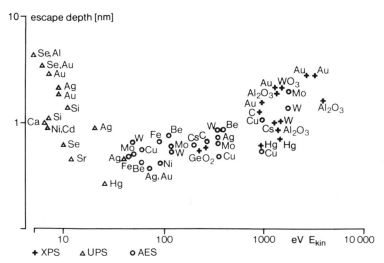

Figure 2. Mean escape depths of photoelectrons and Auger electrons. (Data compiled by Brundle [2].)

tamination, especially of the elements C and O. In an ESCA spectrum the chemically relevant layers are nevertheless analyzed without the contamination layer having to be removed. As the sticking coefficients for hydrocarbons fall rapidly after one or two monolayers have been deposited, the contamination layer does not grow to a thickness of more than a few monolayers during the time taken by the measurement, even if the vacuum is poor ($\approx 10^{-6}$ mbar).

Under certain conditions one can also draw conclusions regarding depth distributions from the relatively thick layer analyzed. It is important that this information be obtained nondestructively.

2.2.2 Structural Investigation and Detection of Compounds

Elements can be detected easily with the aid of Eq. (1) and of the tables published for the photoeffect and Auger effect [3,4]. Identification is almost always certain because in nearly every case an element has several lines in the ESCA spectrum.

If individual ESCA lines are examined with expanded abscissa, chemical shifts that provide information on the binding states of the detected elements are found [1,3,4].

How the shifting of energy levels occurs and how it becomes visible in the ESCA spectrum can be explained qualitatively with references to an oxidation process. As oxidation causes a partial removal of electrons, those that remain are attached to the nucleus more firmly than they were before the oxidation began. This is indicated by an increase in the ionization energy and therefore by a loss of kinetic energy by the photo and Auger electrons.

Sulfur can be taken to illustrate what happens. Figure 3 gives the S2p ionization energies of several sulfur compounds, which are arranged in order of decreasing electron density near the nucleus. According to this criterion the following groups can be distinguished::

1. S ions negatively charged at 162 ± 0.5 eV.
2. Divalent sulfur at 164 ± 1 eV.
3. Quadrivalent sulfur at 166 ± 0.5 eV.
4. Hexavalent sulfur at 169 ± 1 eV.

Shifts within these groups result mainly from differences in the electronegativity of the binding partners. Figure 4 shows that it is quite easy to distinguish between sulfides and sulfites, between sulfinates and sulfonates, and even between thiosulfonates and Bunte salts, for example.

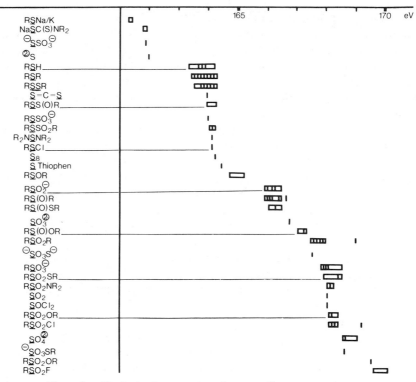

Figure 3. S2p ionization energies of some sulfur compounds.

If the chemical shifts of elements in compounds in which they have approximately the same binding state in each case (e.g., as oxides) are plotted against the atomic numbers [4], the picture obtained for the main groups is somewhat reminiscent of the ionization energy curves: the shifts within a period increase, whereas those within a group decrease (Fig. 5). As far as the subgroups are concerned, the largest shifts are observed for the elements of groups IV and V; then as the atomic number rises, the shifts decrease until they reach zero for elements of subgroups I and II (Fig. 6). In the case of transition metals the same shift is often observed for the same valence state [e.g., Cr(VI), Mo(VI), W(VI)].

There is no comprehensive explanation for the periodicity of the chemical shifts. Obviously, the ionization energy shows a similar variation according to the atomic number. Elements with low ionization energies show small chemical shifts, or none at all, whereas those with high ionization energies show large shifts. On the whole, though, this applies only to the main groups. The large chemical shifts seen for Ti, Nb, Ta, and so on

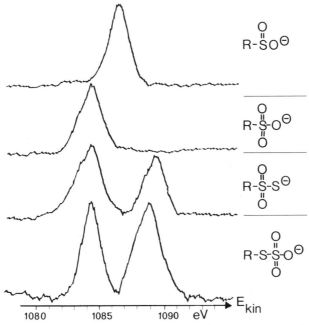

Figure 4. S2p photoelectron spectra of some sulfur compounds.

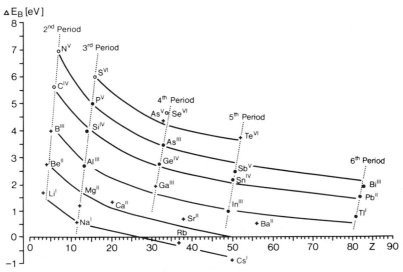

Figure 5. Chemical shifts ΔE_B for oxides of the main group elements versus the atomic number.

43

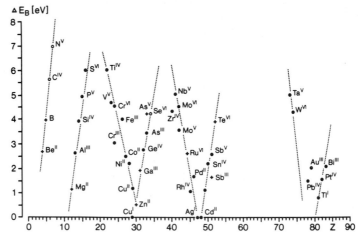

Figure 6. Chemical shifts ΔE_B for oxides versus the atomic numbers of the elements.

(elements of subgroups VI and V) run parallel to the heat of formation of the oxides of these elements.

A special difficulty with organic samples is that carbon is always quadrivalent and lacks marked chemical shifts such as those caused by valency changes in the case of other elements (e.g., S, Mo). The chemical shifts that occur at the C1s level are therefore observed only in consequence of

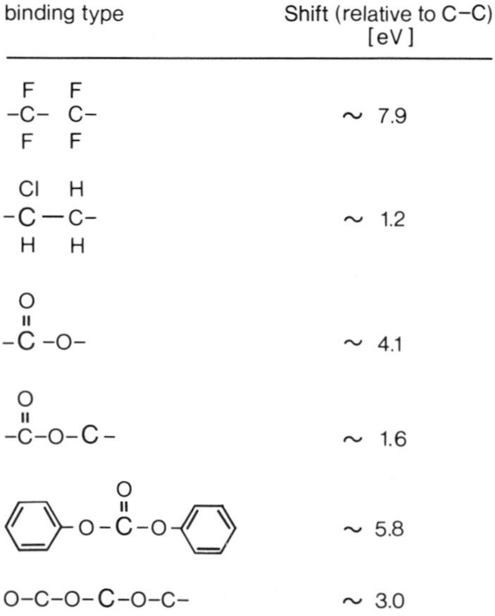

binding type	Shift (relative to C–C) [eV]
F F –C– C– F F	~ 7.9
Cl H –C — C– H H	~ 1.2
O ‖ –C –O–	~ 4.1
O ‖ –C–O–C–	~ 1.6
⬡–O–C(=O)–O–⬡	~ 5.8
O–C–O–C–O–C–	~ 3.0

Figure 7. Chemical shifts for the C1s level.

differences between the electronegativities of the elements bound to carbon; they are particularly marked in the case of such highly electronegative elements as F and O. In contrast, it is generally impossible to distinguish single and multiple C—H— and C—C— bonds according to shifts. Numerical values for chemical shifts of the Cls level are compiled in Fig. 7.

Chemical shifts in the Cls and Ols ranges, as often encountered among high polymers, are reproduced in Fig. 8. It is evident that in carbonyl groups not just C, but O too, can be clearly distinguished from atoms

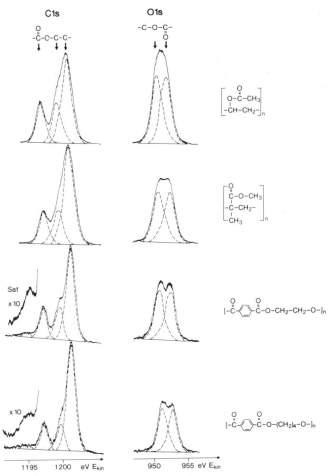

Figure 8. Cls and Ols spectra for some polymers containing oxygen (polyvinyl acetate, polymethyl methacrylate, polyethylene terephthalate, polybutylene terephthalate).

bound in an etherlike manner. Where the binding partner is F the shifts of the C1s line are still more pronounced than when it is O. The summarized result of a large number of investigations is presented in Fig. 9 [5]. It shows that, where F is concerned, effects by the next but one neighbor are measurable also. Thus, for example, the C1s shift of a CHF group differs according to whether the neighboring group is CH_2 or CF_2.

As the chemical shift depends on the electron density of the atom in question, ESCA may be used for the detection of polar structures, as encountered among, for example, azides, amine-imides (Fig. 10), or mesoionic compounds in general. The difference between the N1s ionization energies of amine-imides corresponds directly to the localization of the charges on the individual N atoms [4].

The measured examples show that the full width at half maximum of the lines in the ESCA spectrum is about 1–2 eV. The largest shifts observed were, at the most, 10 times greater. In NMR spectroscopy, for comparison, the same quotient of maximum chemical shift and line width is larger than 1000. From this simple estimation it follows that ESCA cannot discriminate between different types of bonds as precisely as NMR can. In many cases, therefore, additional information is needed if compounds are to be identified with certainty. Such additional information is provided by chemical shifts of the Auger lines [3,6] and by the occurrence of certain satellites in the vicinity of the photoeffect lines.

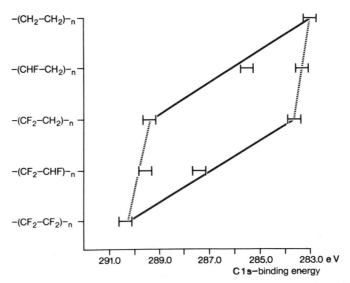

Figure 9. Chemical shifts in the C1s spectrum for fluoropolymers. (After Ginnard and Riggs [5].)

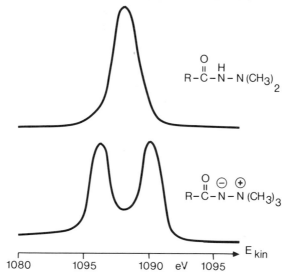

Figure 10. N1s photoelectron spectra of hydrazides and amine imides.

The kinetic energy of the Auger electrons is particularly dependent on the binding state when the electrons are emitted from the valence band. To avoid the necessity of finding the absolute kinetic energies, but nevertheless to enable very slight differences to be determined, Wagner [6] introduced the so-called Auger parameter—the difference between the kinetic energy of a photoelectron line and an Auger line of the same element as a characteristic of the element's binding state.

Satellite peaks are an important secondary source of information in the identification of compounds of transition metals. They are formed in the ionization of an electron from an inner atomic orbital (kinetic energy > 100 eV) if an electron from an outer molecular orbital is simultaneously raised into a higher unoccupied orbital (shake up) or above the vacuum level (shake off) (energy generally < 10 eV). As this amount of energy is lacking to the electron arising from the inner orbital, the kinetic energy of this electron is smaller to this extent. A satellite peak whose intensity depends on the probability of the simultaneous excitation of a second electron may thus appear on the lower kinetic energy side of certain ESCA lines. A well-known illustration of this fact is the characterization of divalent copper by such a satellite peak in the Cu2p spectrum (Fig. 11).

The shake-up satellite peak of aromatic ring systems has still more practical importance, especially in organic chemistry. With π-electron systems the distance between the lowest unoccupied level and the highest

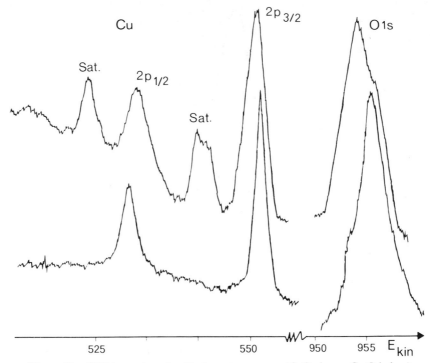

Figure 11.   ESCA spectra of oxide layers on copper (CuO above, Cu$_2$O below).

occupied level is 5–9 eV. Clark et al. [7] estimated the probability of a simultaneous electronic transition in the ionization of the C1s level at 1–5%. They attributed weak lines on the low energy side of the C1s line (distance from the C1s line approximately 7 eV, intensity ≈5% of the C1s line) to the shake-up processes described earlier.

Figure 12 shows the satellite peak of polystyrene. Its intensity is about 5% of the C1s line. The shape of the line suggests a division into two parts corresponding to an electronic transition from the highest or second highest occupied energy level to the lowest unoccupied level. The distance from the C1s line is 6.7 eV. The satellite peak can also be seen in the C1s spectrum of polyethylene terephthalate (PET) and in that of polybutylene terephthalate (PBT) (Fig. 7). It is the most reliable distinguishing characteristic between the very similar spectra of polyethylene terephthalate and polymethyl methacrylate (PMMA).

Spin–spin interactions in molecules whose electron shells are not closed is another possible cause of satellite peaks [4]. In many cases, however, they merely cause line broadening, and they have not yet acquired much practical importance.

These examples show that ESCA is not only capable of detecting elements, the special strength of the method being its ability to characterize compounds. Indeed, the earliest studies by Siegbahn [1] raised very high expectations for ESCA, which appeared to permit the study of binding effects on all the partners involved. Although the original hopes have not been entirely fulfilled, ESCA has remained an important method for the determination of composition and structure. It is even indispensable in the investigation of polar structures and mesoionic compounds.

As examples of structural investigations in the biological field one can mention the studies concerning the thermal stability of chlorophyll[a] monohydrate [8], in which the hydrate oxygen and the rest of the oxygen in the molecule can be distinguished according to the O1s spectrum; the detection of the valence states of nitrogen in bile pigments [9]; and XPS studies of Mg ions bound to the cell walls of Gram-positive bacteria [10]. In electron microscopy, OsO_4 is often used for fixation and staining; the various resultant valence states of the Os have been investigated by ESCA [11]. In structural investigations advantage can be taken of the fact that only a very small amount of the substance is needed. Special preparation techniques (such as vapor deposition or sublimation, evaporation from uncontaminated solvents, electrolytic deposition) must then be used to cover the sample holder with a thin film.

On the other hand, the experience gained in structural investigations is the foundation for the detection of compounds by ESCA in surface analysis. In this case there is no other method that provides such far-reaching structural information in routine work.

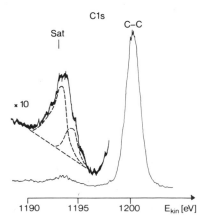

Figure 12. Shake-up satellite peaks in the C1s spectrum of polystyrene.

2.2.3   Surface and Subsurface Analysis

Let us consider the simple case of a continuous oxide layer of thickness d on a metal substrate (Fig. 13). With variation of d the metal intensity I^{met} will decrease according to

$$I^{met} = I^{met}_{\infty} e^{-d/\lambda}$$

where I^{met}_{∞} is the metal intensity without any oxide and λ is the mean escape depth of the photoelectrons or Auger electrons considered.

The oxide intensity will increase according to

$$I^{ox} = I^{ox}_{\infty}(1 - e^{-d/\lambda})$$

where I^{ox}_{∞} is the intensity of a very thick oxide layer.

Whenever the oxide layer on a metal is thinner than the escape depth of the photoelectrons, the oxide and the metal, separated by the chemical shift, appear side by side in the ESCA spectrum. The oxide–metal intensity ratio is then a measure of the oxide layer's thickness. In this way one

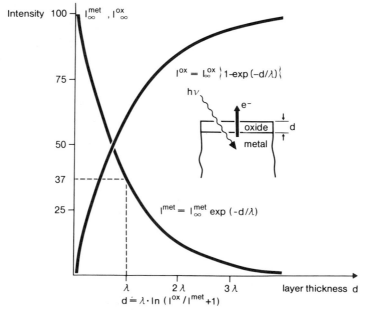

Figure 13.   Intensities of surface and subsurface signals as functions of the thickness d of the surface layer at a given electron escape depth λ.

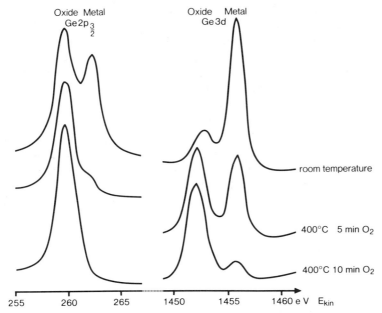

Figure 14. Dependence of the information depths on the kinetic energy of the photoelectrons, demonstrated by germanium samples oxidized to different degrees. At a given oxide layer thickness the oxide–metal intensity ratio in the low energy 2p range is higher than in the high energy 3d range because the layer thickness analyzed by ESCA increases with the kinetic energy of the photoelectrons.

can, for example, observe the growth of the oxide layer on metals exposed to air or H_2O or, quite generally, study the compositions and rates of formation of oxide and passive layers on metallic materials as functions of thermal or chemical treatment.

With multicomponent systems (e.g., steels), the oxide and metal are determined separately for all the alloying constituents. If the intensity ratios of the individual elements in the oxide and metal phase are formed separately, the change of concentration in the oxide, as compared with the metal, can be read off directly from an ESCA spectrum. For such results, however, the oxide must lie substantially undisturbed on the metal. In some cases, in which an element has sharp lines at low and high kinetic energies, one can check nondestructively whether or not this is so by electron spectroscopy: The oxide–metal intensity ratio must vary in accordance with the energy dependence of the escape depth (see Fig. 2). Figure 14 demonstrates this with reference to Ge surfaces oxidized to various extents. As the mean escape depth of the electrons is only about half as great in the case of the Ge $2p_{3/2}$ line as it is in that of the Ge 3d line, the oxide–metal intensity ratios at a given layer thickness differ greatly. If the metal–oxide intensity ratio happens to be independent of the energy, there can be no doubt that an islandlike top layer is present. Angular variation experiments (i.e., investigation of the dependence of the oxide–metal intensity ratio's dependence on the sample's angle of inclination) constitute another technique that enhances surface sensitivity, but it is only well suited to the study of flat surfaces.

The application of the preceding remarks on metal–oxide layers can be extended to layer systems of all kinds. If A and B are a pair of elements or binding states of the same element and A occurs only in the surface layer and B only in the substrate, the intensity ratio of the lines for A and B (or its dependence on the sample's angle of inclination) enables the mean thickness of the surface layer to be estimated. If the lines of B do not occur in the ESCA spectrum, it may be concluded that the covering is continuous and considerably thicker than the mean escape depths of B's photoelectrons.

In many cases the opportunities for obtaining depth information nondestructively are not sufficient, and it is therefore necessary to use methods by which layers are removed. Such methods have the fundamental drawback that an experiment cannot be repeated at the same location. Surface layers can be removed in many ways by selective dissolution or decomposition, but such chemical methods generally imply massive interference with the system, and it is therefore only in exceptional cases that they are suitable for the successive removal of very thin layers. Hence their use is generally restricted to the complete removal of layers that are of little interest. The same applies to such mechanical methods as microtomy. The layer removal technique generally used in surface physics is ion bombardment. But it has to be applied with particular care where organic substances are concerned.

In several publications [12,13] it has been shown that metal oxides are reduced by ion bombardment, even at ion doses that are as low as those normally necessary to remove contamination layers.

Where inorganic chemicals are concerned, ion-induced reactions are often overlooked because the separated binding partners may immediately recombine.

With organic material, however, annealing effects are not observed, the ion-induced effects being simply additive.

This is amply demonstrated by zinc phenyl sulfinate [14], a rubber additive that can be kept unchanged for years under normal laboratory conditions. Under Ar^+ bombardment, however, it is completely destroyed, as can be seen at three points in the ESCA spectrum (Fig. 15): The binding states of S and Zn are changed and the satellite peak in the

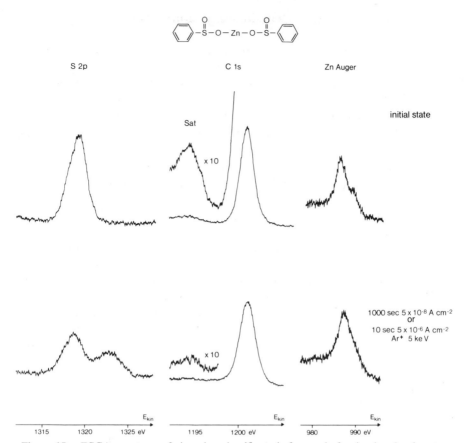

Figure 15. ESCA spectrum of zinc phenyl sulfinate before and after ion bombardment.

C1s spectrum disappears, suggesting destruction of the phenyl residue. The possibility of restoration of the original state is virtually negligible. Consequently, as every hit produces irreversible changes, whether the damage is detectable by ESCA should be only a matter of the dose applied. Figure 15 shows that this is in fact the case: The same effects were detected at 1000 s and 5×10^{-8} A/cm^2 and at 10 s and 5×10^{-6} A/cm^2.

The controlled removal of organic deposits can also be accomplished by plasma etching with activated oxygen. Here also it is obviously impossible to detect the compounds originally present, because, in consequence of the action of oxygen, all the elements are present in the oxidized state. In addition, the formation of residues from substances that produce nonvolatile compounds with oxygen must be expected after prolonged application.

2.2.4   Quantitative Measurements

The intensity of an ESCA line is determined by the following quantities:

1. The probability of the production of a photoelectron or Auger electron (x-ray flux, photoelectric cross section for the electron level under consideration, probability of the Auger process).
2. The probability of loss-free emission from the solid (dependent on the escape depth).
3. The probability of detection (design and mode of operation of the spectrometer; probability of perception by the detector).

In general, these quantities are not sufficiently well known; hence one is often forced to use standards and to determine relative sensitivity factors of the elements experimentally (cf. what is said in this connection in Sections 2.3 and 2.4). The commonly used method is the investigation of a large number of chemical compounds having a common element (e.g., F) [16]. Another possibility is the vapor deposition of extremely pure elements in an ultrahigh vacuum. Figure 16 shows that the relative sensitivities for the most intensive lines of the individual elements, as determined in this way, are not very far apart. (They are within a factor of 10 for most

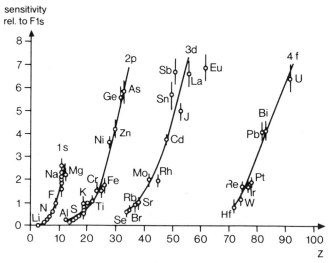

Figure 16.   Relative sensitivity of ESCA for the most intensive lines of various elements, measured with a Varian IEE-15 spectrometer (after Wagner [15]). With other spectrometers and modes of operation the curves for the individual levels may be still flatter.

sample	percentage of C-atoms bound in $-C^1-C^1-$		$-\overset{O}{\overset{\|\|}{C}}-O-C^2-$		$-\overset{O}{\overset{\|\|}{C^3}}-O-$	
	calc.	meas.	calc.	meas.	calc.	meas.
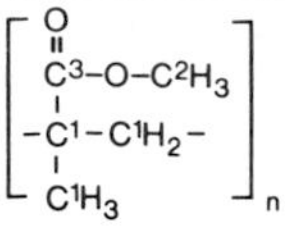	60	63	20	19	20	18
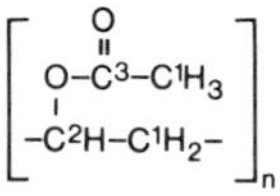	50	52	25	24	25	24
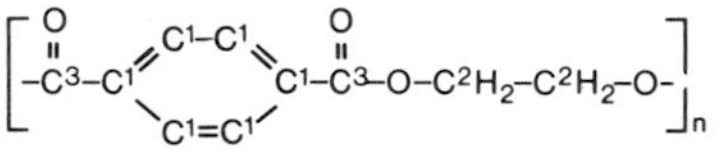	60	60	20	20	20	20
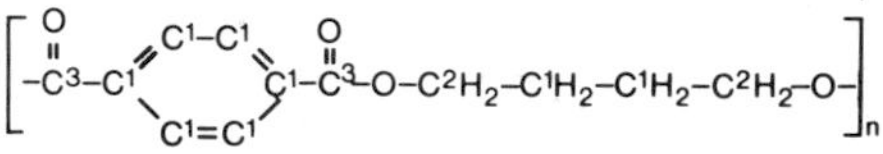	66.7	67	16.6	16	16.7	17

Figure 17. Calculated and measured percentage of C atoms in different binding types in some polymers.

elements.) The sensitivity ratios of the elements have to be individually determined for each type of spectrometer. They are then substantially independent of most of the experimental parameters, such as the x-ray intensity and the geometry of the sample.

An elementary analysis in the normal sense is not possible because ESCA does not detect hydrogen in its compounds.

Where there are chemical shifts in the spectrum of an element (as in Fig. 7) the intensity ratios (e.g., $C1s_{c-c}$: $C1s_{c=0}$) reflect very accurately the quantity ratios for the C atoms in the corresponding binding states (Fig. 17) [12]. That is because the interaction with the x-ray quantum is principally an atomic property. It is unlikely that chemical binding appreciably affects the photoionization cross sections for inner shells. It follows that a calibration performed for an element is generally valid for the compounds of that element too. This represents a decisive advantage of ESCA over other methods, for calibration standards for surfaces are difficult to obtain.

From Fig. 17 it can be seen that polymers having the same shifts in their C1s spectra (e.g., PMMA and polyvinyl acetate, or PET or PBT) can be distinguished by quantitative evaluation of the C1s lines. The data were obtained for freshly cut samples of materials of the highest purity obtainable. With PMMA and polyvinyl acetate, however, the percentage for the C—C bonds is systematically somewhat too high as a result of the slight contamination. In the case of PET and PBT this effect is largely compensated for by the fact that 1–2% of the C—C signal is contained in the satellite. In fluoropolymers the numbers of CF, CF_2, and CF_3 groups can be deduced from the intensities of the various C1s lines.

Concentration data (molar or weight percentages) for copolymers or polymer alloys are best obtainable by plotting a calibration curve with the aid of standards. Normally, polymer surfaces are not very sensitive to hydrocarbon contamination because the sticking coefficients of hydrocarbons are rather low. Nevertheless, for exact measurements attention should be paid to the fact that contamination layers and migration of additives may contribute to the C1s signal in an uncontrolled manner if the nature of the sample is not known. It is therefore advantageous to consider not the entire C1s line, but only a specific contamination-independent part of it (e.g., the line of carbonyl groups or the satellites of aromatic rings) and thus to form the intensity ratio to a line of a heteroatom. In the biological field quantitative determinations of N and S have been carried out on milled cereal grains to enable the protein content to be estimated [17]. The results agreed well with those of the Kjeldahl method. ESCA has also been found suitable for the quantitative determination of various sulphoxide groups in peptides [18].

In considering the sensitivity of ESCA one should bear in mind that this is a method for the analysis of surfaces and that all the data apply only to the analyzed layer, whose thickness is ≤ 10 nm, and never to the bulk of the sample. If an element or compound is present in the analyzed surface layer at a concentration of more than 1%, it can be detected and determined quantitatively. As a rule, indeed, a content of 1% in a single molecular layer is sufficient. Where low concentrations are concerned it is often necessary to accumulate data over a fairly long time (e.g., overnight) in order to obtain a good signal–background ratio. That is because the intensity of the exciting x radiation cannot be intensified as desired in commercial instruments.

2.3 EXPERIMENTAL SECTION

An ESCA spectrometer is a vacuum apparatus with an x-ray tube for stimulation of the photoelectrons and Auger electrons and an energy analyzer and detector system for measuring the number of emitted electrons as a function of their kinetic energy (Fig. 18). A commercial ESCA instrument is shown in Fig. 19.

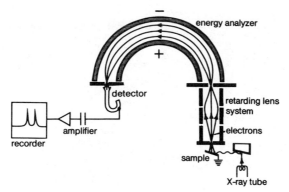

Figure 18. Block diagram of an ESCA spectrometer.

At the center of the instrument above the pump frame there is a spherical recipient, serving as a sample chamber, and a hemispherical analyzer above it. A preparation chamber and the sample airlock are flange-connected to the sample chamber at the right. In addition to the x-ray tube at the back, further devices, all of them directed toward the center of the sphere (sample location), are attached to the spherical recipient. These are an ion gun for removal of surface layers and for excitation in ISS and SIMS, an electron gun for AES, and a quadrupole mass spectrometer for residual gas analysis and SIMS. The data system with the display screen can be seen at the far left.

2.3.1 X-Ray Source

An x-ray source consists essentially of a heating filament and an anode. The electrons emitted by the heating filament are accelerated toward the anode, causing the emission of x rays. The energy distribution of x radiation is characterized by a continuum (*bremsstrahlung*), on which x-ray lines of discrete energy (characteristic radiation) are superimposed. To be ideal for ESCA, an x-ray source should (1) have enough energy to be able to excite inner orbitals of all the elements; (2) have a small full width at half-maximum because this substantially codetermines that of the photoelectron lines; (3) be monochromatic, in order to suppress the *bremsstrahlung* background and undesirable lines (such as those caused by $K_{\alpha3,4}$, K_β, etc.), (4) have a high radiation density at the location of the sample so that high detection sensitivity is possible.

In commercial ESCA spectrometers the anode material is generally either Al or Mg. The K_α doublet of $Al(K_{\alpha1,2})$ has an energy of 1486.6 eV

Figure 19. Commercial ESCA instrument (Leybold Heraeus, Cologne).

58

(0.834 nm) with a full width at half-maximum of 0.83 eV. The Mg $K_{\alpha1,2}$ line with an energy of 1253.6 eV (0.989 nm) has a full width at half-maximum of 0.68 eV. Several other lines of lower intensities, the so-called (x-ray) satellite peaks, are emitted in addition to the $K_{\alpha1,2}$ lines.

In ESCA instruments there is a window between the x-ray anode and the sample (Al window for Al and Mg radiation) that retards the electrons scattered by the anode and absorbs a portion of the *bremsstrahlung*. At a constant x-ray power and a given distance between anode and sample the specific radiation density at the site of the sample depends on the operating voltage of the x-ray tube and on the transmission of the window. Provided the surface of the anode is clean, the maxima of the yield curves for Mg and Al are at 10 kV and 14 kV, respectively. The transmission of a pure Al window with a thickness of 5 μm is 44% for Mg K_{α} radiation and 60% for Al K_{α} radiation.

The use of Al or Mg x radiation with most commercial ESCA instruments represents a convenient compromise with regard to the radiation density that may be obtained at the location of the sample, the full width at half-maximum, and the energy. Other x-ray anodes (e.g., Si, Cr, and Cu anodes) are available for special circumstances (e.g., when the lines of certain elements overlap or an extension of the measurement range is required). The full width at half-maximum can be reduced with an x-ray monochromator that is so adjusted that, for example, only a width of about 0.2 eV is diaphragmed from the Al–K_{α} line [1]. As even the best monochromator crystals reflect only a fraction of the incident radiation, the loss of sensitivity has to be compensated for by increasing the power of the x-ray tube, using a multichannel detector for simultaneous investigation of a larger energy range, or increasing the measurement time.

2.3.2   Energy Analysis and Detection of the Electrons

The first ESCA spectrometers constructed at Uppsala and Berkeley used magnetic energy analyzers. The first commercial instruments were fitted with nondispersive energy filters or dispersive electrostatic analyzers [1]. All the ESCA spectrometers now commercially available are fitted either with hemispherical analyzers or with cylindrical mirror analyzers (CMA) connected with a retarding field. The retarding field–analyzer combination enables the spectrometer to be operated either in such a way that the resolution is constant over the entire energy range (ΔE = const) or in such a way that the retarding factor is adjusted and hence the resolution of the spectrometer is altered as a function of the electron energy E ($\Delta E/E$ = const). As the different modes of operation determine the transmission function of the spectrometer, either the resolution or the transmission can be varied within wide limits to suit the particular problem. After the energy analysis the electrons are detected with an open electron multiplier (channeltron or dynode multiplier). In the case of x-ray excitation this is done by single electron counting.

2.4 COMPARISON WITH CLOSELY ALLIED METHODS

The principal methods for surface analysis of solids [4] are ESCA (XPS), AES, SIMS, and ISS. All have their advantages and disadvantages. ESCA is in general nondestructive, permits more reliable detection of compounds and more reliable quantitative measurement, and is normally free from charging problems in measurements on insulators.

The great advantage of AES appears in microanalysis: the electrons, unlike the x rays of ESCA, can be focused. SIMS is the most sensitive method for surface analysis. Unlike AES and ESCA, it enables hydrogen and its compounds and isotopes to be detected. The information depths are ≤ 10 nm with ESCA, ≤ 5 nm with low-energy AES, and ≤ 1 nm with SIMS. ISS offers the least information depth, only elements in the uppermost layer being detected. If these methods are combined with continuous sputtering of the surface, information on greater depths and depth profiles can be obtained. These general remarks are of course valid for polymers too.

But the surface analysis of polymers requires some special precautions:

1. Polymers are in general nonconducting materials that charge up when they are treated with electrons and ions.
2. Polymers are much more sensitive to electron and ion-induced reactions than metals and oxides.
3. Polymers are very often not UHV compatible because they cannot be baked and contain additives with rather high vapor pressures.
4. Information about chemical bonding is much more important in the polymer field than in inorganic chemistry. The simple detection of C and O is by no means sufficient; one has to know how these elements are connected.

This leads to a somewhat specialized consideration of the methods for surface analysis in polymer application, as was shown in the case of ion-induced reactions in some detail.

With ISS and SIMS one must therefore expect the composition of the surface to be altered by ion-induced reactions and selective sputtering unless the primary ion energy and ion dose are restricted to a very low level (i.e., to establish quasi-static SIMS and ISS conditions). As technical high molecular weight polymers and biological material are unsuitable for baking, it may be impossible to establish UHV conditions. The application of low ion-current densities and low ion doses is often limited because SIMS and especially ISS, having a higher surface sensitivity (smaller information depth) than ESCA, require sputter removal of at least parts of the layer to detect the elements and compounds underneath. Whenever a polymer sample has been subjected to ion bombardment, one has to be very cautious in interpreting any sort of chemical bonding.

SIMS has in general very high sensitivity, and information about chemical bonding can be derived from molecular fragment ions. To compensate for the positive charge induced by the primary ions, polymer samples are simultaneously bombarded with low energy electrons [4]. But where high polymers are concerned, practically every mass number may be obscured by a fragment ion, with the result that unequivocal interpretation of the SIMS spectra becomes more difficult.

Although SIMS and the related technique FAB have been used with great success to detect undecomposed organic molecules, the relevant experiments belong to the field of organic mass spectroscopy, especially as applied to molecules that cannot be vaporized without decomposition. At the moment, however, their application cannot be generalized to include the analysis of the real surfaces, especially where these are polymers. In such cases ESCA is still the most informative method.

All the facts discussed in connection with chemical information, quantitative measurements and in-depth information, are rather independent of the special system under study. This permits generalization and far-reaching estimations for unknown systems without model experiments, which is one of the reasons why ESCA still plays a dominant role in routine surface investigations of technical polymers.

2.5 EXAMPLES OF SURFACE INVESTIGATIONS

The main applications of ESCA are in the study of corrosion and wear [19–21], heterogeneous catalysis [22,23], and polymer technology [16, 24–27].

Metals are of interest to organic chemists and biologists mainly when they are in contact with polymers or biological media. Investigations on metal implants [28] can be taken as an example.

The use of metal implants for osteosynthesis (reconstruction of bones by fixation of their fragments) and alloarthoplasty (replacement of bones and joints by prostheses) has become an indispensable part of orthopedic surgery. But the corrosion of implants by body fluids may result in local inflammatory reactions of the tissue known as metallosis. In particular, when implants are left in the body for long periods of time, the extremely high concentrations of alloy-specific metal ions caused by corrosion may lead to changes in the body's trace element balance. Possible consequences of this excess supply of metals cannot be foreseen yet. Therefore it is of great importance to control the migration of alloy components from an implant into the implant's ambience. The surfaces of metallic implants and especially their passive layers, which play an important role in the inhibition of corrosion, can be investigated by ESCA. This method has been used widely to study oxide layers on metals [19–21]. The experience thus gained is now being applied to metal implants in biological media. While ESCA studies of implants give information on the interface between metal and surrounding tissue, instrumental neutron activation analysis, for example, is able to detect extreme traces of elements in the contact tissue.

A varization osteotomy of the tibia of the left hind leg was performed on 30 one-year-old rabbits of the same strain by removing a wedge-shaped piece of bone with medial base. The fragments were fixed subperiostically in various positions with the aid of stainless steel plates and three to four screws each. A total of 14 normal rabbits (without implants) of the same strain serving as a control group were kept under the same conditions. The rabbits were sacrificed 12 months after the operation.

The thickness of the oxide layer that had been formed on the metal implants was found to be ~3–5 nm on all samples. The oxide layer contained Cr in the trivalent form. The chemical shift of the Fe oxide peak (3.6 eV) was considerably lower than that for Fe(III) in the form of pure Fe_2O_3. It follows that Fe is present in the surface layer both as Fe(II) and as Fe(III).

Ni was observed almost entirely in the metallic state, which indicates that it does not contribute much to the formation of the surface oxide layer. Mo was seen in different chemical forms. When oxidation took place in air, Mo was present in the oxide layer almost exclusively in the hexavalent form. In the animal implant experiment, however, a small amount of Mo(IV) was detected. Quantitative determination was not possible because this oxidizes quickly in air.

The Fe(oxide)–Cr(oxide) intensity ratio shows that the concentration of Cr in the passive layer on the used implants has increased greatly (approximately seven times). But there is also a slight Cr accumulation in the metal immediately beneath the passive layer. So it is concluded that the individual alloying components are dissolved at different rates when the implant comes into contact with water. Fe and Ni are dissolved more quickly than Cr and Mo. The corrosion rates decrease in the order Co > Ni > Fe > Mo > Cr. Obviously, Cr oxides remain on the surface and accumulate there until a passive layer has been formed. Afterward, Fe and Cr are dissolved to the extent that corresponds to the composition of the alloy. Therefore, assuming that the implant is entirely undamaged, the transfer of corrosion products from the implant to the tissue should occur in element ratios corresponding approximately to that of the alloy. If, on the other hand, an implant suffers continuous friction or other mechanical damage sufficient to break the oxide film, there will be a considerably higher total concentration of the alloying elements and a proponderance of Fe and Ni over Cr and Mo.

A combination of ESCA and Ar^+ ion sputtering has been used to determine the surface and subsurface composition of dental enamel [21]. Beneath a mucopolysaccharide coating, untreated enamel was found to have invariant P and Ca concentrations. The surface of enamel treated with acid phosphate–fluoride gel was shown to be almost entirely converted to calcium fluoride. A noncalcium phosphate species was also found to be present in the surface layer. At a depth lower than 60 nm there was a gradual increase in the phosphorus concentration, which approached the value for untreated enamel. The investigation of adhesive bonding of various materials to hard tooth tissues has also been assisted by ESCA [30].

High polymers are encountered in many forms (e.g., as fibers, foams, engineering plastics, or coatings). The problems that can be investigated by the methods of surface analysis, especially ESCA, are correspondingly diverse. They include not merely such typical surface changes as the effects of etching, weathering, and gas discharges, but also the contamination of polymer surfaces and the detection of migration by such polymer additives as lubricants, antioxidants, and antistatic agents, the identification of which is important because they may prevent bonding, printing, or other treatments. Interface problems also arise when there is contact between polymers and pigments, fillers, glass fibers, and so on. Some examples will be briefly reviewed.

Several publications describe investigations concerned with the etching of polytetrafluoroethylene (PTFE) as a means of facilitating the bonding of this material [31,32]. Satisfactory sticking of adhesives can be obtained by, for example, treating the surface with Na in liquid ammonia. In the etched layer only C and O in one binding state each were detected by ESCA, with N, Na, and F not giving a signal. It follows that a layer of the PTFE thicker than the escape depth of the photoelectrons of F or of C in CF_2 groups was decomposed. The less severe treatment with Na naphtha lene tetrahydrofurane gives a considerably thinner etched layer; signals from the PTFE below it are still obtained. The adhesiveness is lost if the plastic is irradiated with UV light in the presence of air. The growth in the intensity of C1s signals from CF and CF_2 groups can be explained by the formation of gaseous products during the deactivation and the associated chemical removal of the etched layer. Heating to 400°C in air or irradiation with UV light in air removes the etched layer entirely, with the result that the ESCA spectrum again shows the signals of the pure PTFE.

Another way in which the surface energy of polymers can be increased is to subject them to corona discharge treatment. This treatment is applied to polyethylene, for example, to make the material easier to print on, easier to bond, and so on. ESCA reveals that this treatment results in oxidation of the carbon atoms at the surface, some even to acid or acid anhydride groups (Fig. 20) [16,33]. Other oxidation products with smaller chemical shifts (ketone, aldehyde, alcohol, and peroxide groups) are represented by almost equal intensities, with the result that no minimum is formed in the region of 1196–1201 eV. Long-lived radicals may also result in the formation of lines within this shift range. Ozone treatment of polybutadiene, on the other hand, results mainly in the formation of ketone groups. Figure 21 shows how the intensity for the oxidation products is reduced during the storage of the treated samples. Volatile products are formed by rearrangement processes at the surface, and some of the oxi-

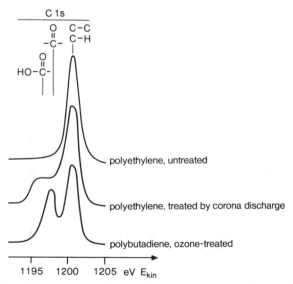

Figure 20.   Effect of surface treatments of polyethylene and polybutadiene.

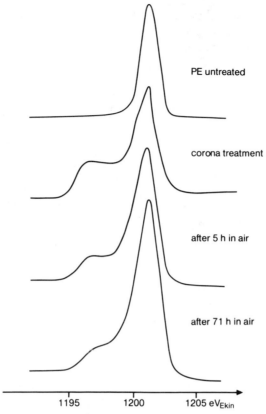

Figure 21. C1s spectra of corona discharge treated polyethylene after storage in air.

dized material may migrate into the bulk. The oxidation products can also be identified by IR and electron spin resonance spectroscopy; to obtain sufficient intensities, however, very high radiation doses are necessary. When ESCA is used as a surface analysis method, effects can be detected, and determined quantitatively, at technically normal irradiation doses. In order to overcome the problems associated with several groups having similar chemical shifts, derivatization techniques for labeling specific functional groups on surfaces have been developed. A review is given in Ref. 34. Some results for other polymers (ABS, polypropylene) subjected to oxygen plasma treatments [35] and noble gas and nitrogen plasma treatments [36] have been obtained.

Briggs et al. [37] have reported on ESCA investigations of polyethylene and polypropylene that were treated with chromosulphuric acid to improve their wettability. Here again, oxidized carbon atoms and the incorporation of SO_3H groups were detected. Polyethylene (PE) that is melted onto aluminum undergoes oxidation—the more so if it is high density, rather than low density, polyethylene [36,39]. Thus the usually good adhesion of extruded PE to aluminum is probably attributable to oxidation processes of this kind. By virtue of its greater surface sensitivity, ESCA enabled the oxidation of PE to be detected, whereas the ATR–IR technique failed to reveal it.

Another application of surface analysis methods is the detection of surface deposits (flame retardant or antistatic agents, fiber finishing agents [40], etc.) and of surface exudations by substances added to polymers, (e.g., lubricants). For example, in a quantitative ESCA investigation Riggs and Beimer [31] found that the amount of an amide lubricant at the surfaces of polyethylene films was proportional to the film thickness.

Another example is the detection and at least semiquantitative determination of calcium stearate, which is often added to polyethylene as a lubricant. An excessive calcium stearate concentration at the surface impedes welding. With ESCA the problem can be solved by measuring the Ca 2p and Ca 2s lines (Fig. 22). The oxygen line is less suitable, because the migration of the Ca stearate begins at an elevated temperature at which at least superficial oxidation of the polyethylene cannot be ruled out.

Surface analysis methods are suitable for the study of the processes involved in the galvanization of plastics, as well as of the possible blemishes. For example, pickling and activation products at the surface can be identified [41]. In the case of ABS polymers the N1s intensity (from the SAN content) reveals the extent to which the surface is homogeneously covered by pickling or activation products (such as Pd).

If plastics are to be used for temporary or permanent organ replacements, it is not sufficient for them to have the necessary physical and chemical bulk properties (such as strength and elasticity). They must also be biocompatible. This means that they must be antithrombogenic, that they must not cause toxic or allergic reactions, that they must not be carcinogenic, that they must not produce changes in plasma proteins and enzymes, that they must not destroy the surrounding tissue, and that they must not cause immune reactions. The antithrombogenicity requirement is the most important one. The adsorption of protein by the plastic, which begins as soon as the surface has been wetted by the blood and which may be more or less pronounced, is of considerable importance in this connection. Model experiments have shown that PTFE is hardly wetted by hemoglobin, which merely forms islands on the surface in this case, whereas a Pt surface is entirely covered [42]. Evidence was found that in the case

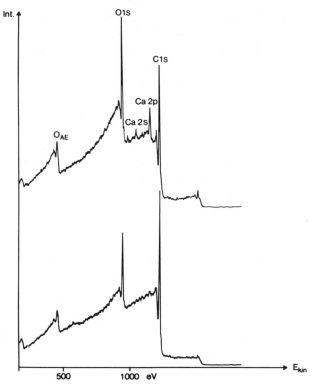

Figure 22.   ESCA wide scan spectra of polyethylene with exuded Ca stearate after exposure to elevated temperature (top) and partial surface oxidation (bottom).

of Pt a denaturation occurs and that it leads to a preferential orientation of
N atoms at the surface.

Several investigations have been concerned with the application of
hydrogel coatings [e.g., coatings consisting of HEMA (hydroxyethyl
methacrylate)] to polymers with suitable mechanical properties in order
to confer biocompatibility on them [43,44]. The good biocompatibility is
attributed to the very high absorptivity for water. ESCA investigations
therefore have been performed on both dried and deep-frozen materials
containing water. Structural changes in consequence of dehydrogenation
and rehydrogenation were observed, and where some systems were concerned, changes in the nature of the surface deposit, which were attributed to some diffusion of hydrogel molecules into the substrate, were
seen. Polyurethanes too show good biocompatibility in many cases. Here,
however, the polymer production and preparation should be carried out

with the greatest care. For solvent-cast films of these polymers it was shown that the side of the film facing the air during preparation differed appreciably from the side in contact with the glass casting surface [45]. The first attempts to correlate blood platelet activation and surface composition have been undertaken [46,47]. It would seem that N in the form of urea or urethane linkages is not necessarily the principal species involved with adsorption of plasma proteins and retention of platelets and that the aromatic rings are possibly more adsorptive. In the case of a polyether urethane–polydimethylsiloxane copolymer, it has been found that "blood compatible" surfaces have a high content of urethane, whereas "blood incompatible" surfaces consist largely of siloxane [48]. Pyrolytic carbon is one of the very few synthetic materials generally accepted as suitable for long-term blood contact applications. ESCA studies showed that approximately 90% of the surface C atoms are not chemically bonded to oxygen. The remaining 10% contain three major types of oxygen functionality: etherlike, quinonelike, and esterlike or carboxylic in nature [49].

Few attempts have so far been made to apply ESCA to biological tissues, bacteria cells, and so on. (A review is given in Ref. 50.) That is partly because such samples are considerably more difficult to prepare than are samples of polymers. Another reason is that protein molecules cannot be adequately characterized, and hence differentiated, by ESCA. Similar difficulties apply to active substances (e.g., those of pesticides from botanical substrates).

Nevertheless attempts have been made to characterize polymer surfaces (untreated and corona-discharged) that are used as substrates for in vitro cell cultures and to correlate the surface properties to the morphology and the properties of the cell culture [48]. A residue of biochemical matter remains on a surface after cells have been cultured on it and then removed. ESCA can give some information on the composition of such cell exudates, on the layer thickness, and on the degree of coverage of the substrate.

In the investigation of the surfaces of bacteria the signals of C, O, and N are generally not very specific. In many cases the quantitative evaluation reveals an influence of the preparation technique, the result depending on whether the sample has been frozen, freeze-dried, or critical-point-dried, for example. Much interest has been shown in the distribution of P (from teichoic acid) in Gram-positive and Gram-negative bacteria [51]. The depth distribution was measured by oxygen plasma etching. Separation of the two membranes in *Escherichia coli* B by their P signal was not achieved. By means of the same technique it has been demonstrated that induction of active DNA synthesis in hepatoma cells is accompanied by a striking increase in calcium concentrations in the cellular interior, whereas the Ca concentrations in the outermost layers of the cell surface are little changed [52]. In an investigation of human red blood corpuscles T1 was found after the removal of an approximately 10-nm-thick surface layer. The signal disappeared after further Ar^+ sputtering. Fe was not found until a roughly 50-nm-thick layer had been removed [53]. Among the changes that occur in a mammalian cell after malignant transformation are alterations in the cell surface [54]. Phenomenological observations should be explicable by analyses, but the preparation of samples without artifacts is exceptionally difficult.

The use of surface analysis methods such as ESCA in the study of polymer surfaces is well developed, but for biological and biomedical applications it is still in its infancy. The examples discussed should have demonstrated, though, that surfaces and, if suitable preparation techniques are used, interfaces too can be characterized.

Where complex problems are involved, surface analysis methods should be used in synergistic combination and with other microscopic and spectroscopic techniques. It is also essential to develop adequate preparation techniques, especially for biological material and in the case of interface analysis. The use of different preparation techniques and the conduct of parallel investigations in cases where several methods can be applied independently are necessary not simply because they increase the supply of information but because the reciprocal checking they permit can reveal systematic errors and preparative artifacts.

REFERENCES

1. K. Siegbahn et al., *ESCA: Atomic, Molecular and Solid State Structure Studied by Means of Electron Spectroscopy,* Almquist and Wiksells, Uppsala, 1967.

2. C. R. Brundle, J. Vac. Sci. Technol. *11,* 212 (1974).

3. C. D. Wagner, W. M. Riggs, L. E. Davies, J. F. Moulder, and G. E. Muilenberg, *Handbook of X-Ray Photoelectron Spectroscopy,* Perkin-Elmer/Physical Electronics Div., Eden Prairie, 1979.

4. R. Holm and S. Storp, Methoden zur Untersuchung von Oberflächen, in *Ullmann's Encyclopädie der technischen Chemie,* Vol. 4, Aufl. Bd. 5, pp. 519–576, Verlag Chemie, Weinheim, 1981.

5. C. R. Ginnard and W. M. Riggs, Anal. Chem. **44,** 1310 (1972).

6. C. D. Wagner, J. Electron Spectrosc. *10,* 305 (1977).

7. D. T. Clark, D. B. Adams, A. Dilks, J. Peeling, and H. R. Thomas, J. Electron Spectrosc. *8,* 51 (1976).

8. N. Winograd, A. Shepard, D. H. Karweik, V. J. Coester, and F. K. Fong, J. Am. Chem. Soc. *98*(8), 2369 (1976).

9. H. Falk, K. Grubmayr, K. Thirring, and N. Gurker, Monatsh. Chem. *109*, 1183 (1978).

10. J. Baddiley, I. C. Hancock, and P. M. A. Sherwood, Nature *243*, 43 (1973).

11. D. L. White, S. B. Andrews, J. W. Faller, and R. J. Barrnett, Biochim. Biophys. Acta *436*, 577 (1976).

12. K. S. Kim, W. E. Baitinger, J. W. Amy, and N. Winograd, J. Electron Spectrosc. *5*, 351 (1974).

13. R. Holm and S. Storp, Appl. Phys. *12*, 101, (1977).

14. R. Holm and S. Storp, J. Electron Spectrosc. *16*, 183 (1979).

15. C. D. Wagner, Anal. Chem. *44*, 1050 (1972).

16. R. Holm and S. Storp, Surf. Interf. Anal. *2*, 96 (1980).

17. J. Peeling, D. T. Clark, I. M. Evans, and D. Boulter, J. Sci. Food Agr. *27*, 331 (1976).

18. D. Jones, G. Distefano, C. Toniolo, and G. M. Bonora, Biopolymers *17*, 2703 (1978).

19. J. Olefjord, Mater. Sci. Eng. *42*, 161 (1980).

20. K. Asami, K. Hashimoto, and S. Shimodaira, Corr. Sci. *18*, 151 (1978).

21. J. E. Castle, Application of XPS analysis to research into the causes of corrosion, in *Applied Surface Analysis*, T. L. Barr and L. E. Davies, Eds., ASTM STP 699, 1978, p. 182.

22. T. E. Fischer, J. Vac. Sci. Technol. *11*, 252 (1974).

23. H. P. Bonzel, Surf. Sci. *68*, 236 (1977).

24. D. T. Clark and W. J. Feast, J. Macromol. Sci.-Rev. Macromol. Chem. *C12*, 191 (1975).

25. D. T. Clark and H. R. Thomas, J. Polymer Sci., Pol. Cnem. Ed. *16*, 791 (1978).

26. D. T. Clark, The modification, degradation and synthesis of polymer surfaces studied by ESCA, in *Photon, Electron and Ion Probes of Polymer Structure and Properties*, D. W. Dwight, T. J. Fabish, and H. R. Thomas, Eds., ACS Symposium Series Vol. 162, Washington, D.C., 1981, p. 247.

27. A. Dilks, Anal. Chem. *53*, 802A (1981).

28. J. Hofmann, R. Michel, R. Holm, and J. Zilkens, Surf. Interface Anal. *3*, 110 (1981).

29. D. M. Hercules and N. L. Craig, J. Dent. Res. *55*, 829 (1976).

30. R. L. Bowen, J. Dent. Res. *57*, 551 (1978).

31. W. M. Riggs and R. G. Beimer, Chem. Technol., 652 (November, 1975).

32. H. Brecht, F. Mayer, and H. Binder, Angew. Makromol. Chem. *33*, 89 (1973).

33. A. R. Blyth, D. Briggs, C. R. Kendall, D. G. Rance, and V. J. I. Zichy, Polymer *19*, 1273 (1978).

34. C. T. Batich, R. C. Wendt, Chemical labels to distinguish surface functional groups using XPS (ESCA), in *Photon, Electron and Ion Probes of Polymer Structure and Properties*, D. W. Dwight, T. J. Fabish, and H. R. Thomas, Eds., ACS Symposium Series Vol. 162, Washington, D.C., 1981, p. 221.

35. J. M. Burkstrand, J. Vac. Sci. Technol. *5*, 223 (1978).

36. H. Jaruda, H. C. Marsh, S. Brandt, and C. N. Reilley, J. Polym. Sci. Polym. Chem. Ed. *15*, 991 (1977).

37. D. Briggs, D. M. Brewis, and M. B. Konieczko, J. Mater. Sci. *11*, 1270 (1976).

38. D. Briggs, D. M. Brewis, and M. B. Konieczko, J. Mater. Sci. *12*, 429 (1977).

39. D. T. Clark, W. J. Feast, K. R. Musgrave, and J. Ritchie, J. Polym. Sci. Polym. Chem. Ed. *13*, 857 (1975).

40. P. M. Soignet, R. J. Berri, and R. R. Benerito, J. Appl. Polym. Sci. *20*, 2483 (1976).

41. K. Richter, G. Kley, J. Robbe, J. Gähde, and J. Löschke, Z. Chem. *18*, 390 (1978).

42. B. D. Ratner, T. A. Horbett, D. Shuttleworth, and H. R. Thomas, J. Coll. Interface Sci. *83*, 630 (1981).

43. B. D. Ratner, P. K. Weathersby, A. S. Hoffman, M. A. Kelly, and L. H. Sharpen, J. Appl. Polymer Sci. *22*, 643 (1978).

44. D. K. Gilding, R. W. Paynter, and J. E. Castle, Biomaterials *1*, 163 (1980).

45. D. J. Lyman, K. Knutson, B. McNeill and K. Shibatani, Trans. Am. Soc. Artif. Int. Organs *21*, 49 (1975).

46. V. Sa da Costa, D. Brier-Russell, E. W. Salzmann, and E. W. Merrill, J. Coll. Interface Sci. *80*, 445 (1981).

47. B. D. Ratner, ESCA and SEM studies on polyurethanes for biomedical applications, in *Photon, Electron and Ion Probes of Polymer Structure and Properties*, D. W. Dwight, T. J. Fabish, and H. R. Thomas, Eds., ACS Symposium Series Vol. 162, Washington, D.C., 1981, p. 371.

48. J. D. Andrade, G. K. Iwamoto, and B. McNeill, XPS studies of polymer surfaces for biomedical applications, in *Characterization of Metal and Polymer Surfaces*, Vol. 2, L. H. Lee, Ed., Academic, New York, 1977, Vol. 2, p. 133.

49. R. N. King, J. D. Andrade, A. D. Haubold, and H. S. Shim, Surface analysis of silicon: alloyed and unalloyed LTI pyrolytic carbon, in *Photon, Electron and Ion Probes of Polymer Structure and Properties*, D. W. Dwight, T. J. Fabish, and H. R. Thomas, Eds., ACS Symposium Series Vol. 162, Washington, D.C., 1981, p. 405.

50. M. M. Millard, Surface characterization of biological materials by x-ray photoelectron spectroscopy, in *Contemporary Topics in Analytical and Clinical Chemistry,* Vol. 3, D. M. Hercules, G. M. Hieftje, L. R. Snyder, and M. A. Everson, Eds., Plenum, New York, 1978, p. 1.

51. M. M. Millard, R. Scherrer, and R. S. Thomas, Biochem. Biophys. Res. Comm. *72,* 1209 (1976).

52. L. Pickart, M. M. Millard, B. Beidermann, and M. M. Thaler, Biochem. Biophys. Acta *544,* 138 (1978).

53. R. G. Meisenheimer, J. W. Fischer, and S. J. Rehfeld, Biochem. Biophys. Res. Comm. *68,* 994 (1976).

54. M. M. Millard and J. C. Bartholomew, Anal. Chem. *49,* 1290 (1977).

LASER MICROPROBE MASS ANALYZER

PETER FRITZ SCHMIDT

Institute for Medical Physics
University of Munster
Munster, Germany

3.1 INTRODUCTION

Mass spectrometry is one of the most versatile analytical methods, because it can be used to obtain qualitative and quantitative information about the organic and inorganic constituents of specimens with a high sensitivity. A further special advantage of mass spectrometry is the ability to elucidate molecular structures.

In principle, a mass spectrometer consists of three sections: a device for producing positively or negatively charged ions from a specimen; a mass analyzer, which sorts out the resulting mixture of ions according to their mass-to-charge ratios (m/e); and a detector, which provides output signals corresponding to the amount of each ionic species. For the determination of elements in biological tissues there are various mass spectrometric techniques [1], but with respect to the scope of microprobe functions—the determination and localization of elements present at trace concentration in small specimen volumes ($\leqq 1 \ \mu m^3$)—there are two recent developments that allow the excitation of microvolumes into an ionized state and the analysis of these ions. These are the ion microprobe, based on secondary ion mass spectrometry (SIMS), and the laser microprobe mass analyzer (LAMMA), based on laser-induced generation of ions. The potentialities of the ion microprobe are shown in the chapter by Burns-Bellhorn in this book. The following is concerned with the applicability of LAMMA in biomedical research. In LAMMA, ions are generated by focusing a pulsed laser beam onto a specimen area of interest. Thus a certain amount of the specimen is evaporated and partly ionized. This microplasma contains neutral fragments, but also atomic and molecular ions representing the constituents of the evaporated material. The mode of ion generation depends on the laser power density in the focus of the optical microscope. Using high power densities, LAMMA is used to analyze the atomic constituents. With low exciting laser power densities more molecular fragments are obtained. The analytical part of LAMMA is a time of flight mass spectrometer, the detector a secondary electron multiplier.

Another possibility for analyzing the ions of laser-evaporated specimen volumes is given by an optical spectrometer: Laser microprobe analysis by emission spectroscopy [2]. This is a well-tried instrument, used mainly in technical applications, but the low sensitivity does not permit the analysis of elements in low concentrations. On the other hand, several attempts have been made to construct an instrument that enables the mass spectrometric analysis of laser-induced plasma, using, for example, magnetic sector spectrometers [3,4]. The detection limit is about 10^{-14} to 10^{-18} g. A Laser Microprobe Mass Analyser LAMMA 500, has been developed [5–9] primarily for biomedical applications. This instrument enables a morphologically controlled analysis of small specimen volumes from histological sections or small particles with a very low limit of detection (10^{-18} to 10^{-20} g), that permits the analysis of elements in trace concentrations at a subcellular level. For the analysis of bulk material, a LAMMA 1000 has been developed.

3.2   THE LAMMA INSTRUMENT

3.2.1   Technical Description

The technical concept of the laser microprobe mass analyzer is realized by a combination of an optical microscope with a laser source, a time-of-

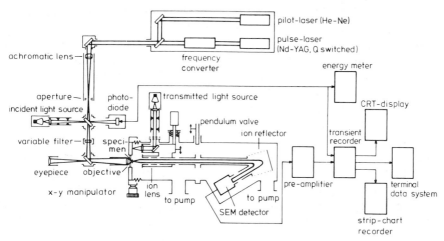

Figure 1.   Schematic diagram of the LAMMA 500 instrument (Leybold Heraeus, Cologne).

flight (tof) mass spectrometer and an open secondary electron multiplier as detector. The basic components of LAMMA 500 are shown in Fig. 1.

The *specimen,* preferably a semithin section, is mounted on a 3.08-mm grid normally used for transmission electron microscopy. The grid fits into a *x–y*-movable stage underneath a thin quartz cover slide that serves as vacuum seal and optical window. For specimen exchange, the specimen chamber can be sealed off from the vacuum in the spectrometer via a pendulum valve. Thus specimen exchange is possible within a few minutes. The *optical microscope* with exchangeable UV-transmitting objectives allows the observation of the specimen with reflecting or transmitting illumination and, at the same time, the objectives are used to focus a very short and intense pulse of a Q-switched, frequency-quadrupoled Nd-YAG-laser ($\lambda = 265$ nm, peak half-width ≈ 15 ns) onto a selected area to be analyzed. The intensity of the laser light can be varied by means of interchangeable filters of definite extinction. In the focus, the field strength of the laser light (power density) is usually between 10^7 and 10^{14} W/cm^2. The area to be analyzed is always indicated by a red spot of a He–Ne laser, which continuously emits light in the red region and which is adjusted to be co-linear to the invisible UV light of the Nd–YAG laser. To select the specimen structure to be analyzed the specimen holder can be manually shifted in the *x*- and *y*-direction by means of two micrometers. In order to perform the analysis, the transmitted light condenser is replaced by an ion optic system and by a short laser pulse a distinct specimen volume is evaporated and partly ionized. In sections or foils the laser pulse causes perforations. The generated ions are analyzed in a time-of-flight mass spectrometer.

The *time-of-flight (tof) mass spectrometer* consists of an accelerating and focusing electrostatic lens underneath the specimen, a field-free drift tube, an ion reflector at the end of this tube, and a second drift tube. An open secondary electron multiplier (SEM) is used as detector. The first dynode serves as ion–electron converter. The ions generated by the laser pulse are accelerated into the tof mass spectrometer in which ions of different m/e ratios are discriminated according to their different flight times. The spectrometer can be used for positive or negative ions, depending on the polarity of the potentials at the electrodes. Furthermore, the tof mass spectrometer allows the registration of a complete mass spectrum for each laser shot either as a spectrum of all positive or negative ions.

The *detector* delivers an analog signal at its output. This is fed to a digital transient recorder of 2048 addresses and a dynamic range of 8 bits. Sample intervals are usually 10 or 20 ns. The selected delay time and sampling rate determine the mass range (max. $m/e = 2000$). The stored mass spectrum is displayed on a CRT or can be either further processed by a computer or recorded by a pen recorder. (For further details see Ref 10.)

3.2.2   Performance of the Instrument

The capabilities of LAMMA are demonstrated by two mass spectra obtained by the analyses of a standard specimen prepared from epoxy resin (Spurr's low viscosity medium), which had been homogeneously doped with the elements Li, Na, K, Sr, and Pb at a concentration of 1 mmole/liter. The section was 0.3 μm thick. The recorded spectra (Fig. 2) are a plot of the ion signal versus time of flight of the ions, respectively, versus mass-to-charge ratio (m/e). This is shown by the following:

After the generations of ions and acceleration by the accelerating lens, (lens voltage U), the ions of mass m have the kinetic energy:

$$eU = \frac{1}{2} mv^2 \tag{1}$$

e, charge; m, mass; and v, velocity of a particular ionic species. It follows that

$$v = \sqrt{2eU/m}$$

taking $\sqrt{2eU}$ as a constant

$$v = k/\sqrt{m} \tag{3}$$

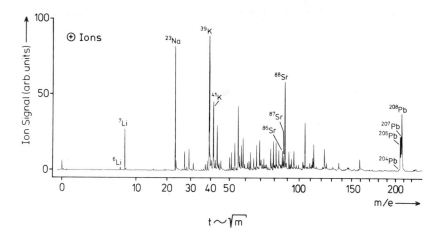

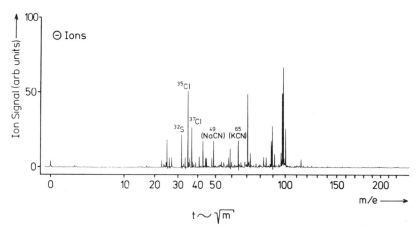

Figure 2. Positive and negative LAMMA spectra of a thin section of "Spurr's low viscosity medium" doped with Li, Na, K, Sr, and Pb as crown ether complexes at a concentration of 1 mmole/liter.

That means that the velocity v of an ion of mass m is proportional to the root of the inverse of its mass. Because $v = s/t$ (s, flight path; t, flight time), and with a constant c ($c = s/k$) it follows that

$$t = c\sqrt{m} \qquad (4)$$

Thus ions of different mass need different times to reach the detector. Typical transient times are 11 μs for ^{7}Li and 60 μs for ^{208}Pb [11]. For calibration, the value c will be derived from those peaks of a mass spectrum that are unequivocally identified [e.g., from the grid of the specimen mounting (Cu or Ni)]. This enables one to label the mass peaks in the mass spectra. In Fig. 2 the ion signals of the elements (isotopes) and some fragments ions of the doped chemical compounds are labeled. The detection of isotopes provides support for the identification of elements, because the amplitudes of the ion signals must correspond to the relative abundances of the isotopes. In addition, characteristic fragment ions from the epoxy resin can be seen (not labeled).

In general, the amplitudes of the ion signal correspond to the number of ions present in the evaporated specimen volume, but as one can see from the spectra, equimolar concentrations of different ions yield different peak heights. This is attributed to the fact that the ratio of ionized to neutral atoms (ion yield) in the evaporated material depends on the ionization potential of a particular ion, (e.g., ionization potential for potassium 4.4 eV and for lead 7.42 eV) and the binding energy of the respective element.

Thus by means of this spectra the feasibility of the laser microprobe mass analyzer can be demonstrated:

Analysis of all elements in trace concentrations.

Discrimination of isotopes (tracer techniques).

Analysis of molecules by fingerprinting or by identification of molecules or molecular fragments.

A further advantage of LAMMA is the speed of analysis, because by use of a tof mass spectrometer, very short measuring times are possible, limited only by the time needed for localizing the area of interest and producing a single laser shot and for pen recording (about 3 min). Therefore LAMMA enables statistically verified results within a relatively short time, provided that the evaluation of the mass spectra is easy.

3.2.3 Interaction of Laser Light and Specimen

Using objectives of an optical microscope, laser light can be focused to small spot sizes with extremely high intensities. This feature is used in LAMMA, where by interaction of a focused laser pulse of high power density (10^7–10^{14} W/cm^2) a small volume of the specimen is evaporated and converted into a plasma that contains neutral fragments, atomic and molecular ions (negative and positive ions, often with similar efficiencies). Thus the area to be analyzed is destroyed. Since the duration of the laser impulse is very short ($\approx$15 ns), heat conduction to the surrounding area of the specimen can be neglected. Interaction is strictly limited to the focus diameter [11]. There is insufficient knowledge concerning the nature of this interaction process, but current research is dealing with this problem [12,13]. This is especially important with respect to quantitative analysis, because the ion yield for the various elements depends on different parameters, such as ionization potential (mainly determining the quantity of positive ions), electron affinity (determining the quantity of negative ions), and matrix effects like strengths of chemical bounding [14].

As seen in the mass spectra of the standard specimens, spectra of biological sections always have two components in the mass spectra: ion signals from the biological specimen itself (inorganic and organic components) and ion signals from the organic materials used for embedding. The occurrence and disappearance of atomic or molecular peaks depend strongly on the laser power density in the laser focus. Therefore the chosen laser intensity is important for the performance of the analysis. A characterization of the influence of different laser power densities to the specimen is possible by the "damage threshold," which means the minimal power density producing a just visible perforation of the specimen. In general, the smaller the exciting laser power density—variable by interchangeable filters—near the damage threshold, the more molecular fragments or molecule peaks are obtained, a mode called "laser-induced desorption" (LID), employed for microprobe mass spectrometry of organic constituents. However, with increasing power densities the destruction of the matrix increases and molecular fragments, ions and atomic ions are produced, a mode employed for the analysis of element composition [15,16]. For elemental analysis of biological sections, laser power densities three to five times above the damage threshold are needed. Thus the excitation of the specimen volume is sufficiently high so that characteristic and reproducible mass spectra are formed. The relative magnitude of the ion signals is independent from the laser power density.

For practical use, as the laser intensity can be experimentally controlled by filters, it is possible to find the optimal power densities by estimating the visible effects and by examining the spectra. The irradiation conditions depend on (1) the specimen or particular specimen structures (influence of absorption); (2) the analytical problem: elements, element concentrations (influence of the ionization potential, high ionization potential, and/or low element concentration needed high power densities in the focus); and (3) the required spatial resolution, which limits the use of too high power densities for high ionization efficiency, because the diameter of the perforation increases with increasing power density.

3.2.4   Resolution in Mass Spectrometry

In mass spectrometry the "mass resolution" is given by the smallest mass difference that can be resolved. It is defined as the ratio $m/\Delta m$. This ratio

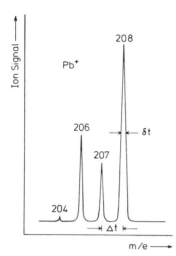

Figure 3. Mass-resolution A demonstrated at the lead isotopes, whereby $A = m/\Delta m = m(\Delta t/\delta t)$ with $\Delta m = 1$, and $\Delta t \sim 1/\sqrt{U \cdot m}\ l$ (l = length of the drift tube) and $\delta t \sim \sqrt{m/U} \cdot \Delta E$ (ΔE = energy distribution of the starting ions, U = accelerating voltage of the lens).

indicates that mass to charge (m/e) value to which adjacent peaks ($m = 1$) are resolved (Fig. 3). In LAMMA the mass resolution is about 600 (determined at mass 208 of the lead isotopes) measured on standard specimens [10]. The equation $m/\Delta m = 600$ means the masses 600 and 601 can be resolved. This is sufficient to discriminate isotopes for all elements and to perform molecular mass spectrometry up to $m/e = 600$ to an accuracy of one mass number.

1.   The width of the ions initial energy distribution is critical for resolution, as a spread in flight time can influence the peak halfwidth δt (Fig. 3) [10]. The starting energy of the ions depends on the laser power density in the focus position, the specimen thickness, and the absorption effects of the specimen. But by using an ion reflector the mass resolution is nearly independent from the primary energy distribution [10].

2.   A further influence is given by the time resolution of the detecting system, the transit time spread of the multiplier, and the frequency bandwidth of the preamplifier and the transient recorder. For the analysis of biological specimens the time resolution is sufficient [10].

3.2.5   Spatial Resolution

For specimen observation, an optical microscope with a magnification of up to 1200 is used, which means that imaging of specimen structures is limited to the resolving power of a light microscope. Additionally, the construction of the optical device allows the observation with phase contrast. The structures to be analyzed should be visible by the optical microscope. For low contrast specimens the correlation between site of analysis and cellular structures is difficult. In this case it is possible either to stain the sections with toluidine blue or to image the specimen by a transmission electron microscope before and after LAMMA investigations, in order to localize the site of perforation [17]. In order to perforate the specimen, the UV wavelength is usually used, which allows a small laser focus on the specimen using the 100× objective. The minimum diameter of the spot is approximately 0.5 μm at a specimen thickness $\leq$0.3 μm. But normally in most applications the perforation has a diameter of typically 0.8–1 μm, in a section thickness of about 0.5 μm. Depending on the thickness of the specimen the evaporated volume is about 10^{-12}–10^{-13} cm^3. Assuming a density of 1, this evaporated volume corresponds to 10^{-12}–10^{-13} g. But larger areas (up to 5 μm diameter) can be evaporated, using higher power densities and/or objectives with lower magnification.

3.2.6 Detection Limits

One of the most important features of laser microprobe mass analysis is the very low detection limit—one of the lowest in analytical instruments—combined with a fairly high sensitivity. The detection limits and the sensitivities for some elements are given in Table 1.

TABLE 1. Detection Limits of the LAMMA Instrument for Some Metals[a]

Metal	Absolute (g)	Relative (ppm)
Li	2×10^{-20}	0.2
Na	2×10^{-20}	0.2
Mg	4×10^{-20}	0.4
Al	2×10^{-20}	0.2
K	1×10^{-20}	0.1
Ca	1×10^{-19}	1.0
Cu	2×10^{-18}	20.0
Rb	5×10^{-20}	0.5
Cs	3×10^{-20}	0.3
Sr	5×10^{-20}	0.5
Ag	1×10^{-19}	1.0
Ba	5×10^{-19}	5.0
Pb	1×10^{-19}	1.0
U	2×10^{-19}	2.0

[a] Atomic constituents dissolved as organometallic complexes in an epoxy resin. The analyzed volume is 1×10^{-13} cm^3 of organic matrix material. (From Ref. 11.)

The limits of detection are mainly determined by the ionization potentials of the elements, by the binding state of atomic ions to the surrounding organic matrix, and by bond strength to inorganic molecules. They are restricted by interference with organic fragment ions (from the specimen itself or from the embedding material) on the same nominal mass values as the elements to be analyzed. This creates problems for the detection of elements in a concentration near the detection limit, especially for elements like Cu [18], Zn, Fe, and Mn. Therefore in the case of frozen-dried and plastic-embedded material, one should make sure beforehand that the chosen plastic does not contain organic molecules that give rise to peaks that may interfere with peaks from the specimen itself. Furthermore, by means of computer processing a subtraction of the organic background can improve detection limit. Another way to reduce the organic background is offered by low temperature oxygen plasma microincineration [19].

3.2.7 Quantitation of Analysis

Quantitation of LAMMA results involves the measurements of the amplitudes of an ion signal and translating this into a concentration. Therefore it is necessary to know the relationship between the amplitude of an ion signal and the corresponding amount of atoms or molecules in the analyzed volume. At present no theoretical model can be realized, because the process of interaction of laser light and matter is partly unknown (see Section 2.3). Additionally, the overall transfer function of the instrument, including parameters such as ionization cross section, ion–electron recombination probability, transmission in the mass spectrometer, ion–electron conversions rate of the first dynode of the multiplier, must be known [14]. For quantitation, therefore, one has to resort to empirical procedures.

As the mass peaks in the recorded spectra correspond to the amount of the particular ionic species in the evaporated matter, quantitative analysis is feasible either if the exact volume of the evaporated material is known or equal amounts of the material are always evaporated. By using laser radiation, the amount of the evaporated volume is sensitive to fluctuations in the laser intensity and to specimen structures with different absorption characteristics. This can result in variations of the evaporated volume with different amplitudes of the ions signals. Therefore the absolute amplitude of the ion signal is not a reliable basis for quantitation. In principle, the estimation of the evaporated volume is possible. By means of electron microscopy the area of the perforation can easily be measured. Problems arise in the exact measurement of the section thickness. There are several methods [20] (e.g., measurement of the electron density in a TEM, microinterferometry, interference microscopy), but these procedures are circumstantial, because the specimen thickness in the region of interest must be determined.

To estimate the precision of LAMMA, perforations on Epon sections of 0.3-μm thickness were performed while keeping all instrument parameters constant. The diameter of the perforations varies less than 10% from shot to shot [8]. LAMMA analysis of special standards—Epon sections doped with different elements in various concentrations—results in a linear correlation between element concentration and the amplitude of the recorded mass signals (Fig. 4). Standard deviations were 5% at higher concentrations (>10 ppm) and 15% at lower concentration levels [9]. Due to the limited dynamic range of the transient-recorder for high concentrations the correlation between element concentration and the ion signal could be nonlinear, but nevertheless unequivocally correlation is possible.

The linear correlation between the amplitude of the ion signal to the concentration of the element in the evaporated sample permits one to

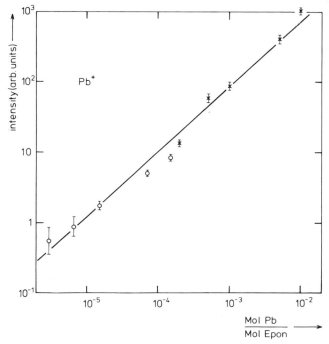

Figure 4. Calibration plot for Pb in standard specimens. Bars indicate S.D. (From Ref. 11.)

establish *concentration ratios* of various elements in distinct specimen structures, e.g., K–Na–Ca ratios from different fibers of an isolated frog sartorius muscle [11]. In the analysis of homogeneous samples, a quantitation by evaporation of a constant volume is feasible, also provided no instrumental parameters are changed. The magnesium content of erythrocytes of 28 patients with renal insufficiency was determined by atomic absorption spectroscopy (AAS). In parallel studies 60 erythrocytes from each patient were analyzed with LAMMA and the average ion signals of Mg were determined. Figure 5 shows a plot of the average intensity of the ion signal of Mg versus the absolute Mg content. The plot indicates a relatively good correlation ($r = 0.75$).

Erythrocytes are relatively homogeneous specimens but biological specimens normally contain structures with different absorption characteristics. That also limits absolute quantitative analysis, because to evaporate the same volume of specimen and standards to obtain a similar ion yield, the specimen and reference standards must be of the same thickness, mass density, and optical density and of comparable element content and distribution. This is difficult to realize.

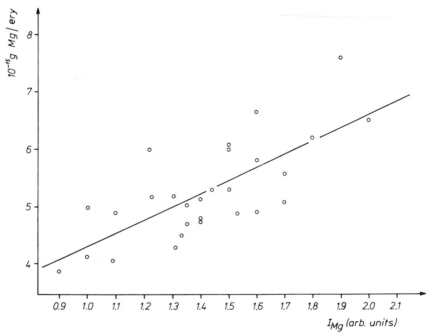

Figure 5.   Plot of the average ion signal of Mg (LAMMA) versus the absolute Mg content (AAS), determined by the analysis of erythrocytes of 28 patients with renal insufficiency.

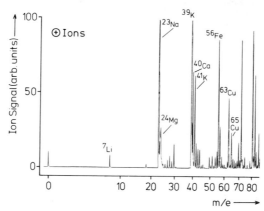

Figure 6.   LAMMA spectrum of control serum (Kontrollogen® L). Content of electrolytes: Sodium, 128 mmole/liter; potassium, 4.6 mmol/liter; calcium, 2.58 mmole/liter; magnesium, 1 mmole/liter. Trace elements: lithium, 0.95 mmole/liter; copper, 18.9 μmole/liter; iron, 18.6 μmole/liter + 0.1 mmole/liter.

But one may correct the variations of evaporated volume by referring to a suitable internal standard, using a matrix element or a doped element of known concentration. This also eliminates the influence of instrumental parameters. It is then possible to estimate concentration ratios by the ion signal of the element of interest relative to the ion signal of the internal standard. This procedure results in a normalization of the spectra that enables one to use the height of an ion signal in the spectra as a basis for quantitation.

In clinical research, information on trace element content is obtained by the analysis of body fluids, and some LAMMA investigations have been made on serum [21]. In general, solutions can be easily doped with an internal standard. Figure 6 shows a mass spectrum of a control serum enriched with 0.1 mmole FeCl. The spectrum demonstrates the high sensitivity and the ability of multielement detection. In order to demonstrate the possibility of relative quantitation in fluids, control sera were enriched with solutions of FeCl of different concentrations. The content of Mg was always constant: 1 mmole/liter serum. A drop of this solution was brought on a grid coated with a thin formvar film and air-dried. Thirty analyses were performed on each concentration and the ratios of Fe (I_{Fe}) and Mg (I_{Mg}) were determined. Figure 7 indicates that the plot between I_{Fe}/I_{Mg} and the Fe concentration is linear [21]. Thus it is possible to determine for each element the efficiency of LAMMA relative to an internal standard ("element specific sensitivity factors") [14]. Furthermore, with knowledge of the absolute concentration of the internal standard and from the

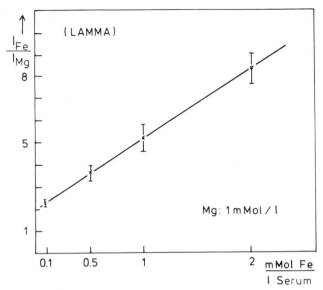

Figure 7. Ion signal intensity ratios of Fe ($m = 56$) and Mg ($m = 24$) plotted versus different concentrations of Fe, obtained from control sera enriched with different FeCl concentrations [2]. As the control serum itself contains iron and this amount is not taken into consideration, the plot does not begin at zero.

ion signals the concentration of the elements to be studied can be determined.

The added internal standard could also be a stable isotope [17] or radioisotope. Investigations using isotope dilution technique (stable isotope ^{44}Ca) to determine the content of small volumes of aqueous solution have been made in order to demonstrate that a cell organelle in crayfish photoreceptor cells was capable of accumulating Ca from extracellular sources [22].

For normal biological specimens it is almost impossible to find natural elements as internal standards, owing to the inhomogeneities in the biological structures. For plastic-embedded specimens insertion of an internal standard by doping the embedded material with metalloorganic compounds is feasible but should be restricted, because specific chemical compounds accumulate in particular specimen structures.

It is possible to deposit thin films of suitable inorganic material directly onto the specimen by evaporation. Metal films like Au and Ag are less useful because their threshold for laser excitation is rather different from that of organic material. Useful standards are thin film deposits of some dielectric materials such as MgF_2, a procedure with reproducible results [17]. A restriction is given by the fact that many of the elements suitable for evaporation are themselves inhomogeneously distributed in the specimen.

Since the organic background from the embedding material or the tissue itself in the spectra are fairly reproducible, this offers the possibility of using organic mass peaks for internal standard. Figure 8 shows the results of measurements on thin Epon sections with various concentrations of Pb [23]. Sections of 0.5-μm and 1-μm thickness were analyzed. Twenty analyses were performed on each section and the ratio of the ion signals of Pb (I_{Pb}) and mass $m = 30$ (probably CH_4N^+) was determined. The plot in Fig. 8 indicates a good relationship, to emphasize the fact that the ratio is independent from the specimen thickness—a presupposition for this procedure. To obtain statistically significant results, it is useful to have more than one mass from the organic background.

Finally, this procedure may offer absolute quantitative analysis, provided that it is always possible to show that the amplitude of the organic peak, acting as an internal standard, is equal in the standard and the specimen to be analyzed or shows an unequivocal quantitative relation [23]. But so far, by using a plot like that in Fig. 8 ("working curve") an estimation of the local element concentration within structures is possible.

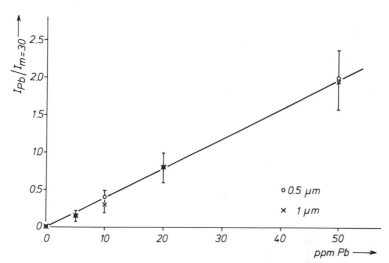

Figure 8.   Calibration plot for the ratio $I_{Pb}/I_{m=30}$ versus I_{Pb} in standard specimens with different concentration of lead. Lead is homogeneously distributed in an Epon matrix; section thickness are 0.5 and 1 μm.

3.3   PREPARATION OF SPECIMENS

The LAMMA sample chamber has been designed for the analysis of small, thin specimens, which have to be mounted on 3.08-mm grids normally used for transmission electron microscopy. Figure 9 shows the schematic diagram of the specimen mounting. The grid is located in a vacuum underneath a thin quartz cover slide. Beyond the specimen mounting the grid separates the specimen from the cover slide to prevent the evaporation of small amounts of the quartz slide while focusing the laser impulse on the specimen.

As the laser and the ion optics are arranged on the opposite sides of the specimen, the nature of the specimens and the specimen preparation has to be performed in such a way that ions formed by laser irradiation can be extracted from the underside of the specimen into the mass spectrometer.

1.   Thin specimens of organic material, metal foils, or coating of evaporated materials, therefore, have to be perforated. This limits the thickness of the specimens. In the case of histological sections the thickness is between 0.1 and 3 μm (optimum thickness, 0.3–0.5 μm). These sections can be mounted onto a blank grid or onto a thin polymer film (Formvar, Pioloform, or Collodium) on a supporting grid.

The preparation of soft biological tissues imposes some problems similar to any kind of microprobe analysis [24]. For the analysis of water-soluble compounds such as the physiological cations, cryotechniques (quench cooling, cryosectioning, and/or freeze drying), plastic embedding in vacuum and sectioning dry on a conventional ultramicrotome are the best preparation methods [25,26]. For the analysis of non-water-soluble elements one can use the normal specimen preparation methods for electron microscopy: fixation in glutaraldehyde, dehydration in ethanol, and plastic embedding.

2.   Soluble compounds in very dilute solution or liquids like serum

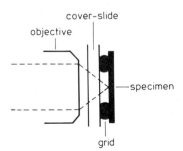

Figure 9.   Schematic diagram of the specimen mounting.

or urine can be inserted in the instrument by attaching a microliter droplet onto one side of the film grid.

3. Microparticles (erythrocytes, bacteria, smears of cells, aerosols) can also be mounted onto a film-coated grid. Powders or microcrystals can be attached to grids by bringing the grid into contact with the particles, so that a few will adhere on the grid. The advantage of this method is that there is no organic background of the supporting foil in the mass spectra.

4. Particles with diameter ≥ 0.2 mm (pieces of renal stones, fibers, hair) can be mounted into a "sandwich grid." These particles have to be analyzed with laser light at grazing incidence on a visible edge. A rough depth profiling can be performed by repeated laser shots onto the same specimen area to create a crater in the specimen.

3.4 APPLICATIONS

The applications of the LAMMA technique are too broad to be listed in detail. A brief survey is given in Section 4.3. The following three applications should demonstrate the potentialities of LAMMA in biomedical research.

3.4.1 Distribution of Lead in Human Aortas

Lead is an environmental contaminant and one of the toxic trace elements in man. In recent years the effect of long-term exposure to small amounts of lead has been of interest, because lead can accumulate in the tissue and can cause toxic poison, probably by paralyzing the activity of specific enzymes [27]. As the level of lead in different soft tissues of humans is in the range of 0.1–2 ppm [28], LAMMA is the instrument able to detect this content of Pb and primarily to localize Pb for determining its cellular and subcellular distribution. One aspect of this field of investigation is the question of whether a relationship exists between the concentration and distribution of Pb in human aortas and the degree of atherosclerosis. By AAS measurements it was shown that lead in the aorta increased with age and presumably with atherosclerosis [29]. In opposition to this finding, other authors show that no correlation between lead content of atherosclerotic and nonatherosclerotic aortas could be ascertained [30]. With LAMMA there is a chance to elucidate these opposing findings. Figure 10a shows the light microscopical micrograph of a morphologically intact human aorta with two LAMMA perforations; Fig. 10b shows the spectra from perforation 1 (muscle cell) and perforation 2 (elastic fiber). The

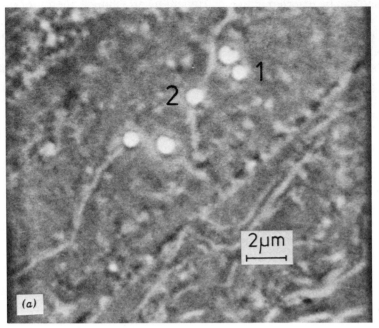

Figure 10a.   Light micrograph (phase contrast) of human aorta (media) with two LAMMA perforations.

spectra indicate that the content of lead in the fibers is relatively high; in the muscle cells the content of lead is just detectable. To compare the lead content, the peak height of ^{208}Pb was referred to mass peak $m = 30$, serving as internal standard [23].

On the contrary, investigations of aortas with atherosclerosis show another cellular distribution near plaques: low lead content in fibers, high lead content in cells [31]. By referring to organic peaks ($m = 30$) LAMMA can determine the relative lead concentration between fibers and cells of the tissues and beyond, to determine concentration profiles across the artery wall in both physiological (Fig. 10c) and pathological conditions (Fig. 10d). Using these results it might be possible to show that the toxicity of lead depends on topochemical localization, because lead can have local toxic concentrations, although integral measurements may indicate a level conforming to normal lead concentrations. By using the working curve (Fig. 8) an estimation of the lead concentration with specific structures is possible. A rough estimation indicates a concentration about 40 times higher than the toxic concentration. Additionally, since LAMMA measures nearly all elements in one analysis, it is possible to

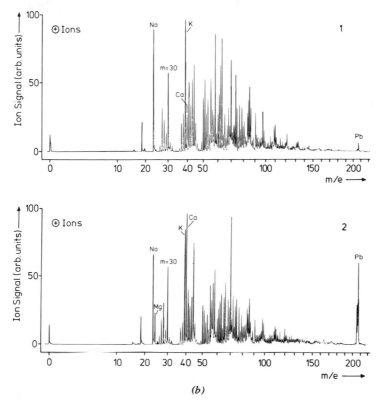

Figure 10b. LAMMA spectra corresponding to the perforations. Perforation 1, muscle cell; perforation 2, elastic fiber.

verify a relationship between the mineral element (e.g., Ca, PO, Mg) and the concentration of Pb.

3.4.2 Lead Distribution in Cellular Elements of Bone Marrow and in the Bone Matrix

Investigations of the subcellular distribution of lead and its turnover are of interest with respect to both its toxicity and possible detoxication mechanisms.

Biopsy material of bone was analyzed with LAMMA to detect lead. The biopsy was made by needle puncture from the iliac crest of a person, who was severely poisoned by lead nitrate over a period of several months. The lead concentration of bone marrow was found to be 26 μg Pb/g wet soft tissue, analyzed by flame spectrophotometry. In sections of

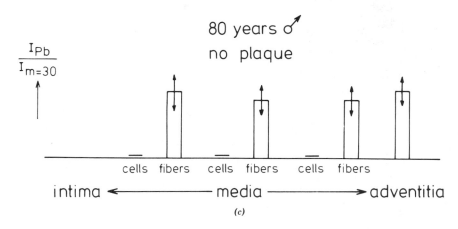

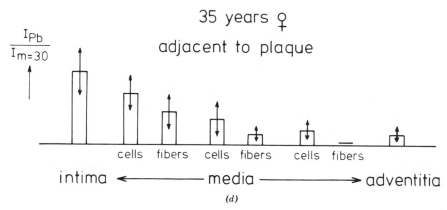

Figure 10c and 10d. Lead concentration in fibers and cells of different regions of human arterial walls of a morphological intact aorta (10c) and an atherosclerotic aorta (10d). (From Ref. 31.)

this tissue LAMMA was able to identify and localize lead in the different cell types of bone marrow [32]. One typical example is shown in Fig. 11, which demonstrates the presence of Pb and some mineral elements in the nucleus (Fig. 11a) and cytoplasm (Fig. 11b) of an erythroblast. In order to compare the different Pb content of all cell types and the subcellular Pb level in the nucleus and cytoplasm it was necessary to refer to an internal standard. The organic mass peaks $m = 30$ serve as internal standards. As seen in all spectra obtained during these investigations, the height of these organic mass peaks is independent of the specific structures (nucleus, cytoplasm). By means of this internal standard it was possible to show that there is a concentration gradient of lead at the subcellular level: The Pb concentration in the nucleus is a factor of 4–5 higher than the Pb concentration in the cytoplasm. These investigations supported the suggestion that a protective mechanism operates within the cells to limit mitochondrial Pb uptake, achieved through the formation of intranuclear inclusion bodies, as a depot of nondiffusible lead.

Moreover, with these investigations it was possible to show for the first time that all cell types reveal fairly identical lead concentrations, as demonstrated in Fig. 12, which shows the relative lead concentration of three different cell types. The standard deviation of these measurements was about 15–20% [23].

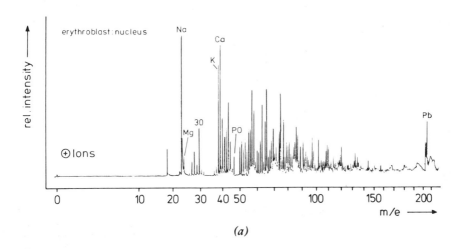

(a)

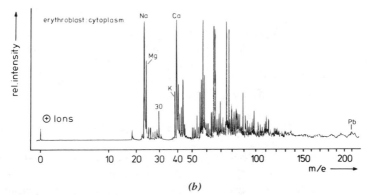

(b)

Figure 11. LAMMA spectra of an erythroblast, nucleus (a) and cytoplasm (b).

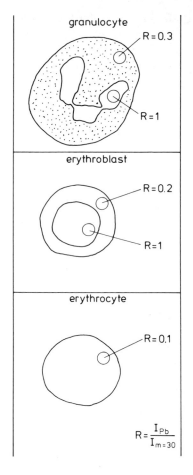

granulocyte

R = 0.3

R = 1

erythroblast

R = 0.2

R = 1

erythrocyte

R = 0.1

$$R = \frac{I_{Pb}}{I_{m=30}}$$

Figure 12. Relative lead concentration in the nucleus and cytoplasm of three different cell types of bone marrow. The gradient is estimated by referring to an organic mass peak $m = 30$ (CH_4N^+).

Furthermore, the distribution of lead in bone matrix could be displayed (Fig. 13). This concentration gradient of lead was also achieved by referring to an internal standard [32]. The results enable the time course of poisoning to be determined. (Specimens from Prof. Dr. Flood, Institute of Anatomy, University of Bergen, Norway.)

3.4.3 A Survey of Further Applications

Many applications are reported that are described in the references given in this chapter. But a brief survey should show further applications to outline the variability in various fields concerning the element analysis in biological tissues: LAMMA was able to detect small amounts of Li present in the brain cells of a rat after a therapeutic dose of LiCl [11]. In patients suffering from renal insufficiency and undergoing intermittent hemodialysis treatment, an accumulation of aluminum in bone was found combined with an increased osteoid volume and a reduced mineralization front (Dialysis osteomalacia). LAMMA was able to localize aluminum and to show a considerable increase of aluminum concentration in osteocytes near the limit between the calcified tissue and the osteoid [33,34]. LAMMA was used to localize uranium deposits [35] and lead deposits [36] in algae after uptake. In the myocardium of a 14-year-old boy suffering from Wilson's disease, LAMMA could demonstrate an inhomogeneous distribution of copper [18]. The selective binding of Li, Na, K, Rb, and Cs to proteins in the A-band of isolated striated muscle cells of frog was demonstrated [37]. In dental research the distribution of fluorine in rat incisor dentine was analyzed [38]. Moreover, it was possible to display the pathway and distribution of constituents of dental material within pulp and periapical tissue following treatment [39]. In order to estimate the cation distribution of the cochlea wall (stria vascularis), it was possible to measure the K/Na ratio across the spiral ligament and stria vascularis (Fig. 14) [40].

Because the identification of specific birefringent renal deposits was difficult by the usual diagnostic histochemical methods, LAMMA was used for an exact analysis. It was possible to show that the renal stones were calcium oxalate [41].

Because of the short analysis time, a further great potential of

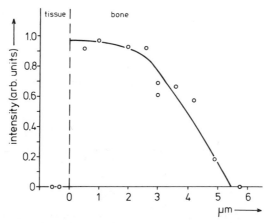

Figure 13.   Relative concentration gradient of lead in bone, achieved by referring to an internal standard. Integral measurement by flame spectrophotometry results in 63 μg Pb/g dry bone matrix.

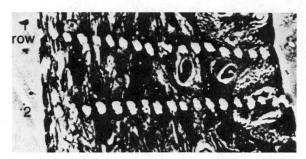

Figure 14a. Light micrograph of a stained section through the lateral wall of the cochlea duct of a guinea pig, with two rows of laser shuts. The schematic drawing indicates individual K/Na ratios determined in each area.

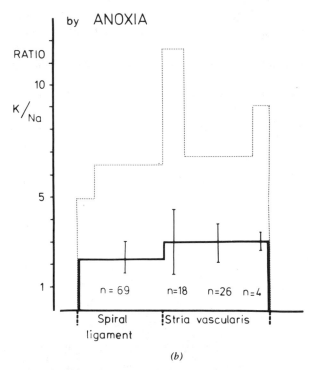

(b)

Figure 14b. Profile of K/Na ratio across spiral ligament and stria vascularis before (black line) and after (dotted line) exposition of anoxia (3 min), n = number of averaged analysis, bars indicate S.D. (From Ref. 40.)

LAMMA is the easy manner of analyzing such microparticles as mineral fibers [42] and aerosols and the classification for chemical composition [14,43].

3.5 ANALYSIS OF ORGANIC MOLECULES

In laser-induced desorption (LID) mode (see Section 2.3), LAMMA offers the possibility of performing mass spectrometry of organic constituents, whereby, in general, the relative intensities of the signals of the molecules and their fragments are reproducible for a given instrument setting, even from thermally labile and nonvolatile molecules [44].

As a typical example, the analysis of leucine demonstrates this ability of organic mass spectrometry. Figure 15 shows a positive ion spectrum of leucine with the parent molecule and typical fragment ions.

By this technique organic mass spectrometry investigations of the influence of electron radiation damage on various organic compounds could be performed to give information on damage effects [45].

First attempts have also been made to detect a specific molecular group out of a complex organic matrix [46]. In patients under chronic dialysis there is a transfer of phthalate plasticizers dissolved out of the plastic blood line into the blood serum. Specific investigations performed with LAMMA could show that in skin samples of such patients phthalate deposits are to be found.

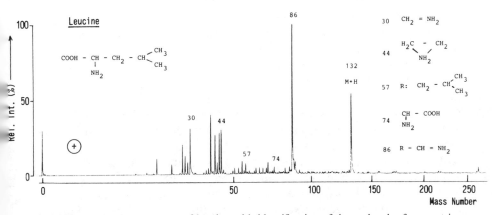

Figure 15. LAMMA spectrum of leucine with identification of the molecular fragment ions.

In tracing isonicotinic acid hydrazide (INH), an attempt was made to detect INH or fragments of INH in treated mycobacterial cells. LAMMA spectra of pure INH were compared with those of INH treated and untreated cells. It could be shown that INH underwent metabolic changes. These primary results show the possibility of detecting pharmaceuticals or their metabolism products in biological tissues [47].

Similar to the detecting of elements, many analyses in organic mass spectrometry have been performed by LAMMA (see Refs. 16, 44, and 48).

3.6　COMPARISON WITH OTHER MICROANALYTICAL TECHNIQUES

As a tool for the analysis of microvolumes, LAMMA is comparable with the following microanalytical techniques producing lateral resolution ≤ 1 μm: Electron probe microanalysis, based on excitation with electrons and detection of characteristic x-ray, and ion microprobe or ion microscope, based on secondary ion mass spectrometry (SIMS) (Table 2).

In making this comparison, atomic absorption spectroscopy (AAS) should be borne in mind. Very often AAS and LAMMA are used as complementary methods. AAS provides quantitative information about specific elements according to integral measurements, while LAMMA gives information about the local distribution of given elements.

LAMMA-Electron Microprobe Analysis

X-ray microanalysis combined with a SEM (scanning electron microscope) or STEM (scanning transmission microscope) is a well-established method for detecting minor elements. It is relatively easy to interpret the results and to establish distribution of a particular element in line scan or in mapping mode. The techniques of quantitation (semi- as well as absolute quantitation) are developed to a relatively high degree of accuracy and precision. But this technique has the following limitations in the analysis of biological and medical specimens [24,49]. (1) Sensitivity for analysis of light elements such as Na, Mg is relatively low. (2) The sensitivity is not sufficiently high for detecting elements in trace concentration: sensitivity of EDS (energy dispersive spectrometer: 700–1000 ppm), of WDS (wavelength dispersive spectrometer: 100 ppm). (3) There is an inability to analyze elements $Z \leq 11$ (EDS) (unless a windowless detector is used) and $Z \leqq 4$ (WDS). (4) It is impossible to discriminate isotopes or to analyze molecules.

LAMMA–Ion Microprobe

LAMMA and the ion microprobe were developed to overcome some of the limitations of electronprobe microanalysis. LAMMA and the ion microprobe are basically different in the excitation and performance of the mass spectrometric analysis. Whereas LAMMA is based on laser-induced evaporation and ionization of the area to be analyzed, which destroys the analyzed area, the ion microprobe technique is based on sequentiae ion sputtering of atomic layers at the surface of the specimens in either the ion microprobe or the ion microscope [50]. A small fraction of the sputtered particles is emitted in an ionized state, the so-called secondary ions. This sputtering enables three-dimensional analysis, restricted only by the fact that different parts of a specimen have different sputter rates [51]. In principle both systems, LAMMA and the ion microprobe, are able (1) to detect all elements with high sensitivity (ppm range), (2) to discriminate isotopes, and (3) to detect molecules or molecular fragments. The mass range of LAMMA is $m/e = 2000$, the mass range of the ion microprobe $m/e = 300$.

The *lateral spatial resolution* is somewhat better in LAMMA (0.5–0.8 μm) than in SIMS (at present state of development the optimum resolution is 1 μm but strongly depending on the sensitivity and the concentration of the element to be detected) [50,51]. Similar to electron microprobe analysis, the ion microprobe is able to image directly the distribution of specific elements; by LAMMA it is possible point by point. The *sensitivity* of both instruments is comparable. Both systems have equal problems created by the interference of organic mass ions with elements at very low concentration. Due to the higher mass resolution of the ion microprobe, in specific instrumental operation conditions up to $m/\Delta m = 5000$, it is possible to separate mass peaks of elements from mass peaks of organic ions, although the increased mass resolution is accompanied by decreased sensitivity [52]. Unlike LAMMA, the ion yield of heavy elements in SIMS decline with increasing Z number [11].

The determination of absolute elemental concentrations may be more difficult with SIMS than with LAMMA, because of the influence of the specimen matrix on the signal intensity of atomic ions [52]. Since laser light is employed for ionization, no problems with charging occur in the analysis of insulating specimens.

TABLE 2. Summary of Characteristics of Electron Microprobe Analysis, Ion Microprobe and LAMMA[a]

	Electron Microprobe Analysis	Ion Microprobe Analyzer	LAMMA 500
Probe	Electrons	Ions	Photons (Laser-beam)
Detection method	Characteristic X-rays EDS or WDS	Secondary ions Mass spectrometer (Double focusing)	Ions ("+" or "−") Mass spectrometer (TOF*)
Resolution of detection system	EDS: 150 eV WDS: 10 eV	$\frac{m}{\Delta m}$: 200 – 10 000	$\frac{m}{\Delta m} = 600$
Lateral resolution	EDS: 50–100 nm (High concentration deposit) WDS: 1 μm	~1 μm	0.5 μm – 1 μm
Depth resolution	1 μm	0.2 nm/s	Sections thickness 0.3 μm – 3 μm
Analyzed specimen volume (Thin sections)	$10^{-12} - 10^{-16}$ g	$2 \cdot 10^{-11}$ g/s	$10^{-12} - 10^{-14}$ g Normally ~10^{-13} g
Detection limits concentration (Relative sensitivity)	EDS: 700 ppm – 1000 ppm WDS: 100 ppm	1 ppm	0.1 ppm – 2 ppm
Minimum detectable mass (Absolute sensitivity)	$10^{-16} - 10^{-18}$ g	$10^{-16} - 10^{-19}$ g	$10^{-18} - 10^{-20}$ g
Elements	EDS: $Z \geqq 11$ WDS: $Z \geqq 4$	All	All

Isotopes	No	Yes	Yes
Molecules	No	Yes Mass range m/e: 1 – 500	Yes m/e: 1 – 2000
Imaging (Spatial resolution)	STEM: 1.5 nm SEM: 7.0 nm	Light microscope Reflection mode, 70×	Light microscope (Transmission 1200×)
Destruction of specimen	No But radiation damage and mass loss	Yes Sputtering	Yes Perforation
Profile of concentration	Yes	Yes	Point by point
Mapping	Direct imaging	Direct imaging	Point by point
Analytical artifacts	Overlapping peaks (EDS) Signal from uncollimated electrons	Spectral interferences Preferential sputtering	Spectral interferences (Superposition with peaks originated from organic material and/or stains in case of low element concentration)
Time required for spectrum acquisition	EDS: minutes WDS: minutes	Depends on mass resolution >10 min	<minute
Quantitation relative absolute	Yes Yes	Yes Yes: being developed	Yes Yes: being developed

^{a} The following abbreviations are used in this table: EDS, Energy Dispersive Spectrometer; WDS, Wavelength Dispersive Spectrometer; STEM, Scanning Transmission Electron Microscope; SEM, Scanning Electron Microscope; m, Mass; TOF, Time of Flight Mass Spectrometer.

101

3.7 CONCLUDING REMARKS

For the purpose of analyzing biological and medical specimens, the usefulness of LAMMA is based on its ability to detect and localize all elements in trace concentrations. The sensitivity for most elements is of the order of ppm, provided no interference with organic mass peaks is present. The absolute amount that can be detected is about 10^{-19}–10^{-20} g. The precision is at present on the order of 20%.

While integral measurement techniques provide useful and needed information about quantitative element concentrations, LAMMA delivers more accurate information on the distribution of elements within various cellular or, with restriction, subcellular structures. Furthermore, by quantitating the analysis it is possible to estimate concentration gradients or, by absolute quantitation, the local content of trace and toxic elements.

Owing to the simultaneous detection of all elements, LAMMA enables studies to be made of the metabolism of individual elements and the interactions between elements (e.g., the determination of concentration ratios between mineral elements and trace elements).

Future investigations will show (1) to what extent elements can be detected at a concentration near the detection limits when superposed by mass peaks of the organic background (this limitation should be partly overcome by suitable preparation methods or computer-aided background subtraction) and (2) to what extent and with what degree of accuracy quantitative analysis is possible.

Because LAMMA can perform mass spectrometry on organic compounds in small volumes, it may also supply more information on the chemical form in which elements are active. Furthermore, because LAMMA is able to analyze molecules, alterations of molecules or the pathway of molecules in tissues can be followed, either by following the molecule itself or by labeling it with isotopes—another very attractive field of investigation.

REFERENCES

1. G. H. Morrison, Mass spectrometry, in *Elemental Analysis of Biomedical Materials,* R. M. Parr, Ed., International Atomic Energy Agency, Vienna, 1980, pp. 201–229.
2. H. Moenke and L. Moenke-Blankenburg, *Laser Micro-Spectrochemical Analysis,* Adam Hilger, London, 1973.
3. J. F. Eloy, Microsc. Acta. Suppl. 2, 307 (1978).
4. J. Conzemius and H. J. Svec, Anal. Chem. *50,* 1854 (1978).

5. R. Kaufmann, F. Hillenkamp, and R. Remy, Microsc. Acta. *73*, 1 (1972).

6. F. Hillenkamp, R. Kaufmann, R. Nitsche, E. Remy, and E. Unsöld, in *Microprobe Analysis as Applied to Cells and Tissues,* T. Hall, P. Echlin, and R. Kaufmann, Eds., Academic, London, 1974, pp. 1–14.

7. F. Hillenkamp, E. Unsöld, R. Kaufmann, and R. Nitsche, Appl. Phys. *8*, 341 (1975).

8. R. Wechsung, F. Hillenkamp, R. Kaufmann, R. Nitsche, and H. Vogt, Scanning Electron Microsc. *1*, 611–620 (1978).

9. H. J. Heinen, R. Wechsung, H. Vogt, F. Hillenkamp, and R. Kaufmann, Acta Phys. Aust. *20*, 257 (1979).

10. H. Vogt, H. J. Heinen, S. Meier, and R. Wechsung, Fresenius Z. Anal. Chem. *308*, 195 (1981).

11. R. Kaufmann, F. Hillenkamp, and R. Wechsung, Med. Prog. Technol. *6*, 109 (1979).

12. N. Fürstenau, Fresenius Z. Anal. Chem. *308*, 201 (1981).

13. U. Haas, P. Wieser, and R. Wurster, Fresenius Z. Anal. Chem. *308*, 270 (1981).

14. R. Kaufmann, P. Wieser, and R. Wurster, Scanning Electron Microsc. *2*, 607 (1980).

15. E. Unsöld, G. Renner, F. Hillenkamp, and R. Nitsche, in *Proc. 7th Int. Mass Spectrometry Conf.,* Florence, Heyden, London, 1977.

16. H. J. Heinen, S. Meier, H. Vogt, and R. Wechsung, in *Advances in Mass Spectrometry,* Vol. 8, A. Quayle, Ed., Heyden, London, 1980, p. 342.

17. W. Schröder, D. Frings, and H. Stieve, Scanning Electron Microsc. *2*, 647–654 (1980).

18. B. Kaduk, K. Metze, P. F. Schmidt, and G. Brandt, Virchow's Arch. A Path. Anat. Histol. *387*, 67 (1980).

19. T. Barnard and R. S. Thomas, J. Microsc. *113*, 269 (1978).

20. W. H. Schröder, European Biophysics Journal (in press).

21. P. F. Schmidt, H. G. Fromme and G. Pfefferkorn, Scanning Electron Microsc. *2*, 623 (1980).

22. W. H. Schröder, Fresenius Z. Anal. Chem. *308*, 212 (1981).

23. P. F. Schmidt and K. Ilsemann, Scanning Electron Microscopy 1983 (in press).

24. P. Echlin and R. Kaufmann, Microsc. Acta. Suppl. 2, 11–45 (1978).

25. R. Kaufmann, F. Hillenkamp, R. Wechsung, H. J. Heinen, and M. Schürmann, Scanning Electron Microsc. *2*, 279–290 (1979).

26. Hj. Hirche, J. Heinrichs, H. E. Schaefer, and M. Schramm, Fresenius Z. Anal. Chem. *308*, 224 (1981).

27. A. de Bruin, *Biochemical Toxicology of Environmental Agents,* Elsevier/North-Holland, Amsterdam, 1976.

28.   E. I. Hamilton and M. J. Minski, Sci. Total Environ. *1*, 375 (1972/1973).

29.   H. A. Schroeder and I. H. Tipton, Arch. Environ. Health *17*, 965 (1968).

30.   O. Pribilla, H. Darmstädter, and Th. Schultek, Z. Rechtsmed. *85*, 127 (1980).

31.   P. F. Schmidt, R. Lehmann, K. Ilsemann, and A. Wilhelm (submitted to Artery).

32.   P. F. Schmidt and H. Zumkley, Eds., in "Spurenelemente," Georg Thieme Verlag, Stuttgart, 1983, pp. 12–24.

33.   H. Zumkley, P. F. Schmidt, H. P. Bertram, A. E. Lison, B. Winterberg, K. Spieker, H. Losse, and R. Barckhaus (submitted to Lancet).

34.   P. F. Schmidt and H. Zumkley, J. de Physique-Colloques (in press).

35.   B. Sprey and H. P. Bochem, Fresenius Z. Anal. Chem. *308*, 239 (1981).

36.   D. W. Lorch and H. Schäfer, Fresenius Z. Anal. Chem. *308*, 246 (1981).

37.   L. Edelmann, Fresenius Z. Anal. Chem. *308*, 218 (1981).

38.   E. Gabriel, Y. Kato, and P. J. Rech, Fresenius Z. Anal. Chem. *308*, 234 (1981).

39.   R. Lehmann, H. Sluka, and P. F. Schmidt, Quintessenz *33*, 613 (1982).

40.   A. Orsulakova, R. Kaufmann, C. Morgenstern, and M. D'Haese, Fresenius Z. Anal. Chem. *308*, 221 (1981).

41.   A. J. Chaplin, P. R. Millard, and P. F. Schmidt, Histochemistry *75*, 259 (1982).

42.   K. R. Spurney, J. Schörmann, and R. Kaufmann, Fresenius Z. Anal. Chem. *308*, 274 (1981).

43.   P. Wieser, R. Wurster, and H. Seiler, Atm. Environ. *14*, 485 (1980).

44.   H. J. Heinen, H. Vogt, and R. Wechsung, Fresenius Z. Anal. Chem. *308*, 290 (1981).

45.   P. Bernsen, L. Reimer, and P. F. Schmidt, Ultramicroscopy *7*, 197 (1981).

46.   K. D. Kupka, W. W. Schropp, Ch. Schiller, and F. Hillenkamp, Fresenius Z. Anal. Chem. *308*, 229 (1981).

47.   U. Seydel and B. Lindner, Fresenius Z. Anal. Chem. *308*, 253 (1981).

48.   Ch. Schiller, K. D. Kupka, and F. Hillenkamp, Fresenius Z. Anal. Chem. *308*, 304 (1981).

49.   P. Galle, J. P. Berry, and R. Lefevre, Scanning Electron Microsc. *2*, 703–710 (1979).

50.   H. Liebl, Scanning *3*, 79 (1980).

51.   R. W. Linton, S. R. Walker, C. R. DeVries, P. Ingram, and J. D. Shelburne, Scanning Electron Microsc. *2*, 583–596 (1980).

52.   M. S. Burns, J. Microsc. *127*, 237 (1982).

ACOUSTIC MICROSCOPY: BIOMEDICAL APPLICATIONS

H. K. WICKRAMASINGHE

Department of Electrical Engineering
University College London
London, England

4.1 INTRODUCTION

Microstructures that occur in biological material have been studied in great detail with conventional techniques such as optical and electron microscopy. In this chapter we discuss a new technique for examining organic substructure based on the use of acoustic waves.

The concept of an acoustic microscope was first put forward for Sokolov in 1949. He realized that the wavelength of sound in water at a frequency of 3 GHz was 0.5 μm and predicted that one day it would be possible to build an acoustic microscope with a resolution comparable to that of the optical microscope. It was not until the early 1970s, when techniques for producing high frequency sound waves were readily available, that Sokolov's proposition was taken seriously.

Several techniques for acoustic microscopy have been proposed, but the scanning acoustic microscope is unique in its image quality and resolution [1]. The use of mechanical scanning has several advantages, the main one being that the lens must perform well only on axis. It is this feature that has made it possible to record high quality acoustic images with submicrometer resolution. The image, which records the mechanical properties (such as density, elasticity, and viscosity) of the sample being investigated, can be obtained either in reflection or transmission. In the following section we present the basic operating principles of the instrument. Several examples of biological images are presented in Section 3. In Section 4 we discuss the contrast mechanisms in reflection and transmission acoustic microscopy. Section 5 is devoted to quantitative measurements, while Section 6 deals with various techniques for improving resolution. Finally, in Section 7 we present some brief concluding remarks.

4.2 PRINCIPLES OF OPERATION

Scanning acoustic microscopy dates back to the work carried out by Lemons and Quate at Stanford University [1]. In those studies they demonstrated a transmission microscope operating in water at an acoustic frequency of 150 MHz and providing a resolution of approximately 10 μm. This group has been so successful that currently their microscope is operating in liquid helium at 2.7 GHz with 90 nm wavelength.

The heart of the acoustic microscope is a sapphire lens (Fig. 1). A short rf pulse (approximately 30 ns in duration) is applied to a piezoelectric transducer deposited on the back surface of the lens rod. An acoustic pulse propagates down the sapphire rod and is focused sharply into the

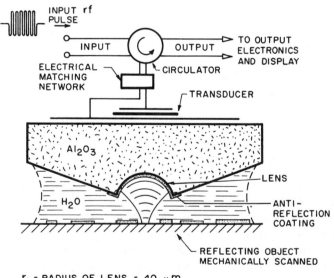

r_0 = RADIUS OF LENS = 40 μm
f = FOCAL LENGTH OF LENS = 1.13 r_0
R = RADIUS OF LENS APERTURE = 0.7 r_0

Figure 1. Basic lens geometry for reflection acoustic microscopy.

coupling liquid (usually water) by the spherical lens. Whereas in optics such a lens would result in severe spherical aberration, in acoustics these aberrations are negligible because of the fact that there is a sevenfold reduction in sound velocity, going from sapphire to water; the focal spot is diffraction limited, and for a well-designed lens its diameter approaches the acoustic wavelength (0.5 μm at 3 GHz in water).

In the reflection acoustic microscope, the object to be imaged is placed at the focus of this lens. Reflected waves return along the incident path and are converted back into an electrical pulse by the transducer. The strength of this pulse is proportional to the "acoustic reflectivity" of the object at the point being investigated. An image is formed by mechanically scanning the object in a raster fashion and using the detected signal to modulate the brightness of a scan-synchronized cathode ray tube (CRT) display. A transmission system can be build using two such lenses arranged in a confocal geometry.

4.3   APPLICATION TO BIOLOGY

The applications of acoustic microscopy can be broadly divided into two areas: biology and nondestructive materials evaluation. In both these areas new and exciting information is emerging. Here we shall concentrate on the biological and organic applications.

During the mid-1970s biological samples were studied extensively using the transmission acoustic microscope. The specimen was attached onto a thin (2 μm) Mylar membrane and scanned in the focal plane in order to form the image.

Acoustic micrographs of biological material show very high contrast without the need for staining. Figure 2 shows a comparison between the optical and acoustic micrograph of a section of human retina demonstrating the high intrinsic contrast [2]. The images marked "amplitude" and "phase" are in essence the acoustic counterparts of optical "bright-field" and "phase contrast" microscopy, respectively. The dark band running across the acoustic image is the pigment epithelium; it contains melanin, which is acoustically absorbing. Red blood cells observed in these images also appear to be acoustically absorbing. The images were taken in transmission at 1 GHz by focusing a sound beam onto a thin (1 μm) section of the specimen.

Frozen sections can be studied in order to determine whether fixation alters the structures that are being imaged. Figure 3 compares optical and acoustic micrographs of the pigment epithelium and choroid and demonstrates the strong acoustic attenuation in the pigment epithelium, Bruchs

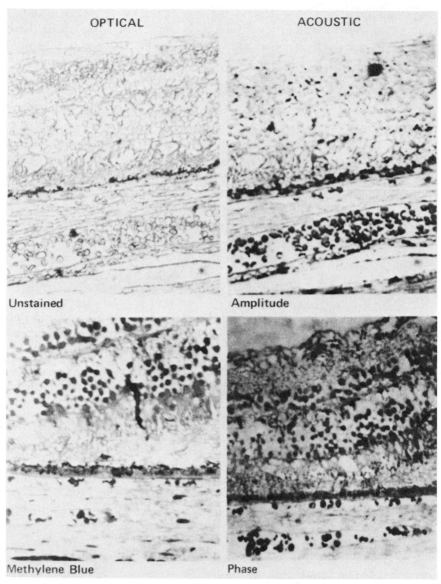

Figure 2. Comparison of acoustic and optical micrographs of human retina (acoustic wavelength, 1.5 μm; magnification 400×).

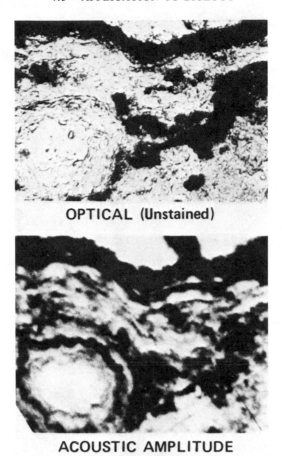

OPTICAL (Unstained)

ACOUSTIC AMPLITUDE

Figure 3. Optical and acoustic micrographs of unfixed human pigmented epithelium and choroid sectioned on a refrigerated microtome; magnification 800×.

membrane, and the choroidal pigment granules. The ring of tissue surrounding the wall of a choroidal arteriole is an elastic lamina that shows up with high contrast.

The fact that one can achieve high contrast without the need for staining suggests that the acoustic microscope could be used to study living cells. Cells in culture may be placed under the microscope and studied under normal and pathological conditions. Physical effects of light or dark on tissues may be studied directly without the need for light to view the sample. Figure 4 shows one such example. The object in this case was a population of Chinese hamster ovary cells, and the water coupling medium was replaced by an appropriate cell culture medium in order to take

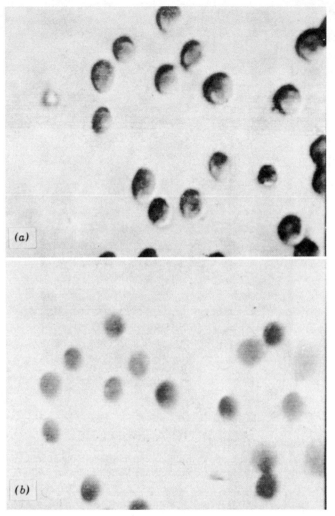

Figure 4. Acoustic micrograph of living Chinese hamster ovary cells. (*a*) Amplitude. (*b*) Phase. Acoustic wavelength, 1.5 μm, magnification 400×.

the images. The transmission acoustic micrographs (1 GHz) show amplitude and phase shifts [3] through the living cells.

The highest resolution acoustic images of biological specimens have been taken using the reflection acoustic microscope. The specimen is mounted on a high acoustic impedance substrate (such as sapphire or quartz) that essentially acts as an acoustic mirror and reflects the transmitted beam [4]. Figure 5 shows a comparison of the optical (stained) and

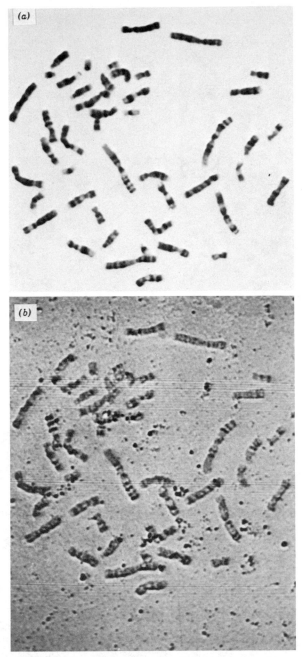

Figure 5. Comparison of (*a*) optical and (*b*) acoustic reflection micrographs of human chromosomes; acoustic image was taken in liquid argon at 2 GHz ($\lambda = 0.4~\mu$m).

111

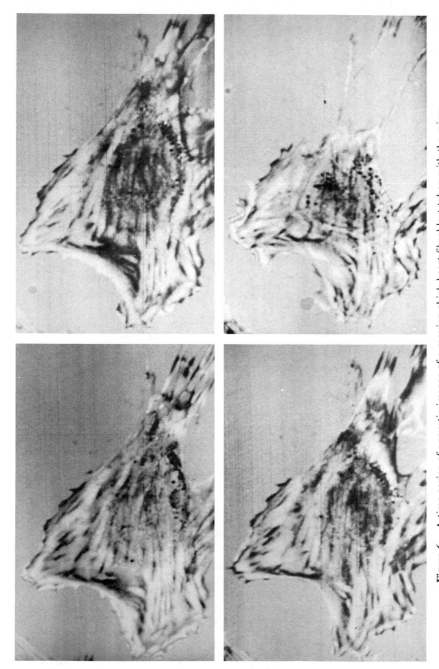

Figure 6. A time series of acoustic images of a moving chick heart fibroblast taken with the microscope focused on the surface of the substrate ($Z = 0$). The dark streaks in the cell periphery show sites of cell–substrate attachment.

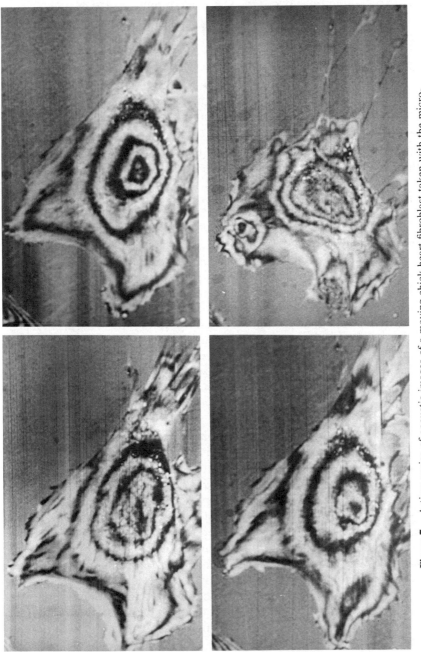

Figure 7. A time series of acoustic images of a moving chick heart fibroblast taken with the microscope focused on the top surface of the cell ($Z = +1.3$ μm). The acoustic interference rings seen in the cell interior can be used to measure cellular impedance.

113

acoustic reflection (unstained) micrographs of human chromosomes. The acoustic images which were taken in liquid argon at 2 GHz ($\lambda = 0.4 \ \mu$m) clearly show up the banding structure [5,6].

Figure 6 shows a sequence of reflection images of a moving chick heart fibroblast cell [7]. The microscope was focused onto the surface of the substrate. The dark streaks in the cell periphery indicate sites of cell–substrate attachment. Figure 7 shows the same sequence of images with the microscope focused onto the cell surface. The acoustic interference rings seen in these images (as we shall see in Section 4.5) can be used to measure cellular impedance (i.e., the product of density and acoustic velocity).

4.4   CONTRAST MECHANISMS

In transmission acoustic microscopy, contrast in biological images can generally be attributed to two factors: attenuation and phase shift. If one is simply using the amplitude of the acoustic wave to form an image, the predominant contrast would be expected to arise because of attenuation changes within the object. However, even in this recording configuration, the system is not totally immune to phase variations. Phase gradients across the object cause the acoustic beam to be "deflected" out of the receiving lens aperture, thereby reducing the detected amplitude. This mechanism of contrast is responsible for the "apparently" high acoustic absorption within human red blood cells [8].

Acoustic attenuation depends on viscosity, density, elasticity, and acoustic frequency, but equations that describe the attenuation of sound in water do not explain the attenuation in complex biological material. Acoustic absorption in tissue is usually attributed to molecular relaxations, which occur as equilibrium states are perturbed by the periodic temperature variation associated with the acoustic wave. Absorption (as well as changes in velocity or phase) arises fundamentally from the time delay required to establish equilibrium between the translational, rotational, and vibrational degrees of freedom of the molecule. Rotational relaxation usually occurs at low frequencies, whereas vibrational relaxation occurs at somewhat higher frequencies. Unfortunately, very little published data are available on the acoustic properties of biological tissue at frequencies in the GHz range. Data obtained with ultrasound diagnostic instruments (which typically operate at frequencies less than 10 MHz) cannot be extrapolated accurately to the GHz range. However, we can say that substances that contain large molecular components, such as collagen, melanin, nucleic acid, and so on, show high acoustic attenuation, probably as a result of the large number of available relaxational modes.

In the phase mode of operation, the acoustic microscope is set up to record phase shifts through the specimen normally resulting from acoustic velocity changes. Recent work has shown that one could further improve the phase sensitivity and contrast by comparing phase shifts through adjacent pixels on the sample [9,10]; in this configuration we have the acoustic equivalent of differential phase contrast optical microscopy.

All the preceding mechanisms of contrast apply equally to reflection acoustic microscopy. However, there is one additional—and very important—mechanism of contrast in reflection acoustic microscopy that is a direct consequence of the substrate that supports the sample. If one stops the raster scan and plots the reflected acoustic signal from the substrate (in the absence of an object) as a function of defocus distance, as measured from the geometrical focus, we find that each substrate material produces a characteristic response (Fig. 8). In particular, nulls in the detected signal are observed spaced regularly along the defocus direction. The spacing between nulls can be directly related to the elastic properties of the substrate. This response, which arises from the characteristic "acoustic reflectivity" of the substrate, can be used to enhance contrast

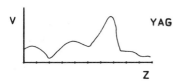

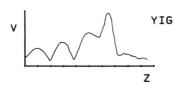

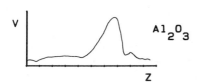

Figure 8. Recorded $V(Z)$ response for various solid surfaces; each division is 3.75 μm.

in biological images by adjusting the focus to correspond to a null response of the substrate. We shall show in the following section how one can use this contrast mechanism to determine the elastic properties of the sample being investigated by carefully choosing the focusing conditions.

The characteristic materials response of the substrate [11] [sometimes called $V(Z)$ response in the literature] has been theoretically studied by a number of authors [12–16]. For smooth, solid substrates the ray model due to Parmon and Bertoni provides the clearest physical picture and accurately predicts the null spacing in $V(Z)$. The basic model can be understood with reference to Fig. 9. Ray 1 is incident at normal incidence, whereas ray 2 is incident at the Rayleigh critical angle. Ray 2 excites a leaky Rayleigh wave at the liquid–solid interface, which reexcites a bulk wave within the liquid, as shown in Fig. 9. Only the ray that phase-matches into the lens will contribute to the detected voltage. A null response will be obtained whenever the phase difference between ray 1 and ray 2 is an odd multiple of π. Application of this principle simply leads to an expression for the null spacing ΔZ in the $V(Z)$ response:

$$\Delta Z = \frac{\lambda_R}{\sin \theta_R} \left(\frac{1 + \cos \theta_R}{2} \right) \tag{1}$$

where λ_R is the Rayleigh wavelength and θ_R is the Rayleigh critical angle. This relationship has been used to measure Rayleigh wave slowness surfaces on isotropic [15] and anisotropic [17] substrates with an accuracy approaching 0.1%.

The Parmon–Bertoni model provides an accurate description of the physical situation in the case of smooth solid substrates. In multilayered substrates (as in the case of a biological object attached to a reflecting substrate) the situation becomes much more complex and it is necessary

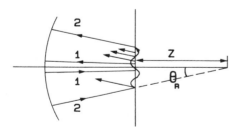

LIQUID   SOLID

Figure 9. Ray diagram interpretation of the $V(Z)$ response.

to resort to a wave approach to calculate $V(Z)$ [13–14]. By decomposing the lens focal distribution into an angular spectrum of plane waves, one can arrive at an integral expression for $V(Z)$ in terms of the "acoustic reflectivity" of the sample being imaged.

$$V(Z) = \int_0^\infty \{P(\lambda \rho R)\}^2 \{U(\lambda \rho R)\}^2 \mathbb{R}(\rho) \exp[jkz\sqrt{1 - \rho^2 \lambda^2}]2\pi\rho \, dp \quad (2)$$

$\mathbb{R}(\rho)$ is the acoustic reflectivity of the sample corresponding to a plane wave having a spatial frequency ρ, p is the lens pupil function, R is the lens radius, and $K = 2\pi/\lambda$ is the acoustic propagation constant in the liquid. U represents the lens input distribution and permits us to simulate the effects of nonuniform lens illumination.

Equation (2) applies to an isotropically layered substrate; the extension to anisotropic situations is straightforward and will not be considered here. It is instructive, however, to use Eq. (2) to determine the null spacing in $V(Z)$ for the case of a smooth isotropic solid surface. In this case the reflectivity function $\mathbb{R}(\rho)$ is essentially constant in amplitude, while its phase changes rapidly (by almost 2π) near $\rho = \rho_R$ corresponding to the Rayleigh critical angle θ_R.

For the sake of simplicity, we shall assume that the lens pupil function P and the lens illumination function U are uniform across the aperture. Referring to Eq. (2), we notice that there will be two regions of P where the integral is likely to have a significant contribution. The first is near $\rho = 0$ when the exponential phase term is approximately unity and the second is near $\rho = \rho_R$ when the rapid changes in $\mathbb{R}(\rho)$ counteract the exponential phase variation. Nulls in $V(Z)$ are obtained when these two contributions are π out of phase, while peaks are obtained when they are in phase. Thus the null condition for $V(Z)$ is:

$$2KZ\sqrt{1 - 0.\lambda^2} - 2KZ\sqrt{1 - \rho_R^2\lambda^2} = (2n + 1)\pi \quad (3)$$

but

$$\rho_R = \frac{\sin \theta_R}{\lambda_R} \quad (4)$$

and

$$K = \frac{2\pi}{\lambda_R} \quad (5)$$

Substituting Eqs. (4) and (5) into Eq. (3) gives the null spacing.

$$\Delta Z = \frac{\lambda}{2(1 - \cos \theta_R)} \qquad (6)$$

In addition, the phase-matching condition for Rayleigh waves requires that

$$\frac{\sin \theta_R}{\lambda} = \frac{1}{\lambda_R} \qquad (7)$$

Substituting Eq. (7) into Eq. (6) gives the final result.

$$\Delta Z = \frac{\lambda_R}{\sin \theta_R} \frac{(1 + \cos \theta_R)}{2} \qquad (8)$$

which is identical to Eq. (1) derived for a smooth solid surface by Parmon and Bertoni using a ray approach. As we mentioned earlier, however, Eq. (2) is quite general and can be used to calculate $V(Z)$ for the case of a biological sample attached onto a reflecting substrate. We shall see in the following section how such $V(Z)$ measurements can be used to determine the elastic parameters of biological tissue on a microscopic scale.

4.5   QUANTITATIVE MEASUREMENTS

Several techniques have been suggested for measuring elastic parameters of biological material on a microscopic scale. One of these [18] is based on the use of the $V(Z)$ response in reflection microscopy. The measurement procedure can best be explained with reference to Figs. 6 and 7. An image taken with the microscope focused on the cell surface shows a set of interference fringes as shown in Fig. 7. Under these conditions the reflection from the substrate surface is weak whereas that from the sample surface is at its maximum. Since the reflection from the sample surface is at any rate quite weak (most biological samples typically have acoustic impedances close to that of water), this focal position represents the optimum condition for obtaining interference fringes. The contrast in the fringes can be further enhanced by choosing a substrate that has an acoustic impedance close to that of the sample. By measuring the depth of contrast in these fringes, we can determine the cellular impedance (i.e., product of density and acoustic velocity), provided we know the acoustic properties of the substrate and coupling liquid (usually water). Figure 6 shows an image taken with the substrate placed at the geometrical focus, and as expected no interference fringes are observed.

Furthermore, by translating the sample toward the lens, we can measure the $V(Z)$ response of the cell–substrate combination in the usual manner. The null spacing in this response then can be related to the longitudinal acoustic velocity within the cell using Eq. (2), provided the cellular thickness and the acoustic properties of the substrate and coupling liquid are known.

Yet another technique is based on the measurement of amplitude and phase shift through a known thickness of the sample in the transmission microscope [19]. One could then use a constrained inversion procedure to calculate the density and longitudinal velocity within the object.

We have measured the acoustic properties of biological tissue by using a modified transmission microscope where one lens is replaced by a plane wave transducer (see Fig. 10). At each pixel point the object is rotated to record the transmission as a function of rotation angle. This record can be shown to be equal to the spatial frequency transmittance function of the object at the point being investigated [20]. For biological samples the transmission reduces to zero at the longitudinal critical angle. Application of Snells law then directly yields the longitudinal velocity. Figure 11 shows an acoustic image at 9.1 MHz of a fixed liver sample (450 μm thick) taken using this system with Flutec (velocity 593 m/s, density 1060 kg/m^3) as the coupling liquid. Figure 12 shows rotational scans at three positions A, B, C marked in Fig. 11. Based on these scans, we have calculated the longitudinal velocities at A, B, and C to be 1680 m/s, 1709 m/s, and 1700 m/s, respectively. These results are within 6% of the macroscopic measurements on unfixed liver tissue reported in the literature.

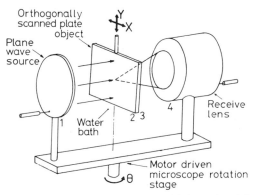

Figure 10. Single-lens acoustic microscope with added rotational degree of freedom.

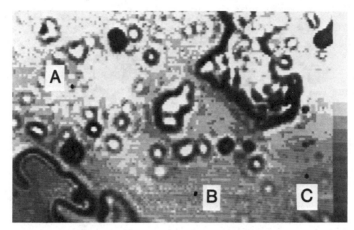

Figure 11. Conventional single-lens image of a 450-μm-thick section of formalin-fixed normal human liver—the field of view is 10 × 6 mm.

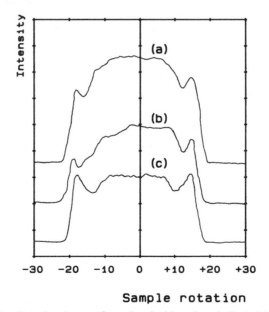

Figure 12. Rotational scans from the pixel locations indicated in Fig. 11.

4.6 TOWARD HIGHER RESOLUTION

By stretching the existing technology to its limits, the reflection SAM has been operated in water at 3.5 GHz, with a corresponding wavelength of 425 nm [6,21]. The major obstacle to improving resolution of the SAM is the high value of the absorption of sound in liquids. Generally, this increases with the square of the frequency. To obtain a wavelength below 425 nm one must find a fluid that has a lower velocity, a lower absorption coefficient, or preferably both. One possibility is to use cryogenic liquids such as argon and helium [22]. Liquid argon has a velocity that is approximately half that of water and comparable attenuation, so that it is capable of improving the resolution by a factor of about 2. Images have been taken at a wavelength of 400 nm in liquid argon at 2 GHz [5]. In liquid helium at 0.1 K, the attenuation is so small that it becomes negligible. The maximum resolution is then set by other factors, such as the accuracies in the scanner. To date, the shortest wavelength that has been reached in a liquid helium microscope is 90 nm at a frequency of 2.7 GHz [23].

The use of cryogenic liquids involves several instrumental complexities as well as restricting the type of sample that can be investigated. We have begun to explore an alternative class of fluid-gases at high pressure. It is well known that the velocity of sound in gases is 5–10 times lower than that in most liquids, although the acoustic absorption is typically 100–1000 times higher. We have shown that the acoustic absorption—at least in the case of monatomic gases such as argon and xenon—varies inversely with pressure. Therefore it should, in principle, be possible to exceed the resolution limit in water using gases at elevated pressures [24]. The results of our initial work in this area, where we demonstrated a reflection SAM operating in argon at 30 bar and 45 MHz, have already been published [25]. The resolution was five times better than that which could be achieved in water at the same frequency.

We can define a coefficient of merit M for a coupling fluid in the SAM by calculating the minimum wavelength that can be achieved for a fixed loss and transit time within the fluid and relating it to the corresponding value for water:

$$M = \left(\frac{\alpha_w}{\alpha}\right)^{1/2} \left(\frac{c_w}{c}\right)^{3/2} \qquad (9)$$

where c is the velocity of sound in the fluid, α is the attenuation coefficient normalized with respect to the square of frequency f, and the subscript w refers to corresponding quantities in water. Table 1 lists the coefficient of merit for several liquids and gases. It is clear that gases such as argon and

TABLE 1   Figure of Merit *M* for Various Fluids

Fluid	Temp. (°C)	Longitudinal Velocity (m/s)	Absorption $\alpha/f^2 \times 10^{17}$ (S²/cm)	*M*
Water	25	1495	22.0	1.00
Water	37	1523	17.7	1.09
Water	60	1550	10.2	1.39
Carbon disulfide	25	1310	10.1	1.81
Mercury	23.8	1450	5.8	2.04
Argon, 40 bar	20	323	412.0	2.00
Argon, 250 bar	20	323	83.0	5.00
Xenon, 40 bar	20	178	953.0	4.00
Carbon dioxide	20	234	350.0	4.04

xenon under pressure can provide substantial improvements in resolution over water. Specifically, argon at 40 bar will provide a factor-of-2 improvement, while xenon or carbon dioxide will improve resolution by a factor of 4. The figure of merit rises with the square root of pressure so that at 250 bar the value in argon is 5.

We are currently working on a system that will use argon at 40 bar as the coupling fluid and a 10-μm-radius lens as the imaging element. We expect to obtain a resolution of 220 nm at 1 GHz. Finally, we are in the process of constructing an instrument that would operate at much higher pressures (around 250 bar) with the aim of attaining resolutions well below 100 nm.

4.7   CONCLUSION

Progress over the past five years has been enormous and the field now attracts worldwide interest. We have covered areas that we think will play an important role in the future development of the acoustic microscope. Examples have been provided where necessary to demonstrate the capability of existing acoustic systems. In many applications the acoustic microscope will be used in conjunction with conventional microscopy techniques (such as optical and electron).

Acoustic images of biological material show high contrast without the need for staining; this brings up the exciting possibility of studying living systems. Progress has been made toward extracting quantitative information from the recorded images, and it will not be long before we see routine recordings of density and velocity images.

Finally, work toward improving the resolving power of the acoustic microscope is progressing along two fronts. Cryogenic liquids such as liquid helium are likely to provide resolutions that are vastly superior to existing SAM's. High pressure gas systems operating at room temperature will, before long, provide resolution superior to the optical microscope. Just how far these developments are likely to take us will be clearer in a few years time.

REFERENCES

1. R. A. Lemons and C. F. Quate, App. Phys. Lett. *24*, 163 (1974).
2. M. F. Marmor, H. K. Wickramasinghe, and R. A. Lemons, J. Invest. Ophthalmol. Visual Sci. *16*, 660 (1977).
3. C. F. Quate, A. Atalar, and H. K. Wickramasinghe, Proc. IEEE *67*, 1092 (August 1979).
4. H. K. Wickramasinghe and M. Hall, Electron. Lett. *12*, 637 (1976).
5. D. Rugar, J. Heiserman, S. Minden, and C. F. Quate, J. Microsc. *120*, 193 (1980).
6. D. Ruger, Proc. IEEE, Ultrasonics Symposium, Chicago, October 1981.
7. These images were taken by John Hildebrand and were furnished by C. F. Quate.
8. H. K. Wickramasinghe, J. Appl. Phys. *50*, 664 (1979).
9. I. R. Smith and H. K. Wickramasinghe, Electron. Lett. *18*, 92 (1982).
10. I. R. Smith and H. K. Wickramasinghe, IEEE Trans. Sonics and Ultrasonics, 321 (1982).
11. R. G. Wilson, R. D. Weglein, and D. M. Bonnel, in *Semi-Conductor Silicon/ 1977*, H. R. Huff and E. Sirtle, Eds., Princeton, N.J., Electrochem Soc. *77*, 431 (1977).
12. A. Atalar, C. F. Quate, and H. K. Wickramasinghe, Appl. Phys. Lett. *31*, 791 (1977).
13. H. K. Wickramasinghe, Electron. Lett. *14*, 305 (1978).
14. A. Atalar, J. Appl. Phys. *49*, 5130 (1978).
15. R. D. Weglein, Appl. Phys. Lett. *34*, 179 (1979).
16. W. Parmon and H. L. Bertoni, Electron. Lett. *15*, 684 (1979).
17. J. Kushibiki, A. Ohkubo, and N. Chubachi, Proc. IEEE, Ultrasonics Symposium, Chicago, 1981, p. 552.
18. C. F. Quate, in *Scanned Image Microscopy*, E. A. Ash, Ed., Academic Press, New York, 1980, p. 23.

19. S. D. Bennet, IEEE Trans. Sonics and Ultrasonics, 316 (November 1982).

20. I. R. Smith, D. A. Sinclair, and H. K. Wickramasinghe, Proc. IEEE, Ultrasonics Symposium, Chicago, 1981, p. 591.

21. V. Jipson and C. F. Quate, Appl. Phys. Lett. *32*, 789 (1979).

22. J. Heiserman, in *Scanned Image Microscopy*, E. A. Ash, Ed., Academic Press, New York, 1980, p. 23.

23. D. Rugar, J. S. Foster, and J. Heiserman, Proc. 12th Int. Symposium on Acoustical Imaging, London, July 1982, to appear in *Acoustical Imaging*, Vol. 12, Plenum Press, New York, p. 13.

24. C. R. Petts and H. K. Wickramasinghe, Electron. Lett. *16*(1), 9 (1980).

25. H. K. Wickramasinghe and C. R. Petts, Proc. IEEE, Ultrasonics Symposium, Boston, November 1980, p. 668.

TRANSMISSION ELECTRON MICROSCOPY

I. PASQUALI RONCHETTI, C. FORNIERI, and G. MORI

Institute of General Pathology
University of Modena
Modena, Italy

One of the principal advantages of electron microscopy is that it provides structural information, down to molecular levels, that can be appreciated and, to a certain extent, interpreted by rather inexperienced people. The same information, when obtained by other physical and chemical approaches, often needs complicated procedures and specific knowledge to be interpreted. In addition, electron microscopy allows the identification of the single components in nonhomogeneous aggregates or systems, whereas other physical methods provide a mean statistical value of a given structure or phenomenon. Therefore, by electron microscopy components can be recognized that, if they are scarce in number, might not be detected by other methods.

Another specific advantage of electron microscopy is that it provides information on the relative position and orientation of macromolecular complexes, as well as their relationships with other components to which they are functionally linked.

For all these reasons it would seem that electron microscopy is the most convenient and the easiest way to study biological systems down to molecular and, if necessary, to atomic level.

Group supported by a grant from the Italian Consiglio Nazionale delle Ricerche.

On the other hand, anyone who has even a little experience in electron microscopy knows that so many artifacts can be introduced during specimen preparation and observation in the microscope that the image obtained can be, however beautiful, meaningless. Actually, in spite of the fact that the technical performance of modern transmission electron microscopes will allow resolution of 0.1–0.2 nm, up to now the specimen preparation methods, the high sensitivity of biological material to electron beam damage, and the high vacuum to which the specimen is exposed in the column of the microscope result in great limitations in the resolution. This, in fact, cannot be better than 1–1.5 nm. Moreover, such great distortion can be induced in the specimen that the final image may actually be misleading. All this is even more relevant if one considers that, in most cases, such artifacts are difficult to recognize and properly evaluate.

5.1 HISTORICAL NOTES

The discovery of electrons by J. J. Thompson in 1897 [1] was followed by fundamental studies in electron optics that led, following the tremendous expansion of physical theory and practice in the first quarter of this century, to the hypothesis of the wave nature of electrons and to the demonstration that electrons can be deviated by an electromagnetic field, as light rays are by glass lenses [2,3].

Since then, studies on electron lenses, performed by Davisson and Calbick [4,5], Brücke and co-workers [6,7] on electrostatic lenses, and Knoll and Ruska [8] on magnetic lenses, led to the definition of their theoretical and practical properties and to the construction in 1932 of the first electron microscope.

An estimate of the resolution limit of the transmission electron microscope with magnetic lenses was made by Ruska and co-workers, who applied the light microscope theory of imaging to electron imaging. The limit of resolution of the light microscope is expressed by Abbe's formula as $d = 0.61\lambda/(n \sin \alpha)$, where d is the resolution limit, 0.61 a constant depending on human visual acuity, λ the wavelength, n the refractive index of the medium between the lens and the object, and α the semiangular aperture of the lens.

By replacing the wavelength of light by the wavelength of 75 KeV accelerated electrons, Knoll and Ruska calculated the maximum imaging aperture α of magnetic lenses and the resolution limit of the magnetic transmission electron microscope. This was found to be 0.22 nm. However, the same authors realized that "whether this high resolving power can be used to make visible structures of this order of magnitude cannot be decided in the present state of knowledge. Further investigations are required, involving detailed studies of the aberrations of the image and an increase in the intensity of the electron source'' [9].

Actually, the resolving power of the first electron microscopes were far from the theoretical limit. A series of studies that also took advantage of the interest of several industrial companies in the development of electronics connected with television led to the design of more perfect lenses [10]. In the early 1960s two points separated by 0.4 nm were observed in nonperiodic specimens by Engel, Koppen and Wolff [11], and two years later Komoda and Otsuki reached the limit of 0.2 nm on linear lattices [12].

To date, several commercial transmission electron microscopes have reached a resolution power of 0.15–0.20 nm for crystalline objects. For biological material such a resolution cannot actually be reached. As will be described later, this is due partly to strong limitations intrinsic to the material, such as high inelastic scattering, partly to the technical procedures for specimen preparation, and partly to radiation damage during observation.

However, as will be discussed later, in biology very often it is not as important to visualize atoms as it is to analyze supraatomic aggregates and their relationships.

5.2 PHYSICAL PRINCIPLES

There are some excellent books that give a complete survey of electron optics and of technical devices applied to electron microscopy [13–17]. In accordance with the aim of the present book, in this chapter a rather elementary and descriptive presentation of the physical principles that are at the basis of the electron imaging is reported.

The electron microscope, like its optical counterpart, is essentially an instrument that produces enlargements of fine details. Better than by their enlargement properties, the characteristics of both optical and electron microscopes can be defined by their resolving power, which expresses the capability, theoretical and practical, of a given instrument to resolve two points or lines at the nearest distance.

For both optical and electronic instruments the resolving power is defined by the formula $d = \lambda/(n \sin \alpha)$; that is, it depends on the wavelength λ and on the angular aperture α of the lens. It is evident that for the best resolution large values of α and short wavelength are required. In light microscopy the use of green light with a wavelength of 500 nm and of an objective lens of wide aperture and corrected for most of the defects, together with the use of oil to increase the refractive index of the material between lenses, allow a resolution limit of about 200 nm. In electron microscopy the wavelength is given by $\lambda = h/mv$, where h is Planck's constant and m and v are the electron mass and velocity, respectively. By considering that $\frac{1}{2}mv^2 = eV$, where e is the electron charge and V the potential through which the electrons are accelerated, and therefore $v = \sqrt{2eV/m}$, it results that $\lambda = h/\sqrt{2emV}$. Inserting values for h, e, and m and considering $V = 100$ kV, $\lambda = 0.004$ nm. Therefore, by using electron waves instead of visible photons, a 10^5 times shorter wavelength can be achieved, and a theoretical resolution in the atomic range can be postulated.

Also in this case the aperture of the magnetic lenses, as well as their properties and aberrations, are important in defining the real image resolution of the instrument.

There are interesting similarities between optical and electron lenses. First of all, in both cases one of the conditions necessary to obtain an ideal lens is that the system has an axis of symmetry. In light microscopy the system consists of glass lenses composed of centered spherical surfaces; in electron optics it is formed by axially symmetric electrostatic or magnetic fields. In the available electron microscopes the electron beam passes through a magnetic field generated by circular conductors carrying a current or by two cylindrical magnet poles with coaxial circular bores.

In both systems lens aberrations, such as spherical aberration, chromatic defects, distortion, astigmatism, and image rotation, have to be considered and possibly corrected for a good image to be achieved.

As pointed out by Ruska in 1966 [18], the resolving power of the electron microscope is so high, because of the short wavelength of electrons, that, in spite of the numerous efforts, significant improvement of imaging has not been obtained during the 30 years since the realization of the first instruments. This is partly due to technical reasons, related to instrument construction and specimen preparation, and partly to theoretical studies pointing to the impossibility of correcting, for instance, spherical aberration of rotationally symmetrical systems, which occur as electrons pass through a magnetic field [19].

The spherical aberration is also called aperture defect, as it is proportional to the cube of the angular aperture of the lens. Therefore, to reduce spherical aberration, the aperture of the objective lens must be small. This works against resolution, which requires a wide aperture of the lens to overcome diffraction deterioration in the image. Therefore, the optimum value for the aperture angle is one that gives practically identical spherical aberration and diffraction errors. Spherical aberration is the most important geometrical aberration of the electron-optical system and it is the same for all points in the image. It has a relevant importance in modern electron microscopes.

Other aberrations include (1) distortion phenomena, in which the image points are shifted from the Gaussian points so that magnification increases or decreases with distance from the axis proportionally to the cube of the point distance from the axis; (2) astigmatism, due to either a nonperfect axial symmetry of the lenses or a misalignment and misorientation; and (3) curvature of field, in which the image of a surface is given on a curved plane. All aberrations depend on the lenses and on the alignment of the instrument and are usually corrected in modern electron microscopes.

One of the most important aberrations is chromatic aberration. This produces a circle of confusion at the image plane that is difficult to correct and actually greatly reduces resolution. This depends on several factors, some of which, such as variations in the intensity of the magnetic field of the lenses and variations in electron energy, can be greatly reduced, whereas others, depending on the energy loss of the electrons due to collisions in the specimen, cannot be overcome. This point will be treated more extensively later.

From theoretical principles, the resolution limit of the electron microscope should be at the atomic level. However, taking into account lens aberrations and diffraction phenomena, the actual resolution limit of electron microscopes for nonperiodic structures is on the order of the nanometer. In actual practice, for biological material it is difficult to demonstrate resolution better than 1.5–2 nm; these limitations mainly depend on characteristics of the specimen, such as thickness of the specimen, lack of contrast between structures, and destruction and distortion of the material under the electron beam.

A schematic drawing of a typical electron microscope is given in Fig. 1. The source of electrons is a hairpin of tungsten wire enclosed in an empty cylinder bearing a negative potential with respect to the filament. The negative bias functions as a lens, which condenses the electron beam below the shield. By passing through the anode, the narrow central beam of electrons is selected and is accelerated through the condenser lens (or lenses). The condenser lenses control both intensity and aperture of illumination at the object. After having crossed the specimen, the unscattered beam reaches the objective lens, where an aperture of 25–100 μm intercepts the scattered electrons. This lens gives the first magnified image of the object and determines the resolution and contrast of the final image. Additional projector lenses are then used to magnify the first image and the final image appears on a fluorescent screen and can be recorded on photographic plates.

The entire system must be perfectly aligned and evacuated to the order of 10^{-4}–10^{-7} mm Hg to allow the beam to reach the screen without being intercepted by gas particles.

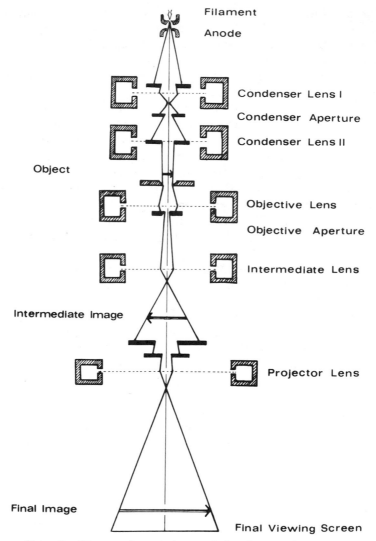

Figure 1.   Diagram of a typical transmission electron microscope.

5.3   ELECTRON–SPECIMEN INTERACTIONS

The electron–specimen interactions can be studied from several points of view and different kinds of information can be achieved: magnification of details in the object, diffraction of crystalline specimens, qualitative recognition, and quantitative measurements of elements by energy dispersive and energy loss spectroscopy. In this chapter the only interactions that are relevant for the final image formation are the electrons that pass through the specimen.

5.3.1 Elastic and Inelastic Scattering

Where the different structures can be distinguished by virtue of their different absorption and deviation of the incident photons, image formation depends on the distribution and quantity of electrons of the incident beam that are scattered by the specimen compared with the unscattered electrons. The amount of the scattered electrons compared with the unscattered ones gives the contrast of the image. Scattering regions in the specimen are visible as darker areas in the image, as electrons that are scattered at angles larger than the objective aperture are prevented from reaching the final screen.

In fact, as a result of the collision of electrons with atoms in the specimen, the incident particles are deviated and lose energy. The collision of electrons with an atomic nucleus in the specimen leads to scattering of the incident electrons with negligible loss of energy, but at a wide angle. These are the elastically scattered electrons, which form a coherent wave and contribute to resolution and contrast. On the other hand, collisions between the incident electrons and specimen electrons lead to a small angle scattering with a relatively high loss of energy. This is the inelastic scattering.

As a practical result, the elastically scattered electrons cooperate to form a high resolution contrasted image, whereas inelastically scattered electrons, of different energy, form a noncoherent wave, which is not focused at a single point in the image, suffers from chromatic aberration of the objective lens, and therefore leads to a severe loss of image contrast and resolution.

The detrimental effect of inelastic scattering is more evident in biological specimens, which have atoms with low atomic number and therefore poor elastic scattering.

Several attempts have been made to subtract the inelastic component of the image-forming wave. Electron spectrometers have been applied between the objective lens and the final screen to select only the electrons with intact energy to form the final image [20–22].

The results show an increase in contrast for biological material; however, resolution is not improved because of a loss of the performance of the microscope. Other attempts were made to correct the final image by filtering elastic and inelastic scattered electrons on the basis that at high resolution contrast and resolution given by elastic scattering depends critically on defocus, whereas the inelastic image is rather independent from defocus [23].

5.3.2 Radiation Damage

Apart from alterations that may occur in the specimen as a result of the vacuum inside the electron microscope column, every object, when exposed to the electron beam, suffers from various kinds of damage. Thin films supported by a circular opening have been calculated to reach temperatures around 550 K [24]. In spite of the difficulties of measuring the temperature of small particles while they are under observation in the microscope, temperatures above 370 K are commonly reached, which means that most organic material is damaged. Thermal effects can be reduced by cooling the specimen chamber and support with liquid nitrogen or helium, or by operating with a low dose of electrons [25–28]. In normal conditions, however, when the specimen is under the electron beam for more than a few seconds, carbonization of biological material has to be taken into account.

Chemical changes that greatly modify the original molecular organization of the specimen have also to be considered. It is very difficult to analyze qualitative and/or quantitative chemical alterations, which most likely depend on the specimen and operating conditions; however, in some cases natural products have been shown to undergo chemical modifications due to neither thermal nor vacuum conditions [29]. In fact, when the specimen is exposed to the electron beam, in a vacuum, a series of excitation phenomena, charge effects, chemical reactions, and/or displacement of atoms might take place [30].

One of the most important processes that alter the morphology of the object is contamination. This is due to an accumulation of material of unknown origin in the areas exposed to the beam. Residual vapors of oil and grease and sublimation and precipitation of material from the specimen itself contribute to the phenomenon, which is undoubtedly favored by the high negative charge of the electron beam. The contamination process was shown to decrease by heating the specimen up to 500 K [31]. However, organic and biological specimens would be completely damaged by these temperatures. Contamination may be reduced by surrounding the specimen with a cold trap [32].

Chemical and displacement phenomena are also partially abolished by the use of stages cooled by liquid nitrogen or, better, by liquid helium [25,30].

Recently the observation of frozen-hydrated biological specimens has been introduced and there is great expectation that this technique will make it possible to observe organic material without any previous chemical treatment, in the natural state and without significant radiation damage [33].

5.4 PREPARATIVE TECHNIQUES FOR BIOLOGICAL MATERIAL

The image in transmission electron microscopy is the result of the ratio between unscattered and scattered electrons during their passage through the object. If the object is too thick, the unscattered fraction is too small and no image is formed. Therefore the thickness of the samples has to be conveniently reduced. As was briefly described in the previous section, biological material has very poor elastic scattering compared with inelastic scattering; therefore it needs to be "stained" in order to increase its contrast and resolution.

Moreover, biological specimens are always highly hydrated and water plays a major role in the molecular and supermolecular organization. In the microscope column a vacuum of at least 10^{-4} mm Hg has to be reached; this greatly modifies the specimen architecture and methods have to be studied to maintain water in the object or to subtract water without affecting specimen organization.

Therefore, it appears evident that biological specimens have to be prepared to be observed in the microscope. For detailed and specific description of various preparative procedures, the reader may refer to reviews and books, such as those edited by Koehler [34], Glauert [35], Hayat [36].

In this report only a survey of the basic methodologies developed to solve general problems is made.

5.4.1 Specimen Contrast and Resolution

Resolution depends both on the instrument and on the specimen. For modern electron microscopes, which reach resolution in the order of 0.1 nm, resolution in biology depends only on the specimen. Two factors have to be considered as playing a major role in limiting resolution of biological objects: low elastic scattering, which also brings about low contrast, and specimen thickness.

Specimen Contrast

The first problem is faced by trying to increase elastic scattering of biological material by staining procedures. Staining can be achieved by linking highly scattering atoms to some chemical group in the specimen or by coating the object with heavy atoms.

In the first case the binding between biological material and heavy atom, usually lead, uranium, and tungsten, is very often nonspecific. The microscopist is accustomed to obtaining conventional patterns, in which all membranes appear as bilayers of electron opaque molecules separated by an electron transparent layer; chromatin results as filaments and granules of a certain dimension with a characteristic packing; mitochondrial matrix consists of more or less aggregated electron opaque granules; myofibrils are built up of a series of thin and thick filaments aligned to form cross-striated structures (Figs. 2 and 3a).

Proteins and nucleic acids do exhibit a certain affinity for uranyl salts; mucopolysaccharides can be stained by iron; tannic acid, added to conventional stains, leads to a heavy electron density of collagen and elastin; however, in the majority of cases staining is absolutely nonspecific.

In some cases a rough indication of the chemical nature of the stained material can be reached. In Fig. 4 toluidine blue O deposits, with characteristic branched appearance, indicate the presence of proteoglycans in the soluble matrix of the aorta wall of seven-day-old chicken. In this specific case, the animal was treated with a chemical that induces lathyrism and toluidine blue O precipitates are present within the abnormally deposited elastin material, indicating that proteoglycan molecules are associated with elastin in this particular pathological condition [37]. However, no indication is obtained on the specific proteoglycan involved. Moreover, the size and shape of the precipitates are so variable that real information on the size and shape of biological molecules involved in the reaction cannot be deduced. Slightly different are cases in which a known enzymatic reaction leads to precipitation of heavy metal salts. Here fairly precise localization of enzymes can be achieved [38].

Immunoelectron microscopy is based on similar principles; in this case the recognition of specific molecules is made through reaction with their specific antibodies, bound either to electron-opaque labels, such as ferritin [39] and gold particles [40] or to enzymes that catalyze reactions leading to electron-opaque precipitates [41]. It is evident that these immunological reactions require previous isolation and purification of the molecules one wants to localize, good preservation of the antigenic properties of the molecules in the specimen, and very careful operating conditions in order to avoid unspecific reactions and misleading results.

"Negative staining" is another procedure that uses salts of elements with high atomic number to increase the scattering properties of biological material. Solutions of uranyl acetate, phosphotungstic acid, or ammonium molybdate are employed to cover and/or to penetrate small isolated

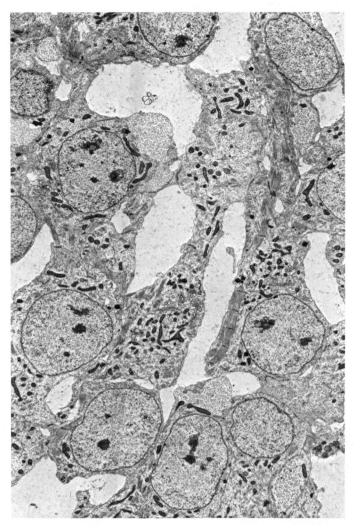

Figure 2. Thin section of chick embryo heart tissue culture in vitro. The sample was chemically fixed with glutaraldehyde and osmium tetroxide, dehydrated in alcohol, and embedded in epoxy resin. Uranyl acetate and lead citrate were used as contrasting stains. The cells are interconnected; nuclei, mitochondria, myofibrils, glycogen deposits, Golgi complexes are clearly seen—their size, shape, and relative position within the cell can be very well appreciated. 4200×.

135

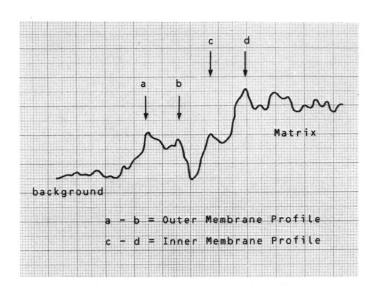

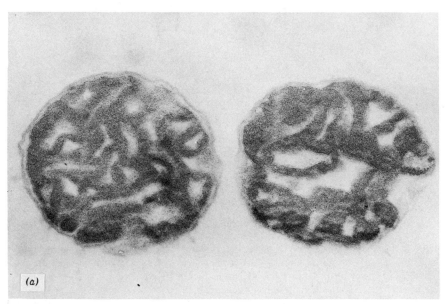

Figure 3a. Above: Thin section of isolated rat liver mitochondria. The organelles were fixed in glutaraldehyde and osmium tetroxide, stained with uranyl acetate and lead citrate and embedded in albumin [52]. The outer and inner mitochondrial membranes, as well as the matrix appear very similar to those of resin-embedded material. Top: Microdensitometric trace of the outer and inner membrane profiles showing their relative thickness and electron densities. With the methodology employed, the specimen was not chemically dehydrated; therefore the membranes retained their lipid content. 80,000×.

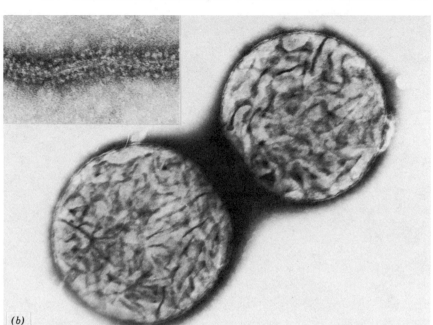

(b)

Figure 3b. Isolated rat liver mitochondria, as those in Fig. 3a, observed after negative staining with 2% ammonium molybdate pH 7.2. Mitochondria lie on an electron transparent film of forvmar and their internal cristae organization is revealed by the stain that penetrates the outer membrane and fills the intermembrane spaces. Inset: Rearrangement of the inner mitochondrial membrane after hypotonic treatment; lipoprotein complexes form long tubules surrounded by ATPase stalked particles. 120,000×. Inset: 200,000×.

objects. After dehydration, a coat of stain lays on the surface or inside the specimen, revealing areas of different electron density. These indicate either regions in which small depressions have collected more negative stain or the presence of chemical groups to which the stain is bound. In the majority of the cases the specimen or part of it is embedded in a highly scattering medium and appears as electron-transparent structure. This technique has been applied with interesting results to isolated structures, such as viruses [42], membranes, plasma lipoproteins [43], collagen fibers (Fig. 4), microsomes, mitochondria (Fig. 3b), and many other specimens. In any case, however, it is impossible to assess if the negative stain is chemically bound or simply forms a coat around the object. Resolution with this method seems to depend mainly on the size of the crystallites formed during dehydration and gives information on the overall organization of isolated biological objects, unless it is associated with immunological procedures for the localization of specific components.

Figure 3c. Isolated rat liver mitochondria, as those in Figs. 3*a* and 3*b*, observed after freeze-fracture. The mitochondria were suspended in 0.25 *M* sucrose containing 20% glycerol, frozen in Freon 22 and liquid nitrogen. On the fracture surfaces a platinum–carbon film was evaporated from an angle of 42°. Convex (*E*) and concave (*P*) fracture planes of the outer mitochondrial membrane show an asymmetrical intramembrane particle distribution. This technique gives useful information on the organization of the internal regions of both natural and artificial membranes. 80,000×.

A disadvantage of negative staining is that the solutions employed are frequently hypotonic with respect to physiological fluids and their pH cannot always be adjusted to neutrality. Therefore, negative stains can dramatically modify and damage osmotically sensitive structures [44]. On the other hand, it has been suggested that the negative stain coat can protect the material, to a certain extent, from exhaustive drying in the microscope [45].

Isolated objects can also be studied by shadow casting. This method was widely employed before negative staining had been developed and consists of deposition under vacuum of a thin metal film (chromium, gold, platinum, tantalum, tungsten, etc.) on the surface of isolated specimens. The metal is evaporated from a given angle with respect to the specimen and a shadow effect is reached. It reveals the surface organization with rather fine details. The electron-opaque areas seen on the final screen

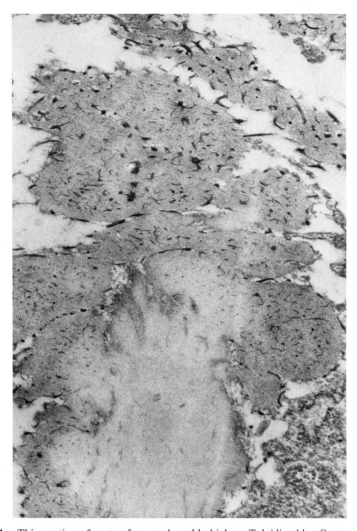

Figure 4. Thin section of aorta of seven-day-old chicken. Toluidine blue O was added to the fixation media and formed precipitates on proteoglycans of the connective tissue. The animal was treated with beta-aminopropionitrile, which inhibits the enzyme lysyl-oxidase; the elastin fiber exhibits lateral knobs formed by scarcely cross-linked elastin molecules with abnormally associated proteoglycans. However, no qualitative or quantitative evaluation of the proteoglycans involved can be made. 48,000×.

represent with certainty surface roughness of the specimen, as all chemical binding can be ruled out in this procedure.

A disadvantage of this technique is that the specimen has to be placed in a vacuum chamber before being coated. Another disadvantage comes from the difficulty of controlling and measuring the real thickness of the metal film [46], although sophisticated thin-film monitors are available for this purpose.

By combining rotation of the specimen and metal evaporation, good resolution and contrast can be obtained for isolated molecules [47]. By scanning transmission electron microscopy, small biological objects can be observed without any staining at all.

Specimen Thickness

The second factor that greatly affects resolution is the thickness of the specimen. In the majority of cases biologists work with voltages in the microscope of 80–100 keV. Under these conditions the electron beam has a limited penetrating range. Therefore a compromise must be reached between electron beam acceleration, specimen thickness, contrast, and resolution. In biology, for the most common operating conditions, the thickness of the specimen must be from a few nanometers up to 100 nm.

Very thin specimens (i.e., isolated molecules) have to be heavily stained to increase their scattering properties or embedded into a highly scattering medium (i.e., the negative stain). The problem is to find out the contrasting approach that interferes as little as possible with the specimen while at the same time giving good resolution.

By increasing the thickness of the object, resolution progressively decreases by increased inelastic scattering and because multiple information comes from partially overlapping surfaces. Therefore, whereas some biological objects are small enough to be observed directly in the microscope (Figs. 3b and 5), others have to be artificially reduced to compatible thickness. In this second case, the specimen is cut into 20–60 nm thin sections, which are stained and observed in the microscope. However, most animal and plant tissues are rather soft and must be embedded in a hard medium before being cut. If the readers are interested in knowing more about embedding procedures and resins employed, they may consult comprehensive reviews and books such as those by Luft [35], Hayat [48], Glauert [49].

Most of the information we have on cellular organelles and their relationships comes from ultrathin sections. In Fig. 2 a typical electron micrograph of animal cells is shown. As pointed out before, with this method the overall shape and size of the cellular components, nucleus, mitochon-

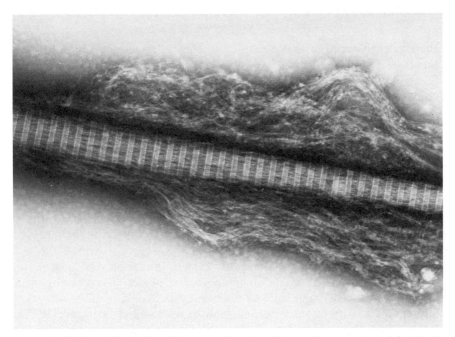

Figure 5. Collagen fiber isolated from rat tail tendon. The specimen was treated for 10 min with 10% glycerol in Tyrode's solution at room temperature and negatively stained with 2% phosphotungstic acid brought to pH 7.2 by NaOH. The fiber is partially swollen and appears built up of about 2-nm-wide fibrils. The characteristic cross striation of the whole fiber is clearly seen. 60,000×.

dria, endoplasmic reticulum, Golgi complex, glycogen, myofibrils, and so on, can be appreciated as well as their relationships. However, this method gives very poor information at the molecular level. Independently from cells and organelles, all the membranes look very similar; chromatin, mitochondrial matrix, materials produced by cells and still retained in the Golgi complex or in the endoplasmic reticulum appear to be identical in all cells from all animals and plants.

In recent years cryotechniques have been developed to preserve the specimen from chemical treatments, embedding resins included. Solid blocks of soft tissues can be obtained by freezing the specimen with liquid nitrogen and keeping it frozen during sectioning. Sections can then be observed in a frozen state or brought to room temperature, stained, and observed in a normal transmission electron microscope. This procedure and the advantages that it brings about will be described briefly in this chapter and more extensively in the chapter by Echlin.

5.4.2 Specimen Preservation

Biological material is very sensitive to the chemical and physical treatments necessary to prepare it for observation in the microscope. The idea of observing living cells or organelles by transmission electron microscopy is at the moment merely utopian. Staining, dehydration, and mainly radiation damage are incompatible with normal biochemical reactions and therefore with life. Moreover, very often they induce dramatic alterations to biological structures.

Fixation

A common procedure to try avoiding damages due to osmotic effects, pH, and dehydration is to treat the object with chemicals that, it is hoped, will suddenly stop all reactions in the living systems, conferring them protection against subsequent damaging effects.

Therefore, "fixation" should fix certain metabolic and structural events and confer some stabilization to the object. Several chemicals have been proposed and used as fixatives: aldehydes, osmium tetroxide, lanthanum and potassium permanganate, and uranyl acetate [35]. The majority of these chemicals bind protein components, functioning as cross-linking agents. Some exhibit affinity for lipids or nucleic acids. Since it was introduced by Sabatini et al. [50], aldehyde fixation, followed by osmium tetroxide, is the most used procedure. However, every biologist knows that different tissues require different fixatives and that concentration, temperature, time, ionic strength, and pH of the fixative are very important and have to be specifically chosen for every material [35,50,51].

Fixation is always performed before embedding and sectioning of tissue blocks or of isolated cells; it helps in maintaining size and shape of osmotically sensitive structures to be observed by negative staining; a slight fixation may be used in some cryotechniques.

Dehydration and Embedding

Dehydration is a very important factor that may destroy the architecture of cells and organelles. Dehydration is essential for embedding in epoxy resins and it consists of a progressive subtraction of water through washing in progressing concentrations of solvents, such as alcohol and acetone. It is evident that lipid and other solvent soluble molecules are removed from the specimen, unless previously chemically "fixed."

Attempts have been made to avoid contact with solvents by using water-soluble resins or inert dehydration with glycerols, or by adopting nonconventional embedding media, such as polymerized glycerols and albumin [49,52,53].

It is worthwhile to notice that, at least in the case of the albumin-embedding method, the structure of isolated mitochondria is very similar to that of normally fixed, dehydrated, and embedded mitochondria (Fig. 3a). The matrix is more homogeneous than in conventionally embedded specimens and the membranes exhibit a regular organization. As far as the membranes are concerned, the outer mitochondrial membrane shows a microdensitometer trace with two identical peaks that correspond to the two electron-opaque layers of the membrane. The inner membrane, on the contrary, exhibits a constant asymmetry, with the matrix peak higher and wider than the peak corresponding to the membrane layer facing the intermembrane space (Fig. 3a). As will be mentioned later, this corresponds to the half-inner mitochondrial membrane richest in intramembrane particles, when observed by freeze-fracture.

Small isolated particles, either shadow cast or negatively stained, undergo a dehydration process during specimen preparation or during observation in the microscope. Devices have been studied to avoid as much as possible collapse of the structures during dehydration. Among these procedures critical-point drying and freeze-drying, based on the sublimation of water, give very good results for viruses, plant and animal organelles, isolated membranes, and cells.

Cryotechniques

In the early 1960s the low temperature procedures applied to electron microscopy started to offer good results and to be applied to all kind of biological material. The first aim was to avoid chemical fixation and embedding and to preserve structures from drastic dehydration; actually, cryotechniques are regarded as the only way to maintain water inside the object so that images of fully hydrated specimens or replicas of fully hydrated macromolecular complexes can be achieved.

Fernandez-Moran and Bullivant were the first to apply liquid nitrogen and liquid helium to fix biological specimens, trying to dehydrate and embed the sample at low temperature to avoid collapsing of the structure [54,55]. This methodology has been developed by several authors with interesting results [53,56]. In some cases it has given rise to new interpretation of biological structures [57].

In cryotechniques one of the most important steps is specimen freezing. Biological material is highly hydrated and at temperatures below 273 K, ice crystals are usually formed that alter the architecture of cells, organelles, and surfaces. It would be ideal if the water molecules could be instantaneously immobilized by very fast freezing. Practically, this is difficult to obtain, as the thermal conductivity of biological material is very low. Even with freezing speeds in the order of 10,000° per second only a very superficial layer in the specimen can be considered frozen, without appreciable ice crystal damage. To avoid water crystallization as much as possible, cryoprotectant agents are currently used. However, there is indication that such molecules interfere with the specimen and "at present it seems unlikely that we have a totally effective cryoprotectant agent that can be said to be entirely artefact-free" [58].

Once the specimen is frozen, it can be (1) frozen-dried, (2) frozen-embedded, (3) cryosectioned, (4) freeze-fractured, (5) observed in the frozen state.

The first two procedures have already been mentioned. It is worthwhile to spend a few words on cryosectioning and freeze-fracturing.

In the first case, ultrathin frozen sections are obtained, on which cytochemical, immunocytochemical, and enzymatic reactions can be performed; the sections are either stained and observed or directly observed in the microscope equipped with a cold stage.

In freeze-fracture, the frozen specimen is placed into a liquid nitrogen or liquid helium refrigerated chamber, where a 10^{-7}–10^{-9} mm Hg vacuum is reached. The specimen is then fractured and on the fracture surfaces a layer of platinum/carbon is evaporated. A second film of pure carbon is evaporated from the top forming a sort of resistant coat on the fractured surface. The biological material is digested and the replica observed in the microscope. This procedure gives interesting results on membrane structure. In fact, the membranes are usually split in the hydrophobic central region and reveal a structure, which is completely different from that obtained by other procedures. In Fig. 3c mitochondria isolated from rat liver, like those of Figs. 3a and 3b, are shown. The inner and the outer membranes have been split in their hydrophobic portions and reveal randomly distributed particles. These should represent proteins or lipoprotein complexes spanning the membranes. Every membrane exhibits peculiar intramembrane particle distribution. In Fig. 3c it appears that the inner membrane is richer in particles than the outer membrane. Generally speaking, the half-membrane of cells and organelles in contact with the protoplasmic space is richer in particles than the half-membrane in contact with the extracellular space or with the internal spaces of cytoplasmic organelles. The distribution of these particles is temperature sensitive; by incubating the specimen at low temperatures before freezing, the particles form clusters leaving large areas of the membrane completely smooth [59].

This is interpreted as migration of proteins or lipoprotein complexes toward more fluid regions. These movements support the hypothesis that biological membranes are fluid bilayers of phospholipids, in which proteins and other constituents are immersed maintaining a certain degree of mobility [60].

By combining electron microscopy with biochemical and physico-chemical data, interesting information on the structure and function of membrane systems can be obtained. By freeze-fracture, in fact, the physicochemical characteristics of a given membrane can be studied. Figures 6a and 6b show the structural modifications induced by temperature on artificial membranes formed of dimyristoyl-phosphatidyl-choline. It is known from scanning calorimetry that the melting point of this phospholipid is around 294 K. In Fig. 6a the liposomes were quenched from 291 ± 2 K and fracture faces exhibited the undulated appearance typical of the premelting phase; whereas in Fig. 6b the liposomes were quenched from

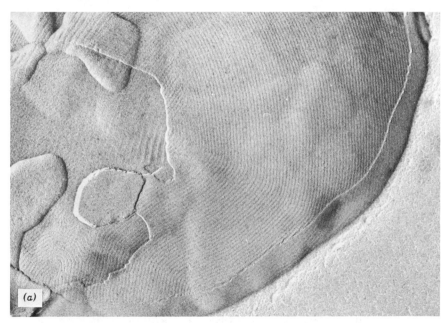

Figure 6a.   Large multilamellar vesicle of dimyristoyl-phosphatidylcholine in water. The sample was equilibrated at 291 ± 2 K and quickly frozen in Freon 22 and liquid nitrogen. In a Balzer BAF 301 freeze-etching apparatus the specimen was fractured at 173 K and the fracture surfaces shadowed with platinum–carbon. The fracture plane reveals the undulated pattern typical of the phospholipid bilayers in premelting phase. 80,000×.

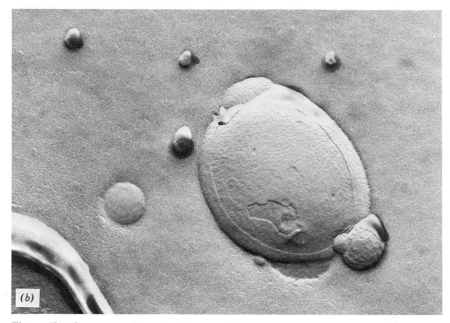

Figure 6b. Same material as in Fig. 6a. The only difference is that the sample was quenched from 300 ± 2 K. The melting point of this phospholipid is 294 K; therefore the membranes were frozen in the melted phase. The undulated appearance in Fig. 6a has disappeared and the fracture surfaces are rather smooth, as it should be above the melting point of pure phospholipid membranes. 80,000×.

300 ± 2 K and fracture faces were rather smooth, as expected for the melted phase. Therefore, freeze-fracture electron microscopy is a rather conservative technique and may only fix precise structural events.

5.4.3   Radiation Damage

As already mentioned, the electron beam interacts with the specimen, inducing excitation, ionization, and displacement of ions, which lead to heating of the sample, bond scission, cross-linking phenomena, and mass loss [61]. These events greatly limit significant information.

In addition, through the effect of local ionization, ions from the surrounding space may precipitate on the specimen and contaminate the sample by forming layers of amorphous materials [28,31,62]. These damaging effects can be, at least partially, reduced for biological specimens by using cold traps in the microscope and cold stages. Recently, cold stages have been studied for the observation of frozen-hydrated biological specimens. The stages are cooled by liquid nitrogen or liquid helium and the specimen, previously frozen, is transferred into the microscope and observed at very low temperature. The temperature of the sample rises when it is irradiated by the electron beam, but it may be maintained with nondamaging limits. In these systems the ionizing effects cannot be overcome; however, the diffusion of radiolysis products is greatly reduced and this reduces the actual damage to the structures. As pointed out by Salih and Cosslett [63], at temperatures below 80 K the molecules retain very little motion, ion displacement may be reduced, and there is the chance of repairing damage, such as radiation bond breaking. At liquid helium temperature no loss of mass from organic material could be detected on electron beam irradiation [27].

In practice, however, in frozen hydrated biological specimens radiolysis of ice cannot be completely avoided and, therefore, free radicals (H$^{\cdot}$ and OH$^{\cdot}$) may produce damage in the sample. A low radiation dose is the only sensible way to reduce these water-mediated alterations [64,65].

5.5 CONCLUSION

Organic and biological materials are not immediately observable in the microscope, but need to be treated in order to increase their contrast and resolution and to be stabilized against dehydration and radiation damages.

However, treatments imply interferences, at atomic and molecular level, between chemicals and specimen or between physical agents and specimen. These intereferences are, in the majority of cases, completely unknown and, in any case, induce modifications in the object that, by simple electron microscopy, it should be impossible to recognize. This is one of the reasons why in biology the real resolution is limited to between 1.0 and 1.5 nm.

The main aim in biology is to correlate structure and function, possibly at molecular level, as the building blocks in biology are superatomic aggregates in the order of 1–10 nm and more. This implies that resolution limit in the transmission electron microscope and suitable techniques for specimen preparation make it possible to obtain useful information on objects whose size is in this order of magnitude. In fact, electron microscopy in biology and organic systems is mainly applied to visualizing supermolecular arrangements and relationships among different aggregates of molecules. In this context transmission electron microscopy, associated with biochemical, biophysical, and chemical studies, has greatly contributed to knowledge of the structure and function of organic material.

REFERENCES

1. J. J. Thompson, Phil. Mag. *4*, 293 (1897).
2. H. Busch, Ann. Phys. *81*, 974 (1926).
3. H. Busch, Arch. Elektrotechn. *18*, 583 (1927).
4. C. J. Davisson and C. J. Calbick, Phys. Rev. *38*, 585 (1931).
5. C. J. Davisson and C. J. Calbick, Phys. Rev. *42*, 580 (1932).
6. E. Brüche and H. Johannson, Ann. Phys. *15*, 145 (1932).
7. E. Brüche and O. Scherzer, *Geometrische Elektronoptik*, Berlin, 1934.
8. M. Knoll and E. Ruska, Ann. Phys. *12*, 607 (1932).
9. M. Knoll and E. Ruska, Z. Phy. *78*, 318 (1932).
10. E. Ruska, Arch. Elektrotechn. *38*, 102 (1944).
11. A. Engel, G. Koppen, and O. Wolff, in *Electron Microscopy*, Proc. Fifth Int. Congr. Electr. Microsc., Philadelphia, 1962, Vol. I, E13, Academic, New York.
12. T. Komoda and M. Otsuki, Japan J. Appl. Phys. *3*, 666 (1964).
13. V. E. Cosslett, *Practical Electron Microscopy*, Butterworths, London, 1951.
14. C. E. Hall, *Introduction to Electron Microscopy*, McGraw-Hill, New York, 1953.
15. B. M. Siegel, *Modern Developments in Electron Microscopy*, Academic, London, 1964.
16. P. Grivet, *Electron Optics*, 2nd ed., Pergamon, Oxford, 1972.
17. P. W. Hawkes, *Electron Optics and Electron Microscopy*, Taylor and Francis, London, 1972.
18. E. Ruska, "Past and recent attempts to attain the resolution limit of the transmission electron microscope," in R. Barer and V. E. Cosslett, Eds., *Advances in Optical and Electron Microscopy*, Vol. I, Academic, London, 1966, pp. 115–179.
19. O. Scherzer, Optik *2*, 114 (1947).
20. R. Castaing, "Secondary ion microanalysis and energy selecting electron microscopy," in U. Valdrè, Ed., *Electron Microscopy in Material Science*, Academic, New York, 1971, p. 103.
21. R. M. Henkelman and F. P. Ottensmeyer, J. Microsc. *102*, 79 (1974).
22. R. F. Egerton, J. G. Philip, P. S. Turner, and M. J. Whelam, J. Phys. Sci. Instrum. *8*, 1033.(1975).
23. D. L. Misell, Contrast enhancement by using two electron micrographs, in *Principles and Techniques of Electron Microscopy*, Vol. 8, M. A. Hayat, Ed., Van Nostrand Reinhold, New York, 1978, pp. 181–204.
24. B. Von Borries and W. Glaser, Kolloid Z. *106*, 123 (1944).

25. R. M. Glaeser and K. A. Taylor, J. Microsc. *112*, 127 (1978).
26. Y. Talmon and E. L. Thomas, J. Microsc. *113*, 69 (1978).
27. E. Knapek and J. Dubochet, J. Mol. Biol. *141*, 147 (1980).
28. A. Howie, Rev. Phys. Appl. *15*, 291 (1980).
29. C. E. Hall, E. A. Hauser, D. S. Le Beau, F. O. Schmitt, and P. Talalay, Ind. Eng. Chem. *36*, 634 (1944).
30. J. Dubochet and E. Knapek, Chem. Scripta *14*, 267 (1979).
31. A. Kumao, H. Hashimoto, and K. Shiraishi, J. Electron Microsc. *30*, 161 (1981).
32. A. E. Ennos, Brit. J. Appl. Phys. *4*, 101 (1953).
33. Y. Talmon, J. Miscrosc. *125*, 227 (1982).
34. J. K. Koehler, *Advanced Techniques in Biological Electron Microscopy*, Springer-Verlag, Berlin, 1973.
35. A. M. Glauert, Fixation, dehydration and embedding of biological specimens, in *Practical Methods in Electron Microscopy*, Vol. 3, Part 1, A. M. Glauert, Ed., Elsevier/North-Holland, Amsterdam, 1975, pp. 1–207.
36. M. A. Hayat, *Principles and Techniques of Electron Microscopy, Biological Applications*, Vol. 8, Van Nostrand Reinhold, New York, 1978.
37. I. Pasquali Ronchetti, C. Fornieri, I. Castellani, G. M. Bressan, and D. Volpin, Exp. Mol. Pathol. *35*, 42 (1981).
38. M. A. Williams, Autoradiography and immunocytochemistry, in *Practical Methods in Electron Microscopy*, Vol. 6, Part 1, A. M. Glauert, Ed., Elsevier/North-Holland, Amsterdam, 1977, pp. 1–234.
39. J. Remacle, S. Fowler, H. Beaufay, A. Amar-Costesec, and J. Berthet, J. Cell Biol. *71*, 551 (1976).
40. D. A. Handley, C. M. Arbeeny, L. D. Witte, and S. Chien, Proc. Natl. Acad. Sci. USA *78*, 368 (1981).
41. L. Y. W. Bourguignon, Cell Biol. Int. Rep. *4*, 541 (1980).
42. R. W. Horne and I. Pasquali Ronchetti, J. Ultrastr. Res. *47*, 361 (1974).
43. I. Pasquali Ronchetti, S. Calandra, M. Baccarani Contri, and M. Montaguti, J. Ultrastr. Res. *53*, 180 (1975).
44. U. Muscatello and V. Guarriero Bobyleva, J. Ultrastr. Res. *31*, 337 (1970).
45. R. W. Horne, J. Microsc. *98*, 286 (1973).
46. J. H. M. Willison and A. J. Rowe, Replica, shadowing and freeze-etching techniques, in *Practical Methods in Electron Microscopy*, Vol. 8, A. M. Glauert, Ed., Elsevier/North-Holland, Amsterdam, 1980.
47. J. Engel, H. E. Oderma, A. Engel, J. A. Madri, H. Furthmayr, H. Rohde, and R. Timpl, J. Mol. Biol. *150*, 97 (1981).
48. M. A. Hayat, *Principles and Techniques of Electron Microscopy: Biological Applications*, Vol. 1, Van Nostrand Reinhold, New York, 1970.

49. J. H. Luft, Embedding media: old and new, in *Advanced Techniques in Biological Electron Microscopy*, J. K. Koehler, Ed., Springer-Verlag, Berlin, 1973, pp. 1–34.

50. D. D. Sabatini, K. Bensch, and J. R. Barrnett, J. Cell Biol. *17*, 19 (1963).

51. J. C. Riemersma, Chemical effect of fixation on biological specimens, in *Some Biological Techniques in Electron Microscopy*, D. F. Parsons, Ed., Academic, New York, 1970, p. 69.

52. I. Pasquali Ronchetti, J. Submicrosc. Cytol. *4*, 205 (1972).

53. D. C. Pease, Substitution techniques, in *Advanced Techniques in Biological Electron Microscopy*, J. K. Koehler, Ed., Springer-Verlag, Berlin, 1973, pp. 35–66.

54. H. Fernandez-Moran, Ann. N.Y. Acad. Sci. *85*, 689 (1960).

55. S. Bullivant, Lab. Invest. *14*, 440/1178 (1965).

56. S. Bullivant, Present status of freezing techniques, in *Some Biological Techniques in Electron Microscopy*, D. F. Parsons, Ed., Academic, New York, 1980, pp. 101–146.

57. F. S. Sjöstrand and W. Bernhard, J. Ultrastr. Res. *56*, 233 (1976).

58. H. Skaer, J. Microsc. *125*, 137 (1982).

59. M. Höchli and C. R. Hackenbrock, J. Cell Biol. *72*, 278 (1977).

60. S. J. Singer, Ann. Rev. Biochem. *43*, 805 (1974).

61. V. E. Cosslett, J. Microsc. *113*, 113 (1978).

62. J. J. Hren, Ultramicroscopy *3*, 375 (1979).

63. S. M. Salih and V. E. Cosslett, J. Microsc. *105*, 269 (1975).

64. I. A. M. Kuo and R. M. Glaeser, Ultramicroscopy *1*, 5 (1975).

65. M. Kessel, J. Frank, and W. Goldfarb, Low dose electron microscopy of individual biological macromolecules, in *Electron Microscopy at Molecular Dimensions*, W. Baumeister and W. Vogell, Eds., Springer-Verlag, Berlin, 1980, pp. 154–160.

AUGER SPECTROSCOPY

A. P. JANSSEN

Standard Telephone & Cables Ltd.
Laser Unit,
Paignton,
Devonshire, England.

6.1 INTRODUCTION

Auger electron spectroscopy (AES) is a technique that allows the chemical composition of a surface to be identified by using an electron beam to excite surface atoms and measuring the energy distribution of emitted electrons from the sample.

The technique has been developed over the last 15 years or so as a means of probing the top few atom layers of a surface. The ability to perform surface chemical analysis with extremely high surface sensitivity arises from the fact that Auger electrons can only escape from a depth of a few nanometers below the surface. Thus details of a specimen's internal composition are not apparent. This surface sensitivity is complementary to a number of other techniques such as x-ray microanalysis, electron energy loss spectroscopy (EELS), ESCA, and laser probe mass analysis. This article will attempt to compare the three electron probe techniques of x-ray, Auger, and electron loss spectroscopy, since these are applicable to either scanning or transmission microscopy—or both.

Because of the sensitivity to only a few atom layers of the surface, samples need to be kept atomically clean by operating under high or ultrahigh vacuum (UHV) conditions (i.e., better than 10^{-7} mbar and often as low as 10^{-11} mbar). Because of the necessity for surface cleaning, Auger analysis has until the present received only very limited attention as an analytical tool for the biological sciences; however, a better understanding of electron scattering processes in solids, particularly for high energy electrons (>1000 eV), may allow the surface cleanliness to be a less demanding constraint.

For many years Auger spectroscopy of surfaces has been carried out using broad, low energy electron beams to obtain information from relatively large areas. In the last few years, however, the technique has been developed as an adjunct to the scanning electron microscope equipped with an energy analyzer to obtain chemical information from extremely small surface areas. At best the spatial resolution of Auger information is comparable with the resolution of the scanning electron microscope (i.e., a few hundred angstroms). The new technique of scanning Auger microscopy has a number of advantages over x-ray analysis and EELS for particular specimens. Bulk samples may be used with very good spatial resolution for analysis, plus an ability to detect all light elements except hydrogen and helium with high sensitivity.

For many applications the sample surface can be removed by ion sputtering to expose clean subsurface layers. The combination of Auger spectroscopy, with its inherent high depth resolution together with ion sputtering, can also yield a chemical composition depth profile. This technique has proved to be extremely valuable in evaluating metallurgical interfaces and semiconductor surfaces.

It is reasonable to expect that as better quantification methods and specimen preparation techniques are developed, Auger spectroscopy will find more widespread application in the biological sciences.

6.1.1 The Auger Process

Auger electron emission, like x-ray emission, results from an initial excitation of the atom by an energetic electron from the incident electron beam, with primary energies ranging from a few thousand volts to several tens of thousand volts. The initial ionization caused by the primary electron is followed by electron transfer from an outer electron shell of the atom to fill the inner shell hole as shown in Fig. 1, which illustrates the Auger process for a silicon atom [1,2]. The atom is now left with an excess energy $E_K - E_{L_1}$, which can be shed by emission of a photon of energy $E_K - E_{L_1}$, which may be in the x-ray range, or else by emission of an electron. The emitted electron may come from any energy level, provided that it can escape into the vacuum. Thus for the illustrated process, the

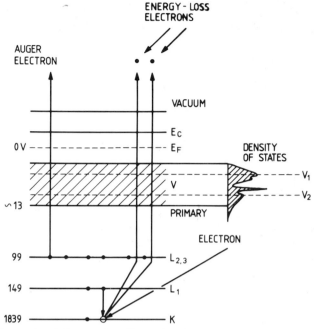

Figure 1.   Energy level diagram for silicon showing the KL_1L_{23} Auger transition. The K electron loss process is also shown.

emitted Auger electron would have an energy approximately given by

$$E_{\text{Auger}} \simeq E_K - E_{L_1} - E_{L_{23}} \tag{1}$$

This particular process is described as the KL_1L_{23} Auger transition. However, a number of transitions are possible involving any two or three separate energy levels including the valence band. Although Auger electron emission leaves the atom doubly ionized, neutralization normally occurs by further transitions involving outer electrons. The probability of de-excitation via a particular transition depends on the electron overlap between shells, and this determines the relative strengths of the Auger peaks. A measurement of the number of emitted electrons over this particular energy range (i.e., 0–2000 eV) will show peaks corresponding to the available Auger transitions for the elements present. Thus the elements present can be identified. The method of detecting the Auger electrons from the background of randomly scattered electrons emitted from the sample is dealt with later. Auger electrons that are of most interest usually have energies from a few tens of eV to about 2 keV, since the probability of ionization of electron levels of these energies is reasonably high for electron beams normally encountered in the scanning microscope. Also, Auger electrons are emitted over this range of energies for all elements except hydrogen and helium. These two elements cannot be detected using this technique, since the Auger process is a three-electron process; therefore beryllium is the lightest detectable element.

6.2 THE AUGER YIELD

In comparison to x-ray analysis, Auger emission is strong for light elements such as carbon and oxygen and in this respect has some advantage for biological analysis. In order to understand the Auger technique more fully and to describe the techniques used for quantification, we shall first discuss the factors governing the yield of Auger electrons from a particular sample. To estimate the Auger current emitted from the sample by an electron beam of current I_p amps, a number of factors have to be taken into account, since these determine the sensitivity of detection for a particular element in different solid environments. The equation for the emitted Auger current can be written [3]:

$$I_A = I_p \cdot N \cdot \phi \cdot \gamma \cdot r \cdot \sec \phi \qquad (2)$$

where I_p is the primary beam current and N is the density of surface atoms. Φ is the ionization cross section, which is a measure of the probability of ionizing a particular electron shell with an incident electron of energy E_p. As the primary energy E_p is increased to the ionization energy E_k of the K shell in the example of Fig. 1, the cross section rises to a maximum at several times E_k. Further increasing the primary energy reduces the ionization cross section. The form of this behavior is shown in Fig. 2 for normalized primary energies [4]. It can be seen that using primary energies much greater than the ionization energy will reduce the Auger signal from the optimum; low primary energies are to be favored, especially for light elements common in biological material.

The term γ is the Auger emission probability (i.e., the probability that deexcitation will occur by electron emission rather than x-ray emission). Since the total probability of deexcitation is 1, γ and the fluorescence yield ω are complementary [i.e., $\gamma = (1 - \omega)$]. Normally, several Auger transitions will be possible from an initial ionization state and the total Auger emission probability should be summed over all transitions, that is,

$$\sum_i \gamma_i = (1 - \omega) \qquad (3)$$

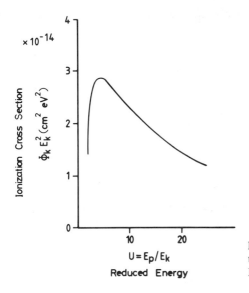

Ionization Cross Section

$\phi_k\, E_k^2\ (cm^2\ eV^2)$

$\times 10^{-14}$

$U = E_p/E_k$

Reduced Energy

Figure 2. Atomic ionization cross-section variation for electrons of energy E_p. Plotted against reduced energy $U = E_p/E_k$.

The Auger emission probability for a given transition is independent of the atom's environment.

In addition to ionization that results directly from impact by the primary electrons of the incident beam, ionization may be caused by electrons scattered in the solid. Scattered electrons will undergo energy loss and simultaneous deflection as they cause ionization until they have insufficient energy to contribute to further ionization. This additional ionization contribution from scattering events near the surface can be accounted for by the backscattering factor r. Numerically this factor varies from ~2 for heavy elements to ~1.2 for light elements and generally increases with primary energy. The backscattering factor for Auger electrons has been measured for a number of elements [5] over a limited range of reduced energy $U = E_p/E_k$ (Fig. 3), and as one would expect, at the ionization threshold there is no contribution to the Auger yield from multiple ionization.

The variation of both the backscattering factor r and the ionization cross section Φ with primary energy E_p is substantial for the range of energies used in the scanning electron microscope, and although the backscattering factor will increase slowly for optimum sensitivity, the decrease in ionization cross section will tend to favor a lower working primary energy. For typical Auger energies of a few tens of eV to 1000 eV, a primary energy of 10 keV would be reasonable. In instruments where Auger spectroscopy is used in conjunction with scanning electron microscopy, considerations of the electron-optical performance (i.e., res-

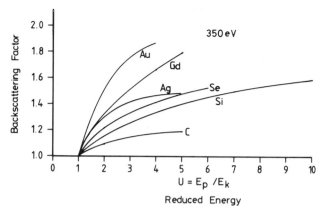

Figure 3. Dependence of the backscattering factor r on reduced energy U for a number of elements. (After Ref. 5.)

olution of the instrument), as well as damage in biological specimens, may be more important factors in determining the primary beam energy.

The factor that is responsible for the widespread use of this technique for surface studies of various kinds and that has allowed very rapid advances to be made in the understanding of solid surface phenomena such as thin film growth, oxidation, and corrosion is the very short escape depth of low energy electrons from a solid. Electrons of less than a few thousand electronvolts energy will lose energy by inelastic collisions within a few interatomic distances. This means that Auger electrons generated deeper than a few monolayers below the sample surface will not escape without energy loss and will therefore not be detected with the characteristic Auger energy E_a of Eq. (1). The average distance traveled by an electron of energy E eV before inelastic collision, more generally called the inelastic mean free path (imfp), has been measured by several authors [6,7] over a wide range of elements. The imfp depends on the particular solid and also on the electron energy, as shown in Fig. 4. For very low energy electrons (less than 10 eV) the number of processes available whereby electrons can lose energy are limited, so that the imfp increases with reduced energy. For energies greater than 50 eV the imfp increases over a range of several monolayers. However, even for energies of several thousand electron volts, the imfp is generally only of the order of 10 atomic layers.

It is more general to replace the effective number of surface atoms cm^{-2}, N in Eq. (2), by $\lambda_e p C$, where λ_e is the Auger electron escape depth, p the atomic density (atoms cm^{-3}), and C is the concentration of a particular element.

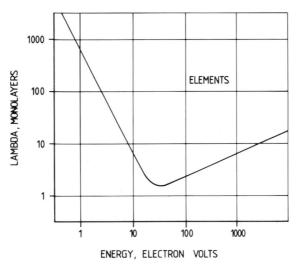

Figure 4. Trend of inelastic mean free path (λ) with electron energy. Experimentally determined values for a variety of elements are scattered about solid line. (After Ref. 7.)

The Auger electron yield also depends on the angle of incidence of the electron beam onto the surface. If the beam direction is tilted toward the surface plane, the path length of the beam within the escape depth λ_e will increase. In addition, there will also be a contribution from scattered electrons, since a higher proportion of forward-scattered electrons can intersect the surface. These effects make the technique sensitive to surface topography in a similar way to secondary electron image contrast formation [5].

One of the main advantages of the Auger technique is its sensitivity for light element analysis. This is illustrated in Fig. 5, which shows the total Auger yields for essentially the total range of elements except hydrogen and helium [8]. The presence of hydrogen in simple compounds may be inferred, however, from shifts in the outer electron levels of other atoms. This results in a shift or shape modification of low-energy Auger peaks from the compound. Elemental yields are plotted in Fig. 5 on a logarithmic scale (after Staib and Staudenmaier; see Ref. 8) for a 10-keV incident beam, where the yield γ is the ratio of number of emitted Auger electrons into 4π steradians to the number of incident electrons. Yields range from 10^{-5} to 10^{-2}, although for the most favorable transitions, yields over the periodic table vary by only a factor of 5.

For biological analysis the enhanced sensitivity for light element analysis compared to the severe drop in x-ray yield for elements lighter than

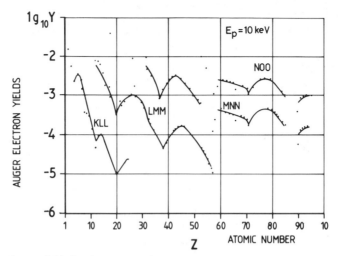

Figure 5. Auger yields for the elements for primary energy E_p = 10 keV. Plotted as $\log_{10}$ of yield Y. (From Ref. 8.)

$Z \simeq 10$ may be a distinct advantage. The detection difficulty for low-Z x rays is compounded in many energy dispersive instruments by absorption in the window placed in front of the detector.

Electron loss spectroscopy is also suited to light element analysis, but yield comparisons with Auger and x ray are difficult, since this depends on the specimen thickness [9]. Without going into unnecessary detail, the three techniques have broadly similar yields (10^{-3}) for reasonably large atomic number Z, with Auger tending to be lowest because of the very small escape depth λ.

The rate at which chemical information is collected for each of these techniques also depends on the collecting efficiency of the detector. A relatively large fraction of available signal can be collected in electron loss spectroscopy, since the loss electrons are emitted within a very small angular cone. However, this is not so for x-ray or Auger spectroscopy.

6.3 INSTRUMENTATION

Auger analysis is performed in the SEM using an electrostatic energy analyzer. There are two types that are generally used for SEM analysis. These analyzers have been designed for relatively large solid angular collection and good background noise discrimination and are shown schematically in Fig. 6.

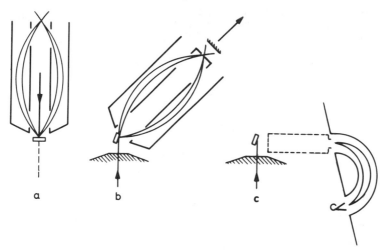

a   b   c

Figure 6. Instrument geometries for electron optical column and electrostatic analyzers used for micro-Auger analysis. (a) Cylindrical mirror analyzer with axial magnetic or electrostatic beam lens system. (b) SEM optical column with retractable CMA to allow use of other detectors and sources. (c) Concentric hemispherical analyzer with magnetic SEM probe lens.

The cylindrical mirror analyzer (CMA) consists of two concentric cylinders with a potential applied between them sufficient to deflect electrons within a small energy range so that they pass through a focusing aperture at the back of the instrument [10,11]. Figures 6a and 6b show the use of this type of instrument with an axial electron gun that gives very good access to the specimen [12] and as an attachment to a conventional electron-optical column (Fig. 6b) [13]. Figure 6c illustrates the concentric hemispherical analyzer, which is capable of greater energy resolution because of the geometric configuration and also because the incoming electrons can be retarded by the input lens [14]. In either case, extremely small electron currents are collected; for a primary current of 5.10^{-9} A the total Auger current may be 10^{-12} A with a collected current of 10^{-14} A from a practical analyzer.

In practice, Auger electrons are detected by electronic differentiation of the energy spectrum. This allows the Auger peaks to be separated from the continuous secondary background current, which is often much larger than the Auger current. This is illustrated in Fig. 7, which shows the cadmium spectrum obtained with a CMA in the configuration of Fig. 6b [15]. The emitted current is shown as a function of the electron energy in Fig. 7b for a number of incident angles. Measurements of the peak intensity for quantitative analysis are made difficult by the large background.

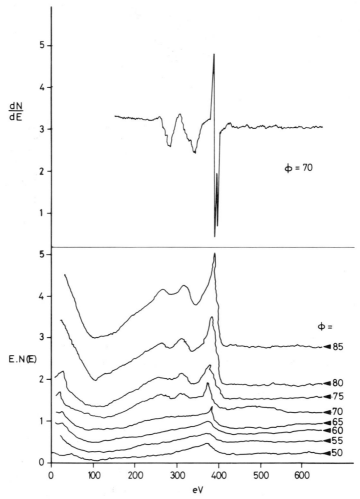

Figure 7. Auger spectra from cadmium showing the energy distribution of emitted electrons (below), for a number of incident beam angles ϕ, and the corresponding electronically differentiated spectrum (above), using a CMA in the mode shown in Fig. 6b with $E_p = 30$ keV. $I_p = 3.10^{-7}$ A.

The Auger intensity can be measured relatively easily from the differentiated signal dN/dE (Fig. 7) because of the suppression of the constant background signal. However, this method of detection is inherently inefficient, since as is seen from the undifferentiated spectra (Fig. 7), most of the Auger current appearing to the low energy side of the main peak edge is not detected. In practice, this tail is modified by a number of loss

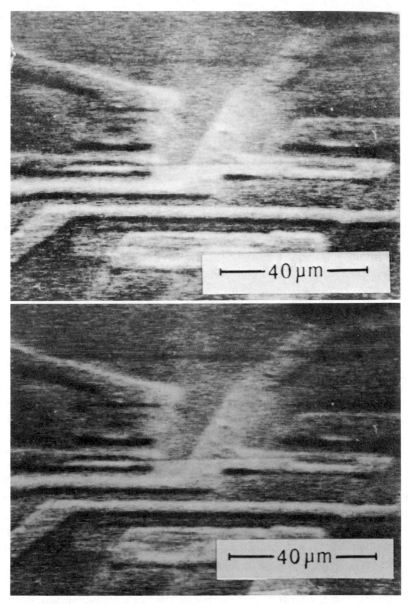

Figure 8. Auger map using the oxygen KLL Auger peak at 510 eV. High concentrations of oxygen can be identified on this image of part of an integrated circuit. Imaging time 500 s.

mechanisms and will change with the presence of other additional atomic species in the sample.

As with x-ray analysis, a chemical picture can be built up using the intensity of a particular Auger peak to show the distribution of an element across the surface. An Auger map using the KLL 510-eV oxygen peak is shown in Fig. 8. The presence of oxygen as SiO_2 is clearly visible in this micrograph of part of an integrated circuit.

6.4   QUANTITATIVE ANALYSIS

The most accurate method of quantitation relies on the use of standard elements or compounds of known composition. As already pointed out, the Auger intensity is very sensitive to surface contamination. Thus it is of critical importance that standards be kept atomically clean. This can be done either by ion sputtering the surface to remove accumulated contamination or by evaporating a layer of material from a vapor source. The ratio of Auger current I_A from the sample to that from the standard I_0 gives a simple first-order measure of the concentration of the element; that is

$$\frac{I_A}{I_0} = \frac{C_A}{C_0}$$

where C_A is the sample concentration and C_0 is the standard concentration. It is then necessary to make corrections for changes in density, backscattering factor, and perhaps the inelastic mean free path. Collectively, these have been shown for certain compounds and alloys to vary linearly with changes in composition.

An alternative method of obtaining a quantitative measurement is to use sensitivity factors [16]. This method assumes that peak intensities for various elements occur with a constant ratio. Thus one standard sample will allow estimates of peak intensities of other elements to be calculated. Values of sensitivity factors for various elements have been measured by Chang [17] and the method is estimated to give an accuracy of better than 30%.

The accuracy of measurement may be decreased by other sample-dependent factors. First (and the most likely error), the surface composition may not be characteristic of the bulk material. This may be due to surface contamination or to surface segregation. The latter means that one or more atomic species comprising the material preferentially tends to migrate toward the surface [18]. Advantage is made of the Auger technique to study these phenomena that otherwise may be troublesome, particularly at elevated temperatures.

Sample cleaning by argon ion sputtering can result in preferential removal of one species with respect to others. Controlled experiments should be made to evaluate the impact of each of the preceding factors [19].

Surface roughness will also affect the Auger intensity. The variation in secondary electron yield with the angle of incidence of the electron beam is responsible for the ability to form images in the SEM. Similarly, the Auger electron yield depends on the primary beam angle of incidence (Fig. 7b), which presents obvious difficulties for rough samples. Methods for largely overcoming these effects rely on comparing the Auger signal to the intensity of backscattered electrons of similar energy that will have a comparable topographical behavior [20].

The accuracy of measurement may also be limited by the noise associated with the available beam current. For beam currents less than 10^{-9} A the signal-to-noise ratio will be approximately 20 for a collection time of 10 s, though this will depend to some extent on the sample, Auger yield, and instrument collection efficiency. Speed and accuracy of analysis of course can be improved by increasing the beam current (normally at the sacrifice of resolution); however, some electron-beam-induced effects may become more significant.

6.5 ELECTRON-BEAM-INDUCED EFFECTS

The electron beam can induce localized adsorption of contamination. This is very commonly seen in most SEMs as a dark area of the surface marking the previous position of the electron beam. The adsorbed material can arise either from the gas phase or from mobile species (organic) loosely adsorbed at the specimen surface. This material can be ionized or fragmented by the electron beam and strong chemisorption will occur. Buildup of contamination depends therefore not only on the chamber pressure but also on the overall sample cleanliness. The extent to which these effects occur does not seem to depend greatly on the beam current density or electron energy and can be completely overcome by operating under sufficiently high vacuum conditions (10^{-10} mbar) and by thorough argon ion sputtering of the sample surfaces. Sample cleaning using either ion bombardment or heating the sample is routinely done for metal and semiconductor samples; however, the effectiveness of ion bombardment for biological material is not clear. From the work done on a UHV SEM it was not found possible to obtain Auger signals other than oxygen and carbon from sections of plant spermatozoids of *Equisetum* and *Phaeceros* [21]. This may have been due to difficulty in removing surface contamination or to internally induced damage and displacement of other species. These specimens were used since they have been shown by x-ray analysis to contain high concentrations of phosphorus and calcium. On the other hand, Hart [22] has shown that stain material apart from carbon and oxygen can be detected. It is likely that more stable materials such as bone and tooth will be more amenable to the Auger technique. Further exploration of the technique with a wider range of biological material may prove necessary to establish the limitations of the technique in this context. However, it is clear already that useful information will only be obtained if satisfactory preparation and cleaning techniques can be developed. Since point analysis of bulk sample, by any means, in the scanning electron microscope involves a static beam focused onto the specimen for a long period, specimen heating may occur. This is generally more severe for biological tissue than for metals and semiconductors but depends critically on the probe diameter, current, and thermal conductivity of the sample. The temperature rise has been calculated by Vine and Einstein [23]. Temperature rises for certain materials in bulk form are shown in Table 1b for operating conditions applicable to the SEM (Table 1a). Biological material would approximate closely amorphous carbon for this purpose.

It is clear that all reasonable probe currents and diameters will probably produce substantial local heating of biological samples such that surface species will be highly mobile and will add to the difficulty of maintaining a clean and stable condition for Auger analysis. These effects will also occur for x-ray analysis, but if lower beam currents can be used, heating effects will be significantly reduced. In addition, x-ray chemical information arises mainly from the subsurface region and is not so sensitive to surface contamination. These arguments are not applicable to electron loss spectroscopy, since in this technique the sample is in the

TABLE 1a   Operating Conditions

Gun	d (nm)	I_p
FEG	5	2.10^{-9}
	20	$1.5\ 10^{-8}$
	100	4.10^{-8}
LaB$_6$	200	3.10^{-7}

TABLE 1b   Temperature Rise (Maximum) ΔT, for Beam Conditions in Table 1a, for Four Different Spot Sizes, 5, 20, 100, and 200 nm

Material	K	p	ΔT for d (nm) (°C)			
			5	20	100	200
Amorphous-C	0.0038	1	71	330	640	(4000)
Amorphous-SiO$_2$	0.0032	0.9	75	360	680	(4300)
Si	0.358	0.9	0.67	3.2	6.1	38
Ag	1.00	0.65	0.18	0.82	1.6	9.8
W	0.489	0.6	0.33	1.6	3.0	19

K is thermal conductivity (cal cm^{-1} s^{-1} °K^{-1}).
p is a power retention factor.

form of a thin film and the majority of the energy from the beam is transmitted rather than adsorbed by the specimen.

Since Auger spectroscopy relies on the detection of relatively low energy electrons, it is particularly sensitive to sample charging. This can result in either complete loss of signal or energy displacement of all Auger peaks by the change in potential of the sample. Samples with resistivities greater than 10^8 Ω cm will show signs of charging in the electron beam and could also result in localized heating sufficient to cause loss of material. Normally, a number of conductive coating techniques are used in electron microscopy to overcome these problems. These, however, cannot be used with an analytical technique that is surface sensitive. For Auger analysis a technique involving charge neutralization with positive ions may be possible. On samples of biological tissues that have been examined at the University of Sussex no charging was evident. This may be because surface carbon layers were always sufficiently conductive.

6.6   COMPARISON OF AUGER WITH OTHER TECHNIQUES

It is useful to compare the Auger technique with two other high-resolution analytical techniques, x-ray analysis and electron loss spectroscopy. This can be done primarily on the basis of applicability to certain materials and types of specimen, quantitative accuracy, and spatial resolution. The main differences between these three techniques are summarized in Table 2. Progress in each of these techniques will inevitably mean that certain limitations will be improved on in the future, particularly with regard to quantification.

TABLE 2

	SAM (AES)	X-Ray	EELS
Elements analyzable	$Z > 3$	$Z > 10$	$Z > 3$
Detectability limits			
Minimum sample vol., nm³	4×10^2	10^8	10^2
Mass, gm	10^{-19}	10^{-16}	10^{-19}
Concentration, ppm	10^3	10^2–10^3	(10^3)
Present quantitative accuracy	$\approx 10\%$	$>2\%$	$\approx 20\%$
Spatial resolution, nm	>30	500	>10
Analysis depth, nm	0.5–2	500	Specimen thickness, <100
Vacuum requirements, torr	$<10^{-8}$ (typically 10^{-9}–10^{-10}	10^{-5}	$<10^{-5}$
Special sample requirements	Usually surface cleaning		Thin film

6.6.1 Spatial Resolution

Auger spectroscopy combines an ability to perform elemental analysis with high quantitative accuracy, high sensitivity for light elements, and high spatial resolution. For thin specimens suitable for transmission electron microscopy the resolution of either the x ray, EELS, or Auger technique will be determined largely by the beam size with some spreading of the electron beam in the specimen. EELS is at a disadvantage to the extent that it is limited only to thin transmission samples. For bulk specimens the detected x-ray or Auger signal comes directly from ionization by the primary beam and also from the cascade of scattered electrons within the solid. The cascade diameter is about 0.5 μm and determines the resolution of x-ray analysis. For Auger, on the other hand, most of the signal arises from ionization by the primary beam with a contribution from cascade processes given by the backscattering factor. Figure 9 shows a comparison between an experimentally measured profile and a calculated Auger intensity profile [24]. These were done for a thin silver layer deposited with a sharp boundary onto a clean tungsten surface. For a beam diameter less than ~0.5 μm the profile consists of a sharp change over a distance determined by the beam diameter and a slow tail due to backscattering that will contribute less than 50% of the total intensity for heavy elements. In the worst case most of the Auger electrons are produced within the probe diameter.

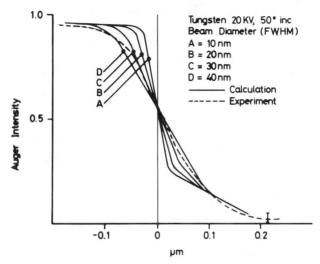

Figure 9. Comparison of measured Auger profiles with calculations for a silver layer on tungsten. The SEM beam is scanned across the sharp boundary of the silver layer.

Degradation of the resolution due to backscattering will be minimal for the light elements of biological samples. Thus details will be seen by micro-Auger analysis with the same kind of resolution as that for SEM.

The extremely shallow depth resolution for Auger spectroscopy has been turned to advantage in conjunction with controlled ion sputtering to obtain depth profiles of metal and semiconductor surfaces. A great deal is now known about oxidation and metal diffusion processes as a result. Depth profiling is now a standard feature of many Auger facilities. Information is generally more reliable from specimens where either the sputter ion beam is defocused to accommodate several square millimeters or where the beam can be rastered over a large area to give a uniform etch rate. Any nonuniformity of structure parallel to the surface plane will also make interpretation of Auger intensity changes more difficult. This technique may be more difficult to apply to biological material since most specimens tend to be inhomogeneous in the surface plane.

6.6.2 Quantitative Accuracy

Much work has been done in recent years to improve the accuracy of all three techniques. X-ray absorption and fluorescence effects can be routinely calculated and corrections made to yield an accuracy of about 2% under controlled conditions. Auger spectroscopy is not complicated by these effects; however, many surfaces have structures and compositions that are not representative of the bulk, in addition to the effects of contamination. Under these circumstances it is extremely difficult to make corrections. Removal of the surface by ion sputtering can introduce uncertainties either because of different species having different sputtering rates or because of the inclusion of the sputtering ions in the surface. Using standard samples it is possible to obtain accuracies of the order of 10%. The use of standard samples for electron loss spectroscopy is complicated by the uncertainty in specimen thickness; however, the accuracy is perhaps comparable with that of Auger analysis.

6.7 SUMMARY AND OUTLOOK FOR
BIOLOGICAL APPLICATIONS

Of the three analytical techniques applied to the electron microscope, x-ray analysis is the most widely used. Routines for background stripping and corrections for absorption by an on-line computer are common. These facilities, together with other data processing techniques, can be applied just as effectively to Auger spectroscopy. The aspect of the Auger technique, namely its extremely high sensitivity to the few uppermost atomic layers of surface, which has been the motivation for its rapid development and wide use in many areas of surface physics and chemistry, appears at present to be a severe limitation to its application to biological material. The difficulties of preparing atomically "clean" or rather "representative" surfaces of biological tissue have hardly been addressed until now. To what extent this can be achieved is unclear. The necessity for surface cleanliness may in the future be deemphasized to some extent for some applications by looking more closely at the structure of the spectrum tail to the low energy side of the main Auger peak. (See Fig. 7.) This tail can extend several hundred electron volts below the main peak and contains Auger electrons that have lost energy in escaping from deeper within the solid. Figure 4 suggests that the detection of high energy Auger electrons >1000 eV with their associated larger iselastic mean free path may also reduce the need for clean surfaces.

As previously explained, the present generation of electrostatic analyzers have a relatively low collection efficiency. This imposes severe restrictions on the speed and accuracy of analysis, particularly with the very low beam currents used in the SEM under optimum resolution conditions. Analyzers are constantly being improved to increase resolving power, transmission, and angular collection efficiency. The concentric hemispherical analyzer (CHA) appears to be particularly suited to electron microscopy because of the flexibility of mode in which it can be operated, and it has an output focal plane that is approximately planar. This allows a multichannel detector to be used to collect a reasonably large part of the spectrum simultaneously, which can speed up the data collection time significantly.

The two prime advantages that Auger spectroscopy can offer the biologist are (1) high sensitivity for light element detection and (2) analysis with spatial resolution approaching the beam diameter. However, as should be apparent from this short survey, the technical difficulties in realizing the potential of the technique for biological application lie not so much in development of instrumentation as in development of suitable sample preparation techniques that will render the surfaces of interest stable and relatively free from contamination by unrepresentative material. It seems likely that for certain types of tissue, such as bone, tooth enamel, or plant silicate structures, Auger spectroscopy may more usefully demonstrate its particular advantages.

It only remains to introduce the surface scientist to the biologist.

REFERENCES

1. C. C. Chang, Auger electron spectroscopy, Surf. Sci. *25*, 53–74 (1974).

2. C. C. Chang, *Characterization of Solid Surfaces,* P. F. Kane and G. B. Larabee, Eds., Plenum, New York (1977).

3. H. E. Bishop and J. C. Riviere, Estimates of efficiencies of production and detection of electron-excited auger emission, J. App. Phys. *40*, 1740–1744 (1969).

4. C. J. Powell, Evaluation of formulas for inner shell ionization cross-sections, Rev. Mod. Phys. *48*, 33 (1976).

5. D. M. Smith and T. E. Gallon, Auger emission from solids, J. Phys. *D7*, 151–161 (1974).

6. C. J. Powell, Attenuation lengths of low-energy electrons in solids, Surf. Sci. *44*, 29–46 (1974).

7. M. P. Seah and W. A. Dench, Quantitative electron spectroscopy of surfaces, National Physical Laboratory report Chem. 82, Teddington, Middlesex, U.K., 1978.

8. P. Staib and G. Staudenmaier, "Quantitative auger micro-analysis, Proc. 7th Int. Vacuum Congress, Vienna, pp. 2355–2358, 1977.

9. D. Joy, D. Maher, and P. Mockel, Quantitative elemental analysis by T.E.M., Proc. Electron Microscopy Conf., Vol. I, Toronto, pp. 528–529.

10. W. Steckelmacher, Energy analyzers for charged particle beams, J. Phys. *E6*, 1061–1071 (1973).

11. D. Roy and J. D. Carette, in *Electron Spectroscopy for Surface Analysis,* H. Ibach, Ed., Springer-Verlag, New York, 1977.

12. L. H. Veneklasen, G. Todd, and H. Poppa, An integral field-emission "microSEM" for UHV surface analysis, Proc. Electron Microscopy Conf., Vol. I, Toronto, pp. 12–13 (1978).

13. J. A. Venables, A. P. Janssen, C. J. Harland, and B. A. Joyce, Scanning Auger electron microscopy at 30 nm resolution, Phil. Mag. *34*, 495–500 (1976).

14. R. Browning, P. J. Bassett, M. M. El. Gomati, and M. Prutton, A digital scanning Auger electron microscope incorporating a CHA, Proc. Roy. Soc. *A357*, 213–230 (1977); and Surf. Sci. *68*, 328–337 (1977).

15. A. P. Janssen, C. J. Harland, and J. A. Venables, A ratio technique for micro Auger analysis, Surf. Sci. *62*, 277–292 (1977).

16. P. M. Hall, J. M. Morabito, and D. K. Conley, Relative sensitivity factors for quantitative Auger analysis, Surf. Sci. *62*, 1–20 (1977).

17. C. C. Chang, General formalism for quantitative Auger analysis, Surf. Sci. *48*, 9–21 (1975).

18. J. Erlewein and S. Hofman, Segregation of tin on copper surfaces, Surf. Sci. *68*, 71–78 (1977).

19. A. Jablonski, S. H. Overbury, and G. A. Somorjai, The surface composition of Au-Pd alloy system, Surf. Sci. *65*, 578–592 (1977).

20. A. P. Janssen, C. J. Harland, and J. A. Venables, A ratio technique for micro-Auger analysis, Surf. Sci. *62*, 277–292 (1977).

21. A. P. Janssen and J. A. Venables, Scanning Auger microscopy—an introduction for biologists, Proc. SEM Int. Conf., Washington (1979).

22. R. K. Hart, Elemental distribution in biological materials by scanning spectrometric microscope, Proc. 8th Int. Elec. Microscope Conference, Vol. 2, Canberra, p. 20 (1974).

23. S. Vine and E. Einstein, Heating effects of electron beam, Proc. IEE III *5*, 921 (1964).

24. A. P. Janssen and J. A. Venables, The effect of backscattered electrons on the resolution of SAM, Surf. Sci. *77*, 351 (1978).

POINT-PROJECTION MICROSCOPY

J. A. PANITZ

Sandia National Laboratories
Surface Physics Division
Albuquerque, New Mexico

Voyages of discovery can be made in new uncharted waters, but also in the familiar bays close to port, provided one has observing apparatus that can see familiar objects with greater detail than that previously possible. [1]

7.1 INTRODUCTION

The desire to see objects in ever finer detail has prompted continual advances in microscopy. Although angstrom resolution has been achieved, and single atoms can be seen under certain circumstances [2,3], large aggregates of atoms—organic macromolecules—have been more elusive.

The morphology of an organic molecule interacting with a metal or a semiconductor surface is of considerable interest. The nature of corrosion-resistant coatings, polymer adhesion, and fundamental properties of metal–molecule interactions could be studied in much greater detail if organic molecules could be directly observed on these substances. If, in addition, a three-dimensional reconstruction of molecule morphology were available, it could lead to a better understanding of the structure of such species, particularly if they were unstained and the imaging procedure was nondestructive.

The electron microscope can be used to observe unstained organic molecules deposited on very thin carbon or dielectric substrates, but image contrast depends on the ability of the molecule to scatter electrons more effectively than the substrate on which it is placed [4]. This means that metallic substrates cannot normally be used. Although cryogenic cooling can reduce the problem of radiation damage by a probing electron beam, volatilization of organic species during imaging remains a serious problem [5]. For example, it has been estimated that 50,000 electrons per square nanometer are needed for statistically reliable imaging in the TEM, at a resolution of 0.5 nm (and a magnification of 50,000×), but only 50 electrons per square nanometer may cause serious damage to the species under examination [5]. Positive ion imaging (with protons) may prove to be less damaging, but high resolution images of organic species have not yet been obtained.

In essence, conventional microscopes have difficulty imaging organic species because it is necessary to support an organic species in the path of a probing beam. The beam must interact with an organic species to ensure adequate contrast in the resulting image, but the interaction must be weak to ensure that radiation damage and heating are minimal. In addition, the imaged species and its support must be isolated from external vibration. Otherwise, any relative movement between the imaged species and the probing beam can be magnified by the imaging process, leading to a loss in image resolution. One way to avoid these difficulties is to image organic species in a microscope that does not require a probing beam of ions or electrons to form an image. Such a microscope is available. It relies on direct point projection of ions or electrons to form an image and has been successfully used to observe a metal surface in atomic resolution. However, the field-electron emission microscope (FEEM) [6,7] and the field-ion microscope (FIM) [8–10], which are prototypic, have never provided reliable images of organic species.

7.2   THE POINT-PROJECTION MICROSCOPE

A point-projection microscope is a simple and elegant device that uses no lenses yet employs charged particles to form highly magnified images in high vacuum. The concept (shown schematically in Fig. 1) is straightforward: Charged particles are accelerated in an electric field that is directed almost radially outward from a highly curved metal (or semiconductor) surface. Since the electric field is divergent, charged particles created at or near the surface will continuously diverge into space. As a result, a highly magnified map of their origin can be obtained by allowing them to

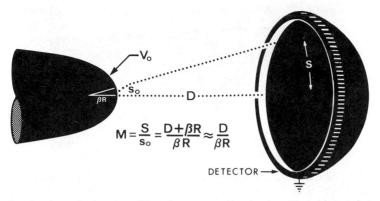

$$M = \frac{S}{S_o} = \frac{D + \beta R}{\beta R} \approx \frac{D}{\beta R}$$

DETECTOR →

Figure 1. A schematic drawing of how image magnification is achieved in a point-projection microscope. Charged particles created at or near the surface of a sharply curved metal specimen (called a field-emitter tip) radially diverge into space as a result of an applied electric field. A detector (placed several centimeters from the tip) records the resulting image. Magnification is high. Image resolution depends on the type of charged particles generated during the imaging process. Under optimum conditions, the image resolution can approach 2 Å.

intercept a suitable detector placed an appropriate distance from the surface. If their velocity component normal to the surface is very large compared to their velocity component parallel to the surface, the detector image will be highly resolved. The idea is to find a means of generating charged particles so that their spatial distribution at the detector will reflect some feature of the curved surface that is to be imaged. This feature might be a species deposited onto the surface or the constituent atoms of the surface itself.

For high magnification and resolution it is necessary to fabricate a surface with a radius of curvature that is smaller than the wavelength of visible light. This can be easily accomplished by polishing the end of a fine wire in an electrochemical bath [11]. The apex of the resulting "tip" can be smoothed by thermal annealing in high vacuum near the melting point of the wire [12]. Thermal self-diffusion will produce a multifaceted end form of minimum free energy that is approximately spherical (Fig. 2). Since the spherical contour is not isolated in space (but is part of an extended equipotential surface), biasing the tip will produce an electric field that is compressed toward the wire axis. If the trajectories of charged particles far from the tip apex are extrapolated back toward its surface, they will tend to intersect at a distance βR inside the tip, where R is the tip radius. The quantity βR can be considered as an "effective" tip radius and $\beta \approx 1.5$ as an "image compression" factor [11]. For an isolated sphere in space, $\beta = 1.0$.

Figure 2. A thermally annealed field-emitter tip viewed in profile in the transmission electron microscope. The tip was prepared by electrochemical polishing of a fine metal wire, followed by heating in vacuum close to its melting point.

The radius of a tip can be accurately determined by viewing the tip in profile in the transmission electron microscope (TEM). Since the tip-to-detector distance can also be accurately measured, the uncertainty in the magnification of a point-projection microscope will be determined by the uncertainty in the image compression factor (which can approach 30%). In order to determine the magnification more accurately, the image compression factor must be determined for each tip geometry. One approach is to determine β from a theoretical calculation employing an idealized tip of comparable shape and dimensions [12]. In practice, it is probably more accurate to image a feature of known size and shape that can then provide an accurate calibration of the magnification under actual operating conditions.

7.3 THE FIELD-ELECTRON EMISSION MICROSCOPE

If a negative bias of several kilovolts is applied to a field-emitter tip in vacuum, an electron current will be emitted from its apex. At a field strength of several volts per nanometer a field-emission current of tens of nanoamperes is typically observed [13]. The current is produced by quantum mechanical tunneling of electrons from the tip into vacuum. Since kilovolt electrons will cause many materials to fluoresce efficiently, the distribution of electrons field-emitted from the tip can be easily observed by placing a fluorescent screen in their path and observing the screen with the unaided eye. The result is the field-electron emission microscope, or FEEM. It has been noted that "in the absence of lenses, illuminating devices, and automatic controls a field-emission microscope is less of an apparatus and more of a direct aid to the eye and brain'' [14].

The magnitude of the field-electron emission current produced by a negatively biased tip is very sensitive to the work function of the tip surface. Relatively small variations in work function over the surface can produce very large changes in the contrast of a FEEM image. For a clean tip, a FEEM image will reflect inherent variations in work function with surface morphology [12,13]. If a contaminant species (such as an organic molecule) is adsorbed onto the tip surface in high vacuum, the work function will change, locally, at the adsorption site. The change in work function will produce a local change in image contrast that can be used to detect the presence of the adsorbed species [12,13].

For more than 30 years, attempts have been made to correlate image changes in the FEEM on adsorption with the known size and shape of the adsorbate. Small organic molecules were usually chosen as the adsorbate because they could be vapor deposited in high vacuum onto a tip that had been previously cleaned by thermal annealing close to its melting point. Since the adsorbed molecules were the major surface contaminant, they were expected to produce the largest change in image contrast.

As early as 1950, high contrast FEEM images were obtained following the adsorption of flavanthrene (Fig. 3) and copper phthalocyanine (Fig. 4) onto a tip surface [14–16]. The resulting images seemed to reflect the known symmetry of each molecule (two- and fourfold, respectively). However, a later study of many other organic species having different shapes and sizes nearly always produced two- and fourfold symmetric images [17]. It became clear that although the contrast of an FEEM image

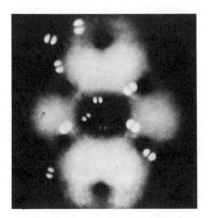

Figure 3. A field-electron emission microscope image at 78 K obtained by depositing flaventhrene, $C_{28}H_{12}N_2O_2$ (a twofold symmetric molecule) onto a tungsten field-emitter tip. (Courtesy of A. J. Melmed, The National Bureau of Standards, Washington, D.C.)

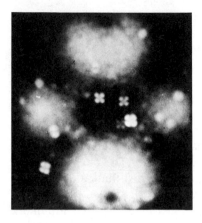

Figure 4. A field-electron emission microscope image at 78 K obtained by depositing copper-phthalocyanine, $C_{32}H_{16}N_{18}Cu$ (a fourfold symmetric molecule) onto a tungsten field-emitter tip. (Courtesy of A. J. Melmed, The National Bureau of Standards, Washington, D.C.)

could reflect the adsorption of an organic molecule on the tip surface, it could not reveal the true shape and size of an individual adsorbate [18,19].

7.4   THE FIELD-ION MICROSCOPE

Several years after its introduction in 1950, the field-ion microscope (or FIM) had achieved a resolution of a few tenths of a nanometer [9,10], about one order of magnitude better than its predecessor, the FEEM. This meant that the individual atoms that formed a metal surface could be unambiguously observed for the first time. In order to achieve atomic resolution, an FIM image probes subtle undulations in the electric field strength above the tip surface when a positive bias of sufficient magnitude is applied. Protruding surface atoms and metallic adsorbates produce field variations that can be detected by field-ionizing gas atoms above the tip surface.

At field strengths of several tens of volts per nanometer, gas phase molecules in the immediate vicinity of the tip apex will ionize with high probability [20]. Ionization occurs when an electron in a gas phase atom tunnels into the tip, leaving a positively charged ion. The ion rapidly accelerates away from the tip surface, radially into space. Since the tunneling probability is a sensitive function of the electric field strength, the local variations in field strength produced by protruding surface atoms will significantly modulate the field-ion current produced by gas phase atoms ionized immediately above the protrusion [11].

The contrast observed in an FIM image reflects the modulation in the field-ion current above the tip surface. To a lesser extent, it reflects inhomogeneities in the supply of gas phase atoms over the tip surface. For a clean, atomically smooth tip, contrast variations in an FIM image will reveal the individual atoms of the surface that occupy kink, edge, and other protruding positions in the crystal lattice (Fig. 5).

At first glance, the FIM would seem to offer several advantages for imaging organic molecules. Image magnification is intrinsically high, and image resolution is of the order of atomic dimensions. Furthermore, since imaging gas ions are formed in space several angstroms from the tip surface, they provide a noninvasive probe of surface morphology. The difficulty with the technique is the very high electrostatic field stress ($P = 4\pi\varepsilon_0 F^2$) that is associated with an imaging field [11]. Even for hydrogen, which has a relatively low imaging field ($F \approx 25$ V/nm), $P \approx 10^9$ dynes/cm^2, or about one-half ton per square millimeter. Under such severe conditions organic species can be severely distorted or even torn apart and removed from the surface as positive ions [21,22].

In order to circumvent the destructive aspects of a high imaging field, several ingenious schemes were developed. At least two procedures involved embedding molecules of interest within a metallic layer that could

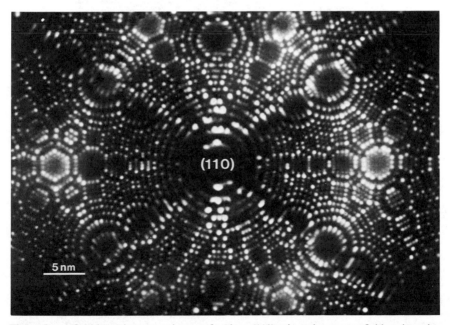

Figure 5. A field-ion microscope image of a clean (110) oriented, tungsten field-emitter tip. The image was taken in 1×10^{-5} Torr of helium at a field strength of 45 V/nm. The tip temperature was 30 K.

survive field-ion imaging [23,24]. In principle, molecules that had been embedded within the layer would be visible as an extended contrast variation in the field-ion image. In general, the images were disappointing. For molecules incorporated within an electrodeposited layer [24], the images were nonstructured, nonreproducible, and unremarkable. When metallic layers were thermally evaporated in high vacuum onto tips presumably covered with biomolecules, image quality improved [23]. A typical image is shown in Fig. 6. Unfortunately, reproducibility was very poor, and the images were not sufficiently detailed to permit useful comparisons with known molecular structures.

The last major attempt to image biomolecules in the FIM was based on a different imaging philosophy. Instead of attempting to shield the molecule from the high imaging field by embedding it within a metallic layer, image intensification was used to observe the earliest stages of image formation [25]. It was hoped that transient images formed prior to field-induced destruction of an uncoated organic molecule might reveal details

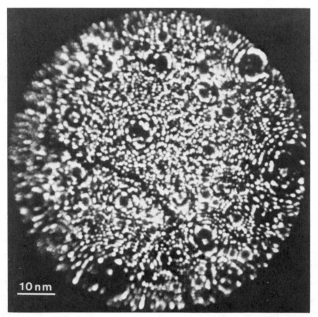

Figure 6. A field-ion microscope image of biological molecules embedded within a platinum deposit on an iridium tip. The tip was dipped into an aqueous solution of DNA (from *Micrococcus luteus*) prior to platinum deposition in vacuum at 400 K. The image was taken in helium at a field strength of 45 V/nm. The tip temperature was 25 K. (Courtesy of F. Hutchinson, Department of Biophysics, Yale University, New Haven, Conn.)

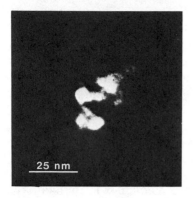

Figure 7.   A transient image recorded during field-ion microscopy of an iridium tip on which a solution containing a single stranded polynucleotide (poly-U) was freeze-dried. The image was taken in hydrogen at a field strength of 5 V/nm. The tip temperature was 5 K. (Courtesy of E. S. Machlin, The Krumb School of Mines, Columbia University, New York, N.Y.)

of its morphology. Since the images were inherently transient, the imaging event was recorded on movie film as the tip voltage was increased. A double blind procedure was then used to select "real" images from those that were thought to reflect random background events. Although a few intriguing image features were recorded (Fig. 7), the transient and nonreproducible nature of the images proved to be major obstacles that could not be overcome.

7.5   MOLECULAR DEPOSITION ON FIELD-EMITTER TIPS

A significant problem with attempts to image organic molecules in the FEEM and FIM was the unknown coverage of molecules on the tip surface. The deposition of molecules is statistical in nature. This means that at the low coverage required to image isolated molecules, the probability of finding a molecule within the field of view of the microscope could be quite small. A species of interest must be placed precisely within the relatively small area of the tip apex that is accessible to imaging. Since the electric field varies with distance from the tip apex, useful images can only be obtained within an approximately circular area having a radius about equal to the tip radius and centered on the tip axis. The resolution of a point-projection microscope decreases with increasing tip radius [13]. As a result, the tip radius (and the imaged area) cannot be made arbitrarily large. For FEEM imaging, a typical tip has a radius of the order of several hundred nanometers [12]. For field-ion imaging, the tip radius is usually about an order of magnitude smaller [11]. This means that unlike conventional microscopes which can examine large surface areas in the hope of finding a deposited molecule, the point-projection microscope has a field of view that is limited to an area only a few hundred nanometers in extent.

It is clear that a severe restriction on the placement of a molecule on the tip surface places, in turn, a severe requirement on the reproducibility and statistical reliability of a given deposition procedure. In order to investigate the deposition problem in detail, it is desirable to use an independent technique to observe the distribution of an appropriate molecule on the tip surface. One useful approach is to examine ferritin-covered tips in the transmission electron microscope (TEM) [26]. Ferritin is a molecule that is uniquely suited for TEM imaging [27]. It is a nearly spherical protein consisting of a hollow shell that is about 13 nm in diameter and about 4 nm thick (Fig. 8). The interior of the shell is accessible through six channels, each about 1 nm in diameter. Ferritin acts as a biological reservoir for iron, which is stored in its interior as a complex ferric hydroxyphosphate polymer. The iron-rich core of the ferritin molecule is easily seen in the TEM [27].

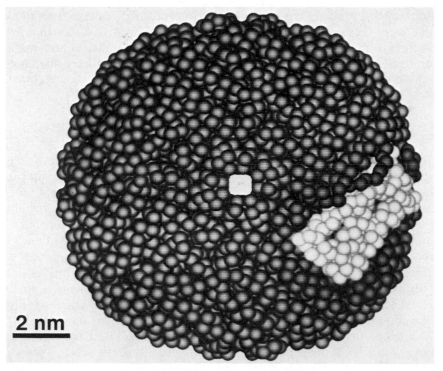

2 nm

Figure 8. A digital reconstruction of the ferritin molecule from x-ray coordinate data. The reconstruction highlights the location of 1 of 24 identical monomers that form its outer protein shell. One of six channels (each about 1 nm in diameter) that lead into the interior of the molecule can be seen in this view. (Courtesy of R. Feldman, The National Institutes of Health, Bethesda, Md.)

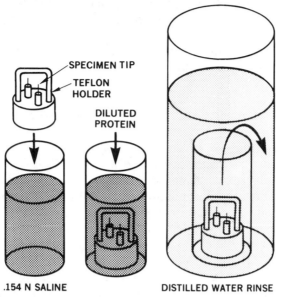

Figure 9. A simple aqueous deposition procedure developed for placing ferritin molecules on a field-emitter tip. The procedure ensures that a tip is initially passed through a molecule-free, air–liquid interface. Similarly, when the tip is removed from solution, it is removed through an air–liquid interface free of denatured molecular species. This ensures that various submonolayer coverages of the molecules can be reliably placed with the imaged area of the tip apex.

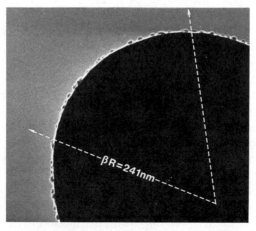

Figure 10. A transmission electron microscope image of a tungsten field-emitter tip exposed to a solution of horse spleen ferritin (10 μg/ml in 0.154 M NaCl for 2 min). Deposition was conducted using the procedure shown in Fig. 9. The iron-rich core of each molecule decorating the tip contour can be seen. The molecule's outer protein shell is invisible.

181

Ferritin can be reliably placed on the surface of a field-emitter tip using a simple aqueous deposition procedure [26] that is shown schematically in Fig. 9. The coverage of ferritin can be controlled on a submonolayer scale within the imaged area of the tip apex. A typical TEM image of a ferritin-coated tip is shown in Fig. 10.

7.6 HIGH-FIELD TOMOGRAPHY

Although in principle any molecule can be placed within the imaged area of a field-emitter tip using the procedure shown in Fig. 9, two obstacles to point-projection imaging remain—the high electric field strength required for imaging and image interpretation. Even if the field strength required for imaging could be made arbitrarily small it is not obvious, a priori, what feature of a deposited molecule would be highlighted in a point-projection image. Since the ionization probability of gas phase species above a deposited molecule might be molecule dependent [28], a point-projection image could look very different from what was expected. Since image magnification and resolution might also depend on molecular conformation [29], a point-projection image could be severely distorted.

A successful high field imaging technique for organic molecules must preserve the inherent simplicity of the point-projection microscope but operate at substantially lower field strengths than the FIM. It must also provide reproducible images that are easy to interpret. During the past four years, a high field imaging technique has been developed that seems to meet these requirements [30–33]. It is a shadowing technique. However, unlike previous FIM techniques (or those developed for electron microscopy), the shadowing species is removed in order to form an image. Since imaging occurs at field strengths that are at least an order of magnitude lower than those encountered in the FIM, organic molecules survive without noticeable degradation [34]. This means that images are reproducible. Images can even be reproduced after repeated exposure of the tip to laboratory ambient temperatures [33], a somewhat surprising (but welcome) observation.

The nature of the imaging process produces a series of contour slice images of the molecule at different elevations above the tip surface [33]. For this reason the imaging process is called *field-ion tomography*. From a series of tomographic images, a three-dimensional picture of the molecule can be reconstructed. An important distinction exists between field-ion tomography and the more familiar computer-assisted x-ray tomography. The latter permits interior detail of an object to be seen. The former does not. Only the exterior shape of a molecule can be obtained with the new imaging technique.

Figure 11 shows schematically one stage in the field-ion tomographic process. A molecule has been deposited onto a field-emitter tip. Its radius is chosen for optimum image resolution. An additional requirement is that the molecule must be much larger than any characteristic surface feature of the tip but much smaller than its apex radius of curvature. This ensures that the molecule behaves like a small dielectric protrusion on an infinite metal plane. As a result, it will not significantly alter the electric field in its immediate vicinity.

After placing the tip in ultrahigh vacuum and cooling it below 30 K, a precise quantity of benzene is condensed onto the tip apex from the gas phase. The benzene is the shadowing species. It completely covers the tip surface, surrounding and embedding each molecule within a frozen, immobile layer. As a potential is applied to the tip, the field strength at the surface of the benzene layer will increase. At some critical field strength the molecules in the outermost layer of the condensed benzene multilayer will field-ionize. The resulting molecular ions will accelerate radially into space. A detector placed several centimeters in front of the tip will record a highly magnified image of the relative position of the benzene molecules in the outermost layer prior to the ionization event. Provided the ioniza-

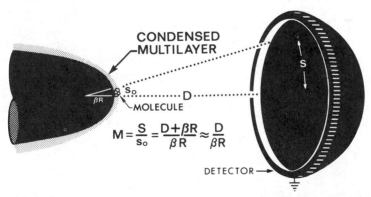

$$M = \frac{S}{S_0} = \frac{D + \beta R}{\beta R} \approx \frac{D}{\beta R}$$

Figure 11. A schematic drawing of one stage in the field-ion tomographic imaging process. A thick layer of benzene condensed onto the cryogenically cooled tip in ultra-high vacuum is gradually removed by generating an increasing electric field strength at the surface of the layer. Benzene ions formed during the process are radially accelerated to a sensitive ion detector where they form a highly magnified image of their relative positions within the layer. As a molecule is exposed by the receding benzene layer, a dark feature appears in the image. The contour of the feature reflects the molecule's contour at its intersection with the remaining benzene layer.

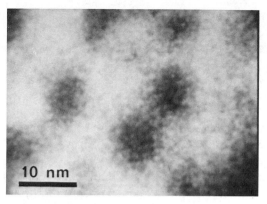

Figure 12. An integrated series of field-ion tomographic images containing an estimated 26 to 28 contour slice images of individual ferritin molecules on tungsten. The image was obtained by the technique shown schematically in Fig. 11 (at a tip temperature of 30 K). The deposition procedure shown in Fig. 9 was used to expose the tip to 10 μg/ml of horse spleen ferritin (in 0.154 M NaCl) for 2 min prior to imaging.

tion probability is high and relatively isotropic, the image will appear bright and relatively contrastless.

As the tip potential is gradually increased, the condensed multilayer will be removed gradually from the tip surface. As soon as a molecule is exposed by the receding benzene layer, a dark region will appear in the detector image. Since the electric field distortion in the vicinity of the molecule will be negligible, the contour of the dark region in the detector image will accurately reflect the contour of the molecule at its intersection with the remaining benzene layer. As the condensed benzene layer is gradually removed, more of the molecule's contour (lying closer to the tip surface) will be revealed. Figure 12 shows a field-ion tomographic image of isolated ferritin molecules on a tungsten tip surface. The image contains an estimated 26–28 molecular contours revealed during a 400-V change in tip potential [35].

7.7 IMAGE RESOLUTION

The resolution of a field-ion tomographic image will depend on the size of the shadowing species (since it delineates the molecule's contour) and the temperature of the condensed benzene multilayer [33]. As the temperature of the tip is lowered, the resolution of the image will increase because the velocity component of each benzene molecule within the condensed layer and parallel to the tip surface will be reduced. At all accessible temperatures, the finite DeBroglie wavelength of the resulting benzene ions can be ignored (i.e., diffraction is negligible). If the condensed benzene multilayer is assumed to be in thermal equilibrium with the temperature of the tip T, the image resolution δ is given by:

$$\delta = \delta_0 + 0.019\beta R \frac{T}{V_0} \qquad (1)$$

where δ_0 is the size of the shadowing species [33]. For benzene $\delta_0 \approx 0.74$ nm and $\beta \approx 1.5$. Typically, $R = 160$ nm and $V_0 \approx 2500$ V. This means that the image resolution at $T = 5$ K should be of the order of 1 nm.

7.8 IMAGE RECONSTRUCTION

The interpretation of any three-dimensional image is made easier if it contains sensory clues related to depth. It is known that the three-dimensional appearance of an object can be enhanced if the object appears as though it were illuminated by an oblique beam of light from above [14]. Our visual experience tells us that under such illumination, the upper half of a protrusion will be highlighted while its lower half will remain in shadow. Since a deposited molecule must protrude from the tip surface, adding highlights to the upper half of features within a series of field-ion tomographic images will enhance the shape of the molecules they represent. The idea is to accentuate the three-dimensional detail inherent in an image without introducing artifacts or subjective bias. A digital image processing procedure has been developed that meets this requirement [36]. It is a variation of a standard digital processing technique known as "unsharp masking" [37].

Processing begins by digitizing an integrated image containing a series of field-ion tomographic images (as shown in Fig. 12) and then smoothing a copy of the image by averaging the pixels within a 5 × 5 pixel window. The final reconstruction is not affected by the size of the averaging window, provided that the window is much smaller than the resolution of the original image. If the pixel-averaged image is shifted vertically downward with respect to the original image by several pixels and then subtracted from it, highlights and shadows will be added in the direction of the shift. The size of the shift is not particularly important. A shift of six video lines is usually adequate. Larger shifts only enhance image contrast; they do not alter image detail.

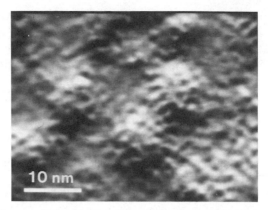

Figure 13. The image of Fig. 12, digitally processed to add highlights and shadows to accentuate each molecular feature. A six video line shift, vertically downward, a 5 × 5 pixel average prior to shifting, and a postfilter using a 15 × 15 pixel window (corresponding to an image resolution of ≈1.5 nm) was used for processing.

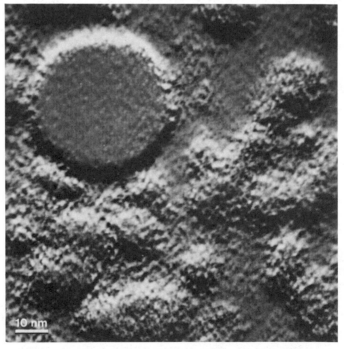

Figure 14. A digitally processed image of ferritin clusters on tungsten taken at 30 K. Four individual field-ion tomographic images were used for image reconstruction, each containing an estimated 5 to 7 contour slice images.

186

Since detail much smaller than the resolution of the image is not physically meaningful, it is convenient to smooth the processed image such that smaller detail is strongly attenuated. The result of processing the image of Fig. 12 is shown in Fig. 13. The features that appear roughly spherical are identified with the protein shell of individual ferritin molecules. The average shell diameter is approximately 13 nm. This is identical to the diameter of the ferritin molecule deduced from an x-ray reconstruction [35]. Figure 14 shows a TEM image of three unstained ferritin molecules on a very thin carbon substrate [27]. Only the iron-rich core of each molecule can be seen. The unstained protein shell is invisible. An estimate of the shell diameter obtained from the average center-to-center distance of each core is about 70% smaller than the x-ray value. This may reflect electron-induced damage during imaging [35].

Figure 14 is a reconstructed image of clusters of ferritin molecules on a tungsten tip. Figure 15 shows the result of adding several more tomo-

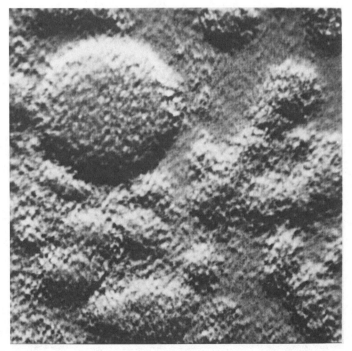

Figure 15.   The image of Fig. 14 with three additional field-ion tomographic images used in the image reconstruction. The large feature in the upper-left-hand corner has been rounded, whereas all other image features remain unchanged. This indicates that the large features extend far from the tip surface, a conclusion supported by TEM observations.

graphic images during image reconstruction. The larger cluster in the upper-left-hand portion of the image has been rounded while all other image features remain unchanged. This indicates that the rounded feature extends far from the tip surface, a conclusion supported by TEM imaging. Figure 16 shows the results of imaging hemocyanin (from *Limulus polyphemus*), and Fig. 17 shows the type of image obtained from a tip on which Poly(dA-dT) DNA was deposited. The central portion of the large image feature appears to be coiled as a double helix. Its size is much larger than double-stranded DNA, a species that would have the appearance of the image feature indicated by an arrow in Fig. 17.

Recently, image reconstruction has been extended to include the generation of true, stereopair images [36,38]. This new feature of high field tomography remains to be explored. It promises to yield even more detailed information about the shape and interaction of unstained organic molecules with metal surfaces.

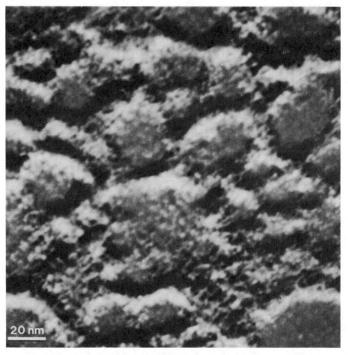

Figure 16. A digitally processed image of a monolayer coverage of hemocyanin (from *Limulus polyphemus*) on tungsten taken at 30 K.

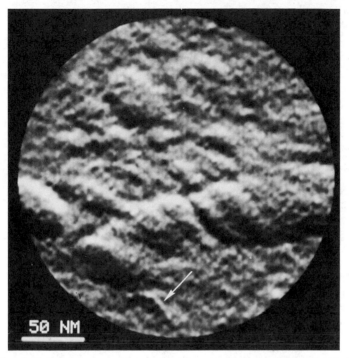

Figure 17. A digitally processed image of a tungsten tip exposed to poly(dA-dT) DNA taken at 30 K. The center of the large feature displays a structure that appears to be a double helix. The structure is much larger than the size of double-stranded DNA molecule, a species that would have the appearance of the image feature indicated by the arrow.

REFERENCES

1. V. L. Fitch, 1980 Nobel Prize lecture in physics, reprinted in Science *212*, 989 (1981).
2. E. W. Müller, Science *149*, 591 (1965).
3. A. V. Crewe, Chem. Scripta *14*, 17 (1978–79).
4. F. P. Ottensmeyer, Science *215*, 461 (1982).
5. A. Klug, Chem. Scripta *14*, 245 (1978–79).
6. E. W. Müller, Z. Tech. Phys. *17*, 412 (1936).
7. E. W. Müller, Z. Phys. *106*, 132, 541 (1936).
8. E. W. Müller, Z. Phys. *131*, 136 (1950).
9. E. W. Müller, Z. Naturforsch. *11a*, 87 (1956).
10. E. W. Müller, J. Appl. Phys. *27*, 474 (1956).

11. E. W. Müller and T. T. Tsong, *Field-Ion Microscopy: Principles and Applications,* American Elsevier, New York, 1969.

12. W. P. Dyke and W. W. Dolan, in *Advances in Electronics and Electron Physics,* Vol. 8, L. Marton, Ed., 1956.

13. R. Gomer, *Field-Emission and Field Ionization,* Harvard University Press, Cambridge, Mass., 1961.

14. T. G. Rochow and E. G. Rochow, *An Introduction to Microscopy by Means of Light, Electrons, X-rays, or Ultrasound* (Plenum, New York, 1978).

15. E. W. Müller, Naturwissenschaften *14,* 333 (1950).

16. E. W. Müller, Life *28,* 67 (June 19, 1950).

17. P. Wolf, Z. Angew. Phys. *6,* 529 (1954).

18. A. J. Melmed and E. W. Müller, J. Chem. Phys. *29,* 1037 (1958).

19. I. Giaever, Surf. Sci. *29,* 1 (1972).

20. M. G. Inghram and R. Gomer, J. Chem. Phys. *22,* 1279 (1954).

21. R. C. Abbott, J. Chem. Phys. *34,* 4533 (1965).

22. J. A. Panitz, Bull. Am. Phys. Soc. *24,* 272 (1979).

23. W. R. Graham, F. Hutchinson, and D. A. Reed, J. Appl. Phys. *44,* 5155 (1973).

24. E. W. Müller and K. D. Rendulic, Science *156,* 961 (1967).

25. E. S. Machlin, A. Freilich, D. C. Agrawal, J. J. Burton, and C. L. Briant, J. Microsc. (GB) *104,* 127 (1975).

26. J. A. Panitz and I. Giaever, Ultramicroscopy *6,* 3 (1981).

27. S. Iijima, Micron *8,* 41 (1977).

28. J. J. Burton and E. S. Machlin, J. Appl. Phys. *43,* 662 (1972).

29. D. J. Rose, J. Appl. Phys. *27,* 215 (1956).

30. J. A. Panitz and I. Giaever, Abstracts of the 25th International Field Emission Symposium (Albuquerque, N.M., 1978); Ultramicroscopy *4,* 361 (1979).

31. J. A. Panitz and I. Giaever, Abstracts of the 26th International Field Emission Symposium (Berlin, West Germany, 1979); Ultramicroscopy *5,* 284 (1980).

32. J. A. Panitz, Abstracts of the 25th Annual Meeting of the American Biophysical Society (Denver, Colo.); Biophysical Journal *33,* 199a (1981).

33. J. A. Panitz, J. Microsc. (GB) *125,* 3 (1982).

34. J. A. Panitz, Ultramicroscopy *7,* 241 (1982).

35. J. A. Panitz, J. Microsc. (GB) (in press).

36. D. C. Ghiglia and M. Flickner, Opt. Lett. *7,* 116 (1982).

37. W. K. Pratt, Digital Imaging Processing (Wiley-Interscience, New York, 1978), p. 322.

38. J. A. Panitz, Abstracts of the annual meeting of the American Vacuum Society (Anaheim, Calif., 1982); J. Vac. Sci. Technol. *20,* 895 (1982).

CHAPTER

8

THE EXAMINATION
OF HYDRATED SPECIMENS
IN ELECTRON MICROSCOPES

V. N. E. ROBINSON

Faculty of Applied Science
University of New South Wales
Kensington, New South Wales, Australia

8.1 INTRODUCTION

When studying any specimen with any analytical instrument, it is desirable to retain the specimen in a condition as close as possible to its natural state, consistent with obtaining accurate analytical information about the specimen. Should it become necessary to change the state of the specimen to obtain more analytical information, it is essential to know the relationship between the natural and "as analyzed" states of the specimen. The natural state of most biological specimens is hydrated at temperatures above 0°C. In this state their water vapor pressure (up to 50 mbar) is not compatible with the low pressure (less than 10^{-4} mbar) required to generate and control the electron beam in an electron microscope. Thus before these specimens can be examined in an electron microscope, special precautions must be taken to make the microscope and the specimens compatible.

Three totally separate principles have been developed to make high

191

vapor pressure hydrated biological specimens compatible with the low pressure requirements of electron microscopy.

1. *Remove the water*—usually done by controlled drying techniques such as freeze-drying and critical-point-drying.

2. *Lower the water vapor pressure*—achieved by freezing the specimen, usually with liquid nitrogen or liquid-nitrogen-cooled solutions.

3. *Construct a specimen environment*—contain the water vapor in a special environment around the specimen and so prevent it from interfering with the microscope.

Most of the studies on biological materials have been based on examining fixed, stained [in the case of transmission electron microscope (TEM) studies] and dehydrated specimens. These studies have well documented a vast number of biological structures. For the most part the changes induced between the "natural state" and the "analyzed state" are sufficiently well understood that misinterpretations of image information are quite rare. Thus, for the most part, hydrated biological materials can be successfully examined after dehydration, that is, according to principle (1) mentioned earlier. Methods and techniques used in this principle have been extensively reviewed elsewhere and will not be further mentioned in this chapter.

However, there are a large number of applications in which it is essential to study biological specimens in the hydrated state. These include such diverse fields as verification of shrinkage and other structural changes induced during dehydration, observations of dynamic biological activity, the observation of structures that cannot be dehydrated without almost total structural collapse, and the study of materials in contact with water and indeed just the study of hydrated structures to show information that cannot be revealed in the dehydrated state. It is for these reasons that there is an interest in the direct observation of hydrated specimens in electron microscopes. It is one of the major aims of this chapter to indicate the many different techniques that have been employed to study hydrated specimens, the results that have been achieved and point out which techniques could best be employed for the different types of results expected.

The early attempts to look at hydrated specimens in a TEM were performed using closed wet cells to contain the water vapor [1,2]. The biological structures to be examined were placed between two film-containing grids, the edges were sealed, and the composite grid was placed in the TEM in the normal manner. More recently, differentially pumped aperture limited chambers have been constructed in the electron optical column of a TEM [3]. These chambers use apertures to contain the water vapor and other gases in the vicinity of the specimen and differentially pump them to prevent the water vapor from interfering with the remainder of the electron optical column. Both of these principles have been extended to scanning transmission electron microscopes (STEMs) [4,5] and scanning electron microscopes (SEMs) [6–8].

8.2 PRINCIPLES OF OPERATION

8.2.1 Design Features

Closed Cells

The essential principle of a closed cell is that a water-impervious, electron-transparent film is employed to retain the water vapor in a closed environment, separating it from the electron optical column. Figure 1 illustrates the incorporation of this principle for (1) TEM, (2) STEM, and (3) SEM. These films are often required to withstand pressure differentials up to atmospheric pressure and are usually supported on TEM grids to give a stronger film. In TEM the environment is enclosed between two films. The beam passes through the upper film, is scattered by the sample and its environment, and passes out of the lower film to contribute to image contrast in the normal manner. In STEM and SEM only one film is used to separate the environment from the electron optical column. In STEM the detector, an unbiased scintillator type or solid-state detector, is located in the environmental chamber. The specimen is mounted on the underside of the film. The beam passes through the film, reacts with the specimen, and carries on to the detector. In SEM, the detector, also an unbiased scintillator type or solid-state detector, is located above the film. The specimen is placed on, or as close as practical to, the film. The beam passes through the film, some of the electrons are backscattered and pass back out through the film to be detected by the detector.

Several different materials have been used for these films. These include collodion–carbon films (120 nm collodion, 10 nm carbon), titanium films (50 nm), Formvar and nitrocellulose (30–44 nm). These films have been rested on or glued to the grid (e.g., using araldite). The composite grid has then been either glued to a plate for further support or clamped directly to the grid support mechanism.

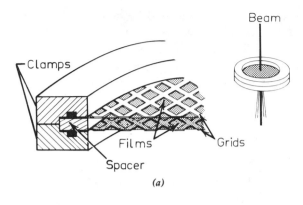

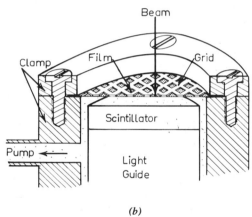

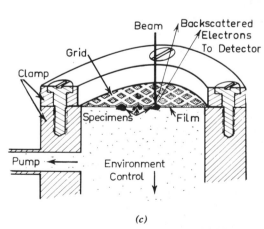

Figure 1. Schematic illustrations of the essential features of closed cell hydration chambers in (*a*) TEM, (*b*) STEM, and (*c*) SEM.

Differential Pumping

In these systems the vacuum in the electron optical column was isolated from that in the specimen chamber, with the only connection being through a small objective lens aperture, and in the case of TEM, a post specimen aperture as well. These principles are illustrated in Fig. 2, for TEM, STEM, and SEM. The combination of aperture and pumping system should be chosen such that optimum microscope operation is achieved. This generally involves a compromise between aperture size, pumping capability, and desired field of view. Specific details are outlined in the following section.

8.2.2 Design of Vacuum Pumping System

When closed cells are used to study hydrated specimens, little or no modification to the microscope vacuum system is required. The film contains the water vapor, preventing it from interfering with the electron beam. The specimen is inserted into the microscope in the normal manner and, provided the film does not rupture, there is no problem of the water vapor interfering with the vacuum system. Some designs, depending on the strength of the film, must be pumped down carefully such that there is never more than a certain maximum pressure across the film.

However, the situation with differentially pumped, pressure-limiting aperture systems is totally different. Here gas can leak away from the specimen environment, through the aperture, and may scatter the beam or interfere with the electron beam generation and/or detection system. It is to avoid these possibilities that the gas must be pumped away as quickly as possible.

Figure 3 shows the percentage of electrons scattered out of the beam by nitrogen gas, at various pressures, for different working distances, extrapolated from the results of Moncrieff et al. [9]. It can be easily seen from this that the amount of beam scattering increases with the beam path length and gas pressure. Although no definite value for the maximum amount of beam scattering that can be tolerated has been determined, suffice it to say that the less the amount of scattering, the better will be the image. Where high pressures are involved (i.e., in the electron optical column), the pressure must be kept as low as possible.

Dimensions of the order of 300 mm and longer are commonly encountered in the electron optical columns of electron microscopes. It is easy to measure the pressure in the specimen chamber and the electron gun of a microscope. However, because of conductance problems, the pressure in the column could be considerably higher than that in the gun and column vacuum manifold.

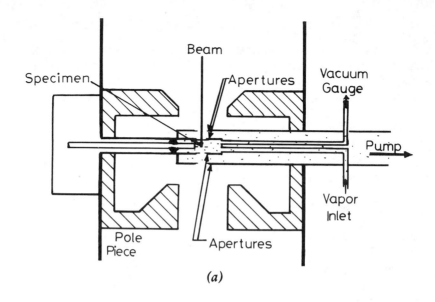

(a)

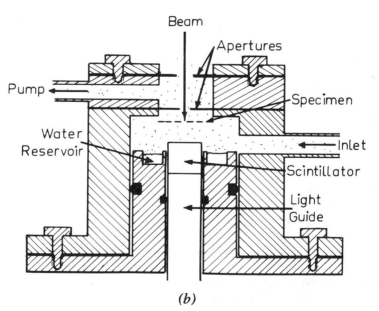

(b)

Figure 2. Schematic illustrations of the essential principles employed for the incorporation of differentially pumped, aperture-limited hydration chambers in (*a*) TEM, (*b*) STEM, and (*c*) SEM.

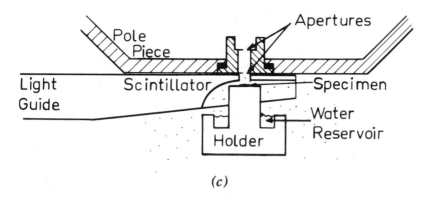

(c)

When designing a vacuum system for a microscope fitted with a hydration chamber, it is necessary to have the conductance above the final aperture as large as possible. Essentially, this means that the conduction path should be as short as possible and through an orifice of as large a diameter as possible.

Techniques to determine the pressure at various positions in the electron optical column through which the beam passes and to which a gauge is inaccessible have been described by Danilatos and Robinson [10]. The calculations are based on the speed of exhaust of gas molecules from the specimen chamber through the final apertures and the conductance of the vacuum system components. These can be determined using the techniques published by Dushman [11]. Danilatos and Robinson have shown in specific detail how to calculate pressure differentials and gradients at all points throughout the electron beam path, and their paper should be referred to for all specific details of the calculation technique.

Their calculations have shown that quite high pressures can be present in the column, particularly just above the final aperture, even though quite low gun pressures are measured. As a specific example, a 10-mbar cham-

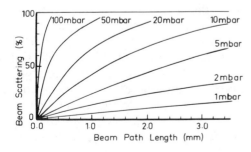

Figure 3. Curves showing the percent of beam scattering versus path length for different vapor pressures of nitrogen gas. (From the data of Moncrieff et al. [9].)

ber pressure across a 136-μm single aperture resulted in an above aperture column pressure of 6×10^{-2} mbar but a gun pressure of less than 1×10^{-5} mbar. With this magnitude of column pressure for over 200 mm, beam scattering in the column can be quite substantial.

The pressure differential that can be maintained across a single pressure limiting aperture depends on the diameter of the aperture and the vacuum conductance on the high vacuum side of the aperture. In the experimental situation described by Danilatos and Robinson [10], a single 100-μm final aperture could sustain a pressure gradient of a factor of approximately 10^3. When pressure gradients above this are required, a double differentially pumped aperture system is required. Such a system has two advantages. High chamber pressures can be tolerated for the same electron optical column pressure, or larger diameter apertures, and therefore larger fields of view can be used for the same specimen chamber pressure. Thus wherever possible, a double differentially pumped aperture system should be employed in any aperture-limited hydration chamber. The apertures should be placed as close together as possible with rotary vacuum pumping between the two. The only disadvantages of a double differentially pumped aperture system are those that relate to the mechanical requirements of accurate fabrication, assembly, and aperture alignment.

8.2.3   Beam Scattering by Gas Molecules

Scattering of the electrons by gas molecules in an electron microscope has been considered by Moncrieff et al. [9]. By considering the elastic and inelastic scattering cross sections of single atoms and appropriately combining them for diatomic molecules, they were able to derive curves for the cross-sectional areas for scattering of electrons by nitrogen gas. These curves indicated that if a 25-keV electron was scattered by a gas molecule (N_2), it would be scattered through an angle greater than 10^{-4} rad, with a typical scattering angle being greater than 10^{-2} rad. Thus once most scattered electrons have traveled more than 0.1 mm, they are over 1 μm away from their beam original trajectory. In the case of SEM and STEM, this means that any scattered electron will be so far outside the original electron beam spot that it will contribute only to a background signal. This signal merely reduces contrast and can be subtracted to obtain normal contrast if desired. Beam scattering does not alter the spot size and thus does not significantly affect resolution. Scattering merely reduces contrast, and when it becomes excessive, there is no contrast in the image.

When it is borne in mind that all contrast observed in an electron microscope (TEM, STEM, or SEM) is due to differential scattering between a feature and its background, it can be easily realized that it is essential to keep the scattering of the beam by the gas surrounding the specimens down to an absolute minimum. This can be achieved by minimizing beam scattering in the electron optical column (as discussed in the previous section) and operating with as low a pressure times working distance factor as practical. As a general rule of thumb, beam scattering in the column of less than 10% is desirable and up to 60% can still be useful. Beam scattering in excess of 60% in the electron optical column causes too much loss of contrast in the final image to be useful.

8.3   CONTRAST AND RESOLUTION

8.3.1   Special Considerations in Contrast Mechanisms

In all electron microscopy—TEM, STEM, and SEM—image contrast is produced by the specimen scattering the electron beam. This has special problems when hydrated biological specimens are studied in an EM. The first is that electron scattering increases with atomic number and that the atomic number of organic materials is quite low. This means that the contrast in the image of any unstained biological specimens will be quite low and will result in images displaying a much lower resolution than the instrument is capable of displaying on more ideal higher atomic number specimens. The second major problem is that the elements in water, H and O, have the same atomic number as some of the elements (H, C, N, O) in organic specimens. Thus the contrast of hydrated organic specimens is even lower than that of dehydrated organic specimens, because with hydrated specimens there is the added factor that contrast now also depends on the differential scattering between water and the organic specimen. As these materials have very similar atomic numbers, scattering differences are quite small and thus contrast is low.

To overcome this small scattering difference and obtain reasonable contrast from hydrated specimens, two factors are necessary. First, beam scattering by the water vapor–gas environment around the specimen must be reduced to as small a value as practical. That is, lowest compatible pressure, shortest working distance. Some examples of the pressure working distance versus beam scattering for 25-keV electrons are shown in Fig. 3. Second, the water surrounding the specimen must be kept as thin as possible, preferably 100 nm or less. This poses no problems at all when hydrophobic specimens are studied in an SEM. The water is naturally repelled from the surface, leaving no water against which differential scattering must occur. However, with hydrophilic specimens in an SEM and all specimens examined in TEM and STEM, thinning of the surrounding water layer is essential. This should be achieved without dehydrating the specimen in any manner.

8.3.2 Limitations and Resolution

As mentioned in the previous section, contrast is formed by the differential scattering of the beam by the water and the sample. This, of course, is superimposed on the additional scattering produced by the gas molecules and the mechanism employed to retain them. In the case of closed cells this scattering is the sum of the scattering induced by the membrane and the gas molecules. With differentially pumped systems, it is due only to scattering by the gas molecules. The additional scattering by the membrane in a closed cell is often counteracted by the shorter beam path length through a higher vapor pressure region and can result in less beam scattering than in a differentially pumped system.

Closed cells have the advantage over differential pumping in requiring less microscope modification and can sometimes produce less beam scattering. They have the disadvantages of being liable to membrane bursting and enclosing a small volume of liquid, making dynamic experiments difficult. As a general rule, closed cells have been used in TEM and STEM studies where hydration was required and where a dynamic gas situation was not required. Differentially pumped systems have been used almost exclusively in SEM, as well as in TEM and STEM, when control and variation of the gas pressure was required. Where either system can be used, the choice between closed-cell and differential pumping should be made on the basis of the practicalities of construction, a good, thin, strong membrane versus an adequate pumping system. Other considerations that can affect the choice of a closed-cell versus differential pumping are that closed cells, which can withstand up to atmospheric pressure, are commercially available for TEM applications, whereas differentially pumped systems have to be individually constructed. It is anticipated that this situation could change with differentially pumped systems soon being commercially available for SEM and STEM studies. Against the commercial availability of closed cells must be assessed the greater versatility of differentially pumped systems.

The studies of Moncrieff et al. have shown that scattering by the gas molecules does not affect resolution. It merely degrades contrast (by adding a uniform background intensity that can be subtracted away). Beam scattering only causes a problem when it is so excessive that either there is no contrast left at all or, after subtracting the background, the image is so noisy (grainy) as to display insufficient information. It is thus expected that resolution obtained from hydrated specimens should be approximately the same as that achieved from similar nonhydrated specimens. This has been verified by Fukami [12], who has shown a resolution of better than 1.5 nm from TEM images of stained hydrated biological tissues. Similarly, Lyon et al. [5] have displayed a STEM resolution of better than 25 nm in a 26-mbar water vapor environment. Swift and Brown [4] obtained a STEM resolution of better than 100 nm for specimens at atmospheric pressure. However, here high image contrast, and subsequently good resolution, was achieved because of the high electron scattering power of the specimens. Most biological material in its natural state consists of low atomic number (and also low density) constituents, which have a low electron scattering factor. It is this low electron scattering power of unstained biological specimens that reduces the available resolution. This can be conveniently expressed in SEM in an empirical relationship of

$$M_{\mu m} = 1000 \, \rho \left(\frac{20}{R}\right) \cdots \tag{1}$$

where $M_{\mu m}$ = maximum useful magnification, ρ is specimen density in g/cm^{-3}, and R is the guaranteed resolution (in nanometers for the particular microscope operating parameters of accelerating voltage and working distance) of the SEM. Most biological specimens have a density close to 1 g/cm^{-3}. The use of non-optimum aperture positioning and accelerating voltage usually between 15 kV and 25 kV gives $R \simeq 7$–10 for the newer-generation SEM's. Thus in the SEM, good images from hydrated biological specimens could be expected at magnifications of 2000–3000×. This figure should be expected to be slightly higher for STEM and TEM. But again, it must be re-emphasized that this is a limitation due principally to the low electron scattering power of the biological specimen, not the presence of the water vapor surrounding the specimen.

8.4 RESULTS

TEM and STEM studies generally reveal information about the bulk structure of thin specimens. This section deals principally with results obtained using an aperture limited differentially pumped hydration chamber in a SEM to study the surface appearance of bulk hydrated specimens. Most of the early studies of hydrated biological samples were performed on keratin fibers (wool and hair) at temperatures close to but above 0°C. These showed significant increases in scale edge protrusions of hy-

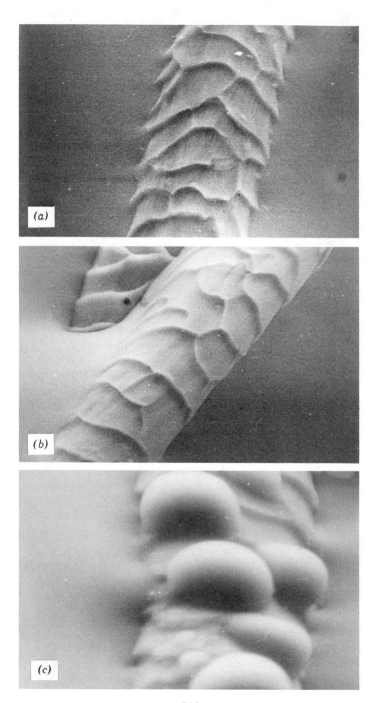

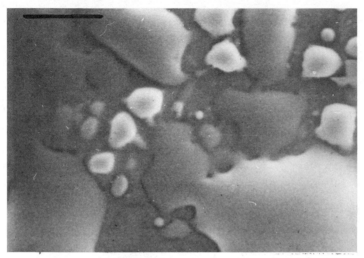

Figure 5.   Micrograph of water droplets on a plastic surface: 6.5 mbars; 15-kV accelerating voltage; magnification marker represents 50 μm.

drated fibers over dehydrated fibers (see Fig. 4*a*). This information has subsequently been employed in studies in wool science to help explain the enhanced differential friction effect of wet wool fibers and add further insight into the problem of wool felting and garment shrinking. Similarly, longitudinal striations in wool fibers, which are very noticeable in dehydrated fibers, are still present in the hydrated fibers. This suggests that cuticle and cortex swelling in the hydrated state are approximately uniform, again yielding additional information about the nature of water absorption and properties of the wool fibers. Similar results have been obtained at room temperature, indicating that there is no significant temperature effect, between 0°C and 20°C, associated with water absorption of wool fibers. This is consistent with the wool fiber property of regain being approximately equivalent at these two temperatures. Figure 4*b* shows formaldehyde cross-linked wool fibers in water. Note that these fibers do not show the same degree of scale edge protrusion as untreated fibers, and this in turn can be correlated to the observation that these fibers do not felt to the same degree as untreated fibers.

Figure 4.   Micrographs of wool fibers partially immersed in water. (*a*) Untreated fiber: note the prominent scale structure and the longitudinal striations; 6.5 mbars; 18-kV accelerating voltage. (*b*) Formaldehyde cross-linked fibers: note the less prominent scale 17-kV accelerating voltage; 6.5 mbars. (*c*) Condensation droplets on untreated wool fiber 15-kV accelerating voltage; 6.5 mbars. Magnification markers represent 20 μm.

Other studies have been performed on the differences in appearance of hydrated and dehydrated chemically modified keratin fibers. These studies have yielded useful information concerning the swelling and wettability of the fibers after treatment. The variation of water contact angle along a fiber can be used to evaluate microvariations in surface state of a fiber. This can be seen in Figs. 4a and 4b, as well as Fig. 4c, which shows water condensation droplets on the surface of a wool fiber.

Figure 5 illustrates water droplets on the surface of a plastic material considered for applications in biomedical engineering. The shape, varia-

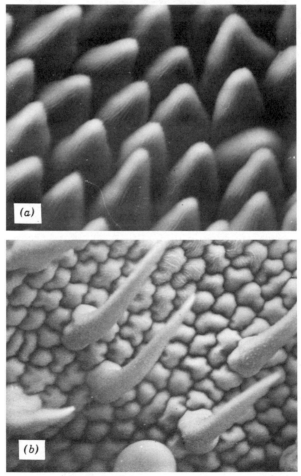

Figure 6.   Micrographs of flower (a) petal and (b) stem structures: 8 mbars; 15 kV; magnification markers represent 20 μm.

tions, and contact angles can yield information concerning the expected behavior of this material in biological applications. Additionally, in the materials science field, Neal and Mills [13] have shown how a water-absorbent material (Spongia) absorbs water, dynamically displaying how a front of water moved through the sponge. They have also dynamically displayed an ice–water interface moving across the field of view. These types of results have served to illustrate how microscopy of hydrated structures can be used to reveal information about water in contact with other surfaces.

Studies have also been made of many other specimens, ranging from soft botanical tissues—flower petals, stamen, leaves, and so on—to animal tissues such as insects and internal mammalian organs such as heart muscle, kidney, and lung tissues [14]. The studies of soft botanical tissues (see Fig. 6) have provided some interesting observations. First, many critical point and freeze-dried structures have shown a considerable amount of fine detail on the surface. When similar structures have been examined fresh, there has been no sign of this fine detail on the surface [15]. The most probable explanation of this was that in its natural state, osmotic pressure would tend to keep the outer membrane slightly expanded, lightly stretched, like any fragile membrane that is subjected to slight internal pressure. During dehydration, when this osmotic pressure is removed, the membrane relaxes a little, possibly settling onto a substructure or folding against itself, to reveal the fine surface detail so often

Figure 7. Micrograph of mandible of a live ant, 20 kV, 25 mbars. Magnification marker represents 50 μm.

observed in dehydrated specimens. Figure 6 illustrates two types of cells. Figure 6*b* shows several different cells in which there is no external surface structure to be observed on the cells. Figure 6*a* shows striated conical petal cells, which display considerable fine surface detail.

Figure 7 shows detail on the mandible of a live ant. In this situation there seems to be no advantage gained from examining the specimen in its hydrated state. The image displays considerably less detail than can be achieved from a dehydrated sample and displays no additional information in the hydrated state.

8.5 CONCLUSIONS

Techniques have been developed that enable hydrated biological specimens to be examined in transmission, scanning transmission and scanning electron microscopes. Those techniques have the advantage that specimen preparation to examine hydrated specimens is simpler than that required to dehydrate specimens. However, they do have a number of disadvantages. These include:

1. The equipment required to construct and operate an electron microscope in the hydrated specimen mode is generally not commercially available (apart from TEM closed cells) and must be individually constructed.
2. Microscope operation is usually more difficult.
3. Image contrast is considerably reduced, showing itself as greatly reduced available image resolution (over conventionally prepared—fixed, dehydrated, and stained or metal coated—specimens).

However, for microscopists who are prepared to persevere with these difficulties, there are a number of advantages to be gained from studying hydrated specimens. These advantages are related to the examination of specimens free from preparation artifacts. In the wool fiber examples illustrated in Section 4, it was shown how hydrated fiber appearance could be related to the felting characteristics of wet wool fibers. The example with plant materials served to illustrate the variability of cell structure with hydration. Some hydrated cells have a smooth external appearance, while others have a highly structured appearance. Only by examining these structures in the hydrated state can we be certain of the appearance and shape of these structures in their natural state. Similarly, until such time as a biological sample has been examined in the hydrated

state, there is no information available as to how much preparation artifact has been introduced during dehydration.

REFERENCES

1. I. M. Abrams and J. W. McBain, J. Appl. Phys. *15*, 607 (1944).
2. E. F. Fullham, Rev. Sci. Instrum. *43*, 245 (1972).
3. V. R. Matricardi, G. G. Hausner, and D. F. Parsons, Proc. 28th Ann. Meet-ing, EMSA, Ed. C. J. Arceneaux, Claitor's, Baton Rouge, p. 542, 1970.
4. J. A. Swift and A. C. Brown, J. Phys. E: Sci. Instrum. *3*, 924 (1970).
5. N. C. Lyon, E. Gasiecki, and D. F. Parsons, Scanning Electron Micros-copy/1976, Part I, Proc. 9th Ann. SEM Sym., O. Johari, Ed., IIT Research Institute, Chicago, p. 101, 1976.
6. V. N. E. Robinson, J. Microsc. *103*, 71 (1975).
7. V. N. E. Robinson, Scanning Electron Microscopy/1975, Part I, Proc. 8th Ann. SEM Symp., O Johari, Ed, IIT Research Institute, Chicago, p. 51, 1975.
8. R. Takahashi, Scanning Electron Microscopy/1977, Part I, Proc. 10th Ann. SEM Sym., O. Johari, Ed., IIT Research Institute, Chicago, p. 71, 1977.
9. D. A. Moncrieff, P. R. Barker, and V. N. E. Robinson, J. Phys. D: Appl. Phys. *12*, 481 (1979).
10. G. D. Danilatos and V. N. E. Robinson, Scanning *2*, 72 (1979).
11. S. Dushman, *Scientific Foundation of Vacuum Technique*, Wiley, New York, 1949, chap. 2.
12. A. Fukami, Electron Microscopy 1976, Proc. 6th Europ. Cong. EM, Y. Ben-Shaul, Ed., Tal International, Jerusalem, p. 63, 1976.
13. R. J. Neal and A. Mills, Jr., Scanning *3*, 292 (1980).
14. V. N. E. Robinson, Electron Microscopy 1976, Proc. 6th Europ. Cong. EM, Y. Ben-Shaul, Ed., Tal International, Jerusalem, p. 85, 1976.
15. G. D. Danilatos, J. Microsc. *121*, 235 (1981).

SCANNING ELECTRON-STIMULATED DESORPTION MICROSCOPY

H. F. DYLLA

Plasma Physics Laboratory
Princeton University
Princeton, New Jersey

J. H. ABRAMS

Department of Surgery
University of Minnesota
Minneapolis, Minnesota

9.1   INTRODUCTION

Much effort has been devoted in recent years to the development of techniques, such as Auger electron spectroscopy (AES) and electron stimulated desorption (ESD), for probing the structure of solid surfaces with incident electron beams. A useful extension of these techniques is the combination of the spectroscopic measurement with a finely focused, scanning electron beam to map the spatial variation of the observed surface properties. The development of scanning Auger microscopy has proceeded to the point where the technique provides a reliable means of mapping the elemental composition of a surface with a resolution of less than 1 μm [1]. In contrast, the process of electron-stimulated desorption of ions and neutrals from surfaces is well known and has been applied to the study of chemisorption in many gas–solid systems [2,3], but there have been few studies reported in the literature on spatially resolved ESD effects [4–10].

In this chapter we demonstrate that ESD provides a useful contrast mechanism for scanning electron microscopy, because ESD signal sensitivity depends on the coverage and binding energy of adsorbed species. The application of scanning ESD (SESD) to biological surfaces is particularly interesting: important surface chemical variations, such as relative hydrophobicity and degree of carbon saturation, can be mapped using the spatial variation in the ESD of common low mass adsorbates (H_2O, C_2H_4, CH_3OH, etc.).

In the case of biological materials, adsorption of low mass molecules with different chemical properties can enhance observation of surface differences without disturbing the biological structure. Although enhancement of contrast by use of an extraneous molecule is similar in idea to the fixing of biological specimens with heavy metals in transmission electron microscopy [11], the use of such low mass adsorbates with biological compatibility will, it is hoped, disturb the specimen to a far lesser degree. For SESD the adsorbate that is used as a surface stain may be a naturally occurring molecular constituent of the surface structure of the biological specimen, such as H_2O or an alkali–metal solute from the normal aqueous environment, or a specific molecular stain applied to a specimen by liquid or gas-phase exposure previous to ESD examination.

In this review we will describe qualitative ESD theory and the experimental apparatus necessary for SESD, examine basic ESD spectroscopic studies necessary for interpretation and optimization of SESD micrographs of biological materials, and, finally, provide two examples of SESD micrographs from biological specimens. In the first example, the variation in carbon–hydrogen surface chemistry from an aortic cross section will be shown. In the second, mapping of alkali and halogen-containing adsorbates across cell surfaces at a spatial resolution of less than 1 μm will be presented.

9.2 LITERATURE SURVEY OF RELATED STUDIES

The first documented discussion of ESD concerned the appearance of background ion peaks attributed to ionization of chemisorbed gases on electrode surfaces [12] in mass spectrometer ion sources. Subsequent studies of ESD effects on several well-defined chemisorption systems (O_2/

Mo, O_2/Ni, CO/W) by a number of investigators [13–17], led to the proposal of a model for ESD by Redhead [16] and Menzel and Gomer [17] that could account for many of the experimentally observed features of ESD from chemisorbed species. A comprehensive review of ESD phenomena was published by Madey and Yates [2]. An examination of the surveyed literature reveals that virtually all the experimental work probed the adsorption and desorption of simple gases (H_2, O_2, CO, etc.) on well-characterized metal surfaces. In the context of such studies, ESD proved useful as a primary analysis technique as well as a complementary to other surface spectroscopies for probing adsorbate structures. The measurement of ESD product intensities, as a function of the adsorbate coverage, substrate temperature, or electron bombardment conditions, generated useful information on both the chemical state of the adsorbate and the nature of the binding site.

The relative simplicity of ESD spectra from atomic or diatomic adsorbates on well-characterized metal surfaces, in comparison to the more complicated spectra from higher mass molecular adsorbates and organic substrates, has apparently discouraged experimental efforts with these more complex surfaces. There are few citations in the literature to ESD from organic molecules or organic surfaces. A series of ESD investigations has been published [18–22] that present desorption spectra and desorption probabilities from metal surfaces contaminated with uncharacterized hydrocarbon films. These studies were performed for the purpose of conditioning vacuum-vessel surfaces in electron storage rings and magnetic fusion devices. Recently, several studies have been reported where the ESD behavior of specific organic molecules adsorbed on metal surfaces was examined. Included in these investigations were CH_3OH [23], C_2H_6, cycloparaffins [24], and neopentane [25]. These studies are representative of the trend evident in the application of other surface spectroscopic methods to more complex molecular species that are relevant to catalytic studies. The most recent developments in basic ESD studies include formulation of two additional ESD models by Knotek and Feibelman [26] and Antoniewicz [27] and the study of the related phenomenon, photon-stimulated desorption using synchrotron radiation, for the excitation source. The reader is referred to the companion chapter by Knotek [25] for a detailed discussion of these developments.

The historical development of scanning ESD as a technique is interwoven with the progression of basic ESD studies. The first spatially resolved ESD measurements were demonstrated by Rork and Consoliver [4]. They used a relatively broad (~0.1 mm) electron beam from a CRT-type electron gun for the scanning electron probe and a magnetic-sector mass spectrometer for the detection of ESD ions. Since this first study, a number of low resolution SESD studies have been reported in the literature [5–9]. In all but one of the preceding studies, ESD ions were imaged after desorption from a simple chemisorbed (or physisorbed) species on a metallic substrate. Systems studied included O^+ on W [5], CO^+ from Au [6], and H^+ on Nb [8]. One unusual measurement reported in these early SESD studies presented a low resolution SESD image of a fingerprint on a Ni substrate reported by Joshi and Davis [7]. The ESD spectrum originating from the oils of the fingerprint showed a relatively large ESD ion yield of high mass molecular fragments. This result provided experimental evidence that ESD from complex organic molecules would be informative and interesting.

The authors of this chapter have attempted to apply ESD specifically to the study of organic and biological surfaces and to push the resolution of SESD considerably beyond the low resolution studies in the literature. A brief description of SESD studies on a model biological surface has been previously published [10]. Although more complete in scope, the present contribution represents only preliminary efforts to develop a new microscopy for biological surfaces.

9.3 EXPERIMENTAL AND THEORETICAL CONSIDERATIONS OF SESD

9.3.1 Experimental Apparatus

The apparatus that is required for SESD measurements includes a scanning electron beam, a detector for ESD particles, that may include mass and energy analysis of the particles, and a thermally and vibrationally isolated sample support in an ultrahigh vacuum environment.

The SESD micrographs and related measurements reported in this review were made in a scanning Auger microprobe. The microprobe, a PHI Model 590 Microprobe [28], contains instrumentation for several types of surface analytical measurements. A cylindrical mirror energy analyzer with a concentrically mounted scanning electron beam allows Auger electron spectroscopy (AES) and scanning Auger micrographs to be made. Ion extraction optics, ion energy analyzer, quadrupole mass filter, and an off-axis, broad focus ion gun are present for secondary-ion mass spectrometry (SIMS). An electron gun with a wide operating range in terms of current–voltage characteristics is useful for SESD studies. The scanning electron beam in the 590 Microprobe employs a finely focused (160-nm, 10^{-9}–10^{-7} A at 0.5–10 kV) beam that is projected onto the sample for standard scanning electron micrographs. The beam can be

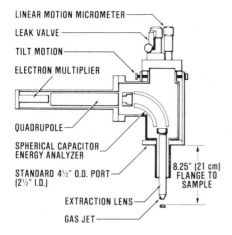

LINEAR MOTION MICROMETER

LEAK VALVE

TILT MOTION

ELECTRON MULTIPLIER

QUADRUPOLE

SPHERICAL CAPACITOR ENERGY ANALYZER

STANDARD 4½″ O.D. PORT (2½″ I.D.)

EXTRACTION LENS

GAS JET

8.25″ (21 cm) FLANGE TO SAMPLE

Figure 1. Diagram of the ion collection optics used for the scanning electron stimulated desorption measurements in this review.

defocused to allow higher current ($>10^{-6}$ A) measurements that are required for low signal AES and ESD applications. The ion collection optics shown in the schematic diagram in Fig. 1 are designed for SIMS studies [29] but are well suited for the detection of ESD ions. The use of the electrostatic ion energy filter, coupled to the mass filter and ion detector, allows improvement of the signal-to-noise ratio for detection of ions produced at the sample surface by electron or ion bombardment. Since both SIMS and ESD produce ions with kinetic energies of several eV [30,2], the energy filter discriminates against thermal ions produced by gas-phase ionization in the neighborhood of the target.

Other types of ion detection systems were employed in the SESD studies reviewed in the previous section. A magnetic sector mass spectrometer was used in the work of Rork and Consoliver [4] and Lichtman and Campuzano [5], and a cylindrical mirror energy analyzer (CMA) was used for the work reported by Larson et al. [9]. The latter type of analyzer will improve signal intensities for the measurement of total ESD ion yields, because high ion transmission efficiency (~10%) exists for such systems [31]. In comparison, typical collection efficiencies of mass spectrometer optics are less than 10^{-2}. However, for the detection of mass-analyzed ESD signals, time-of-flight analysis has been added to CMA-based systems [9], and the efficiency advantage decreases rapidly with the complexity of the mass spectrum.

The ion collection efficiency for the ion optics shown in Fig. 1 has been estimated to be (0.25 ± 0.1)% for mass 56 ions emitted at an iron alloy target surface, sputtered by an argon ion beam [29]. Since a quadrupole mass filter exhibits decreased ion transmission with increasing ion mass

[32], the net ion collection efficiency for this system is expected to be somewhat larger than 0.25% for low mass ($m < 56$ amu) ions and smaller than 0.25% for high mass ($m > 56$ amu) ions.

The low-level signal intensities that are encountered with SESD studies affect the design of the apparatus in several other areas. In addition to requiring optimal ion collection, SESD studies benefit greatly from pulse-counting electronics with a large dynamic range ($>10^6$ cps). For imaging ESD signals, the simplest arrangement couples intensity modulation of a CRT display with the analog output of a count-rate meter. The optimal method of recording SESD images would involve digital recording and storage of both the spatial (scanning beam position) and ion intensity data. Digital signal processing allows more control of image intensity and contrast as well as long scan-period imaging and multiple-image signal averaging.

The use of long scan periods ($>10^3$ s) for low signal level SESD micrographs requires that the electron optics, ion optics, and the specimen support be mechanically and thermally stable, in order that micrographic spatial resolution be governed only by the counting statistics. Finally, the vacuum environment has stringent requirements for SESD. As with any surface-specific measurement, ultrahigh vacuum ($<10^{-9}$ Torr) conditions are necessary to stabilize the surface conditions during the course of the measurement. If ESD measurements are performed in the vacuum conditions of a typical SEM (10^{-5}–10^{-6} Torr), the ESD signals would be dominated by the decomposition of residual gases on the specimen surface, and not dependent on the specific surface properties of the specimen.

9.3.2 Theoretical Considerations

The signal-to-noise level of ESD intensities and the limit of spatial resolution of SESD micrographs are related. We can quantitatively evaluate the relationship between sensitivity, signal intensity, and spatial resolution, using a simple physical description of the ESD process. In general, ESD refers to the process by which an adsorbed species is removed from a surface by impact of low energy (<2 kV) electrons. The desorbed particle may leave the surface in an ionic, neutral, or excited state. The observed effects are believed to be the result of electronic excitations and not thermal or momentum transfer effects because ESD is observed at low incident power levels (10^{-5} W/cm^2), where thermal excitation would be negligible, and desorption by direct momentum transfer would be possible only for very weakly bound ($E < 0.1$ eV) adsorbate species, a consequence of the low mass of an electron. By analogy to similar electronic excitation processes in gas-phase collision phenomena, an ESD cross section, Q_{ij}, can be defined [16,17] that relates the number of desorbed species n_i (per unit time and area) from an adsorbate state j with coverage n_j (cm^{-2}):

$$n_i = n_j Q_{ij} \frac{J}{q} \tag{1}$$

where J/q is the incident electron flux given by the current density J(A/cm^2) divided by the electronic charge q(1.6 × 10^{-19} coulomb).

Only two cross sections need be considered for most systems: the total desorption cross section Q and the ionic desorption cross section Q^+. Measurements of the total ESD cross section from chemisorbed molecules on metal surfaces [2] have values in the range of 10^{-18}–10^{-21} cm^2, several orders of magnitude smaller than the analogous gas-phase electronic excitation processes: ionization, dissociation, and the excitation cross sections for free-molecular collisions [33] (typically 10^{-16} cm^2). Only in the case of weakly bound adsorbates, where the effect of the surface on the electronic configuration of the adsorbate is minimal, have large (10^{-17}–10^{-16} cm^2) ESD cross sections been measured [34,35]. The measured values of the ionic cross section Q^+ from gas–metal systems are typically 10^{-2}–10^{-4} times smaller than the total cross section. Thus for practical considerations, the total cross section can be equated to the cross section for neutral desorption.

Consideration of the difference in cross-section measurements between gas phase and surface phase excitation processes led Menzel and Gomer [17] and, independently, Redhead [16] to a one-dimensional, classical model of ESD. In this model, desorption is the result of electronic excitation of an adsorbate–substrate complex. Electron impact causes a Franck–Condon transition from the ground state (the surface bond) to a repulsive portion of the potential energy curves for the ionic state or to any of the higher-energy neutral states of the complex (Fig. 2). The cross section for this primary excitation is believed to be comparable in magnitude to similar excitations that occur as the result of electron impact of free molecules. On excitation the neutral or ionic particle accelerates away from the surface and would desorb if the presence of the surface were ignored. The increased density of electronic states within the substrate allows a number of competing recapture processes to occur with high probability. First, an initially excited ionic particle can be neutralized by electron tunneling from the substrate to fill the ground state hole caused by the primary excitation. Depending on the distance of the ion

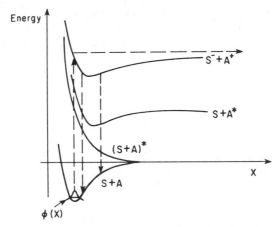

Figure 2. Potential energy curves for the lowest-lying states of a substrate (S)–adsorbate (A) complex. In the Menzel–Gomer–Redhead model of ESD, electron impact causes a transition from the ground state $\phi(x)$ to a repulsive portion of the potential energy curves of the high-energy states.

from the surface at the time of neutralization, the reneutralized ion is recaptured in the ground state or escapes as a neutral. Second, the efficiency of neutralization of an ion near a metal or semiconductor surface is found to be very high from evidence obtained from ion-neutralization spectroscopy measurements [36,37]. Therefore, even if excitation to the ionic state were the predominant transition during ESD, the larger number of desorbed neutrals compared to ions is qualitatively understood. In the context of the Menzel–Gomer–Redhead model, the magnitude of the total cross section can be affected by any deexcitation process, and these processes are more probable with the increased localized density of states available in the substrate. The observed large variation in ESD cross sections with binding energy is interpreted as a consequence of the sensitive variation in the electron-tunneling rate with localized bond geometry (barrier width).

To maximize the sensitivity of SESD, adsorbate–substrate configurations should be chosen with maximum values of Q and Q^+/Q. As the value of Q increases, the interaction of the incident electrons with the absorbate increases while the interaction with the substrate decreases. Since it is experimentally more efficient to detect only the ionized fraction of the desorbed particles, the maximum value of Q^+/Q increases the detected signal intensity.

9.3.3   SESD Sensitivity-Resolution Limits

We can now formulate an expression for the signal level expected in SESD. From Eq. (1), the total number of detected ESD ions (S) from a sample area d^2, where d(cm) is equal to the incident electron beam diameter, is given by

$$S = \epsilon n_0 d^2 (1 - e^{-\Delta t/\tau}) \tag{2}$$

where Δt is irradiation time, $\tau = q/JQ$ is the desorption time constant, n_0 is the initial adsorbate coverage (cm^{-2}), and ϵ is the detection efficiency. For the detection of ionic signals, ϵ is equal to the product of the fraction of desorbed particles that are ionized at the surface (Q^+/Q) and the ion collection efficiency of the detection system.

Assuming that the primary source of noise will be the shot noise in the comparatively small number of detected particles, the signal-to-noise ratio can be expressed as:

$$\frac{S}{N} = \frac{S}{S^{1/2}} = (\epsilon n_0 f)^{1/2} \, d \tag{3}$$

where the quantity $f = 1 - \exp(-\Delta t/\tau)$ is the fraction desorbed of the initial adsorbate coverage.

Since Q is much larger than Q^+, one might expect larger ESD signals from desorbed neutrals where the full cross-section Q is applicable. However, this potential gain is diminished by the difficulty in detecting neutral species and by the additional noise source caused by ionization of residual gas molecules. Including background ionization, the signal-to-noise ratio for desorbed neutral signals is:

$$S/N = \frac{\epsilon n_0 d^2 f}{(\epsilon n_0 d^2 f + BP_B \, \Delta t)^{1/2}} \tag{4}$$

where P_B is the background partial pressure of the molecule of interest and B is a constant related to the ionizer efficiency. Equations (3) and (4) are plotted in Fig. 3 for several values of the detection efficiency ϵ and desorption fraction f. The straight lines in Fig. 3 correspond to desorbed ion signals or to neutral signals where the background pressure of the desorption species is negligible. For the instrumental parameters of the PHI 590 scanning Auger microprobe used for the measurements in this chapter [1], and for ESD ionic cross sections of 10^{-19} cm^2 for the adsor-

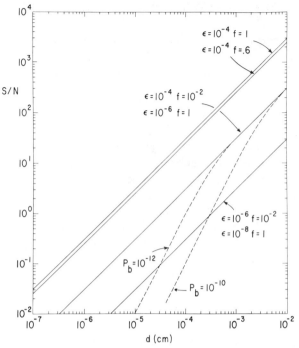

Figure 3. Signal-to-noise ratio compared with resolution, where d = minimum resolvable diameter, for scanning electron stimulated desorption microscopy where only shot noise is considered. The curves are plotted for various values of the total detection efficiency, ϵ, and the desorption fraction $f = [1 - \exp(-\Delta t/\tau)]$, assuming a full monolayer initial coverage and a scan time per distance d, of $\Delta t = 2$ ms. Two curves are shown for the detection of desorbed (28 amu) neutrals in the presence of a background pressure ($P_B = 10^{-10}, 10^{-12}$ Torr). (From Ref. 10.)

bate, the estimated spatial resolution limit is $\simeq 2 \times 10^{-5}$ cm for SESD micrographs of minimal quality ($S/N \simeq 5$). Measurements of ESD cross sections are considered in Section 9.4.1.

9.4   BASIC MEASUREMENTS

From the discussion in the previous section, ESD ion intensities are proportional both to the surface coverage of the ESD active adsorbates and to a factor (Q^+/Q) that describes the surface ionization and escape probability for the species. An analogous situation exists in the description of secondary ion formation during low energy ion bombardment [30] used for secondary ion mass spectrometry (SIMS). Since the ionic intensities generated from a surface under low energy electron or ion bombardment are not simply related to the surface coverage, knowledge of the surface ionization and escape probabilities is necessary to maximize the utility of either ESD or SIMS as surface probes and imaging techniques. During the development of SIMS, investigators have adopted a semiempirical approach [38], using measurements on standard samples to establish elemental sensitivity factors that account for the first-order variations in ionization probability. For the case of molecular SIMS, the recent cataloging of spectra from several well-characterized surfaces of biologic materials [39–42], is useful for qualitative analysis (i.e., species identification), just as the existing literature catalogs of gas-phase mass spectra have been for four decades.

The analogous studies for ESD, the cataloging of ESD probabilities, desorption cross sections, and standard spectra are available for only a few well-studied cases of simple gases on clean metals. This section presents several examples of ESD measurements that are required to improve the utility of ESD for the analysis and imaging of organic surfaces. The first series of measurements was performed on monolayer-thick films of polymethylglutamate (PMG) that were cast on silicon substrates using the Langmuir-Blodgett technique [43]. The second series of measurements was performed on a multilayer soap film formed by collapsing a soap bubble onto a similar silicon substrate.

ESD examination of monolayer or ultrathin films demonstrates the surface specificity of the technique by distinguishing the properties of such films relative to the substrate. In addition, complementary analysis by Auger electron spectroscopy (AES) can be performed using similar electron beam parameters to yield film thickness and atomic composition. Silicon substrates were chosen for these studies, because optically flat samples of high purity are readily available from semiconductor suppliers, and the thin (3–5 nm) passivation oxide layer represents a fairly inert surface for deposition of organic films. The Si substrates, examined by AES previous to film deposition, showed only a small residual surface carbon concentration ($\approx$6 at. %) that is due to hydrocarbon adsorption from atmospheric exposure. No other impurities were detectable by AES. ESD spectra of the substrates showed only a H$^+$ intensity of ~10 cps at an incident electron beam current of 1 μA. This signal is also assumed to be the result of residual hydrocarbon contamination.

9.4.1 PMG Monolayers

PMG was chosen as a model surface for these ESD studies for several reasons. Because PMG is a polymerized amino acid, its functional groups

(—NH$_2$, —COOH) are representative of a protein surface. In addition, the conformation of solvent-cast PMG monolayers has been studied with IR spectroscopy [44], and the structure of the film is found to be preserved after deposition of the cast film onto solid substrates. Studies of PMG monolayers by Loeb and Baier [44] have shown that PMG monolayers can be cast on H$_2$O in either the α-helix or β-sheet conformation, depending on the relative concentration of two solvents, chloroform and pyridine. For the purpose of ESD studies the α-helix conformation, which results from solution in pure chloroform, was chosen to enhance the hydrophilicity of the resulting film. When PMG monolayers were cast onto H$_2$O, the films were deposited onto the substrates by slowly lifting the substrate through the film. The films were air-dried at room temperature and received no other treatment before introduction into the vacuum system for surface analysis. The presence of the PMG monolayer on the substrate was checked by AES. Attenuation of the low energy (92 eV) silicon Auger transition provided an estimate of the film thickness $\simeq 2$ nm, and the detected C/N elemental composition ratio ($\simeq 6$) agreed with the stoichiometry of the methylglutamate monomer.

Because of the α-helical conformation of the deposited PMG films, substantial water of hydration is expected to be retained by adsorption on and within the film. For this reason, similarly prepared PMG films were used as model hydrophilic surfaces in an earlier study of SESD by the authors. To demonstrate the use of ESD as a contrast mechanism for mapping the surface chemistry of biologic surfaces, a grid of a hydrophobic material (pure C) was evaporated over the PMG monolayer. Figure 4a shows a standard electron micrograph of the C/PMG sample, and Fig. 4b shows the same area of the sample imaged with H$^+$ ions desorbed with a scanning 2.0 kV electron beam. The desorbing H$^+$ ions are seen only from the PMG areas and not from the pure C areas of the sample. Because of the relatively small ion collection efficiency of the instrument used in the C/PMG study, only H$^+$ (at a count rate of ~300 cps) was detected in the ESD spectrum. For the instrument used in the present study, the ion collection efficiency has been substantially improved (approximately tenfold), and the ESD spectra that are obtained from PMG monolayers are more informative (Fig. 5). In addition to H$^+$, peaks are visible at masses 16, 17, and 19 amu in the positive ion spectrum, corresponding to H$_2$O-related fragments, O$^+$, OH$^+$, and H$_3$O$^+$.

The small number of peaks in the ESD mass spectrum shown in Fig. 5, compared to a SIMS spectrum or vapor-phase spectrum obtained from volatilized PMG, deserves comment. Qualitative ESD theory dictates that the ESD spectrum should be dominated by the species with the largest product of coverage and surface ionization probability. Thus H$_2$O-

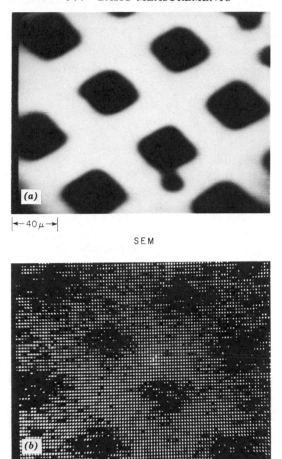

SEM

SDM (H+)

Figure 4. (*a*) Scanning electron micrograph of a test sample consisting of a grid of carbon evaporated over a monolayer of polymethylglutamate (PMG) cast onto a stainless steel substrate. (*b*) Scanning electron stimulated desorption micrograph (SDM) of the same area imaged with desorbing H+ ions. (From Ref. 10.)

related fragments in ESD spectra of the hydrated PMG films are not surprising. Although some contribution to the PMG spectrum may be the result of electron-induced decomposition of PMG functional groups (such as —OH, —COOH), this interaction is apparently not dominant in the presence of the water of hydration: the ESD spectrum lacks organic fragments, CH_n, CN, and so on. The substantial attentuation in intensity with increasing mass shown in Fig. 5 is also consistent with ESD theory: The

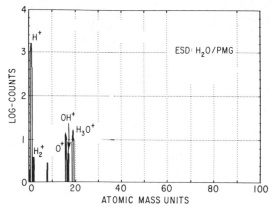

Figure 5. ESD positive ion mass spectrum from a PMG monolayer cast from a water–chloroform solution onto a silicon substrate. Fragment ions from the water of hydration are evident. Electron beam parameters: 2.0 kV, 0.5 μA.

escape probability for an ESD ion produced at or near a metal (or semi-conducting) surface decreases exponentially with increasing mass of the ionized fragment. Heavy mass fragments have a slower velocity for a given excitation energy and thus have a higher probability of neutralization from electron tunneling from the substrate. The various factors that affect the surface ionization efficiency and neutralization probability warrant further investigation as a possible means of increasing overall ESD ion intensities. A dependence of ESD intensity from organic samples on the substrate material is expected in view of recent SIMS results. An effect in the variability of surface ionization with substrate has been seen in the SIMS studies of Benninghoven et al. [39,40] for thin layers of amino acids. The effect may be due to the secondary-electron emission properties of the sample–substrate structure. A substantial fraction of the surface ionization caused by low energy electron or ion bombardment at incident particle energies that are much larger than the threshold energies (5–50 eV) is due to secondary electron emission. Therefore, the use of a substrate material with high secondary-emission characteristics, or the use of a thin sample layer to maximize the interaction of the substrate secondaries with the organic sample, should enhance ESD ion (or second-ary ion) intensities. The effect of sample thickness has been qualitatively demonstrated. For thick (>0.1 m) PMG films that are cast on silicon substrates from solution, the ESD spectrum is attenuated by more than an order of magnitude in comparison to the spectrum from the PMG monolayer. A similar effect in the SIMS yield from amino acid layers has been seen in the study by Colton et al. [45].

Figure 6 shows the time dependence of the ESD ion intensities from the H_2O/PMG sample at an incident beam current of 5 μA. The current density ($J = 2$ mA/cm^2) was sufficient during this measurement for the depletion of one or more ESD active species to be evident. If a simple exponential decay can be fit to the time dependence of the ion intensities $i^+(t)$, then the characteristic time for the decay τ can be used to extract the total desorption cross section for the ESD active species from the solution to Eq. (1):

$$i^+(t) = i^+(0) \exp(-t/\tau) \qquad (5)$$

where τ is equal to q/JQ.

The time dependence of the H^+ and H_3O^+ ion intensities can be fit with two characteristic times, which suggests that at least two ESD processes are responsible for the generation of each species. The initial rapid drop in H^+ intensity yields a total ESD cross section of $Q^+ = 7.7 \times 10^{-17}$ cm^2. The subsequent, much slower decay in H^+ intensity may represent a second H^+-generating ESD process with a total cross section that is approximately 10^2–10^3 times smaller than the value of the initial state. However, it is also possible that the H^+ signal, after removal of the high cross section-species, approaches a steady state intensity that represents an equilibrium between ESD of H^+ (from H_2O) and adsorption of H_2O from

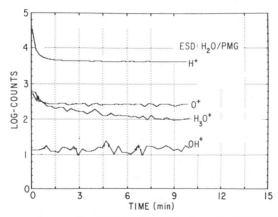

Figure 6. Time dependence of the ESD ion intensities from the H_2O/PMG sample with an incident 2.0-kV, 5-μA electron beam.

the gas phase. Such a steady state situation has been discussed by Redhead [16], who has shown that the steady state ESD ion current i_S^+, is given by:

$$i_S^+ = \nu(p)S(n)A_b q \left(\frac{Q^+}{Q}\right) \tag{6}$$

where $\nu(p)$ is the pressure dependent specific arrival rate for the adsorbate species from the gas phase (in molecules cm^{-2} s^{-1}); $S(n)$ is the sticking coefficient of the adsorbate that may depend on the coverage n; and A_b is the electron-bombardment area. Values for all the parameters in Eq. (6) have not been measured for the H_2O/PMG system, but estimates can be made for comparison of the steady state current predicted by Eq. (6) to the measured values from Fig. 6. The total pressure in the vacuum system for the PMG measurements was $\simeq 1 \times 10^{-9}$ Torr. Assuming that the partial pressure of H_2O vapor was approximately one half of the total pressure and that the sticking coefficient is unity, Eq. (6) predicts $i_S^+ \simeq 10^{-16}$ (Q^+/Q) A. Comparison of this value to the measured steady state H^+ current, which is the intensity of H^+ from Fig. 6 divided by the estimated H^+ ion collection efficiency ($\simeq 10^{-2}$), gives $i_S^+ \simeq 10^{-13}$ A. The measured and predicted values agree if the ratio of ionic to total cross section is $(Q^+/Q) \simeq 10^{-3}$, which is typical of published values.

Thus the redeposition argument seems a plausible explanation for the observed steady state ESD ion currents from the H_2O/PMG system. For the purpose of scanning ESD, redeposition of the active ESD species from the gas phase is useful for maintenance of signal level. To minimize contamination of the vacuum system and possible interference from gas-phase generation of ions, a localized gas jet directed at the sample surface (see Fig. 1) would be more advantageous for redeposition than reliance on the partial pressure in the vacuum system.

The initial ESD ion intensities from H_2O/PMG, before any significant decay of the signal has occurred, are shown in Fig. 7 as a function of the incident electron beam current. The essentially linear dependence of the H^+ and H_3O^+ intensities with electron beam current shows, first, that the ion production reaction for these ions is a first-order process, and second, that the adsorbate coverage is not subject to depletion during the time interval required for these measurements, 10 s per point. The nonlinear dependence of the OH^+ signal may indicate that the ionic cross section for the generation of these ions is small and that the signal level is given by the steady state current due to redeposition. It should be noted that the desorbed ion current predicted by redeposition equilibrium [Eq. (6)] is independent of incident current.

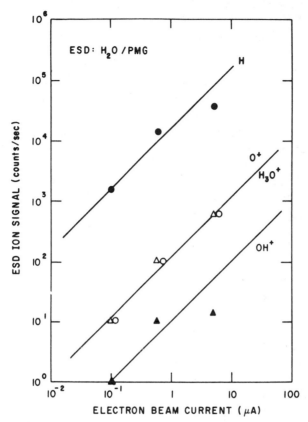

Figure 7. Dependence of the ESD ion intensities from the H_2O/PMG sample with incident electron beam current. The beam voltage was 2.0 kV.

9.4.2 Adsorbates as SESD Surface Stains

A final series of measurements was performed on the PMG monolayers to examine the variability of the ESD spectra with the exposure to different adsorbates. The authors propose that the use of low mass adsorbate molecules of differing chemical reactivity will demonstrate, in SESD measurements, variations in surface chemistry. The results from the exposure of identically prepared PMG monolayers to a series of nonaqueous adsorbates were initially surprising and informative.

The test adsorbate molecules were common reagent-grade solvents that, as a group, spanned a range in polarization from highly polarized to completely covalently bonded. They included formaldehyde (37% H_2CO in H_2O solution), 1-propanol ($CH_3CH_2CH_2OH$), carbon tetrachloride, and

n-heptane ($CH_3(CH_2)_5CH_3$). The PMG monolayer samples were prepared as described in the previous section: The films were cast onto H_2O and then deposited onto clean Si substrates. Following a brief air-drying period, each sample was rinsed in a different solvent, then again briefly air-dried before introduction into the vacuum system for ESD measurements.

Unexpectedly, the resulting ESD spectra from these PMG samples were all identical to the spectrum shown in Fig. 5 for the film, which was simply deposited from H_2O and received no subsequent solvent exposure. No evidence of the solvent exposure was seen with ESD from the PMG samples. We interpret these results to indicate that no significant adsorption of these solvent molecules occurred. The ESD spectra were identical, because in all cases the PMG surface is covered with the H_2O adsorbed during the casting process. Apparently, this water of hydration

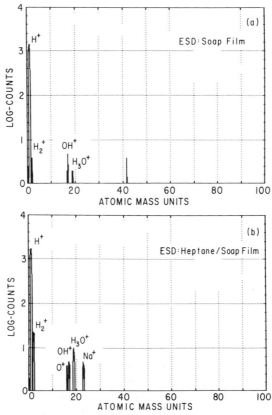

Figure 8. (*a*) ESD ion spectrum from a soap film deposited onto a silicon substrate. (*b*) ESD spectrum from the soap film after exposure to heptane. Electron beam parameters: 2.0 kV, 0.5 μA.

completely dominates the surface chemistry of the PMG monolayers, to the exclusion of significant surface reactions with other adsorbates. The fact that no trace of Cl, or Cl-containing fragments, from the original chloroform solvent was seen in the ESD spectra of the PMG films (or in AES spectra) is additional evidence for the dominance of H_2O in the surface chemistry. On the basis of these results, we expect that the ESD behavior of any highly polar biologic surface (proteins, sugars, etc.) to be dominated by adsorbed H_2O.

To investigate the ESD behavior from a biologic surface that was not dominated by water of hydration, a set of samples was prepared by collapsing an ordinary soap film (Ivory Snow) onto clean Si substrates. Figure 8a shows the ESD from the soap film taken with the same instrumental parameters as the previous ESD spectra from PMG surfaces. Except for a large H^+ peak, no significant high mass positive ions were observed. A lack of high mass ESD ion intensities is believed to be a manifestation of the thickness of the soap film, estimated to be in the range of 0.5–1 μm.

A significant enhancement in the ESD spectrum was produced by exposing an identically prepared soap film to heptane previous to introduction in the vacuum system, Fig. 8b. Following exposure to heptane, significant mass peaks due to H_2^+, OH^+, H_3O^+, and Na^+ are evident in the spectrum. We interpret the H_2^+ ion as being an ESD product of the heptane adsorbate, whereas the remaining high mass ions are from the soap film itself (presumably from $-COO^- Na^+$ and its probable water of hydration). This interpretation is based on the time dependence of the ESD intensities from the heptane–soap film shown in Fig. 9. The H_2^+ intensity initially decays by more than an order of magnitude, whereas the H^+ intensity initially decays but then slowly begins to rise before the H_2^+

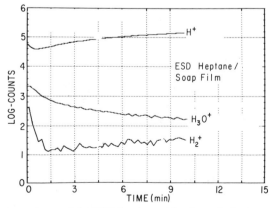

Figure 9. Time dependence of the ESD ion intensities from the heptane–soap film sample with an incident 2.0-kV, 5-μA electron beam.

TABLE 1   Measurements of the Total ESD Cross Section Q, and Ionic Desorption Probability $P^+ = Q^+n$, from Several Organic Surfaces

System	Desorbed Species	Q (cm²)	P^+ (ions/electron)	Ref.
H₂O/PMG	H⁺	$(7.7 \pm 0.8) \times 10^{-17}$	10^{-7}	this work
	H₃O⁺	$(1.8 \pm 0.5) \times 10^{-18}$	10^{-9}	
Heptane/soap	H₂⁺	$(3.6 \pm 0.4) \times 10^{-18}$	10^{-9}	this work
	H₃O⁺	$(1.4 \pm 0.5) \times 10^{-18}$	10^{-8}	
C₆H₁₂/Ru	H⁺	$(8.0 \pm 2.0) \times 10^{-17}$	—	Ref. 24

intensity reaches a steady state. This behavior could be explained by ESD removal of the H_2^+-producing adsorbate, heptane, and subsequent exposure of the ESD active species or adsorption site that produces the predominant H^+ signal. The behavior of the H_3O^+ peak is interesting in that it appears to be unaffected by the time behavior of the H^+ and the H_2^+ peaks. The total cross section that is derived from Fig. 9 for ESD of H_3O^+, $Q \simeq 1.4 \times 10^{-18}$ cm^{-2}, is equivalent, to within experimental error, to the total cross section derived from H_3O^+ from PMG.

These measurements of the interactions of organic surfaces and simple adsorbates on the PMG and soap film samples are examples of the basic measurements required from a wide variety of materials to maximize the usefulness of SESD. From consideration of the signal-to-noise arguments given in Section 9.3, in order to maximize the signal intensity of SESD micrographs, ESD products should be chosen with total Q values that maximize the desorption fraction $f = 1 - \exp(-\Delta t/\tau)$ per given scan time, Δt; and ratios of Q^+/Q that maximize the fraction of desorption product ionized at the surface. Values of the total cross section can be extracted from moderate-current exposures, as illustrated in this section. However, accurate ionic cross sections are not likely to be derived from organic surfaces, because of the inability accurately to predict or measure adsorbate coverages on such surfaces. As an alternative, ion desorption probabilities $P^+ = Q^+n$ can be measured to gauge relative ionic signal strengths. Table 1 lists total Q and ionic desorption probabilities, derived from the measurements in this section and from the literature, for ESD from organic surfaces.

9.5   SESD MICROGRAPHS OF BIOLOGICAL SURFACES

We have discussed the basic mechanisms of SESD, have considered parametric constraints that set limits on the signal-to-noise ratio and resolution of SESD micrographs, and have presented examples of basic measurements of the primary ESD parameters from homogeneous organic surfaces. In this section we present examples of ESD spectra and ESD micrographs taken from two biological samples: a cross section of a canine aorta and sheep red blood cells.

9.5.1 SESD: Canine Aorta

We chose the cross section of a large blood vessel as an interesting material for SESD studies because of the large-scale (0.1–1 mm) juxtaposition of tissues of very different structures. The examined samples were prepared as thin (approximately 6 μm), frozen sections in a cryomicrotome using CCl_2F_2 as the refrigerant and OCT compound to facilitate sectioning [46]. The sections were deposited onto clean Si substrates, briefly air-dried, and then kept at $-12°C$ until examination. A sample was mounted directly onto the introduction probe of the PHI 590 microprobe without further preparation. Figure 10 is a series of standard secondary electron micrographs (SED) and a scanning Auger micrograph (SAM) that show the morphology of the canine aorta cross section. Two different tissue structures are visible in the low magnification (100×, 200×) secondary electron and carbon Auger electron micrographs: The smooth muscle tissue of the vessel wall appears flat, and the external layer of connective tissue appears rough.

Figure 11a shows the ESD spectrum that is obtained from the aorta cross section with 500 eV incident electrons. To integrate the ESD signals from a large area of the sample, the electron beam was rapidly scanned across the same area imaged in the SED micrograph of Fig. 10a. In addition to the expected strong H^+ and H_2O related peaks (OH^+, H_3O^+), a succession of hydrocarbon peaks from parent molecules of the form $C_nH_m^+$ is seen through the entire range of the mass scan. The attenuation of $C_nH_m^+$ peaks with increasing mass is probably the result of two effects. First, the probability of ionic neutralization increases with increasing mass, as discussed in Section 3. Second, the mass discrimination of quadrupole-type mass spectrometers is well known [32]. For comparison with the ESD spectrum, a SIMS spectrum (Fig. 11c) was obtained from a different but similarly structured area of the aorta, using 5 kV argon ions in the primary ion beam. Except for the usual expected alkali peaks (Na^+, K^+) and increased fragmentation of the hydrocarbons, the ESD and SIMS spectra are similar in appearance.

The surface specificity of the ESD spectrum obtained from the aorta sample was demonstrated by examining the ESD spectrum from a second sample that was prepared in a similar manner with one important difference. Prior to introduction to the microprobe the sample was rinsed in

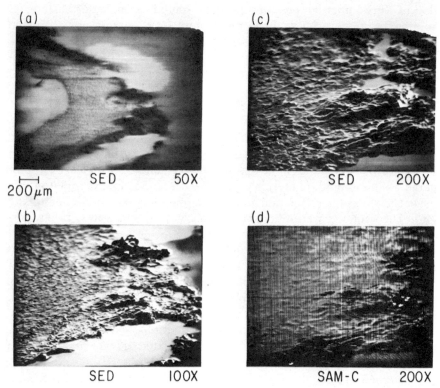

Figure 10. Secondary electron (SED) micrographs (*a, b, c*) and a carbon Auger micrograph (*d*) of a cross section of a canine aorta. The smooth muscle tissue of the vessel appears flat, and the fatty-connective tissue appears rough.

CH$_3$OH. Figure 11*b* shows the resulting ESD spectrum that was obtained using the same instrumental parameters as with the first aorta sample. The surface species that were responsible for the C$_n$H$_m^+$ emission were removed by the CH$_3$OH exposure, and the only remaining features in the ESD spectrum are the H$_2$O-related peaks (H$^+$, O$^+$, OH$^+$, H$_3$O$^+$).

The spatial origin of the C$_n$H$_m^+$ ions in the Fig. 11*a* ESD spectrum becomes evident when specific ion signals from the spectrum are used to image the sample surface. Figure 12 shows a low magnification (50×) SED image of the aorta sample and scanning desorption micrographs (SDM) of the same sample area taken with H$^+$, C$_2$H$_3^+$, and C$_2$H$_5^+$ ESD ion emission. The SDM images were scanned at 100 lines per frame in a period of 150 s per frame, and a simple linear relationship between ESD ion intensity and CRT brightness was used to form the image. The micro-

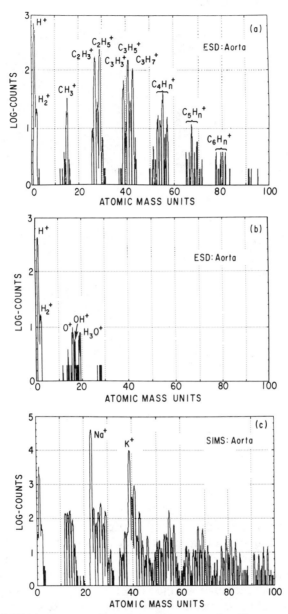

Figure 11. (*a*) ESD ion spectrum from the aorta cross section. The spectrum was obtained by scanning the same area of sample as shown in Fig. 10*a* with a 500-V, 2-μA electron beam. (*b*) ESD ion spectrum from the aorta cross section after the sample was rinsed in methanol. (*c*) SIMS spectrum from the aorta sample obtained with a primary A⁺ ion beam of 5 kV, 15 nA.

231

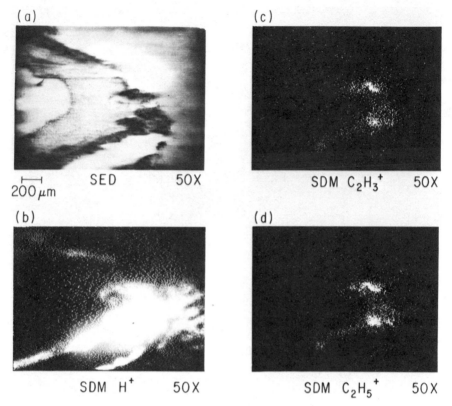

Figure 12. (*a*) Secondary electron (SED) micrograph of the aorta cross section. Scanning electron stimulated desorption micrographs (SDM) of the same sample area imaged with (*b*) H^+, (*c*) $C_2H_3^+$, and (*d*) $C_2H_5^+$ ion emission. The SDM images were scanned with 100 lines per frame over a period of 150 s.

graphs show that the predominant $C_nH_m^+$ and H^+ ion emission originates from the fatty connective tissue of the sample.

The intensity of the higher mass hydrocarbons in the ESD spectrum from the aorta sample, compared to the lack of such high mass peaks seen in the ESD spectrum of the soap film, presumably formed from a fatty acid, is puzzling. We hypothesize that the surface of the fatty connective tissue of the aorta is covered with loosely bound, fat-related oils. The low binding energies of these species, evident from their removal from rinsing in CH_3OH, would enhance the ESD desorption probabilities. These oil molecules, present in the adventitial tissue layer, provide a natural contrast mechanism for distinguishing muscle from connective tissue.

Comparison of the surface chemical information that is contained in the SDM images with the information contained in the carbon Auger image of the same sample (Fig. 10*d*) is interesting. Because of the higher signal intensity, the carbon Auger image provides more topographic information. However, the signal that is used to form the image (the intensity of the carbon Auger transition at 272 eV), shows none of the variations in carbon–hydrogen surface chemistry that occur across the fatty connective tissue–smooth muscle tissue boundary. It is unlikely that any of the subtle variations in Auger peak shape that are the result of different chemical states [47] can be used for enhancing the information content of Auger images. Except for several special cases where large Auger peak energy shifts occur (e.g., with the oxidation of certain metals [48]), the use of Auger peak intensities in Auger imaging is generally limited to providing the spatial variation of the elemental composition.

9.5.2 SESD: Red Blood Cells

To demonstrate the capabilities of this technique at more interesting subcellular dimensions, individual red blood cells (RBC) were chosen as the sample for the second set of SESD micrographs. The samples were prepared from whole sheep blood that had been preserved with EDTA. A blood smear was prepared on a silicon substrate, allowed to air-dry briefly, and then mounted in the microprobe without further preparation. The sample was examined in the standard secondary electron mode to find an area of the sample where the individual red blood cell density was small (i.e., 1–10 cells per 5000× field of view). Figure 13 shows a high

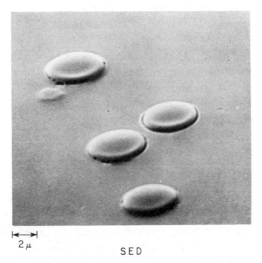

$\overset{\longleftrightarrow}{2\,\mu}$

SED

Figure 13. Secondary electron micrograph of sheep red blood cells deposited onto a silicon substrate.

magnification (5000×) secondary electron micrograph of such an area of the sample. The major axis diameter of an individual red blood cell is seen to be $5.0 \pm 0.5 \ \mu$m.

After secondary electron imaging of the RBC sample, the electron beam energy was lowered to 2 kV, the beam current increased to 1 μA, and the raster area increased to obtain an ESD spectrum (Fig. 14a) averaged over a larger area of the sample (25 μm)^2. The spectrum shows the ubiquitous H$^+$ and H$_2$O-fragment ions, O$^+$, OH$^+$, and H$_3$O$^+$. However, in comparison to the previously examined biologic surfaces, certain distinguishing characteristics are seen: strong peaks corresponding to Na$^+$ and Cl$^+$ and a peak at 14 amu, which could be CH$_2^+$ or N$^+$. Further, some fraction of the larger than expected 17 amu peak may be due to NH$_3^+$, which suggests that N$^+$ may indeed be part of the peak at 14 amu.

A SIMS spectrum taken with 2 kV Xe$^+$ ions scanned over a similarly sized area of the RBC sample (Fig. 14c) also shows strong saline-related peaks (Na$^+$, Cl$^+$, Na$_2^+$). Unfortunately, the hydrocarbon (C$_n$H$_m^+$) fragmentation spectrum, which has interferences at higher masses (silicon at C$_2$H$_n^+$, and K$^+$, Ca$^+$ at C$_3$H$_n^+$) lends no additional evidence for the ambiguity between CH$_2^+$/N$^+$ and OH$^+$/NH$_3^+$ in the ESD spectrum. The process of obtaining a SIMS spectrum from the RBC sample did reveal an interesting property of the ESD-active species responsible for the ion peaks appearing in the original RBC ESD spectrum (Fig. 14a). After the spectrum of Fig. 14a was obtained, the RBC sample was lightly sputtered with 2 kV Xe$^+$ ions. Assuming that the sputtering rate of pure carbon is applicable, the incident Xe$^+$ dose would have removed the equivalent of 3–4 atomic layers ($\approx$2 nm). Figure 14b shows the ESD spectrum obtained from the RBC sample after this light sputtering. All peaks except H$^+$ have disappeared and the H$^+$ is severely attenuated, an observation that suggests the ESD-active species are surface species (adsorbates) and not predominantly decomposition products of the biologic substrate.

Fresh RBC samples were prepared for the attempts at SESD micrographs of single cells. Figure 15a shows a secondary electron image (SED) at 2000× of a group of red blood cells and a SESD micrograph of the same sample area imaged with desorbing H$^+$ ions (200 lines per frame with a 150 s frame period). The SESD micrograph was taken before the SED image in order to minimize the electron exposure. As a result, the SESD image actually shows the oblate shape of the red blood cells more clearly than the Fig. 15 SED image. Through the course of several exposures that were required for adjustment of the electron beam conditions for the SED image, the red blood cells became distorted, probably as the result of electron beam damage.

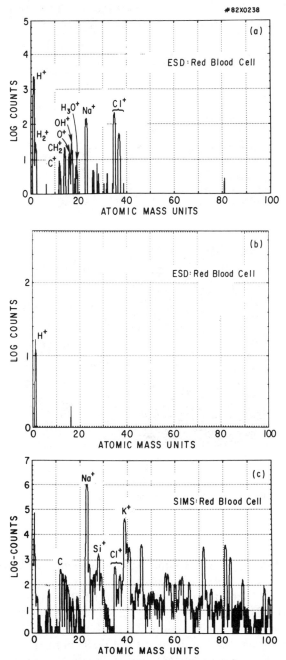

Figure 14. (*a*) ESD ion spectrum from the red blood cell sample with a 2.0-kV, 1-μA electron beam. (*b*) ESD spectrum from the RBC sample after light sputtering with Xe$^+$ ions. (*c*) SIMS spectrum from RBC sample obtained with a primary Xe$^+$ ion beam of 2 kV, 10 nA.

235

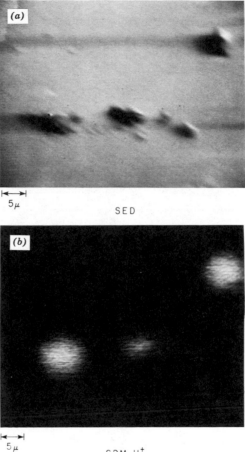

Figure 15. (a) Secondary electron (SED) micrograph of a group of red blood cells, (b) scanning electron stimulated desorption micrograph (SDM) of the same sample area imaged with desorbing H^+ ions. The SDM image was scanned with 200 lines per frame over a period of 150 s.

Figure 16 shows a SESD H^+ micrograph of a single red blood cell from a new area of the sample, along with line scans across the approximate center of the cell that show the variation in ESD ion intensity (displayed linearly) for H^+, Na^+, and H_3O^+ ions. The line scans show that the spatial resolution is definitely subcellular and probably better than 1 μm for these scanning conditions, H^+ ion intensities of ~80 counts per image point of 0.2 μm width. A more accurate estimate of the resolution attained in these micrographs would require the measurement of the variation in ion intensity across a known concentration gradient. It is interesting to speculate

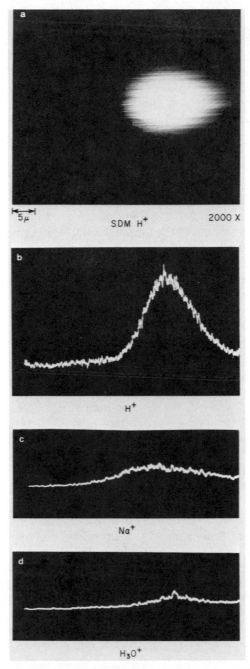

Figure 16. (*a*) Scanning electron stimulated desorption micrograph (SDM) of a single red blood cell imaged with desorbing H⁺ ions. Diametral line scans of the (*b*) H⁺, (*c*) Na⁺, and (*d*) H₃O⁺ ion emission across the cell surface.

237

on the cause of the observed concentration profile of ESD ions, which is at a maximum at the center of the scan across the cell. Since whole blood (rather than washed cells) was used to prepare these samples, the ESD active species (Na, Cl, H_2O, NH_x, etc.) are likely components of the blood plasma that coalesced and adsorbed onto the cell surface as this liquid component dehydrated in the microprobe vacuum system.

9.5.3 Future Studies

Biologic surfaces present interesting problems for further study by SESD. One area of intensive investigation is the composition of cell surfaces, which are implicated in diverse functions. For example, surface properties of cells are important in the binding of hormones to hormone receptor complexes [49]. Cell membranes represent important means for manipulating cell function with pharmacologic agents [50]. Activation of neutrophils by various stimuli is mediated both by direct stimulation of the neutrophil membrane and by indirect ligand-receptor stimulation. Stimulation appears to produce changes in the distribution of membrane-associated calcium [51]. The pathologic behavior of malignant tumor cells, which includes continued proliferation, altered adhesion to other cells, and different migratory properties, is mediated by surface changes. Migration of neoplastic cells and invasion by malignant cells require alteration of the relationship between the cell surface and the extracellular environment [52]. Indeed, because cellular surface structure is involved in those altered physiologic properties related to neoplastic behavior, such as cell growth, division, differentiation, and immunogenicity, Nicolson suggests that analysis of the cell surface would allow assessment of the efficacy of specific molecular therapy in neoplastic diseases [53].

SESD appears capable of providing surface-specific information about normal and pathologic cells. As described in Section 4.2, the use of biologically compatible molecules as surface stains to enhance surface contrast and, if related to substances that normally bind to transport sites and receptor complexes, to label physiologically important sites would have great utility. An interesting surface stain molecule would be D_2O, in which cells have been incubated, to identify those aspects of cellular physiology that have surface expression. The use of labeled antibodies, directed against a cell surface component of interest, might demonstrate physiologically important areas of cell surfaces. With the antibody technique, one might use a label (e.g., bromine) bound noncovalently to those amino acids comprising the antibodies that contain protonated amino groups, lysine and arginine. Bromine should be detected with sensitivity comparable to chlorine (compare ESD of sheep RBC in Fig. 14a) at a mass where few competitive species are expected. Evidence for the feasibility of such methods comes from several investigations. From work with sulphonated stilbene compounds, with high affinity and specificity for erythrocyte anion transport system binding sites, an estimated $(0.4–4) \times 10^6$ sites exist per cell [54]. Studies using a radiolabeled antigen–antibody inhibition assay demonstrate that mouse spleen cells contain a comparable number $(0.5–1) \times 10^5$ of an entirely different surface component, surface immunoglobulin [55]. A simple estimate suggests that 3×10^4 Br labels can be expected per square micrometer of cell surface, a number that should be within the presently demonstrated limits of detection by SESD.

ACKNOWLEDGMENTS

The authors wish to acknowledge the financial and technical support of these studies by the Physical Electronics Division of the Perkin-Elmer Corporation. The able assistance of C. Hovland, B. Phillips, and A. Hirt of Perkin-Elmer is gratefully acknowledged. This work was inspired by the previous biological studies of Prof. J. G. King and his colleagues. The authors thank D. Manos and S. Cohen for useful discussions and G. Smith for preparing the manuscript.

REFERENCES

1. N. C. MacDonald, C. T. Hovland, and R. L. Gerlach, in *Proc. 1977 Conf. on Scanning Electron Microscopy,* Vol. 1, IIT Research Institute, Chicago, 1977, p. 201.

2. T. E. Madey and J. T. Yates, J. Vac. Sci. Technol. *8,* 525 (1971).

3. D. Menzel, Surface Sci. *47,* 370 (1975).

4. G. Rork and R. E. Consoliver, Surface Sci. *10,* 291 (1968).

5. D. Lichtman and J. Campuzano, Japan. J. Appl. Phys. Suppl. *2,* 189 (1974).

6. H. F. Dylla and J. G. King, Bull. Am. Phys. Soc. *21,* 241 (1976).

7. A. Joshi and L. E. Davis, J. Vac. Sci. Technol. *14,* 1310 (1977).

8. H. Poppa and E. Bauer, Surface Sci. *97,* L309 (1980).

9. L. A. Larson, F. Soria, and H. Poppa, J. Vac. Sci. Technol. *17,* 1364 (1980).

10. H. F. Dylla, J. H. Abrams, C. T. Hovland, and J. G. King, Nature *291,* 401 (1981).

11. P. B. DeNee, R. G. Frederickson, and R. S. Pope, in *Proc. 1977 Conf. on Scanning Electron Microscopy,* Vol. II, IIT Research Institute, Chicago, 1977, p. 83.

12. A. J. Dempster, Phys. Rev. *11*, 316 (1918).

13. G. E. Moore, J. Appl. Phys. *32*, 1241 (1961).

14. D. A. Degras, L. A. Petermann, and A. Schram, Nat. Symp. Vacuum Technol. Trans. *9*, 497 (1962).

15. L. A. Petermann, Suppl. Nuovo Cimento Ser. I, *I*, 601 (1963).

16. P. A. Redhead, Can. J. Phys. *42*, 886 (1964).

17. D. Menzel and R. Gomer, J. Chem. Phys. *40*, 1164 (1964).

18. D. Lichtman and R. B. McQuistan, Prog. Nucl. Energy Ser. IX, *4*, 95 (1965).

19. A. W. Jones, E. Jones, and E. M. Williams, Vacuum *23*, 227 (1973).

20. A. G. Mathewson, in *Proc. Int. Symp. on Plasma Wall Interactions,* Pergamon, Oxford, 1977, p. 917.

21. A Mathewson and M. H. Achard, in *Proc. 7th Int. Vac. Congress and 3rd Int. Conf. on Solid Surfaces,* R. Dobrozemsky, F. Rudenauer, F. P. Viehbock, and A. Breth, Eds., Vol. II, Vienna, 1977, p. 1217.

22. S. A. Cohen and H. F. Dylla, J. Vac. Sci. Technol. *14*, 559 (1977).

23. R. Stockbauer, E. Bertel, and T. E. Madey, J. Chem. Phys. *76*, 5639 (1982).

24. T. E. Madey and J. T. Yates, Surf. Sci. *76*, 397 (1978).

25. M. L. Knotek, "Electron and Photon Stimulated Desorption," in *Analysis of Organic and Biological Surfaces,* P. Echlin, Ed., Wiley-Interscience, New York, 1984.

26. M. L. Knotek and P. J. Feibelman, Phys. Rev. Lett. *40*, 964 (1978).

27. P. R. Antoniewicz, Phys. Rev. *B21*, 3811 (1980).

28. PHI Model 590 Scanning Auger Microprobe, manufactured by the Physical Electronics Div., Perkin-Elmer Corp., Eden Prairie, Minn. See Ref. 1 for specifications.

29. B. F. Phillips and R. L. Gerlach, in *Microbeam Analysis,* D. B. Wittry, Ed., San Francisco Press, San Francisco, 1980, p. 85.

30. A. Benninghoven, Surf. Sci. *35*, 427 (1973).

31. H. Niehus and E. Bauer, Rev. Sci. Instrum. *46*, 1275 (1975).

32. W. E. Austin, A. E. Holme, and J. H. Leck, in *Quadrupole Mass Spectrometry and Its Applications,* P. H. Dawson, Ed. Elsevier, Amsterdam, 1976, p. 121.

33. H. S. W. Massey and E. H. S. Burhop, *Electronic and Ionic Impact Phenomena,* Oxford, London, 1952.

34. D. A. Degras and J. Lecante, Nuovo Cimento, Suppl. *5*, 598 (1967).

35. W. Jeland and D. Menzel, Surface Sci. *42*, 485 (1974).

36. H. D. Hagstrum, Phys. Rev. *119*, 940 (1960).

37. D. D. Pretzer and H. D. Hagstrum, Surf. Sci. *4*, 265 (1966).

38. O. P. Leta and J. H. Morrison, Anal. Chem. *52*, 514 (1980).

39. A. Benninghoven, D. Jaspers, and W. Sichtermann, Appl. Phys. *11*, 35 (1979).

40. A. Benninghoven and W. K. Sichtermann, Anal. Chem. *50,* 1180 (1978).

41. H. Kambara and S. Hishida, Org. Mass Spectrom. *16,* 167 (1981).

42. L. K. Liu, K. L. Bush, and R. G. Cooks, Anal. Chem. *53,* 109 (1981).

43. K. B. Blodgett, J. Am. Chem. Soc. *57,* 1007 (1935).

44. G. I. Loeb and R. E. Baier, J. Colloid Interface Sci. *27,* 38 (1968).

45. R. J. Colton, J. S. Murday, J. R. Wyatt, and J. J. DeCorpo, Surface Sci. *84,* 235 (1979).

46. Lab-Tek Products, Napierville, Ill.

47. D. R. Jennison, J. Vac. Sci. Technol. *17,* 172 (1980).

48. L. E. Davis, N. C. MacDonald, P. W. Palmberg, G. E. Riach, and R. E. Weber, *Handbook of Auger Electron Spectroscopy,* Perkin-Elmer Corp., Eden Prairie, Minn., 1976.

49. M. D. Hollenberg, Pharmacol. Rev. *30,* 393 (1979).

50. V. T. Marchesi, Pharmacol. Rev. *30,* 371 (1979).

51. G. Weissman, J. E. Smolen, and H. M. Korchak, New Eng. J. Med. *303,* 27 (1980).

52. R. O. Hynes, Biochem. Biophys. Acta *458,* 73 (1976).

53. G. L. Nicolson, Biochem. Biophys. Acta *458,* 1 (1976).

54. A. L. Hubbard and Z. A. Cohn, in *Biochemical Analysis of Membranes,* A. H. Maddy, Ed., Wiley, New York, 1976, p. 427.

55. E. Rabellino, S. Colon, H. M. Gray, and E. R. Unanue, J. Exp. Med. *133,* 156 (1971).

OPTICAL MICROSCOPY
IN THE NEAR INFRARED

BARRY G. COHEN

Research Devices, Inc.
Berkeley Heights, New Jersey

10.1 INTRODUCTION

Infrared microscopy is a specialized branch of optical microscopy. It deals with the magnification and observation of small objects, either illuminated by or emitting radiation in the near infrared (750–1200 nm) [1]. In this sense, IR microscopy is different from thermography [2], which is the detection and measurement of thermal (blackbody) radiation emitted from warm objects, and different from IR photography [3], which uses sensitive films or papers but does not produce a "real-time" image. IR microscopy does, however, allow the measurement of temperature in certain cases, and normally allows photographic recording of images. The applications of IR microscopy fall into four primary classes:

1. Study of materials that are opaque in the visible but transparent in the near IR (e.g., semiconductors).

2. Study of materials that are transparent in the visible, but show absorptions or other contrasts in the near IR (living organisms, blood cells, etc.).

3. Study of materials that absorb visible energy and re-emit it as fluorescence or luminescence in the IR (chlorophylls, impurities in crystals, etc.).

4. Study of materials that are photosensitive in the visible, but not in the IR (photoreceptor cells, photographic materials, photosynthetic organisms, etc.).

In almost every case the use of IR microscopy is indicated by a special or unusual property of the system or material under study. The theory and procedures of normal visible microscopy generally apply, except for the requirements of the longer wavelengths, and the necessity for conversion to a visible image for the observer's use.

The limiting contrast and resolution of an infrared microscope are also proscribed by the laws of optics and the particular material under study. Since IR microscopy is only used when some material or system property requires its use, it is generally fruitless to attempt direct comparisons. The longer wavelengths used may reduce the ultimate resolution of the instrument by as much as a factor of 2. However, the special properties of infrared may enhance the visibility of surface and internal structures. Thus the net result of the use of IR microscopy is often a striking increase in visibility.

10.2 MICROSCOPE STRUCTURE

As shown in Fig. 1, an infrared microscope contains an objective system that operates in a manner similar to visible objective systems. Two separate illumination systems are utilized, one for light transmitted through the sample and the other for light reflected from the sample. There is a conversion system that takes the IR image and converts it into a visible image, and a viewing or image presentation system that may include photography, television display, and so on. The microscope also includes the various necessary focusing and adjusting components, such as mechanical stage, diaphragms, filters, condensers, and so on.

In addition to normal bright field and dark field illumination, the infrared microscope is frequently equipped with infrared polarizers in the illuminator and a polarizing analyzer in the image converter system. This allows straightforward observation of infrared birefringence due to elastic strain or other nonuniformities in materials.

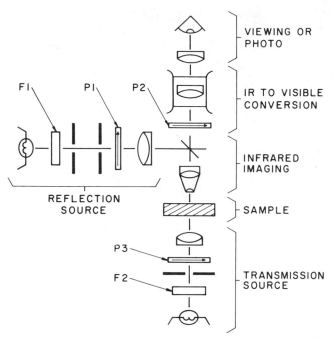

VIEWING OR PHOTO

IR TO VISIBLE CONVERSION

INFRARED IMAGING

REFLECTION SOURCE

SAMPLE

TRANSMISSION SOURCE

Figure 1. Basic components of the infrared microscope showing the major functional groups, including infrared filters F1 and F2, and polarizing elements P1, P2, and P3.

It is useful to have a working knowledge of the wavelength ranges that are accessible and of interest for various materials. The normal visible range is 400–750 nm. The image converter tubes used in IR instruments have a spectral response that extends through this range and beyond, to about 1200 nm in the IR. These instruments use internal filters to remove the visible light, thus allowing operation in the range 750–1200 nm. This is a wider range of wavelengths than is normally visible to a human eye, and a great many interesting optical phenomena exist in this range.

10.3 APPLICATION TO BIOLOGICAL AND GEOLOGICAL SAMPLES

10.3.1 Photosensitive Organisms

The infrared microscope allows many different studies to be performed, particularly in the area of photosensitive materials. Since many living

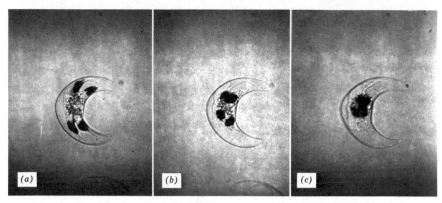

Figure 2. A marine organism, *Pyrocystis lunula,* photographed in the chlorophyll red absorption band at 673 nm. (*a*) A "dark-adapted" cell whose chloroplasts are distally contracted into the extreme ends of the spindle-shaped plastids. (*b*) After 5 min exposure to 500 nm at 100 μW/cm^2. the chloroplasts have moved in a contracted mass toward the center of the cell. (*c*) Continued exposure ultimately contracts the entire chlorophyll content into a single mass at the cell center.

organisms are light sensitive, they cannot be observed with an ordinary microscope when they are in their dark phase. Furthermore, if their responses to light are rapid and depend on the wavelength of excitation, the observation of a particular structure or behavior may be an artifact of the microscope illumination, rather than a characteristic of the system or organism.

The response of the photosynthetic and photoluminescent marine organism *Pyrocystis lunula* to visible light excitation has recently been measured by use of an infrared microscope [4]. The action spectrum for chloroplast motion, as a function of excitation wavelength, indicates which component of the organism triggers the observed reaction.

Figure 2 shows the behavior of the marine alga *P. lunula* while observed in the infrared. The "dark phase" (Fig. 2*a*) is not observable for more than a very short interval under visible light, since the cell rapidly transforms to the "light phase" (Fig. 2*c*) when illuminated. Using the IR microscope, accurate measurements of the velocity of chloroplast motion can be made.

10.3.2 Infrared Fluorescence

The identification or location of certain compounds that exhibit fluorescence is a common technique. Normally, visible fluorescence is performed by exciting a sample in the near ultraviolet and observing the visible emission that occurs. Infrared fluorescence is the same technique except the excitation is done with visible light, and the IR microscope is used to detect fluorescence in the IR spectrum. Many materials that do not fluoresce in the visible, do so in the infrared. This allows simple determination of their presence and amount.

Figure 3 is a freshwater diatom *Melosira* that shows very little contrast in visible light but emits strongly in the infrared. Since fluorescence is a "dark-field" procedure, it is easy to determine the presence of the fluorescent material and to measure the amount of emission under standard excitation conditions.

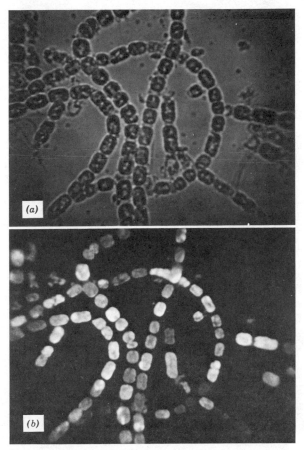

Figure 3. (*a*) The diatom *Melosira* viewed with normal transmitted visible light. Intracellular contrast is low. (*b*) The same cells in IR fluorescence. Nonuniform emission and high contrast is observed.

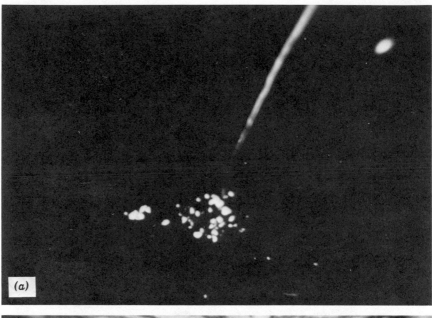

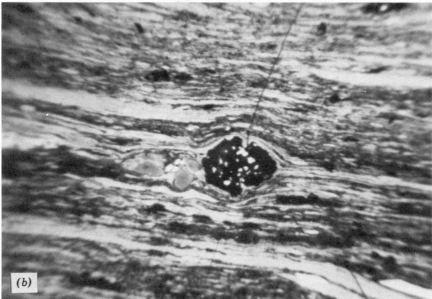

Figure 4. (*a*) A section of medium volatility bituminous coal is opaque when viewed in visible light. (*b*) Transmitted infrared. The sample becomes very transparent allowing determination of included spores, algae, and vitrinite materials. IR fluorescence may also be observed. (200×; wavelengths, 605–1200 nm.)

248

10.3.3 Sediments and Coal

Infrared transmission, reflection, and fluorescence measurements have been well known in the study and characterization of peats, lignites, and coals [6]. The infrared microscope shows in great detail the structure and carbonization of materials. The advantage of infrared microscopy for these studies lies in the relatively greater transparency of their organic compositions to the IR. Thus relatively thick sections can be examined with much less loss of the carbonized components. Furthermore, the infrared fluorescence can be observed directly on bulk or powdered samples.

Figure 4 is a coal sample observed in visible and infrared transmission. The presence of the carbonaceous material prevents any significant visible transparency but the infrared transmission is much higher, and considerable detail is observed.

10.4 APPLICATIONS TO SOLID STATE DEVICES

Some difficult problems occurring in solid state materials and devices may be simply analyzed with the IR microscope. Defects in semiconductor devices can cause low yields or premature failures. These defects may be associated with internal flaws in the material or with faulty processes, such as diffusions or metal-coating layers [7,8]. Defects that intersect a surface may be covered by layers of metal coating. Coating layers themselves may suffer from problems such as pinholes or nonuniform thickness.

Most of the problems are undetectable by visible light microscopy or by scanning electron microscopy, as the materials are opaque to the radiation employed.

By operating the IR microscope at wavelengths just longer than the absorption edge of the host materials (1100 nm for Si, 950 nm for GaAs), internal defects and structures may be directly observed.

10.4.1 Precipitates and Decorated Defects

Precipitates, being localized defects, often exist completely immersed in a material. Since the host material is transparent to the IR radiation, precipitates can readily be observed. Usually the orientation and morphology of the precipitate can be studied allowing an experienced investigator to determine its composition.

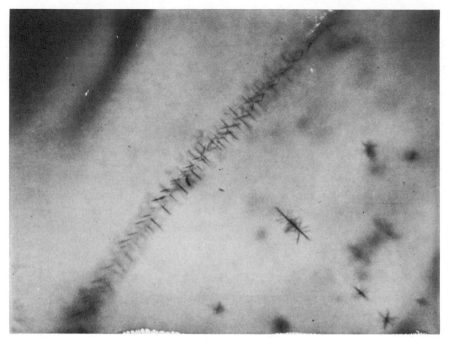

Figure 5. Copper decorations in silicon. (BAR = 50 μm). Material defects can be inferred from the location of the copper precipitates.

A special case of this kind of analysis involves the intentional precipitation of a heavy metal such as copper, gold, or nickel on an otherwise invisible defect such as dislocation or a stacking fault in silicon [9]. Copper precipitates, a few micrometers in size, can be readily formed. The location of a dislocation can be inferred from the array of copper precipitates that form along it (Fig. 5). When using the infrared microscope, the line of precipitates may be following laterally and in depth all the way through the material. This procedure allows the location of defects, as delineated by the precipitates, to be correlated with devices that have been made on the silicon wafer.

10.4.2 Strain Birefringence Centers

Isolated strain patterns may occur whenever a localized defect exists in a material. The strain patterns shown in Fig. 6 were found in a wafer of bulk silicon—prior to any device diffusions. Each of the strain patterns is apparently associated with a defect in the silicon, but at most of the birefringence centers nothing is visible in normal transmission. Either the

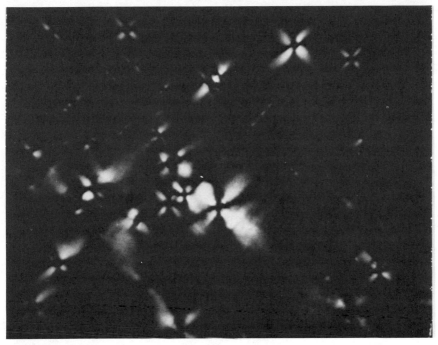

Figure 6. Transmitted IR with crossed polarizers (100×) reveals strains in the vicinity of point defects immersed in a silicon slice.

defects are too small to be resolved or they consist of material with no infrared absorption and very little optical discontinuity. The observations in Fig. 6 were made in transmitted, polarized infrared.

10.4.3 Microcracks

In the manufacture of semiconductor devices, one of the last operations usually involves the separation of a wafer into individual chips. The separation is usually done by scribing and breaking the slice. Whether the scribing is done with a laser, a diamond point, a saw blade, or any other technique, small microcracks may be generated around the edges of the chips. At some later time, a microcrack may propagate across the chip, causing it to fail.

Even when a microcrack extends to the surface of a chip, it will be very difficult to observe. The trace of the plane of the microcrack in the device surface will be a very fine line. It may be obscured by metal-coating layers or other surface patterns. In infrared transmission, the whole projection of the crack plane will be observable, making it straightforward to detect and identify these defects.

10.4.4   Metal-Coating Defects

Using ordinary visible light and metallographic techniques, it is almost impossible to detect small changes in the thickness of a metal layer, or small defects such as pinholes or thin spots. Since the semiconductor material is quite transparent in the IR, and the IR transparency of typical metals, such as aluminum or gold on silicon, is frequently greater than their visible light transparencies, these defects can easily be found in transmission.

Pinholes in an otherwise opaque metal layer are easily detected, even at low magnification. Although a pinhole diameter may be considerably smaller than a wavelength, making it impossible to find by reflection techniques, a relatively large amount of energy can still diffract through the defect during transmission.

The diffraction effect is particularly noticeable when the defect takes the form of thin lines where the metal is coated over an oxide step on the devices. The thin line at the step acts as a slit that passes appreciable infrared energy. The location of the defect is easily found.

10.4.5   Die Attach Bonds

When a completed device chip is bonded to a metal or metallized ceramic header, uniformity of the contact between the semiconductor and the metal is very important [12]. This is a critical problem in power devices where the best possible thermal conduction is required between the chip and the heat sink. X-ray examination is difficult since the x rays must penetrate the header and the chip, and the thin contact region shows very little contrast.

By using the IR microscope with special filters in the reflection illuminator, the interface between the semiconductor and the die attach metallization may be examined directly and nondestructively. In the case of a gold–silicon eutectic bond, for example, it is often possible to observe not only the uniformity of the bond, but the grain size of the regrowth region. In cases of nonuniform bonding, one would expect that considerable elastic strain would exist in the chip, which could cause leaky junctions or even catastrophic mechanical failure. This strain may be observed when it exists by polarizing the reflection source and adding a crossed analyzer in front of the image converter tube.

An example of this technique is shown in Fig. 7. A mounted silicon transistor is observed in reflected infrared (Fig. 7a). The surface metalli-

Figure 7. A silicon transistor, gold eutectic bonded to a metal header. (BAR = 0.2 mm). (*a*) Reflected, penetrating, infrared; the regrowth of the Au–Si bond is observed in the left portion of the chip. (*b*) Crossed polarizers; the bright areas correspond to highly strained silicon.

zation and diffusions are slightly out of focus, since the image plane of the IR microscope is set at the back of the chip. Under cross polarizers (Fig. 7b) the bright, highly strained regions, associated with the large-grained regrowth areas of the Au–Si eutectic bond, may be easily seen.

10.4.6 Beam Leads and "Flip Chips"

The most recent technique of mounting semiconductor devices involves making direct contact between beam leads or solderable pads on the face

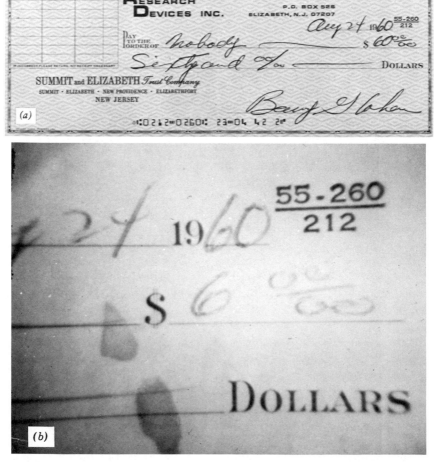

Figure 8. (a) Visual appearance of an altered check. (b) The same check under the IR microscope shows the great contrast of the two inks used.

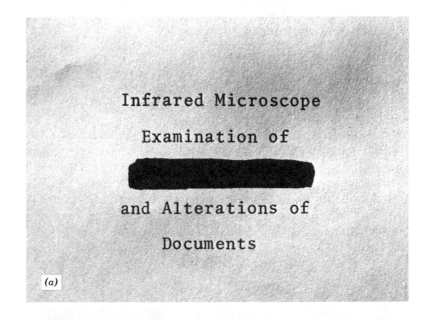

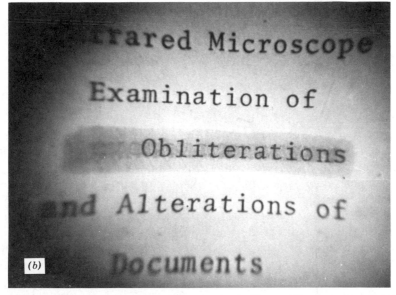

Figure 9. (*a*) Visual appearance of a document that had been obliterated with a dark black ink. (*b*) The obliterating ink is transparent in the infrared and the underlying writing can easily be read.

of the chip and matching connecting patterns on the header [13]. In both cases the chips are mounted with their active faces toward the header. Examination of the chips after bonding can be readily performed in the infrared.

10.5   APPLICATION TO FORENSIC STUDIES

Infrared examination of forensic materials is an established technique [14]. The IR microscope makes these investigations rapid and convenient.

One major application of IR microscopy concerns the examination of documents for alterations or obliterations. The reading of bleached or burned documents, the detection of gunpowder particles embedded in colored or blood-stained clothing, and similar investigations are well within the normal range of use of the IR microscope.

The document shown in Fig. 8 is an example of the use of IR microscopy for determining that a check had been altered. The characteristic of the inks used to originate and to alter the check are such that they appear identical in the visible, but very different in the infrared.

In addition to alteration detection, IR microscopy is often used to read obliterated writing as shown in Fig. 9; obliterations can take the form of an overlay of pigment as in the figure, or can be a chemical process such as burning or bleaching. Even in those cases, the original writing can often be detected and deciphered.

10.6   CONCLUSION

IR microscopy is a generally useful procedure that finds wide use in many diverse fields. It is a normal extension of visible light microscopy into a wavelength range where the human eye is insensitive.

Many naturally occurring materials have extremely interesting properties in the near infrared that may be observed in an IR microscope. As these instruments find wide use in such diverse fields as semiconductor devices, biochemistry, criminalistics, geology, paleontology, ophthalmology, and so on, we may expect a rapid expansion in the number of known applications.

REFERENCES

1.   B. G. Cohen, U.S. Patent No. 3,624,400.
2.   R. F. Leftwich, The Infrared Microimager, Proc. SPIE *104*, 104 (April 1977).

3. W. Clark, *Photography by Infrared,* 2nd Ed., John Wiley and Sons, New York (1946).

4. B. G. Cohen, Action Spectrum for Chloroplast Migration in the Marine Dinoflagellate *Pyrocytis lunula,* Johns Hopkins University, Baltimore, Maryland (1976).

5. E. Swift and W. R. Taylor, Bioluminescence and Chloroplast Movement in the Dinoflagellate *Pyrocytis lunula.*

6. W. D. Ian Rolfe, Uses of Infrared Rays; *Handbook of Paleontological Techniques,* pp. 344–350, W. H. Freeman and Company, San Francisco (1965).

7. C. G. Peattie et al., Proc. IEEE 62:149 (Feb. 1974).

8. G. L. Schnable and R. S. Keen, *Advances in Electronics and Electron Physics,* Vol. 30, Academic Press, N.Y. (1971).

9. W. C. Dash, J. Appl. Phys. *27,* 1193 (1956).

10. J. S. Ahearn, et al., Phys. Stat. Sol (a) 38:315 (1976).

11. J. E. Johnson, Die Bond Failure Modes, IEEE Proc. on Device Reliability (1974).

12. B. G. Cohen, Infrared Microscopy for Evaluation of Silicon Devices and Die Attach Bonds, Proc. SPIE, Vol. 104, p. 125 (April 1977).

13. M. P. Lepselter, B.S.T.J. 45 (1966).

14. D. A. Crown, *The Forensic Examination of Paints and Pigments* Charles C. Thomas, Springfield, Illinois, pp. 176–178 (1968).

SECONDARY ION MASS SPECTROMETRY

MARGARET S. BURNS

Departments of Ophthalmology and Biochemistry
Albert Einstein College of Medicine
Montefiore Medical Center
New York, New York

11.1 INTRODUCTION

Secondary ion mass spectrometry (SIMS) is, in a sense, the intellectual offspring of the electron probe since Georges Slodzian, who designed one of the first SIMS instruments, was the student of Raymond Castaing, credited with designing one of the first electron probes. SIMS is not well known in the biological field but has been used extensively in the electronics industry for elemental analysis of semiconductors, in geochemistry for analysis of isotopic composition of minerals, and in metallurgy for compositional mapping. Although studies on ion bombardment physics had been carried out for many years, the use of ion bombardment as an analytical tool dates from the mid- to late 1960s when commercial instrumentation designed by Liebl or Slodzian became available [1–3]. Today there are several available commercial instruments as well as instruments designed for specific use in individual laboratories. Since most of the biological applications have been made using a direct imaging instrument of Slodzian's design, this chapter will use the Cameca IMS 3f to illustrate the

This work was supported by NIH Grant #EY 02093 and Research to Prevent Blindness, Inc., New York.

technique and applications. Most of the early biological applications of SIMS came from Galle's laboratory and are referenced in a recent review [4]. The IMS 3f is a second-generation instrument and has the most sophisticated technical specifications of the commercially available instruments.

11.2 ANALYTICAL TECHNIQUE

Secondary ion mass spectrometry is so named because a beam of ions (the "secondary" ions) are caused to erupt ("sputter") from the sample surface that is being bombarded by a focused beam of ions (the "primary" ions). The energetics are such that some of the atoms that are removed are ionized as either positive or negative species. One can think of SIMS as bombardment by a cannonball as compared to the stiletto-like penetration of electrons in an electron probe. Two factors that determine the number of ions that may be measured is the rate at which atoms are sputtered from the specimen surface, the *sputtering yield,* and the amount of ionization of the sputtered atoms, the *ionization yield.* Unlike the physics of the electron probe, there is as yet no comprehensive theory that can be used to predict the energy exchanges occurring between the primary ion beam and the sample surface. Recent work by Garrison et al. [5] uses computer modeling of bombardment of crystalline metal systems, in which the bond distances and energies are well known, to illustrate some of the collision processes that may occur. The primary ion can penetrate the specimen and interact with atoms beneath the surface, may be trapped within the specimen, may initiate collision cascades of specimen atoms to combine with each other and to ionize, forming ions from more than one atomic layer (Fig. 1). Although the collision processes in organic and biological materials are undoubtedly similar, the absence of large-scale order makes it more difficult to predict either sputtering or ionization yields.

Empirically, however, it is known that there are many interrelated factors that influence both the sputtering and the ionization yield [1]. I will mention some of these factors and relate them to analysis of biological materials. The sputtering yield (the number of atoms removed per incident ion) increases with increasing primary ion energy up to a plateau that is characteristic of the ion–target combination. In commercial instruments the primary ion beam may be varied from about 5–20 keV, although in practice a fixed ion energy is used for analytical purposes. The sputtering yield is also a function of the atomic number of the bombarding ion, reaching a maximum for the noble gas specie in each group of elements,

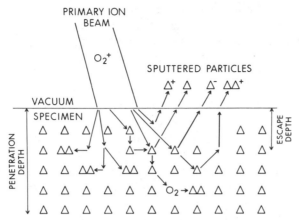

Figure 1. Schematic diagram of possible primary ion interactions with atoms of specimen.

and gradually increasing as the atomic number increases. However, the chemistry of the incident ion may affect the ionization yield. For example, oxygen bombardment is routinely used to enhance positive secondary ion yields, and cesium bombardment is used to enhance negative secondary ion yields [2]. Whether or not oxygen enhances ion emission from biological specimens is not known. Since tissue specimens have such a high intrinsic oxygen content, it is not clear whether there would be enhancement of ion yield by oxygen as compared with argon bombardment. In fact, almost all the studies of biological specimens have used positive oxygen bombardment, perhaps for the same reason I did—to optimize ion yield of what was expected to be an insulating sample with charging problems (*vide infra*). In recent work on organic material, noble gas bombardment has been used to optimize available chemical information in the spectra [6–9]. The nature of the target material also influences the sputtering yield and there is a rough correlation that relates increased sputtering yield to the number of electrons in the atomic d shell or to the heat of sublimation, which is also related to electronic configuration [1]. Whatever the controlling parameters, it is clear that the chemistry of the specimen being bombarded is a factor in both sputtering yield and ionization yield.

This dependence of the sputtering and ionization processes on the electronic and chemical properties of the material being analyzed is usually covered by the umbrella term of "matrix effects." First, the ionization of any given atom is related to its ionization potential as a positive ion or its electron affinity if discussing a negative ion. Therefore there is an intrinsic difference in the ability of any given atom to ionize, and this is further modulated by the interactions of the primary ions and adjacent atoms and the energy (work function) of the specimen surface the ions must overcome in order to escape. These differences can give rise to ion yields for different atoms that vary as much as four orders of magnitude, but are less dramatic for elements in an insulating matrix [2]. There is little evidence of a matrix effect in biological systems, but there are some reasons why variations in ion yields may not be found. One is that biological specimens are uniform in concentration of the major elements present: C, H, N, and O. Therefore they provide a relatively uniform chemical matrix for the elements of interest such as Na, K, Ca, Mg. These elements are present in highest concentration in tissue and this is only about 0.2% by weight. Most of the matrix effects that have been studied involve large variations in local concentrations of a specie, such as the difference in aluminum emission from pure aluminum or from an oxide. Since the oxygen content of tissues is high and relatively uniform and to the extent that this enhances and stabilizes ion yields, it should decrease a "matrix effect" in tissues. A common cause for ion emission that is not proportional to concentration is when different crystal faces are examined, since the work function varies depending on the atomic density of the face. Since there are very few examples of true crystallinity in soft biological specimens, one would not expect to see "matrix effects" from this source. One report dealing with mineralization of egg shell noted differential carbon emission from the organic carbon matrix and the crystallizing apatite shell [10]. In this example there are gross surface and crystalline differences between the two areas, which may account for the anomalous emission. There may be chemical and electronic properties of biological specimens that give rise to anomalous ion yields, but they have not yet been reported.

Other operating parameters that affect the sputtering and ionization yields in addition to the energy, mass, and chemistry of the primary beam and the properties of the target itself are the angle of incidence and current density of the primary ion beam. Since the angle of incidence is not readily variable in commercial instruments, we will say only that it is important in controlling the characteristics of the collision cascade in surface layers. The primary ion current density is the critical determinant of the sputtering rate, the number of atoms removed per unit time. Since the sputtering rate is a property of the sputtering yield (atoms removed per incident ion), it is obvious that the greater the fluence (the number of ions arriving at the surface per unit time), the greater will be the rate of removal. In practice, biological samples have limited useful current densities due to local heating, combustion of the sample, and rate of destruction. Primary beam currents of about 100 nA are focused in a

Gaussian distribution to a 50 μm beam diameter, giving a current density of the order of 10^{-4} A/cm². This is an order of magnitude less than the current density routinely used to analyze electronic or metallurgic samples. However, it is still in the range of "dynamic SIMS" and above the very low fluences used to generate molecular organic ions by "static SIMS" in which the probability of more than one bombardment event in any spot is small [6].

The analysis conditions chosen will depend on the question being asked. A high sputtering rate may be appropriate to locate an element present at low concentration since more ions per unit time will be collected. However, the in-depth position of that element will be less clear than if a low sputtering rate were used. As always, resolution and sensitivity must be played off against each other. SIMS instruments operated with a fine primary ion beam may be properly called ion microprobes and can be thought of as analogous to an electron microprobe [1]. The direct imaging instrument is more difficult to understand because the area and point analysis are achieved by a combination of the ion optical system and mechanical apertures. A large beam of about 50 μm is rastered over an area as large as 500 μm, producing a crater as material is sputtered sequentially from the surface (Fig. 2). This is done in order to provide a

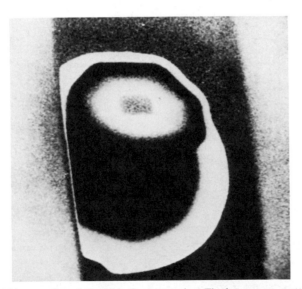

Figure 2. SEM image of a gelatin film after sputtering. The low power sputtered crater is seen as a rectangular dark area close to the center of the field. Differential secondary electron emission from the sample is probably due to effects of the primary beam on the surface.

large flat area in the center of the crater which will be analyzed. One can obtain anomalous secondary ions from sloping surfaces. This is because a critical factor is the evenness of the electrical field between the source of secondary ions (the specimen surface) and the first lens, the immersion lens (Fig. 3). A distortion of the field here can deflect ions totally, or skew them so that the apparent location is misleading. The entire unfiltered secondary ion beam is directed through the transfer optics, which enables the ions to be directed to the entrance slit of the electrostatic analyzer with little loss of signal at varying magnifications. The actual magnification of the image is dependent on ion optical conditions selected and is in the range of 100×–500× [12].

The primary slit and apertures are used to select a portion of the secondary ions, coming from the center of the crater, to be mass analyzed. This is done by two sequential mass analyzers, an electrostatic and a magnetic analyzer. The ions that leave the sample surface are not monoenergetic but have a range of energies, the lowest of which is equal to the sample surface potential. For some applications, knowledge of the

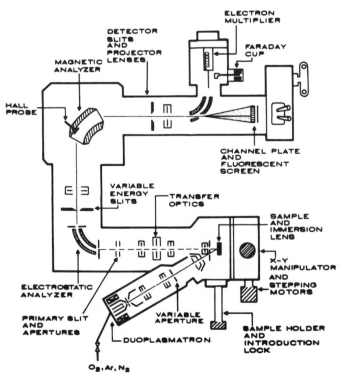

Figure 3. Schematic diagram of ion microanalyzer.

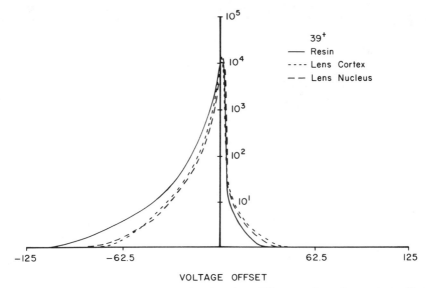

Figure 4. Energy distribution of potassium ions from different regions of a specimen. Since the energy maxima has not shifted, charging of the specimen is probably not significant.

ion energy distributions, which are indicators of the sample surface conditions or chemistry of the sample may be important [13] (Fig. 4). The variable energy slits are used to select one portion of the ion energy curve for analysis. The total secondary ions are separated in the two analyzers to individual ions of a given mass–charge ratio. For example, calcium may be ionized as a singly charged ion, Ca^+, with a mass–charge ratio (m/e) of 40, or as Ca^{2+}, $m/e = 20$. The energetics of the bombardment under dynamic SIMS conditions are such that the singly charged specie predominates, usually by a factor of at least 100 [2]. Therefore the secondary ions are usually atomic and not molecular in nature, unlike many other forms of mass spectrometry. There is much current interest in SIMS as a technique for producing volatile species from underivatized molecules [6–9]. The ions produced are a function of the composition of the specimen surface, as well as ionization effects that were discussed earlier. Figure 5 shows relative ion yields in a specific matrix to illustrate this point. Those ions of physiological interest, Na, K, Ca, are readily ionized as positive species. In complex organic systems, the problem of mass interferences could be substantial. For example, at $m/e = 40$, C_2O^+ could be present in addition to calcium 40. However fine tuning of the selected mass can be achieved up to a mass resolution of 10,000. Thus calcium ($m/e = 39.96259$) could readily be discriminated from C_2O ($m/e = 39.99491$).

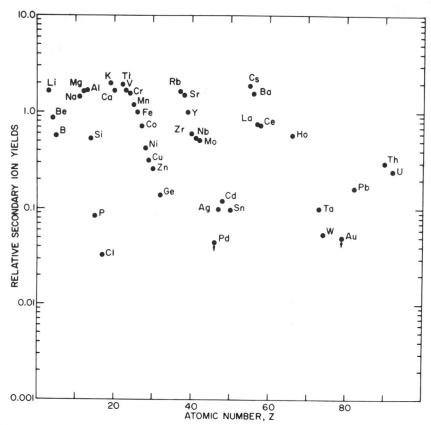

Figure 5. Relative positive secondary ion yields from an aluminum oxide matrix. (From Ref. 2.)

In fact, resolution of organic polyatomic species from elemental ions does not require high resolution, usually no more than about 2000. The use of a mass spectrometer has two substantial advantages. First, all isotopes of an element can be analyzed independently, making tracer experiments possible. Second, the signal to noise ratio is very high. The background count rate in the IMS 3f is of the order of 1 count every 6 s. For comparison, a sodium signal in a typical tissue is of the order of 100,000 to 1,000,000 counts per second.

There are two detection systems available: ion counting through the use of an electron multiplier (for low count rates) or a Faraday cup for higher count rates. Alternatively the ions, still retaining their relative spatial location, as they were on the sample surface, are converted to elec-

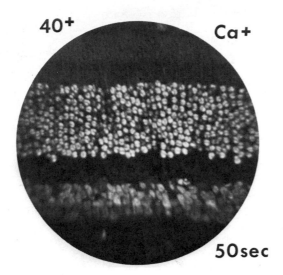

Figure 6. Calcium, 40^+, ion image of glutaraldehyde-fixed cat retina. The bilobed nuclei are apparent in the center field. The lobes are separated by 0.5–1 μm. (From Ref. 4.)

trons at a channelplate, and this is displayed on a fluorescent screen that may be directly viewed or photographed on a camera that is outside the vacuum system. This mode of detection forms the ion image, with an optimum resolution of less than 1 μm [12] (Fig. 6). The field of view is most often 150 μm, although with selection of different apertures and optical parameters, a field of 25 μm may be imaged at higher magnification. Ion counting may be used in conjunction with scanning the magnetic field to select sequentially ions of different m/e ratios, thus forming a mass spectrum of the secondary ions (Figs. 7–8). Alternatively, the count rate can be monitored in a given area, compared to standards, and converted to concentrations [14–15]. Because of the complexity and lack of understanding of all the parameters governing ion production, direct ion counts are usually referenced to a standard [16].

Since ion generation results from the sequential destruction of the surface, any of the preceding modes of data collection can be followed in time to form a "depth profile" or analysis of the specimen in the third dimension. For example, the location of subsurface layers of ions are crucial in controlling the properties of semiconductors, so monitoring boron counts while sputtering a silicon wafer is used to form a depth profile of the location and quantity of boron in an implanted layer. Sequential ion images could be used to build up a three-dimensional struc-

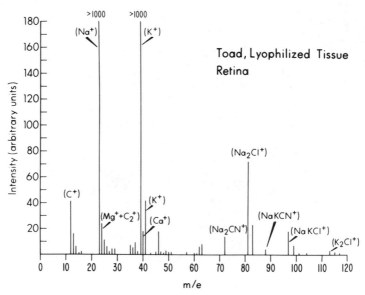

Figure 7. Positive secondary ion mass spectrum using a primary ion beam of O_2^+. The alkaline earth elements are prominent and form polyatomic species with each other and with chlorine.

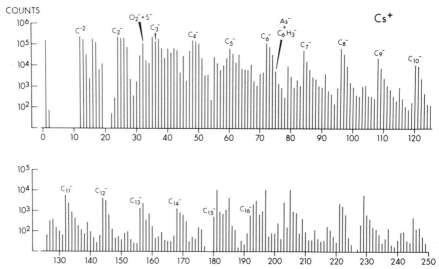

Figure 8. Negative secondary ion mass spectrum of lyophilized mouse liver using a primary ion beam of Cs^+. Carbon and hydrocarbon species dominate the negative spectrum to such an extent that detection of other elements such as sulfur (32^+) or arsenic (75^+) is compromised.

268

ture, or recorded on videotape to form a visual three-dimensional tour through the specimen [17–18].

11.3   EXPLANATION OF USE

The ion optical system requires a good vacuum and therefore the sample must be dry. The vacuum systems are typically 10^{-7}–10^{-8} Torr in the specimen chamber, maintained by ion pumping. Different investigators have used fixed, dehydrated or freeze-dried specimens, embedded or not embedded to achieve this condition [10,19–21]. There is no need to go into problems of ion loss and ion translocation during tissue preparation, which in this, as in so many other techniques, has complicated (or made meaningless) the interpretation of the results. The continued use of chemically dehydrated specimens for elemental analysis of biological tissues is to be avoided in all cases unless one has independent evidence of lack of loss and lack of translocation during processing [22]. Let us assume that one has an appropriately dehydrated specimen. The other parameter which is essential and specific for SIMS analysis is that the specimen be flat relative to the immersion lens, in order not to have artifactual ion emission and location. For this reason I prefer to use dried material that is directly embedded in epoxy resin. The cut surface can be briefly sputtered, to remove any surface contamination, and the underlying tissue then analyzed. The size of the sample can be up to 25 mm in diameter, and of any thickness since only the surface layers are sputtered and analyzed at any given moment in time. The thickness is limited by the predilection for thicker specimens to charge, although thin specimens of 1 μm or less frequently do not present charging problems [10]. The samples are mounted on a conductive surface; gold foil, silicon wafers, and tantalum films among other materials have been used. The higher the mass or the more unusual the element used for the substrate, the less likely it is to confuse specimen analysis. The upper surface is sometimes covered with a conductive grid or solid coating that is removed during the course of analysis [23–25]. There is no theoretical reason why frozen hydrated specimens could not be analyzed, and, in fact, ion emission may be enhanced in the presence of frozen water [26], but the technology of a cold stage or cold transfer stage has not been available.

Typically, one proceeds as follows in the analysis of an unknown specimen: A mass spectrum is taken to get a general idea of the composition of the specimen (Figs. 7 and 8), and to verify if an element of interest is present in the sample. In organic specimens it is necessary to confirm the presence of the element and differentiate it from hydrocarbon peaks by

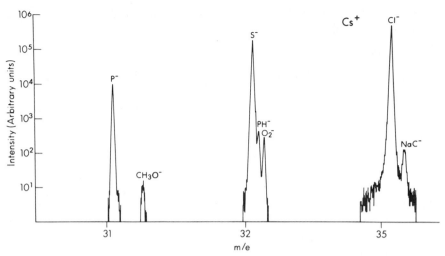

Figure 9. High mass resolution spectra of three masses showing interferences by hydrocarbon species. Note that the intensity scale is logarithmic, so the contribution of PH^- and O_2^- to the S^- is less than 1%.

carrying out a high mass resolution spectrum of the elements of interest [27], particularly when analyzing negative ion spectra, which always include carbon polyatomic species (Fig. 9). It takes about 20 min to generate either a full mass spectrum up to mass 300 or a high mass resolution of an individual element with good mass resolution. If the main question is simply where an element is located, the ion images of individual elements may be taken. Since the camera is outside the vacuum system, no time is taken in venting the instrument, and exposures of $\frac{1}{2}$ sec (for Na or K) or up to a few minutes (for Ca) are usually sufficient for adequate images. The presence of the channelplate introduces a background due to spontaneous ionization events, so exposures longer than a few minutes are not useful. If there are significant interfering species, ion images, which can be taken only in the low mass resolution mode, are not useful and may show spurious locations (Fig. 10). Analysis of images by digital image-processing techniques would have large advantages for analysis of complex and heterogeneous specimens [28].

If quantitative data are required, standards of gelatin or epoxy doped with various ions of interest have been used to generate standard curves [14] or, alternatively, ion implantation in the specimen has been used to form an internal standard [15] (Figs. 11 and 12). Quantitative analysis is then done under the same operating conditions as the standards were run. Apertures are used to select the analysis area, generally 30–50 μm. With

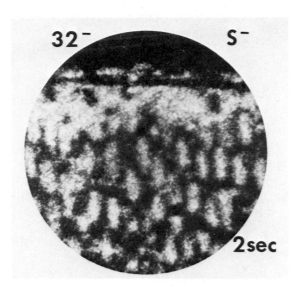

Figure 10. Negative secondary ion image of toad photoreceptor outer segments at $m/e = 32^-$. Although both sulfur, phosphorus, and oxygen species are present, the sulfur intensity is so much greater (see Fig. 9) that the ion image reflects the location of S^-.

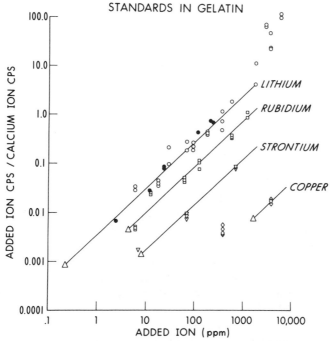

Figure 11. Standard curves of elements doped in gelatin films. The higher ionization yield of lithium compared to copper is evident from the lower detection limit. (From Ref. 14.)

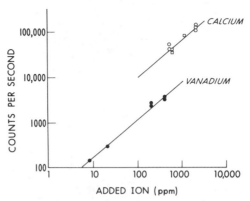

Figure 12. Standards may also be prepared in epoxy resin, as is used for embedding specimens. (From Ref. 14.)

sputtering rates of 2 Å/s and counting times of 0.1–10 s, a sample volume of about 200 μm^3 is analyzed. If significant mass interferences are present, one can count in high mass resolution conditions. It is pertinent that the instrument is adjusted to optimize the ion images as compared to high mass resolution and for mass spectra, so it is not practicable in terms of time or reproducibility to change between operating modes for quantitative analysis. Data generation is extremely fast, and it is easy to accumulate large quantities of numbers and images in a short time.

11.4   RESOLUTION

There are several different ways to specify resolution with SIMS. The mass resolution of the technique is defined as the mass of interest divided by the differences of two resolvable peaks, $M/\Delta M$. In low mass operating conditions a mass resolution of about 300 in used. This means that mass 30.0 can be resolved from 30.1, or mass 299 from 300. This is not adequate for differentiating many hydrocarbons from elemental peaks, so the mass resolution may be increased to as high as 10,000, but at the cost of losing signal and decreasing sensitivity. In practice, a mass resolution of 5000 is adequate to separate most elements from the hydrocarbon interferences, the most serious problem in biological specimens. Because Na and K are usually prominent in biological material, polyatomics such as NaCl can be a significant interference, even though they are at a level of a small fraction of the parent peak, if they occur at masses where the elemental peak is small, such as trace elements. For example, the presence of nickel (mass 58) at trace levels could not be demonstrated without resolving the presence of NaCl$^+$ at the same nominal mass.

The lateral spatial resolution of the ion optical system is optimally 0.3 μm, but for practical purposes a spatial resolution of 1 μm is reasonable, particularly for elements present in higher concentrations. In Fig. 13, the K image clearly shows some of the individual pigment granules, which are 1 μm or less in diameter. The Ba image shows emission from the same areas, and with the same general distribution, but with less spatial resolution since the concentration is less and more material was sputtered to obtain the image and the ionized particles came from different lateral locations at different depths.

The depth spatial resolution is similarly dependent on the concentration of the species and the ionization probability of the species (Table 1). Other factors that influence the depth resolution are the penetration of the bombarding ion, which can cause mixing in the specimen, and the current density of the primary ion, which controls the sputtering rate. The measured sputtering rates for biological tissue are about 2 Å/s. The interrelationships of these different parameters are shown in [2]. Under optimum conditions, depth resolutions of 30 Å have been obtained but for practical

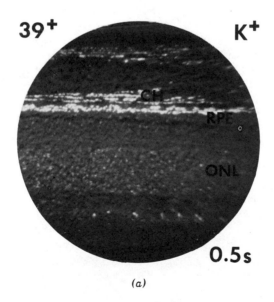

(a)

Figure 13a. Potassium image of cat retina showing high emission from pigment granules in the choroid (CH) and retinal pigment epithelium (RPE). Emission is less in the outer nuclear layer (ONL).

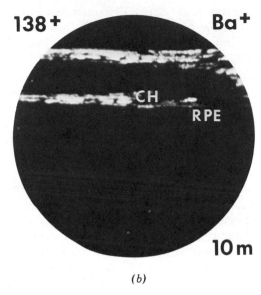

(b)

Figure 13b. The principal barium isotope, 138⁺, is also present in the choroidal pigment granules but not the RPE.

TABLE 1 Depth resolution calculated from quantitative analysis of lithium in gelatin [14] and relative ionization yields [2], assuming a concentration of 1 ppm for each element and no mass interferences [4].

OPTIMUM DEPTH RESOLUTION

ELEMENT	CONCENTRATION	MINIMUM DEPTH
Lithium	0.14 mmole/kg dry weight	0.16 Å
Sodium	0.04 mmole/kg dry weight	1.1 Å
Potassium	0.025 mmole/kg dry weight	1.7 Å
Calcium	0.025 mmole/kg dry weight	14. Å
Zinc	0.016 mmole/kg dry weight	470. Å
Rubidium	0.012 mmole/kg dry weight	70. Å
Barium	0.007 mmole/kg dry weight	160. Å

purposes depth resolutions of 100 Å are more likely to be obtained in biological tissue for favorable elements. A complicating factor is that sputtering does not occur evenly throughout a specimen, particularly if there are different densities present (Fig. 14). Thus, at a given time of analysis, ions would be coming from several different depths, and degrade the spatial resolution [29].

The accuracy and precision of quantitative analysis has been tested for two different sets of standards on a first-generation SIMS instrument [14–15]. Gelatin standards doped with various atoms gave unnormalized standard curves with a precision of ±15%. When implanted ions were used to form an internal standard, the precision was improved to ±5%. The only practical application of quantitative analysis has shown that the accuracy, as compared to atomic absorption or flame photometer values, is within 20% for potassium and calcium, but that sodium is more variable [30]. One problem with knowing how to estimate accuracy is that there may be considerable heterogeneity in a biological specimen that is not apparent when one carries out a bulk analysis on a whole tissue as compared to analysis of a few cells with SIMS.

The sensitivity of SIMS is very high for the alkaline earth elements, most important to cell physiology. Because of the variable ionization

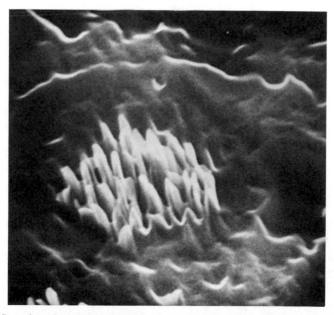

Figure 14. Scanning electron micrograph after 1 h of sputtering shows a cell with the cytoplasmic rim etched more than the nucleus.

TABLE 2 Estimation of minimum detectable concentrations of elements compared to whole blood values [4]. The calculations did not take into account mass interferences. Elements may be concentrated in different organs at higher levels than blood, and would then be detectable.

| Elements | mmole / kg dry weight | |
	Minimum Detectable Concentration	Concentration in Whole Blood
Li	0.0004	0.03
Ba	0.0004	0.013
Mg	0.0018	17.50
V	0.0018	0.003
Fe	0.008	71.40
Co	0.028	0.002
Cu	0.08	0.14
Ni	0.17	0.0005
Zn	0.24	0.94
Pb	0.28	0.007
Cd	2.40	0.0004
Hg	2.40	0.00025

rates the sensitivity is different for different elements (Table 2), and is not likely to be useful for trace elements such as Cu, Mn, and Zn, unless there is a biological mechanism that concentrates and localizes the trace elements.

11.5   COMPARISON WITH OTHER TECHNIQUES

SIMS is truly a surface technique, but with one exception has not been used in biological applications as a surface tool. The problem is, again, the preparation of a specimen that is truly clean on the surface, without artifactual contamination. In general, the applications have used cross sections of material and have taken advantage of the imaging mode, rather than the surface analysis capability.

SIMS appears to be similar in sensitivity to LAMMA analysis, and the ionization processes may be similar, since the spectra generated are quite alike. The currently available laser probes are quadrupole mass analyzers, which have less resolving power than the IMS 3f and so the spectra undoubtedly have mass interferences. For example, Mg is probably a combination of Mg^+ and C_2^+. The main difference is using a fine probe beam to ionize a spot, rather than a large area. This is advantageous in the examination of fine structure or small particles, which can be done in SIMS by image analysis, but the lateral resolution is of the same order of magnitude.

The electron probe and SIMS are highly complementary techniques. Electron probes have greater sensitivity for heavier elements, and SIMS for light elements. The electron probe has much greater lateral spatial resolution, but no depth resolution. It also takes much longer to collect data over large sample areas and has appreciable background counts. The requirement in electron probes for thin specimens in order to take advantage of the high spatial resolution gives problems during sample preparation that are much simpler in SIMS since the inherent spatial resolution is less.

A superb solution would be to take advantage of the complementary nature of these techniques and establish a laboratory with several types of instrumentation as a central resource.

11.6 ARTIFACTS

SIMS is a destructive technique. The mechanism by which ions are produced involves destruction of the material being analyzed at a relatively slow rate of 2 Å/s under the conditions used for analysis. At higher current densities the samples are visibly charred, and there is undoubtedly some rearrangement of elements within the specimen. Whether this is a serious deterrent to analysis has not been studied. The constancy of ion images suggests that the degree of mixing or potential translocation during analysis is not a serious problem.

A potentially serious artifact could be the occurrence of matrix effects in analysis of biological tissue. If there are conditions within the sample that give enhanced emission of an element in one area versus another, a false relative location would be found. One paper has described differential carbon emission from a mineralized or organic material, in which carbon emission from the mineral layer was less intense even though the concentration of C was higher [10]. There have not been any systematic studies looking for matrix effects in soft biological tissue.

11.7   APPLICATIONS

A few applications, mostly drawn from work in my laboratory, will be mentioned to give an idea of the types of studies that are possible with SIMS. A more comprehensive bibliography has recently been published [4].

Mass spectra are generally useful for defining the overall composition of material, which in the case of soft biological tissue is generally Na and K [20]. A potential application would be to use the spectrum to define the chemical composition of tissue, but the spectra of common biological compounds in the dynamic mode of SIMS analysis is so similar that differentiation in a complex biological matrix is not feasible (Fig. 15). However, SIMS has had a surge of interest recently when operated in the static mode in producing parent ions from nonderivatized and nonvolatile compounds, an important advance for organic mass spectrometry [6–9]. This mode of analysis has not yet been tried on complex biological tissue to see if individual molecules within a specimen can be identified.

Ion images have been used extensively to locate specific elements within a field (Fig. 13). When a simple yes or no answer to the question, "Where is this element localized?" is desired, an ion image is the quickest response. Beryllium was found in lung tissue, and lanthanum localized in plant specimens [32–33]. Mineralized specimens showed specialized deposition of Sr in radiolaria [23,25]. The advantage of the ion image is that a large area may be surveyed at one time. If an element is present and shown to be free of mass interferences [34], an ion image of the element is a positive result. If the element is not seen, it does not mean that it is not present, it may be that the ionization is not within detectable limits.

The use of stable isotopes for tracer experiments in biological specimens has just begun [13] and offers a valuable new tool for studying ion fluxes in tissue. One example of this was to microinject barium into one cell of *Fundulus* eggs while recording from its electrically coupled partner (Fig. 16). Fifteen minutes after injection, most of the barium is still in the injected cell, but a trace was seen in the coupled cell.

The ion image contains much information that can be readily appreciated visually, but to convert that to meaningful physiological form it must be changed into quantitative data. This approach is being pursued by digital image-processing systems and offers a great advance in the use of the ion image [18,28,35].

Applied quantitative analysis in biology has been tried in two systems. Lodding and his co-workers [36] used apatite standards to devise an analytical scheme for fluorine in enamel. I have applied quantitative analysis using epoxy-doped standards to study formation of cataract in rat lens

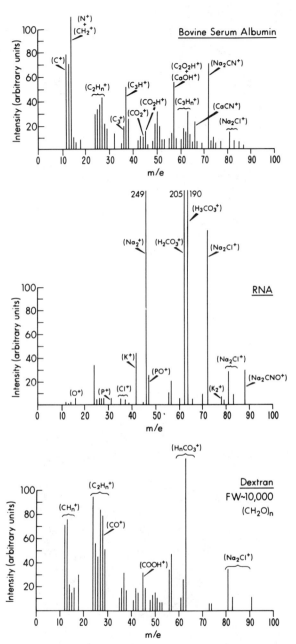

Figure 15. Mass spectra of different types of biological polymers. Although differences between the spectra are seen, it is unlikely that individual components in a complex tissue could be differentiated.

279

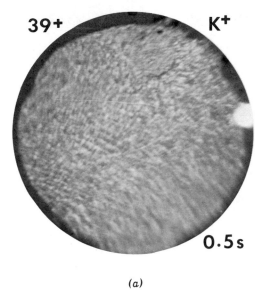

(a)

Figure 16a. Potassium image of two coupled Fundulus eggs. No differentiation of the cells is seen.

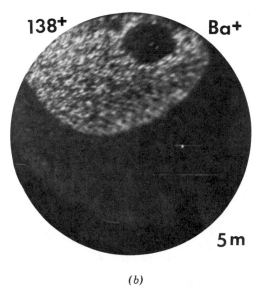

(b)

Figure 16b. Barium image shows up clearly the microinjected cell with a vacuole that excludes barium. The second cell has a small amount of barium 15 min after injection.

280

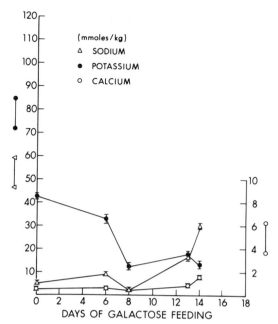

Figure 17. Ion counting of galactose-induced cataract in rat lens nucleus showing a loss of potassium and gain in sodium and calcium as the cataract developed.

(Fig. 17). The agreement with bulk analysis by other methods was quite good, and the precision was within ±15%, acceptable for this type of problem. An interesting phenomenon is that when carrying out local area analysis of tissue there is more variability than might be expected from previously available bulk measurements.

Depth profiling has been used to study the gradient of fluorine in the surface of teeth [37]. The problem with true surface analysis in biological systems is, of course, preparation of the specimen, and an examination of biomaterials and their interaction with tissue would be a fruitful area for application of this aspect of SIMS analysis.

ACKNOWLEDGMENTS

The ideas expressed here have been developed in conversations with Pierre Galle, Annick Quettier, David File, Vaughn Deline. I thank Dr. Otto Friedheim and Dr. David Spray for permission to use preliminary data, and to a superb staff who make my work possible, Noel Roa, Mary Ellen Murphy, Judith Channer, Benjamin Day, and Patricia Lynch.

REFERENCES

1. G. Carter and J. S. Colligon, *Ion Bombardment of Solids,* Elsevier, New York, 1968.
2. J. A. McHugh, Secondary ion mass spectrometry, in *Methods of Surface Analysis,* A. W. Czanderna, Ed., Elsevier, Amsterdam, 1975.
3. S. J. B. Reed, Ed., Review issue on secondary ion mass spectrometry, Scanning *3,* 57 (1980).
4. M. S. Burns, J. Microsc. *127,* 237 (1982).
5. B. J. Garrison, N. Winograd, and D. E. Harrison, Phys. Rev. *B18,* 6000 (1978).
6. A. Eicke, W. Sichtermann, and A. Benninghoven, Organic Mass Spectrom. *15,* 289 (1980).
7. S. E. Unger, T. M. Ryan, and R. G. Cooks, Surf. Interface Anal. *3,* 12 (1981).
8. J. A. Gardella and D. M. Hercules, Anal. Chem. *53,* 1879 (1981).
9. H. T. Jonkman and J. Michl, *J. Am. Chem. Soc. 103,* 733 (1981).
10. C. Quintana, A. Quettier, and D. Sandoz, *Calcif. Tissue Int. 30,* 151 (1980).
11. K. F. J. Heinrich, in *Microprobe Analysis as Applied to Cells and Tissues,* T. Hall, P. Echlin, and R. Kaufmann, Eds., Academic, New York, 1974, p. 75.
12. G. Slodzian, Surf. Sci. *48,* 161 (1975).
13. M. S. Burns, *Microbeam Analysis,* San Francisco Press, San Francisco, 1982, p. 138.
14. M. S. Burns-Bellhorn and D. M. File, Anal. Biochem. *92,* 213 (1979).
15. D. Zhu, W. C. Harris, and G. H. Morrison, Anal. Chem. *54,* 419 (1982).
16. D. E. Newbury, *Scanning 3,* 110 (1980).
17. M. S. Burns-Bellhorn, Secondary ion mass spectrometry: future application to biological samples, in *Microbeam Analysis in Biology,* C. Lechene and R. R. Warner, Eds., Academic, New York, 1979, p. 129.
18. W. C. Harris and G. H. Morrison, *Microbeam Analysis,* San Francisco Press, San Francisco, 1982, p. 227.
19. M. S. Burns, D. M. File, K. T. Brown, and D. G. Flaming, Brain Res. *220,* 173 (1981).
20. K. M. Stika, K. L. Bielat, and G. H. Morrison, J. Microsc. *118,* 409 (1980).
21. M. Truchet, *Microsc. Acta,* Suppl. 2, 355 (1978).
22. A. J. Morgan, Preparation of specimens, changes in chemical integrity, in *X-Ray Microanalysis in Biology,* M. A. Hayat, Ed., University Park Press, Baltimore, 1980, p. 65.
23. P. Galle, Biomedical applications of secondary ion emission microanalysis, in *Secondary Ion Mass Spectrometry SIMS II,* A. Benninghoven, C. A.

Evans, R. A. Powell, R. Shimizu, and H. A. Storms, Eds., Springer-Verlag, New York, 1979, p. 238.

24. M. B. Bellhorn and R. K. Lewis, Exp. Eye Res. *22*, 505 (1976).

25. R. Lefevre, Scanning *3*, 90 (1980).

26. T. Sato, M. Nambu, Y. Omori, and N. Hayakawa, Proceedings of the United States-Japan Joint Seminar on Secondary Ion Mass Spectrometry: Fundamentals and Applications, National Science Foundation, Osaka, Japan, 1978.

27. M. S. Burns, Anal. Chem. *53*, 2149 (1981).

28. B. K. Furman and G. H. Morrison, Anal. Chem. *52*, 2305 (1980).

29. M. E. Farmer, R. W. Linton, P. Ingram, J. R. Sommer, and J. D. Shelburne, J. Microsc. 124, RPI (1981).

30. M. S. Burns and D. M. File, Quantitative analysis of galactose cataract formation by SIMS, submitted to J. Microsc.

31. M. S. Burns, unpublished data.

32. J. L. Abraham, R. Ross, N. Marquez, and R. M. Wagner, SEM ITTRI Proceedings, 1976, p. 501.

33. A. R. Spurr and P. Galle, Localization of elements in botanical materials by secondary ion mass spectrometry, in *Secondary Ion Mass Spectrometry SIMS II*, A. Benninghoven, C. A. Evans, R. A. Powell, R. Shimizu, and H. A. Storms, Eds., Springer-Verlag, New York, 1979, p. 252.

34. M. Truchet and J. Vovelle, Calcif. Tissue Res. *24*, 231 (1977).

35. W. Steiger and F. G. Rudenauer, Anal. Chem. *51*, 2107 (1979).

36. A. Lodding, J. M. Gourgout, L. G. Petersson, and G. Frostell, Z. Naturforsch. *294*, 897 (1974).

37. L. G. Petersson, H. Odelius, A. Lodding, S. J. Larsson, and G. Frostell, J. Dent. Res. *55*, 980 (1975).

CONFOCAL SCANNING
LIGHT MICROSCOPY

G. J. BRAKENHOFF

Department of Electron Microscopy and Molecular Cytology
University of Amsterdam
Amsterdam, The Netherlands

12.1 INTRODUCTION

The newly developed confocal scanning light microscope (CSLM) possesses in comparison with conventional light microscopy a number of attractive properties of which the most important are improved resolution, the possibility of image processing and the availability of a number of unique imaging modes. When operating with high aperture immersion optics, resolutions are theoretically possible (depending on wavelength) down to 60 nm. Until now a resolution of 130 nm has been demonstrated with fidelity of imaging down to this improved resolution limit.

These resolution capabilities of CSLM systems are of special value for the visualization of live biological specimens in their natural watery environment. Although electron microscopy (EM) possesses superior resolution, in practice only dehydrated, chemically fixed specimens can be examined. Some observations in EM under hydrated conditions have been reported by Parsons [1] with the so-called wet cell method. This process is unfortunately technically rather difficult while in addition during observation a high, probably lethal, radiation dose is delivered to the specimen. This leaves high resolution immersion CSLM possibly as the only technique that can deliver morphological data on live specimens below the lowest limits of standard light microscopy. The scanned x-ray imaging technique as presently under development by Spiller [2] may eventually offer better resolution than the 60 nm lower limit mentioned earlier. However, because of signal-to-noise considerations and the consequent necessary number of quanta to be absorbed by the specimen, it can be expected that the delivered radiation doses will surpass the survival limits for many, if not all, living specimens. For nonliving specimens we will also indicate some areas of application where the use of the CSLM technique is of value. These are mostly related to the scanning aspect of the CSLM and of course its higher resolution.

Historically, scanning light microscopy dates back to the first attempts by Roberts and Young [3] based on a flying spot approach. To achieve improved resolution at a given aperture the scanning aspect has to be combined with the confocal principle—first suggested by Minsky [4]—as explained below. Lemons and Quate [5] incorporated the confocal arrangement in one of their first acoustic microscopes. The first actual demonstration of improved imaging with high aperture confocal optics was reported by Brakenhoff et al. [6]. Confocal microscopy has become a practical proposition mainly thanks to the present-day availability of lasers; there is no other light source that so conveniently permits the focusing of sufficient light intensities through the small illumination pinholes necessary in this technique.

12.2 THE CSLM PRINCIPLE

The basic idea behind the CSLM is rather simple. As explained in Fig. 1 one and the same spot in the specimen is both optimally illuminated as well as imaged on the detector by the confocal optics. The illumination intensity distribution $I(u)$ and the detection sensitivity distribution $D(u)$ can be described by the relation [6]

$$I(u) \text{ resp } D(u) = \text{const} \frac{J_1^2(u)}{u^2} \tag{1}$$

with J_1 the first-order Bessel function and u an optical unit related to the transverse coordinate x (see Fig. 1) by the relation

$$u = \frac{\text{N.A.}2\pi x}{\lambda} \tag{2}$$

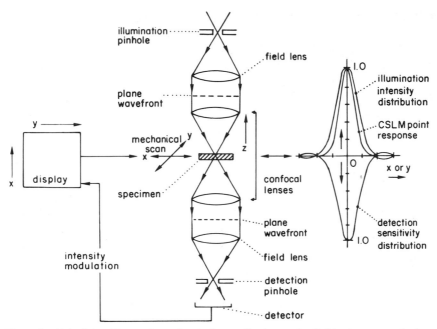

Figure 1. Principle of image formation in the confocal scanning light microscope. A plane wave generated by a point light source and a field lens is brought to a focal point in the specimen by the top confocal lens. The lower confocal lens is so positioned that the backward-projected image of the detector pinhole exactly coincides with this focal point. If the optics are diffraction limited, both the illumination intensity distribution and the detection sensitivity distribution will be Airy-disklike patterns as sketched in the righthand figure. Now the point response in the CSLM is equal (see text) to the product of the illumination and detection sensitivity distributions resulting in a sharpening of the response and therefore improvement of resolution. The specimen is scanned mechanically through the common focal point. The image is generated on a CRT display running in synchrony with the mechanical scan and modulated by the detector signal.

N.A. is the numerical aperture of the confocal optics employed and λ the wavelength of the light used. The confocal point response $C(u)$ is now given by Sheppard and Choudhury [7] calculated in the paraxial approximation as

$$C(u) = \text{const} \, \frac{J_1^4 u}{u^4} \tag{3}$$

thus being equal to the product of the illumination and detection sensitivity distributions.

The imaging in conventional light microscopy being described by $D(u)$ [Eq. (1)], it can be calculated that an increase in linear resolution in the

CSLM will result by a factor of 1.4. As this happens in both directions transverse to the optical axis, and because the depth of field is also reduced by the same factor [6], one image point in the CSLM will correspond to a volume element in the specimen that is smaller by a factor of $(1.4)^3 \approx 3$ than in conventional microscopy. This will result in a proportional increase of information that can be extracted from the specimen.

It should be pointed out that CSLM makes special sense when optics of maximum possible aperture are employed (e.g., immersion objectives of N.A. = 1.3 or 1.4). Because when one is working at lower numerical apertures and higher resolution is desired, it is easier to employ higher numerical apertures than to go to confocal scanning microscopy, unless of course the scanning aspect of this type of imaging is to be exploited (see below). Apodization with annular diaphragms in either or both illuminating and detection paths may be employed to obtain still higher resolutions [6], but at the price of increased strength of the ghost images. This makes the images thus obtained less trustworthy and limits the use of such apodization techniques.

12.3   THE CSLM INSTRUMENT

We will now describe the apparatus presently in use in our institute. For an alternative approach see Wilson [8]. A schematic diagram of the optical layout is shown in Fig. 2 together with a corresponding photograph of the instrument. The two confocally arranged immersion objectives employed (Zeiss Planapochromats, 100×, N.A. = 1.3) were chosen "corrected for infinity" in order to have more freedom in the handling of the input and output beams; this makes, for instance, the indicated interferometer setup possible. Actually, an even simpler system results when using objectives corrected for 170 mm. Then the respective pinholes are just positioned at this optical distance from the objectives and the field lenses can be omitted.

To obtain confocal conditions we have to make the back projection of the detector pinhole in the specimen coincide with the focal spot of the top confocal lens (see Fig. 1). To be able to accomplish this the top confocal lens has been mounted on a precision slide, permitting vertical movement, while mirror 7 can be tilted in two directions. Both the vertical slide movement as well as the tilt movements possess a mechanical preadjustment combined with a piezoelectric drive for fine tuning. As a criterion for proper confocal adjustment we used the transmission of maximum power through the detection pinhole, while looking through an open area of the specimen. The specimen to be imaged is placed on a scan table in the object plane between the two immersion objectives, which form the confocal arrangement (see Fig. 3). The use of cover glasses of a specific thickness both for specimen support and specimen covering is dictated by the optical requirements of the objectives being used. Figure 3 shows the situation when operating at visible wavelengths; in ultraviolet, quartz objectives are used with 350-μm-thick cover glasses. Because of the mechanical scan rather small "cover" slides were selected in order to keep the mass of the object to a minimum. The top objective lens could be raised about 20 mm to enable the object to be placed on the scan table. The optical axes of the confocal lenses were chosen to be vertical to avoid the effect the combined influence of gravity and mechanical scan movement might have on unwanted displacements of the immersion oil between specimen and objectives. In practice we did not experience any trouble of this kind up to the largest scan amplitudes used (≈ 1 mm).

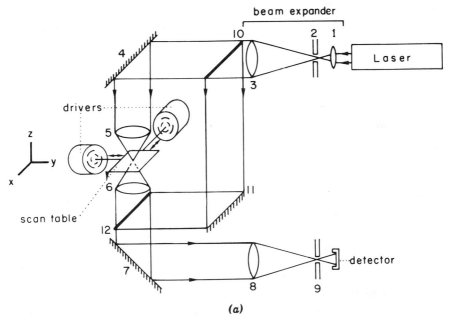

(a)

Figure 2a. Optical layout of the CSLM. A laser beam is focused on a small pinhole (diameter, 5 μm) by lens 1. The high intensity point light source illuminates, with the help of field lens 3 ($f = 200$ mm) and mirror 4, the confocal lenses 5 and 6. These are 100× microscope immersion objectives (N.A. = 1.3). After passing through the object placed on a scan table between the confocal lenses, the light is focused on the detector pinhole 9 (diameter, 5 μm) by field lens 8 ($f = 200$ mm). 4 and 7 are beam steering mirrors. In addition the beam splitters 10 and 12 and mirror 11, shown in the optical layout, when present permit the operation of the microscope in the interference mode.

Figure 2b. Photograph of the CSLM. The lasers (not visible) are mounted on the back of the optical board. Their beams are made to enter lens 1 with the help of two mirrors (not visible). The numbers in the photograph correspond to those in Fig. 2a. ST indicates the scan table, D the detector.

The mechanical scan movement of the scan table is effected in the x–y plane under suitable mechanical guidance by two drivers, modified from loudspeaker systems. Along the optical axis—in the z direction—the position of the scan table is determined by a small ruby attached to the table, skating in the x–y direction over a finely polished hard-metal plate. With a spring a small adjustable contact pressure is maintained over the slightly lubricated interface between ruby and metal plate. This contact pressure is critical: if too small, the table will lose contact, resulting in an undetermined z position; if too large, the scan table will ground to a halt at high magnifications (small scan amplitudes). By proper adjustment stable scans can be obtained with scan fields down to 100 by 100 nm, with scan position repeatability of better than 10 nm. This is considerably better than required in view of the resolutions of 100–200 nm expected from optical considerations.

The scan format is conventional, a fast line scan with the image repetition frequency depending on the number of lines per frame. Typical values are 50–100 lines per frame during searching and focusing, and 4000 during recording on photographic film. The fast line movement (frequency range 50–200 Hz) is sinusoidal, from which the approximately linear part

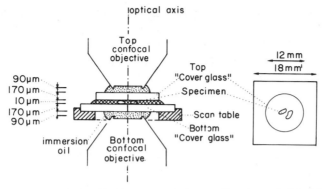

Figure 3. The specimen embedded in an appropriate medium is sandwiched between two thin (170 μm) "cover glasses" and moved during image formation through the confocal point by the mechanical movements of the scan table. The righthand figure shows a top view of the object.

is used for image formation. A piezoelectric sensor senses the actual table movement and after suitable amplification drives directly the x-axis of the display, thus avoiding image distortion due to the sinusoidal nature of the scan and mechanical resonances of the scan table plus suspension system. The slow sawtooth-type y-axis movement is driven directly by the scan generator. The remaining controls are similar to the ones known from scanning electron microscopy with of course extra controls for aligning the optical elements.

12.3.1 Light Sources

As the resolution attainable in CSLM is proportional to the wavelength, there is an advantage to using as short a wavelength as possible. An underlimit is present at about 200 nm due to absorption in the (quartz) immersion optics. At N.A. = 1.3 a resolution of 57.5 nm can then be expected. At lower wavelength reflection optics have to be used that have a much lower numerical aperture, which nullifies a potential gain in resolution from the wavelength aspect [compare Eq. (2)].

For image formation, contrast, as associated with a specific property of the specimen, is the most important factor. This then determines through the particular absorption band(s) of the features of interest, the wavelength to be used. We may then state that at this wavelength CSLM can always produce images at higher resolution than conventional microscopy.

Because for adequate operation of a CSLM, a few milliwatts of continuous wave (or very high repetition rate pulsed) laser light suffice, we can say that nowadays with dye lasers and doubling techniques about every wavelength between 200 nm and 800 nm can be reached. These techniques are expensive but gas lasers fortunately offer usable output at isolated lines at a much lower price. We mention the HeNe laser ($\lambda = 633$ nm) HeCd laser ($\lambda = 442$ nm and 325 nm) and the argon ion laser ($\lambda = 529$, 514, 497, 488, 476, 457 nm and others). We have until now used the first two types of lasers.

12.4   IMAGING TESTS AND EXAMPLES

Confocal microscopy only provides an effective gain in resolution with respect to conventional microscopy when operated at maximum optical apertures. It is therefore crucial to test whether the theoretical expectations can in practice be realized in high aperture immersion systems, where the ray trajectories traverse the optics at angles up to 60°. An unequivocal imaging test is the response to a point object as predicted by Eq. (1) for conventional microscopy and Eq. (2) for CSLM. We used as a test object small (5–10 nm) holes, which can be found in evaporation-deposited gold films. When such a hole is scanned through the confocal point, the desired point response can be recorded directly as a function of the x-axis position (Fig. 4a). The result may be directly compared with theory after x-axis normalization with the help of Eq. (2). We observe with respect to conventional microscopy a sharpening of the response as measured from the width at half intensity (w.h.i.) by a factor of 1.4 and we found, in addition, that the confocal response is very close (within 6%) to the one expected at optimal diffraction limited performance. In absolute values we found the following w.h.i's of the point responses as a function of wavelength: at $\lambda = 663$ nm a w.h.i. of 196 nm (184 nm); at $\lambda = 442$ a w.h.i. of 150 nm (127 nm), and at $\lambda = 325$ nm a w.h.i. of 130 nm (98 nm, N.A. = 1.25). The theoretical values are given in parentheses. In the same way, but scanning the point object *along* the optical axis, the depth of focus can be determined. This value is also reduced by a factor of 1.4 in CSLM; we obtained, for instance, at $\lambda = 633$ nm a depth of focus of 0.75 μm versus 1.08 μm in conventional microscopy. For further details see Ref. 6. The imaging of a waffle-type grating replica (Fig. 4b) confirms the point resolutions found in CSLM and shows fidelity of imaging down to these improved resolutions.

In Fig. 5 CSLM images of a *Drosophila hydei* polytene chromosome are presented. In a joint trial study, the results of which were reported by Grond [9] it has been shown that because of the increased resolution, up to 60% extra bands could be identified by CSLM in comparison with

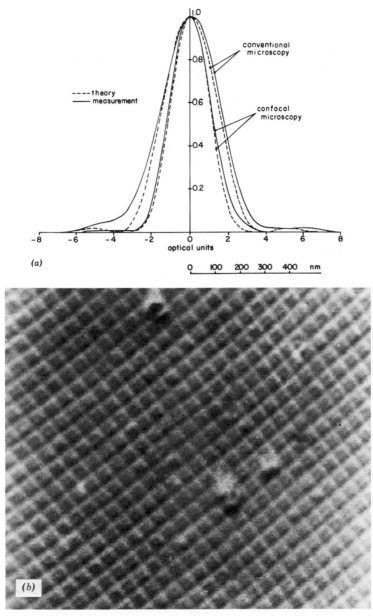

Figure 4. (*a*) The improved CSLM imaging properties are demonstrated in the response to a point object at λ = 633 nm. The response is sharper than possible in conventional microscopy and because of the high quality immersion optics used (N.A. = 1.3) very close to the theoretically expected diffraction-limited performance. (*b*) A replica of a waffle-type grating, ruling distance 463 nm imaged by CSLM at 325 nm. The grating lines are clearly resolved without spurious structure being present with an apparent linewidth in accordance with the width of the observed point response at half maximum of 130 nm.

293

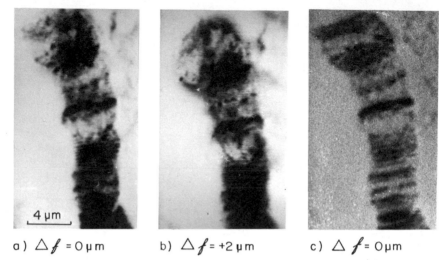

a) $\triangle f = 0 \mu m$ b) $\triangle f = +2 \mu m$ c) $\triangle f = 0 \mu m$

Figure 5. (*a* and *b*) CSLM images at λ = 442 nm of a chromosome preparation of *Drosophila hydei* observed at two levels 2 μm apart (*c*) For comparison a white-light conventional microscope image of the specimen at a level corresponding with the one in (*a*). The images from this three-dimensional structure demonstrate the resolution and excellent depth separation (due to the reduced depth of focus) attainable with CSLM in such objects.

standard microscopy. Also, the sectioning capabilities of CSLM, thanks to the reduced depth of focus, are clearly demonstrated in Fig. 5. This property enables one to obtain from a certain thickness of material 1.4 times more independent images than in standard LM and points to an application of CSLM for the investigation of embedded tissue material at resolutions less good than EM but better than conventional LM. The worse resolution is compensated for in CSLM by two aspects: First, in EM we would have to section the material in a great number of ultrathin sections, from whose images the structures can later laboriously be reconstructed; in CSLM fewer but thicker sections would suffice. Second, we also have more freedom in the choice of plane to be imaged. This plane does not, for instance, need to be perpendicular to the optical axis, nor flat as in conventional microscopy. Because the scan movement of the specimen can be transverse as well as parallel to the optical axis, the scan can be made to follow the curved or inclined plane in which the features of interest are located. We intend to incorporate this rather unique imaging mode of CSLM into a new instrument presently under construction.

As mentioned earlier, a CSLM system is especially valuable for the observation of live specimens. Originally, the main motivation for us to develop the technique was the study of the bacterial nucleoplasm during growth and replication. From the morphology as observed in EM after fixation and dehydration we have not been able to make reliable deductions about the live conformation, while Zernike phase contrast microscopy just lacks sufficient resolution. Figure 6 shows a CSLM imaging example—though not technically perfect, owing to some instabilities in the system—together with a standard microscopy comparison image. In addition to the resolution in the image, we would like to draw attention to the fact that, thanks to the electronic contrast expansion, it has been possible to obtain an *absorption contrast* image of a live bacterium. This means that the CSLM contrast enhancement capabilities actually open up a new observation mode for weakly absorbing specimens from which otherwise an image could be formed only on the basis of their refractive index characteristics by phase contrast microscopy. This phase informa-

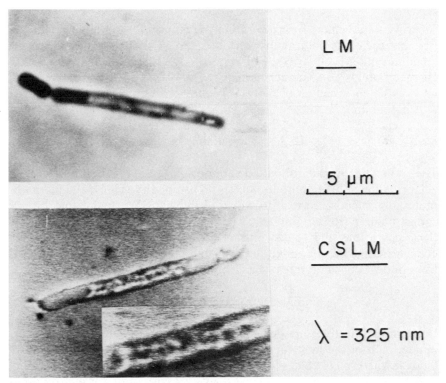

Figure 6. (*a*) Live *Escherichia coli* bacterium imaged by CSLM at $\lambda = 325$ nm. The image is contrast enhanced from an original 7–8% modulation depth. The insert shows the center part of the bacterium at a higher magnification. (*b*) A comparison image of a bacterium, visualized by normal white light, Zernike phase contrast microscopy.

tion can also be obtained in CSLM by incorporating the confocal arrangement in an interferometer setup and recording it quantitatively [10].

12.5   LIVE SPECIMEN PREPARATION AND IMAGING CONDITIONS

The observation of live specimens in the CSLM requires both that certain physiological conditions be met to ensure (temporary) survival and that some constraints set by optical considerations be satisfied. Based on our experience with the preparation of live bacteria we will discuss the following aspects: (1) immobilization, (2) embedding liquids, and (3) refractive index effects on image formation.

12.5.1   Immobilization

To prevent movement from a certain image frame during viewing, a bacterium has to be immobilized to a certain extent. To effect this we adapted a technique pioneered by Tsutsui et al. [11], in which cells are fixed to charged polylysine-coated glass slides. The bacteria, when attached, still have sufficient freedom of motion to allow growth and replication after attachment and are subsequently embedded in a suitable medium.

12.5.2   Embedding Liquids

The embedding medium should be nonpoisonous, stay in solution without gel formation at the required salinity and osmolarity values, and permit or preferably support normal growth and division. We found that bovine serum albumin (BSA) solutions as earlier used by Marquis [12] can satisfy these conditions and enable one to adjust, by variation of the solution concentration, the refractive index of the medium to the value required from optical considerations (see Section 12.5.3).

12.5.3   Refractive Index Aspects

Immersion objectives are designed under the assumption that the whole optical path of the light rays traverses a medium of a certain refractive index n with typical values for n of 1.51 or 1.52 for glass objectives and of 1.47 for the quartz types. However, of necessity during visualization of the live specimen, part of the trajectory of the light ray will be through the water-based embedding medium and the live object itself. Because of the irregular nature of the surface between the live object and the medium, the optical effects of the refractive index change over the interface cannot be predicted exactly; hence it is advisable to match as closely as possible the embedding medium refractive index to the value of the bacterial cytoplasm (typically $n = 1.39$). This can be very effectively done with the BSA-immersion technique as described by Binnerts et al. [13].

It can be shown [14] that the optical effects on image formation of a medium between the glass slides with a different refractive index do not seriously affect the image formation, provided the thickness of the medium between the glasses does not exceed 10–20 μm. The optical aspects mentioned above are also valid for image formation in conventional microscopy, which, however, because of its lower resolution capabilities, is less sensitive to this type of fault.

12.6 RESOLUTION AND RADIATION DAMAGE

A fundamental resolution limit in image formation with electromagnetic radiation (light, x-ray radiation) is set by quantum aspects together with the statistics involved in the imaging. To image a feature of transmission T with respect to a transparent background ($T = 1$) with a signal to noise ratio S/N, we have from statistical considerations that on the average n events or quanta will be needed with

$$\bar{n} = \left(\frac{S}{N}\right)^2 \frac{(1 + T)}{(1 - T)^2} \tag{4}$$

With $E_q = hc/\lambda$ the energy per quantum (h = Planck's constant, c = velocity of light), and d the feature size we arrive at an energy per unit surface E_s that has to be deposited in the feature (transmission T) during imaging of

$$E_s = \frac{\bar{n}E_q(1 - T)}{d^2} = \left(\frac{S}{N}\right)^2 \left(\frac{1 + T}{1 - T}\right) \frac{hc}{\lambda d^2} \tag{5}$$

As d has to be larger than the Rayleigh limit R_L of the optical system with numerical aperture N.A., we have

$$d > R_L = \frac{0.61\lambda}{\text{N.A.}} \tag{6}$$

Substituting from Eq. (6) in Eq. (5) and putting E_s equal to E_m, the maximum energy per surface unit the specimen can tolerate, we find that the minimum feature size d_{min} that can be imaged is given by

$$d_{min} = \left(\left(\frac{S}{N} \right)^2 \left(\frac{1 + T}{1 - T} \right) \frac{0.61hc}{\text{N.A. } E_m} \right)^{1/3} \tag{7}$$

To get an idea of the implications of Eq. (7) we will calculate d_{min} as applicable to typical circumstances during viewing of a live specimen. From reported values of 37% survival of UV-radiated eukaryotic cells at doses of 0.5 J/m^2 [15] and our own observations on bacteria (1–5 J/m^2 for 10% survival), we estimate that E_m for biological systems will be in the range of 0.1–10 J/m^2. Taking then the imaging of a biological system with a transmission $T = 0.9$, a desired $S/N = 5$, and optics with a N.A. $= 0.5$, we find from Eq. (7) that putting $E_m = 1 \text{ J/m}^2$ a value for $d_{min} = 40$ nm. This value is not too far from the lower resolution limit of high aperture immersion CSLM (at N.A. $= 1.3$ and $\lambda = 200$ nm) of 57.5 nm and indicates that CSLM might very well be an optimal technique for viewing live specimens. Applying Eq. (7) to the x-ray imaging techniques presently under development [2] and assuming a comparable radiation sensitivity at these wavelengths, we can conclude that the higher resolution capabilities of x-ray microscopy cannot be used for the imaging of live specimens because of radiation damage.

12.7 SCANNING ASPECTS AND CONCLUDING REMARKS

To emphasize the importance of the scanning concept to light microscopy, we would like to make a tabulation of its specific properties.

General to all scanning approaches are the following:

1. Data are collected quantitatively, making image processing possible.
2. Specimen exposure is limited to the point of interest.
3. More sensitive detectors are available than the photographic plate. The quantum efficiency of photomultipliers is 0.1 to 0.3; of the photographic plate typically 0.01.

Although scanning can be accomplished either by flying spot methods [3] or by mechanical scanning, the following properties are inherent to the latter:

4. The optics used need only be optimal on the optical axis, and because less correction is required should be inherently less expensive to produce.

5. Optical conditions are identical for each point in the image plane.
6. Both amplitude and phase of any point in the image can be recorded with unambiguity at the same moment.
7. No field-of-view limitations are imposed by the optics. Scan amplitude, which can be set at will, determines the area imaged.
8. The imaged plane does not need to be flat or perpendicular to the optical axis.
9. The magnification is continuously variable through the scan amplitude setting.

With confocal optics we have in addition:

10. Essentially improved imaging.
11. Larger dynamic range due to virtual absence of glare [6].

These properties come in addition to the availability of various imaging modes. Apart from the already mentioned $x–y–z$ scan mode, we would like to mention the possibilities of stereo- and differential imaging, the latter in relation to both phase and amplitude [10]. Basically, these imaging modes are connected with the fact that more than one confocal imaging point can be used to "probe" the specimen. Depending on the way the detector outputs corresponding to these points are combined, various imaging conditions result.

CSLM can also operate in the reflection mode [8] and most of the specific points of scanning microscopy mentioned are applicable in this mode. It is, however, not very suitable for imaging very transparent or weakly reflecting specimens as, for instance, live objects. Applications reported in this mode seem to be mostly on metallurgical specimens and for the inspection of microelectronic circuits.

The illumination and the detection do not need to be at the same wavelength, as, for example, in fluorescence microscopy. It can in fact be shown [16] that in the fluorescence mode, CSLM under some circumstances can give further improvements in resolution.

In comparison with conventional microscopy, CSLM has the disadvantages of not having an immediately observable image (but a scanned image on a CRT) and a definitely more complicated operation. To alleviate these difficulties we are presently building a CSLM instrument that will operate under microcomputer control. Some of the functions to be included will be automatic confocal alignment, a "constant"—fast repetition rate—viewing image (generated from computer memory) independent of the mechanical scanning and facilities for the realization of the

$x-y-x$ scanning mode. Because for every magnification change and specimen position change, a different setting of the controls is required, this $x-y-z$ scan mode can be much more conveniently realized with digital than with analog electronics. In conclusion we think in view of the preceding considerations that CSLM is a technique of great potential and may take its place, especially when operated under microcomputer control, under the imaging techniques employed inside and outside biology.

REFERENCES

1. D. F. Parsons, "Biological applications of electron-microscope environmental chambers," in Proc. 6th Europ. Congr. Electron Microscopy, Jerusalem, 1976, Vol. 2, p. 79.

2. E. Spiller, in *Scanned Image Microscopy*, E. A. Ash, Ed., Academic Press, New York, 1980, p. 365.

3. F. Roberts and J. Z. Young, Proc. IEE *99*, Part. IIIA, 747, 1950.

4. M. Minsky, U.S. Patent 3013467, Microscopy Apparatus, Dec. 19, 1961 (filed Nov. 7, 1957).

5. R. A. Lemons and C. F. Quate, Appl. Phys. Lett. *24*, 163 (1974).

6. G. J. Brakenhoff, P. Blom, and P. Barends, J. Microsc. *117*, 219 (1979).

7. C. J. R. Sheppard and A. Choudhury, Otica *24*, 1051 (1977).

8. T. Wilson, Appl. Phys. *22*, 119 (1980).

9. C. J. Grond, J. Derksen, and G. J. Brakenhoff, Exp. Cell Res. *138*, 458 (1982).

10. G. J. Brakenhoff, J Microsc. *117*, 233 (1979).

11. K. Tsutsui, H. Kumon, H. Ichikawa, and J. Tawara, J. Electr. Micr. *25*, 163 (1976).

12. R. E. Marquis, J. Bacteriol. *116*, 1273 (1973).

13. J. S. Binnerts, C. L. Woldringh, and G. J. Brakenhoff, J. Microsc. *125*, 359 (1982).

14. G. J. Brakenhoff, in *Scanned Image Microscopy*, E. A. Ash, Ed., Academic Press, New York, 1980, p. 183.

15. V. M. Maher, R. D. Curren, L. M. Oulette, and J. J. McCormick, in *Fundamentals in Cancer Prevention*, Magee et al., Eds., University of Tokyo Press, Tokyo/University Park Press, Baltimore, 1976, p. 363.

16. J. Cox, C. J. R. Sheppard, and T. Wilton, Optik *60*, 391 (1982).

CATHODOLUMINESCENCE MODE SCANNING ELECTRON MICROSCOPY

D. B. HOLT

Department of Metallurgy and Materials Science
Imperial College of Science and Technology
London, England

Light is emitted by many organic molecules under electron bombardment. This is known as cathodoluminescence (CL). The molecules may be present naturally and give autoluminescence, or fluorescent dyes may be added to give CL at selected sites. Spectral analysis can be used to identify the luminescent molecules. Video display of the CL signal in scanning electron microscopes shows the distribution of those molecules, with a high spatial resolution.

The CL is of low intensity and electron bombardment produces damage in organic and biological specimens. These are the factors that mainly limit the spatial resolution and spectral, biochemical, and microanalytical sensitivity of the CL mode of the SEM. The instrumentation developed to maximize signal detection efficiencies and to minimize damage, especially low temperature stages, is discussed.

The biological and medical applications of the technique are briefly outlined. An introductory bibliography of the relevant literature is given.

A summary of the advantages of the technique suggests that it merits greater use than it has had thus far. The availability of commercial cold stages and spectroscopic CL detection systems should remove the major obstacle to its wider application to biological specimens.

13.1 INTRODUCTION

One of the effects of electron bombardment of many types of specimen is the emission of light. This "cathodoluminescence" constitutes the "signal" forming the basis of the "luminescence mode" of the scanning electron microscope (SEM). The specialized terminology in this field requires some explanation.

There are five modes of scanning electron microscopy. The kinetic energy of the incident beam electrons of the SEM is dissipated into several other forms in interactions with the material of the specimen. These include (1) electrons emitted back into the space above the specimen, (2) x rays, (3) visible light, (4) the generation of currents or e.m.fs (voltages) in the specimen or, if the specimen is sufficiently thin, (5) transmitted electrons. Each of these types of energy can be detected or transduced into an electrical current. This is an electrical signal that can be amplified and electronically processed. It can then be displayed as a video signal on a synchronously scanned cathode ray tube to produce a television-type image. The five forms of energy give rise to the five modes of operation of SEMs: (1) the emissive, (2) electron probe microanalysis (EPMA), (3) luminescence or cathodoluminescence (CL), (4) conductive or charge collection (CC), and (5) scanning transmission electron microscope (STEM) modes [2,3].

The terminology concerning luminescence also requires definition. The distinction first made in science was that between the light emitted by matter "immediately" after excitation, that is, after the supply of some form of activating energy, and light that was emitted over a noticeable period after excitation. The former was referred to as *fluorescence* and the latter as *phosphorescence*. In the case of inorganic solids it was found that a continuous range of decay times occurred for different luminescence emission mechanisms. There was no significant difference between fast and slow emission phenomena. Solid state physicists therefore abandoned use of these terms. Instead they adopted prefixes to indicate the type of energy used to excite the emission. Cathodoluminescence is the light emitted as the result of cathode ray (electron) bombardment. Cathodoluminescence is the most widely and intently observed of all technological phenomena as it is responsible for the emission of light by television screens.

The terms *fluorescence* and *phosphorescence* remained in use by scientists concerned with the luminescence of organic molecules. In this case a significant distinction arises because "permitted" singlet–singlet state transitions occur rapidly. "Forbidden" triplet–singlet state transitions occur slowly [4,5].

The technique of cathodoluminescence (CL) mode scanning electron microscopy has received much developmental effort because of its value for the study of optoelectronic semiconducting materials [2,6]. There has also been considerable interest in the polymer and biological applications of the CL mode [7–9]. Widespread adoption of the technique has been prevented by the unavailability of CL mode detection systems for attachment to SEMs. This is now changing.

The examination of CL in a scanning electron beam instrument can be carried out in two ways just as in the case of the x rays used in microanalysis. That is, micrographs giving a spatial resolution of a micrometer or two can be obtained with a particular wavelength of emission and spectra can be obtained for point analyses. These possibilities were first explored by McMullen (see Ref. 10). This led Bernard et al. [1] to construct a CL

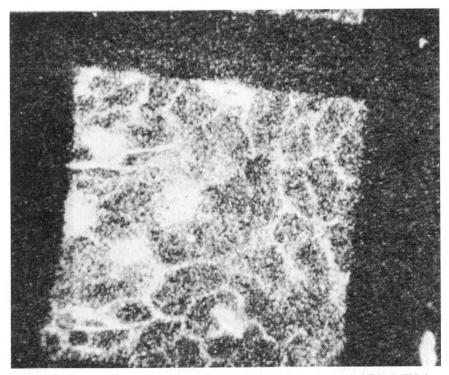

Figure 1. The first SEM CL micrograph of a biological specimen ever published. This is a spinach leaf stained with thioflavin T. $I_b = 10^{-9}$ A, $V_b = 27$ kV, exposure time 30 s, 200×. (After Ref. 12.)

facility for an SEM and Davey [11] one for an electron microprobe. Pease and Hayes [12] first used the SEM CL mode to observe biological specimens. They recorded CL micrographs using the autoluminescence of unstained biological samples and the CL produced by staining with thioflavine T. This was known to produce fluorescence, that is, photoluminescence (PL), when excited with ultraviolet light. They published Fig. 1 showing that in a spinach leaf this dye was selectively absorbed at the cell walls. The combination of spectral analysis of the CL emission with CL micrographs, Pease and Hayes suggested, could be used to examine biological specimens for functional chemical groups in a manner similar to optical fluorescence microscopy.

Giles [13] and Herbst and Hoder [9] reviewed the field of biological CL mode studies 9 and 11 years later, respectively. Both made essentially the same point, namely, that the potential of the technique "has not yet been fully realized" (Giles). This is still true although steady progress is now being made as can be seen from the bibliographies of Refs. 14 and 15 and the reviews of Refs. 16 and 38. In this chapter the reasons for this will be discussed. These include the drawbacks and limitations of the technique, some of which can be partly overcome by instrumental developments. Another reason is the small effort devoted to the SEM CL technique. It is hoped that this chapter may help to stimulate additional interest in this promising field.

13.2 FLUORESCENT AND PHOSPHORESCENT CL OF ORGANIC MOLECULES

The emission of light by organic molecules is due to electron transitions between excited state and ground state energy levels. The emitted CL spectra therefore contain information related to the quantum mechanical energy level diagrams of the molecules. The data constituting this information include the wavelengths or, equivalently, the photon energies of the peaks of the emission bands, their characteristic decay times and efficiencies that is, their relative intensities. Thermal broadening makes difficult this potential means for identifying the molecules concerned. This broadening blurs the sharp emission lines of the quantum mechanical theory. However, if the form of the spectrum of the particular compound is known, it can often be used to recognize the molecules at a particular site when that spectrum is found. This requires sufficient spectral resolution to distinguish similar spectra. Lowering the temperature, preferably to the liquid helium range (i.e., the vicinity of 4.2 K, the boiling point of He at 1 atm of pressure), helps. It decreases thermal broadening to sharpen the spectra somewhat. More important, in the case of the CL of organic materials, is the fact that it generally increases the intensity of the emission and reduces the rate of electron beam damage of the specimens and the consequent changes in the intensity and shape of the spectra [8,17].

Molecules can absorb or emit energy by three mechanisms. These are (1) changes in the electronic energy state, (2) changes in the vibrational energy of the molecule, and in some cases (3) changes in the rotational energy of the molecule. Each of these forms of energy are quantized, that is, they can only take one of a discrete set of energy values. They can be considered separately but they interact.

The separation of the rotational energy levels is very small. Hence rotational spectra are found in the far infrared region.

(For a photon the energy is given by

$$E = h\nu \qquad (1)$$

where ν is the frequency and h is Planck's constant. For electromagnetic radiation

$$\nu\lambda = c \qquad (2)$$

where λ is the wavelength and c is the velocity of light. Hence

$$\nu \, \alpha \, \frac{1}{\lambda} \qquad (3)$$

That is, low frequencies and, by Eq. (1) low photon energies, correspond to long, "far infrared" wavelengths. Higher frequencies and photon energies correspond to shorter, "near infrared" and visible wavelengths of emission.)

The energy spacings between the vibrational energy levels are larger. The rotation–vibration spectra of molecules are found in the near infrared region. The electronic energy levels have still larger spacings. Consequently, vibration and electronic transitions give rise to visible spectra. The term *vibronic* has been coined to denote the system of combined vibrational and electronic energy levels and the spectra to which it gives rise. The rotational energies, if any, are negligible on this scale. Thus molecules have energy level diagrams like that shown schematically in Fig. 2, for purposes of the discussion of visible CL.

SEM CL mode studies have been carried out almost exclusively in the visible region. Much use is made in chemical analysis of infrared spectros-

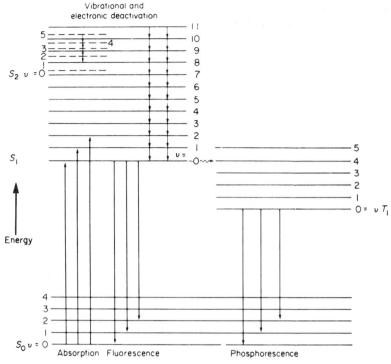

Figure 2. A schematic energy level diagram of an organic molecule. The singlet states are labeled S and the triplet states T. The values of v indicate vibrational levels of the ground and excited electronic states.

copy. This information cannot as yet be used in the interpretation of CL mode spectra of biological specimens. The development of infrared CL detection systems for SEMs has begun, however, and initially promising results have been reported in work on silicon [18,19]. The intense interest in silicon and in materials for near infrared fiber optical communications will ensure that the development of instrumentation for SEM CL work in this wavelength region will continue.

However, thus far, all the work in the biological field has concerned the visible, vibronic CL spectra of molecules to which we now turn. Excitation of molecules by the incidence of high energy electrons in SEMs tends to increase all the three forms of the energy of molecules and even to ionize or dissociate them into smaller units or cause cross-linkage into larger units. The latter processes are the mechanism of the damage, which is a severe problem. However, in favorable cases at least, some of the molecular excitation results in raising the collective "configuration" of the electrons from the ground state to an excited state. These are of two types called singlet (S) and triplet (T) states as marked in Fig. 2. The energy states of the molecules are specified by the values of configurational quantum numbers. The important one in this context is **J**, the total angular momentum quantum number. For some states this has only one value. These are the singlet states. Alternatively, **J** can have three values and these are the triplet states. This means that singlet states have only one energy value but triplet states are triply degenerate. That is, the single energy of a triplet state can be split into three levels in the presence of, for example, a magnetic field.

Excitation, that is, the absorption of energy by the vibronic system, results in a transition from the singlet ground state to a singlet excited state (Fig. 2). The system can then relax to the lowest excited states by vibrational and electronic deactivation that is, emission processes. There are two possibilities for luminescence. The system can make an emissive, radiative transition of the S $\rightarrow$ S type. These luminescent transitions take place relatively rapidly, with decay times less than or equal to 10^{-7} s, and are referred to as fluorescence (Fig. 2). Alternatively, an intersystem crossing can take place from an excited singlet to an excited triplet state. Luminescent transitions from such states are of the T $\rightarrow$ S type and have longer decay times, greater than 10^{-7} s. Such emission processes are referred to as phosphorescence as indicated in Fig. 2.

13.3   THE CATHODOLUMINESCENCE SPECTRA OF ORGANIC COMPOUNDS

The results of the early work were reported by De Mets [7,21] and Falk [20]. One general but not invariable rule was found to be the following. Compounds containing unsaturated covalent bonds, those known as π bonds, tend to be luminescent. The luminescent yield, that is, the efficiency of conversion of excitation energy into emitted light, was often particularly high in compounds with five- and six-membered ring structures. Beam damage was reduced at lower specimen temperatures. The CL spectra closely resembled the photoluminescent (PL), ultraviolet-excited spectra, so PL spectra can be used to give at least tentative identification of compounds from their observed CL spectra. The rule that molecules containing π-bonded ring structures are luminescent was supported by Giles' CL studies of a number of synthetic fiber polymers. He reported that (poly)ethyleneterephthalate, which is known by the tradename Terylene and which contains six-membered unsaturated rings, had a CL yield that was several orders of magnitude greater than polyamide or poly-

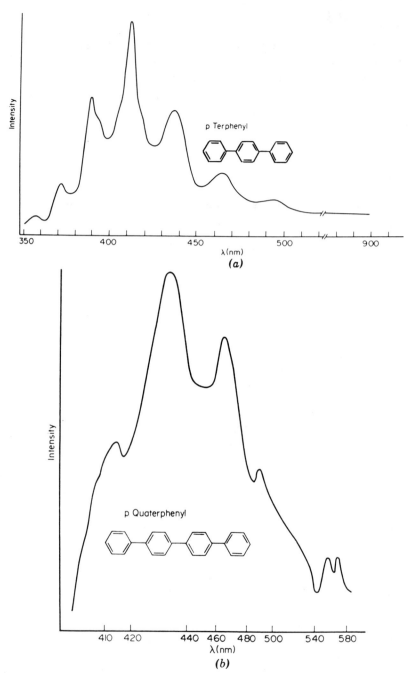

Figure 3. Cathodoluminescence spectra of polyphenyls. (*a*) p. terphenyl crystalline plate-lets, purissimum grade at 25 K and $I_b = 4 \times 10^{-10}$ A. (*b*) p. quaterphenyl microcrystalline platelets, purissimum grade at 20 K, $I_b = 10^{-9}$ A. (After Ref. 8.)

acrylic fibers such as those known as Nylon and Acrilan, which contain only linear π bonds.

A related observation of more biological interest was made by Bond et al. [22]. They observed a relatively intense CL emission within the cells of *Eucalyptus* stem fibers. They ascribed this to local concentrations of phenolic substances. They confirmed this by staining with Toluidine blue.

The CL spectra of adenosine, albumin, cytosine, DNA, guanine, lysozyme, thymine, trypsin, tryptophan, and tyrosine were also determined [23–26]. In this work also it was found that beam damage could be reduced by the use of cold stages.

Systematic studies of the CL spectra of series of organic compounds were reported by De Mets et al. [8]. Figure 3 contains their spectra for

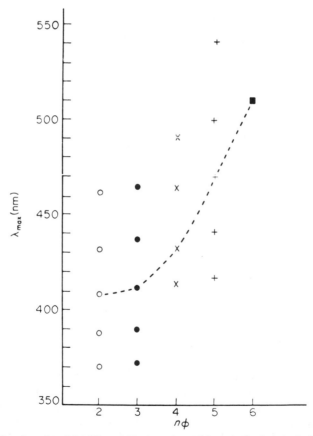

Figure 4. Wavelengths of the CL emission maxima of the p. polyphenyls, indicated as $n\phi$. The dashed curve indicates the wavelength shift of the main emission peak. (After Ref. 8.)

two of a series of polyphenyl compounds. The family resemblance of the spectra of members of this series is well illustrated here. In general there was a broad emission band with a series of peaks scattered along it. Figure 4 shows the steady increase in the wavelengths of the five main peaks for the first six of these compounds. The curve for the main peak wavelength of the central doublet is indicated by the dashed curve. Such data can be used to identify polyphenyl molecules by their spectra and to determine the number n of the particular member of the series involved. De Mets et al. also published series of spectra for fused-ring hydrocarbons (anthracene, perylene, etc.) and phenyl derivatives of (poly)alkenes (tr. stilbene, 4,4'-diphenylstilbene, etc.) together with a plot of the peak wavelengths for the first three members of the latter series, in the manner of Fig. 4.

The distinction must be made between autoluminescence and CL due to fluorescent stains. Autoluminescence is emitted by organic molecular constituents of the biological specimen. Its analysis should identify the molecules or structures involved. If damage can be limited or corrected for and if the intrinsic CL yield of the substances involved is known, the concentration of the molecules in the volume emitting can, in principle, be obtained from the observed emission intensity. This would be analogous to the way in which the concentrations of elements are found from their intensities of emission in x-ray microanalysis. The difficulty of this possible method of analysis is compounded by (1) the low intensity of CL emission in the SEM, especially from organic substances, and (2) beam damage problems as well as the lack of systemic interpretive data of the kind shown in Figs. 3 and 4.

13.4   BEAM DAMAGE AND SEM OPERATING CONDITIONS

Giles compared the CL from efficient inorganic phosphors such as ZnS powder for cathode ray tube screens and GaP for light-emitting diodes (LEDs) with that from organic molecules. He reported that the CL from even the most luminescent organics was orders of magnitude less than that from the inorganic phosphors under the same electron beam and detector system conditions.

The effect of continued electron beam bombardment is generally found to be a decrease in the CL emission intensity. Pease and Hayes referred to it as "poisoning" [8]. It has been repeatedly reported [7,27,28] and is now usually referred to as beam damage or degradation. It is believed that the mechanism of this degradation is bond scission and cross-linking. These processes involve intermediate ionic and free radical states produced by beam ionization of the molecules [29]. When this occurs, so much energy is transferred to the molecule that one or more electrons are lifted right up out of the range of bound states represented in diagrams like that of Fig. 2.

Colebrooke and Windle [30] studied beam damage in (poly)methylmethacrylate (PMMA also known under the tradenames Perspex in Britain and Lucite in the United States). They found the beam effect to be approximately inversely proportional to the beam accelerating voltage V_b and directly proportional to the beam current I_b. Heidenreich et al. [31] put forward a semiempirical model of these effects that fitted the observations of Colebrooke and Windle. The interest of Heidenreich et al. and many subsequent workers arose from the fact that PMMA is the simplest of the "resists" that can be used in electron beam lithography to produce large-scale integrated circuits with components of submicron dimensions. The beam polymerization is then required to harden, or the beam scission to soften, the areas to be protected or exposed, respectively.

The basic idea is that the rate of damage is dependent on the beam energy dissipated per unit volume of the specimen. The incident beam electrons dissipate their energy by scattering interactions that deflect and slow them. They therefore spread out into a droplet-shaped volume like that shown in Fig. 5a. This can often be approximated by a sphere or portion of a sphere as shown in Fig. 5b. The penetration range of the incident beam and the volume of the energy dissipation volume increases with the beam accelerating voltage V_b. Hence for a given beam current, an increase in beam voltage leads to a reduction in the average energy dissipated per unit volume. Conversely, an increase in beam current for constant voltage results in an increase in the energy density deposited since more incident electrons are arriving per unit time and the dissipation volume is constant.

To obtain the maximum CL emission for the minimum damage therefore Giles recommended the use of a high beam voltage and low beam current and a fast scan speed. The latter means that the beam is incident at each point for a minimum time. Factors that limit the values of these operating parameters that can be adopted are as follows. The increasing beam dissipation volume means a decreasing resolution and increases the depth of the subsurface layer examined. A minimum beam power $I_b V_b$ is needed to get sufficient intensity to detect and a sufficiently slow scan speed is needed to obtain a high enough signal-to-noise ratio to record a clear, stable micrograph. Trial and error is required to define optimum operating conditions for a particular SEM, CL detection system and type of specimen.

Figure 6 is an observation by Giles showing the effect of beam damage in reducing the CL spectrum emitted from an area of a synthetic fiber on

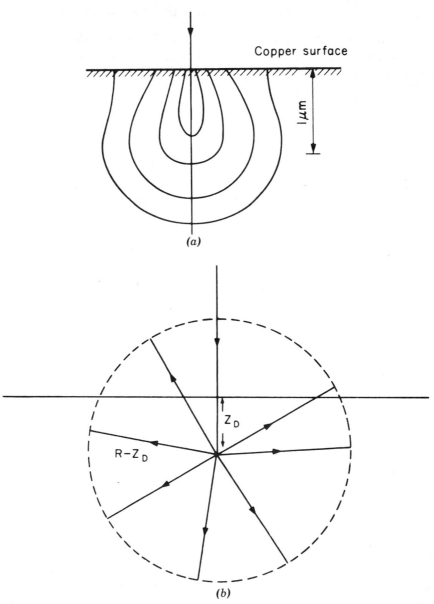

Figure 5. (*a*) Contours of equal energy loss under a 29-keV electron beam impact point, calculated by the Monte Carlo method for a copper target by Bishop [32]. (*b*) The simple Archard [33] diffusion model for the beam energy dissipation volume. The beam is assumed to dissipate its energy at a point at a depth Z_D whence the energy diffuses out into a spherical volume. $Z_D = 4OR_B/7Z$, where R_B is the beam penetration range and Z is the (average) atomic number of the material. In some simple cases $Z_D = R_B/2$ and the sphere is just inside the solid and in others $Z_D = R_B$ and the energy dissipation volume is a hemisphere with the specimen surface bisecting the sphere.

312

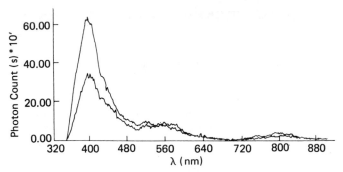

Figure 6. CL spectra of polyethylene terephthalate (Terylene fiber, type 111H). The upper curve was obtained on first sweeping the monochromator through the wavelength range and the lower curve was obtained on immediately repeating the observation. (After Ref. 28.)

two immediately successive recording runs. It is the reduction in intensity together with possible changes in the form of the spectrum that makes damage such a serious problem in CL observations of organic materials and biological specimens.

Figure 7 is a result of De Mets et al. [8] showing the effect of reducing the specimen temperature to liquid helium levels in sharpening the emission bands and revealing more detail.

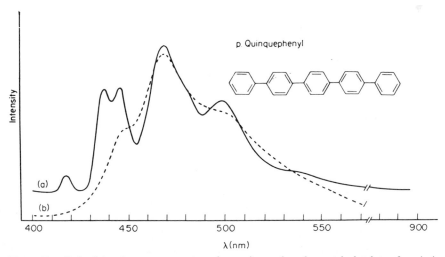

Figure 7. Cathodoluminescence spectra of p. quinquephenyl crystal platelets of purissimum grade (a) at 6 K and (b) at room temperature. Both were recorded with $I_b = 4 \times 10^{-10}$ A and the intensity scale units are the same for both but the zero level of spectrum a has been raised for greater clarity. (After Ref. 8.)

13.5 CATHODOLUMINESCENCE DETECTION SYSTEMS

The beam powers available in scanning electron microscopes are small. The power usable in examining biological specimens is further limited by considerations of beam damage. For Fig. 1, for example, $I_b V_b$ was 27 μW. The power in the emitted CL spectrum due to the low luminescent efficiencies of organic molecules is always many orders of magnitude lower than that in the beam. This means that the highest possible fraction of the small numbers of photons emitted per second must be collected and detected.

Most modern CL detection systems incorporate some form of semiellipsoidal mirror [34]. The design originated by Steyn and Giles [28] is shown in Fig. 8. The semiellipsoidal mirror of Fig. 8a is made of aluminized epoxy and acts to bring the light emitted in all directions from the specimen at the first focus, F_1, via the second focus F_2, through a collimating lens into the fiber optic light guide. The latter is a specially made "transformer" that changes in cross section from a circle to match the

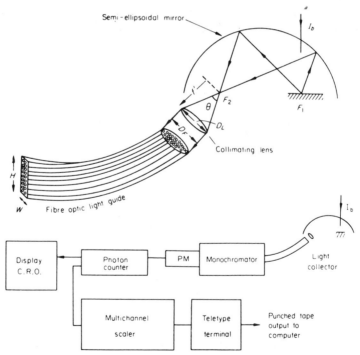

Figure 8. Components of a spectroscopic detection system. (a) The light collector. (b) Schematic diagram of the whole system. (After Ref. 28.)

lens at the input end to a rectangle to match the monochromator entrance slit at the output end.

The monochromator (Fig. 8b) disperses the light so one narrow "pass-band" of wavelengths, $\lambda \pm \Delta\lambda/2$, at a time is detected by the photomultiplier, PM. The output of the PM is a train of pulses, each due to a single photon. A pulse-counting circuit measures the count rate, which is an indicator of the CL emission intensity. One wavelength band can be displayed on the cathode ray oscilloscope (CRO) to form a "monochromatic" CL SEM micrograph. Alternatively, the whole spectrum can be detected and displayed as video signals to produce a TV-like "panchromatic" or "integral" CL micrograph. Panchromatic micrographs cannot distinguish the spatial distributions of different luminescent substances but can be used when no one color band is strong enough to produce a clear micrograph.

Alternatively, the electron beam can be left stationary, or scanned over a small selected area, while the monochromator runs through the visible range of wavelengths. This gives a spectrum via the multichannel scaler and computer in the system of Fig. 8.

The evolution of light collection systems of higher efficiencies is outlined by Giles [13] and Herbst and Hoder [9]. Many different types of CL detection system have been developed [16,35]. An alternative approach that is of interest for organic CL studies is the use of beam chopping and phase-lock amplification. This turns the electron beam of the SEM on and off in regular pulses and is said to "chop" it. Phase-lock amplification gives the amplified difference between the PM output with the beam on and off. This eliminates the incandescent (thermoluminescent) light emitted by the hot tungsten filament in the electron gun, which may not be negligible compared with the weak CL of biological specimens. To minimize PM noise so low intensity CL can be detected reliably, cooling and magnetic focusing of the PM are desirable. The relatively new GaAs : Cs (caesiated gallium arsenide) photocathode photomultipliers, which have relatively high and roughly constant efficiencies for counting photons across the whole visible range of wavelengths, are probably to be preferred also.

The most important additional item of equipment for organic and biological CL studies is a cold stage. Although a liquid helium stage was used by De Mets et al. [8] in the early 1970s, and liquid nitrogen stages are in use, no other such stage appeared in the literature for nearly a decade. Then the importance of liquid helium temperatures for the interpretation of the Cl of inorganic materials was stressed [36] and the value of liquid helium temperatures in increasing CL intensities and simplifying damage in biological specimens were demonstrated [37] at the same conference in

1980. At a conference in early 1981 on semiconducting materials the existence of five new liquid helium stages was announced. Of perhaps the most significance here is the fact that one of them is now available commercially together with a minicomputer-controlled spectroscopic CL detection system. The SEM stage is based on a continuous flow liquid helium cryostat shown in Fig. 9. The specimens are attached to copper mounts that are inserted into a cooled sliding plate, to allow some specimen movement. The specimen is enclosed in a cooled, thermal shielding box. This reflects heat radiation and prevents vacuum pump oil contami-

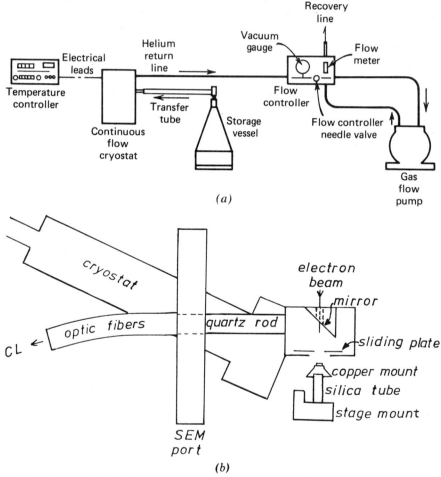

Figure 9. (a) The Oxford Instruments temperature-controlled, continuous flow liquid helium cryostat. (b) Schematic diagram of the liquid helium stage for SEM CL observations mounted on the end of the cryostat. (From Ref. 35.)

nation in the SEM specimen chamber from reaching the specimen. A system of mirrors, a lens, and a quartz rod take the light out to the port in the side of the specimen chamber.*

Hörl and Roschger [37] reported that cooling unstained human lymphocytes and erythrocytes to liquid helium temperatures resulted in an increase by a factor of about 8 in the light emission, which rapidly decreased. Large increases in luminescence efficiency are found in many cases also on cooling inorganic materials. The increase in CL spectral detail on cooling to liquid helium temperatures was clear in the work of de Mets et al. (see Fig. 7).

It is desirable to be able to measure other parameters of the CL emission from biological specimens in order to unambiguously interpret the molecular mechanisms responsible. The most obvious of these parameters is the decay time for continued emission after electron bombardment ceases. Hörl and Roschger observed faint horizontal stripes in their liquid nitrogen temperature micrographs. These they ascribed to a triplet, phosphorescence emission with a half-life of the order of 30 μs. Thus low temperature stages may make decay times measurable so that at least fluorescence and phosphorescence transitions can be reliably distinguished.

It would also be desirable to extend observations into the infrared, where vibration–rotation spectra might be observed. This requires totally different detection systems: semiconductor detectors instead of photomultipliers and Fourier transform spectrometers instead of grating monochromators, and so on. The feasibility of such techniques has been demonstrated [19] and progress in this field should be relatively rapid because of its value for the study of fiber optical communications materials and devices. As in the visible, so in the infrared region, the lower CL emission intensities of organic and biological specimens and their damage sensitivity, however, will present added difficulties.

13.6 SPECIMEN PREPARATION

The subject of specimen preparation lies well outside the authors' field of competence. Fortunately, workers in the field have reviewed the literature recently [9,16, and especially 38]. Interested readers are referred to those reviews for detailed accounts of the subject matter of the remainder of this chapter as well as the full literature references. Here we will give only a summary of the main conclusions of work in the field.

---

* This SEM liquid helium CL stage and detection system is available from Oxford Instruments Ltd., Osney Mead, Oxford, U.K.

The essential requirement is that the preparation method must not alter either the autoluminescence or the luminescent staining in the specimen. Staining is usually carried out first, followed by fixation, dehydration, and coating as normally used in preparation of SEM biological specimens. Unfortunately, some fixatives can increase autoluminescence or decrease the luminescence of a dye. Glutaraldehyde is a good fixative for vaginal cells and mesenteric tissues but tests are needed to determine the best treatment in other cases.

Critical point drying has been used successfully in CL work as have cryostat sections mounted either on polished aluminum stubs or nylon foils stretched across aluminum rings. The use of a cold stage together with a transfer line and cold finger requires only the prior rapid freezing of the specimen. Freeze-drying requires neither fixing nor dehydration. Rapid freezing in liquid nitrogen and freeze-drying is used. Again the importance of cold stages for SEM CL work is clear. [See Chapter 21 for further details.]

The usual embedding compounds are generally luminescent but epon is less so than the others. Coating with gold or carbon to avoid charging reduces the emission of light. Often charging is not a problem using CL specimen preparation and SEM operating conditions. It is best to carry out the CL observations first and subsequently coat the specimen if necessary for high resolution secondary electron microscopy.

13.7   BIOLOGICAL AND MEDICAL APPLICATIONS

Bröcker and Pfefferkorn [38] classified the applications of the CL mode of the SEM. This classification is used in Table 1.

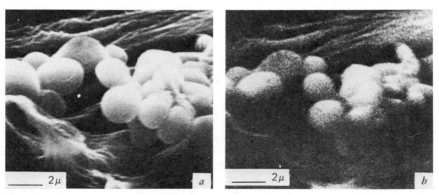

Figure 10.   A clump of spores of *Pityriasis versicolor* in human skin (*a*) secondary electron and (*b*) CL mode SEM micrographs. (After Ref. 43.)

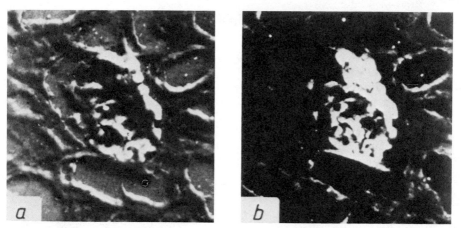

Figure 11. Uncoated cryostat section of rat kidney (*a*) secondary electron and (*b*) CL SEM micrographs. Specific CL immunostaining of fibrin deposition in glomeruli is visible. (After Refs. 44 and 45.)

An example of specific autoluminescence is shown in Fig. 10, where clearly it is the "spores" that are luminescent.

Examples of cathodoluminescent staining are shown in Figs. 11 and 12; see Table 1 for brief explanations. For further details of this work and of the remainder of the literature see the reviews of Refs. 9 and 38 and the bibliographies of Refs. 14 and 15.

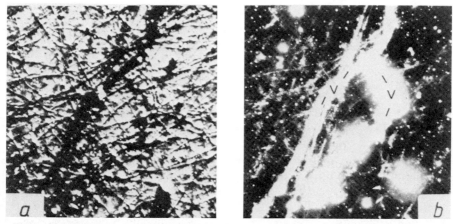

Figure 12. Rabbit mesenteric tissue after intravenous injection of the dye brilliant sulfoflavine complexed with albumins and critical point drying (*a*) primary (backscattered) electron and (*b*) CL SEM micrographs. The latter shows luminescent halos of the microvessels marked v. (After Ref. 47.)

TABLE 1 Types of Applications of the CL Mode in Biology and Medicine

Autoluminescence	CL Staining
Direct analysis of the luminescent constituents of specimens. This has included:	1. Demonstration of splits, cracks, or fissures through the penetration of luminescent solutions followed by drying. Demonstration of the depth of penetration of, e.g., dental filling materials [38]
Distribution of luminescent herbicides over leaf surfaces [39]	
Distribution of calcification of human aortae [40]	
CL of localized Xanthones, etc., in the lichens *Lecanora, Buellia,* and *Launera* [41]	2. Immunocathodoluminescence, e.g., fluoresceinisothiocyanate (FITC) labeled anti-rat-fibrinogen serum was used to study the deposition of fibrin in rat kidneys after the infusion of an endotoxin [44,45]—Fig. 11
Identification of nucleic acid and other biological molecules in situ with high resolution [23,24,42]	
CL of hypheae and spores of *Pityriasis versicolor* in human skin [43]—see Fig. 10	3. Staining of cells, tissues, and organs by luminescent dyes (e.g., Refs. 12, 44–46]—see Fig. 12
Localized CL in wood and plant tissue [22]	
	4. The production of luminescent precipitates may be possible

After Bröcker and Pfefferkorn [38].

13.8 THE ADVANTAGES OF THE CL MODE OF THE SCANNING ELECTRON MICROSCOPE

The difficulties of the technique were recounted at some length in Sections 13.3–13.5, together with the instrumentation developed to minimize them. To balance our account we will close by summarizing the attractive features of the CL mode for biological and organic studies.

The most directly comparable technique to the CL mode of the SEM is fluorescent (ultraviolet-excited emission, i.e., PL) light microscopy. The CL mode has a number of advantages over fluorescent light microscopy. These include the fact that some substances emit CL but not photoluminescence (PL). In the CL mode electronic amplification can be used to enhance weak emission, and various forms of signal processing such as differentiation can be used to enhance micrographic contrast. Moreover, computer data processing of the CL signal and pattern recognition or image analysis can be applied to obtain quantitative results rapidly. This has not yet been seriously exploited in biological studies, or indeed in other fields to any great extent, but its value for the future evolution of SEM techniques as routine laboratory methods is great. The results of the CL mode are available simultaneously with those of other modes of the SEM. It is often useful to compare the results with those of the emissive mode, as in Figs. 10–12, of the x-ray mode that is, of electron probe microanalysis and of scanning transmission electron microscopy when these are available.

In principle, CL mode micrographs can have higher resolution than light micrographs as they are not limited by the wavelength of light. Despite some exaggerated claims it is doubtful whether such resolutions have been attained in CL mode observations of biological specimens. In the case of highly luminescent, damage-resistant specimens, however, there is no reason why present and future advances in instrumentation should not lead to submicron resolutions. The large depth of field, well known in the case of secondary electron SEM micrographs, is also available in CL mode pictures.

The conclusion of these considerations must be that the CL mode of the SEM is of greater potential use in biological and medical research than has so far been realized. The commercial availability of cold stages and spectroscopic CL mode detection systems removes a major obstacle to the wider use of this technique.

BIBLIOGRAPHY

CL Mode SEM Studies of Organic and Biological Specimens

Bröcker, W. (1976). Biologisch-Medizinische Anwendungen der Kathodolumineszenz am Raster-Elektronenmikriskop, Microsc. Acta 78, 105–117.

Bröcker, W., and G. Pfefferkorn (1979). Applications of the cathodoluminescence method in biology and medicine, Ref. 38.

Giles, P. L. (1975). Cathodoluminescence, Ref. 13.

Herbst, R., and D. Hoder (1978). Cathodoluminescence in biological studies, Ref. 9.

de Mets, M. (1974), Ref. 7.

de Mets, M., K. J. Howlett, and A. D. Yoffe (1974). Cathodoluminescence spectra of organic compounds,'' Ref. 8.

Pfefferkorn, G., W. Bröcker, and M. Hastenrath (1980). The cathodoluminescence method in the scanning electron microscope, Ref. 16.

CL Mode with Biological and Medical Applications

Bröcker, W., and G. Pfefferkorn (1978). Bibliography on cathodoluminescence, Ref. 14.

Bröcker, W., and G. Pfefferkorn (1980). Bibliography on cathodoluminescence, Part II, Ref. 15.

Theory of the Luminescence of Organic Molecules

Becker, R. S. (1969). *Theory and Interpretation of Fluorescence and Phosphorescence*, Wiley, New York.

Carmier, M. J., D. M. Hercules and J. Lee, Eds. (1972). *Chemiluminescence and Bioluminescence*, Proc. 2nd Conf. Chemilumin., Univ. Georgia, Plenum, New York.

Finkelnburg, W. (1950). Molecular physics, in *Atomic Physics*, McGraw-Hill, New York, pp. 346–417. Most readable introduction known to the author.

Harris, D. C., and M. D. Bertolucci (1978). *Symmetry and Spectroscopy: An Introduction to Vibrational and Electronic Spectroscopy*, Oxford University Press, New York.

Herring, P. J. Ed. (1978). *Bioluminescence in Action*, Academic, London.

Howarth, O. (1973). *Theory of Spectroscopy: An Elementary Introduction*, Nelson London. Basic quantum mechanics of emissive transitions.

Lumb, M. D. (1978). Organic luminescence, in *Luminescence Spectroscopy*, M. D. Lumb, Ed, Academic, London, chap. 2.

Luminescence

A central role is played by the *Journal of Luminescence,* which has also published the proceedings of most of the biennial International Conferences on Luminescence.

SEM

Central roles are played in this case by the journal *Scanning: International Journal of Scanning Microscopy and Related Methods* and the Proceedings of the Annual SEM Symposia. These are published as (Scanning Electron Microsc.) "SEM 1981," "SEM 1982," etc., by SEM, Inc. whose address is the curious "AMP O'Hare, IL 60666, U.S.A."

REFERENCES

1. R. Bernard, F. Davoine, and P. Pinard, C.R. Acad. Sci. *248,* 2564–2566 (1959).

2. P. R. Thornton, *Scanning Electron Microscopy,* Chapman and Hall, London, 1968.

3. D. B. Holt, M. D. Muir, P. R. Grant, and I. M. Boswarva, Eds., *Quantitative Scanning Electron Microscopy,* Academic, London, 1974.

4. R. S. Becker, *Theory and Interpretation of Fluorescence and Phosphorescence,* Wiley, New York, 1969.

5. M. D. Muir and P. R. Grant, Cathodoluminescence, in Ref. 3, pp. 287–334.

6. D. B. Holt, Quantitative scanning electron microscope studies of cathodoluminescence in adamantine semiconductors, in Ref. 3, pp. 335–386.

7. M. de Mets, Cathodoluminescence of organic chemicals, in M. A. Hayat,

Ed., *Principles and Techniques of Scanning Electron Microscopy*, Van Nostrand Reinhold, New York, 1974, chap. 1.

8. M. de Mets, K. J. Howlett, and A. D. Yoffe, J. Microsc. *102*, 125–142 (1974).

9. R. Herbst and D. Hoder, Scanning *1*, 35–41 (1978).

10. K. C. A. Smith and C. W. Oatley, Br. J. Appl. Phys. *6*, 391–399 (1955).

11. J. P. Davey, Ph.D. thesis, Cambridge University, 1965.

12. R. F. W. Pease and T. L. Hayes, Nature *210*, 1049 (1966).

13. P. L. Giles, J. Microsc. Biol. Cell. *22*, 357–370 (1975).

14. W. Bröcker and G. E. Pfefferkorn, in Scanning Electron Microsc. 1 (1978), AMF O'Hare, Chicago, pp. 333–351.

15. W. Bröcker and G. Pfefferkorn, Scanning Electron Microsc. *1*, 298–302 (1980), AMF O'Hare, Chicago.

16. G. Pfefferkorn, W. Bröcker, and M. Hastenrath, Scanning Electron Microsc. *1*, 251–258 (1980), AMF O'Hare, Chicago.

17. E. M. Hörl and P. Roschger, in Scanning Electron Microsc. *1*, 285–292 (1980), AMF O'Hare, Chicago.

18. S. M. Davidson, T. J. Cumberbatch, E. Huang, and S. Myhajlenko, in *Microscopy of Semiconducting Materials, 1981*, A. G. Cullis and D. C. Joy, Eds., Institute of Physics, Bristol and London, 1981, pp. 191–196.

19. T. J. Cumberbatch, S. M. Davidson, and S. Myhajlenko, in *Microscopy of Semiconducting Materials, 1981*, A. G. Cullis and D. C. Joy, Eds., Institute of Physics, Bristol and London, pp. 197–202.

20. R. H. Falk, Cathodoluminescence of herbicides, in *Principles and Techniques of Scanning Electron Microscopy*, M. A. Hayat, Ed., Van Nostrand Reinhold, New York, 1974, chap. 2.

21. M. de Mets, Microsc. Acta *76*, 405–414 (1975).

22. E. F. Bond, D. Beresford, and G. H. Haggis, J. Microsc. *100*, 271–282 (1974).

23. W. A. Barnett, M. L. H. Wise, and B. L. Jones, J. Microsc. *105*, 299–303 (1975).

24. P. V. C. Hough, W. R. McKinney, M. C. Ledbetter, R. E. Pollock, and H. W. Moos, Proc. Natl. Acad. Sci. U.S.A. *73*, 317–321 (1976).

25. P. V. C. Hough, in Scanning Electron Microsc. 1 (1977), Chicago, pp. 257–264.

26. R. Mohan, R. Steiner, and R. Kaufmann, in Proc. Int. Conf. Microprobe Anal. in Biol. Med., Hirzel, Stuttgart, 1977; quoted in Ref. 9.

27. M. D. Muir, P. R. Grant, G. Hubbard, and J. Mundell, Scanning Electron Microsc. Proc. 4th Ann. SEM Symposium, IITRI, Chicago, 1971, pp. 401–408.

28. J. B. Steyn, P. Giles, and D. B. Holt, J. Microsc. *107*, 107–128 (1976).

29. A. Colebrooke, M.Sc. thesis, Imperial College, University of London, 1972.

30. A. Colebrooke and A. H. Windle, in *Scanning Electron Microscopy: Systems and Applications 1973*, W. C. Nixon, Ed., Institute of Physics, London and Bristol, pp. 132–135.

31. R. D. Heidenreich, L. F. Thompson, E. D. Feit, and C. M. Melliar-Smith, J. Appl. Phys. *44*, 4039–4047 (1973).

32. H. E. Bishop, Proc. Phys. Soc. *85*, 855–865 (1965).

33. G. D. Archard, J. Appl. Phys. *32*, 1505–1509 (1961).

34. E. Hörl and E. Mügschl, in Proc. 5th Eur. Congr. Elect. Microsc. (EMCON 72), Institute of Physics, London and Bristol, pp. 502–503.

35. D. B. Holt, in *Microscopy of Semiconducting Materials, 1981*, A. G. Cullis and D. C. Joy, Eds., Institute of Physics, Bristol and London, pp. 165–178.

36. D. B. Holt and S. Datta, Scanning Electron Microsc. *1*, 259–278 (1980), AMF O'Hare, Chicago.

37. E. M. Hörl and P. Roschger, Scanning Electron Microsc. *1*, 285–292, AMF O'Hare, Chicago.

38. W. Bröcker and G. Pfefferkorn, in Scanning Electron Microsc. *2*, 125–132 (1979), AMF O'Hare, Chicago.

39. F. D. Hess, D. E. Beyer, and R. H. Falk, Weed Sci. *22*, 394–401, (1974).

40. W. Bröcker, H. J. Höhling, W. A. P. Nicholson et al., Path. Res. Pract. *163*, 310–322 (1978).

41. D. Hoder and A. Mathey, Microsc. Acta Suppl. *2*, 271–280 (1978).

42. W. A. Barnett, E. L. Jones, and M. L. H. Wise, Micron 6, 93–100 (1975).

43. J. P. Carteaud, M. D. Muir, and P. R. Grant, Sabouraudia *10*, 143–146 (1972).

44. W. Bröcker, E. H. Schmidt, G. Pfefferkorn, and F. K. Beller, Scanning Electron Microsc. IITRI, Chicago, 1971, pp. 243–250.

45. E. H. Schmidt, H. Bröcker, W. Wagner, et al., Am. J. Pathol. *81*, 43–48 (1975).

46. R. Herbst, E. M. Hörl, and A. M. Multier-Lajous, Beitr. Elektronenmikr. Direktabb. Oberfl. *6*, 169–176 (1973).

47. W. Bröcker, G. Hauck, R. Blaschke et al., Microsc. Acta Suppl. *2*, 260–270 (1978).

THREE-DIMENSIONAL ANALYSIS
OF SURFACES

PETER G. T. HOWELL and ALAN BOYDE

Department of Anatomy and Embryology
University College London
London, England

14.1 INTRODUCTION

Three levels may be ascribed to the study of surfaces in three dimensions: the first is simple stereoscopy, the second is photogrammetry, and the third is stereophotogrammetry. These represent three discrete aspects of the same subject, and may be considered to form a logical progression from stereoscopy through photogrammetry to stereophotogrammetry, as the interest in three-dimensional analysis tends to move away from simple observation toward measurement and reconstruction in three dimensions. The field of interest should not be confused with the science and art of stereology, in which the bulk properties of an object, such as relative volume fractions of the constituent phases within a sample, are estimated from essentially flat two-dimensional sections (see Chapter 19 in this volume).

This chapter deals with the answers provided by analyses based on imaging with a scanning electron microscope (SEM) operated at low ac-

The authors' work in the field of three-dimensional analysis of surfaces has been generously supported by grants from the Science Research Council.

celerating voltage and using secondary electrons to give the least depth information from the sample, and using oriented pairs of images to deduce information about the position of the surface. The general approach is the same under all circumstances concerning unfamiliar surfaces about which nothing is known in advance.

Because it may be important to know the orientation of a surface facet to be subjected to any SEM-based method of surface-zone chemical analysis (e.g., EDX), the information given here is of significance for such other methods of surface analysis. It is important that three-dimensional imaging techniques be employed whenever any rough surface is viewed.

Historical Note. Stereo techniques were first introduced into transmission electron microscopy in 1940 by von Ardenne and into scanning electron microscopy in the late 1950s. Wells (1957) applied these techniques to the study of fibers. Stereo analysis has been reviewed most recently by Howell (1975), Boyde and Howell (1977) and Wergin and Pawley (1980).

14.2   LEVEL 1: STEREOSCOPY

Stereoscopy is seeing objects in three dimensions and is the result of stereoscopic vision. It requires, first, stereopsis—the individual ability to see objects in three dimensions. This is an innate ability that one either possesses or does not possess; approximately 10% of the population do not preceive their surroundings stereoscopically! (These persons obtain information about the third dimension in their everyday worlds by such visual clues as size and shape or by generating motion parallax through walking at a constant speed up to an object.) Second, a pair of images is required that provides the observer with two slightly different views of the same subject. The third requirement is a means by which these images may be viewed.

14.2.1   Methods for Recording Stereopairs

How may the stereoscopic images be created so that the surface may be viewed in its full three-dimensional detail? The stereoscopic images must provide two aspects of the same area of the sample that differ only slightly in their viewing point, such that when the two single images are observed, the brain fuses them into a composite image that is interpreted as a three-dimensional model of that surface. In the scanning electron microscope there is normally a mechanical stage that enables the operator to rotate (or tilt) and translate (or shift) the sample about three mutually orthogonal axes so that the greater part of the sample surface may be viewed from any desired orientation. The electron beam acts as the line of sight, or viewing point, and the apparent illumination is provided by the secondary electron detector.

The raster in the SEM is formed by a bundle of rays that arise from a single point (PC, the perspective center) and diverge across the sample surface from here (Fig. 1a). This then represents a perspective geometry. With increased magnification the angle of scan is significantly reduced so that for all intents and purposes the individual rays of the raster may be considered to be parallel—and a parallel projection can be assumed for the image formation (Fig. 1b). At either high or low magnification stereopairs may be created by tilting the sample, while at low magnification it is also possible to move the sample across the field of view to create the stereopair. Because of its universal application the *tilt-only* method is the method of choice for taking stereopairs.

Tilt-Only Method

The tilting mechanism of the SEM stage is used to alter the gross inclination of the sample with respect to the electron beam and may be used to

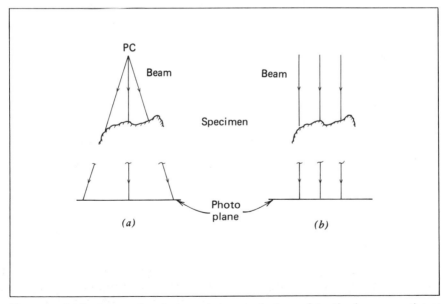

Figure 1. (a) In the scanning electron microscope the image formation is a perspective or central projection. This is especially important at low magnifications where the angle of scan is significant. (b) At higher magnifications the angle of scan across the specimen is greatly reduced so that above 500× a parallel projection may be assumed for the image formation.

create the stereopair by altering the sample orientation. Recording a stereopair proceeds as follows:

Step 1. Initially the sample is set up to show the selected area of the sample. The tilt axis should lie parallel to one of the sides of the cathode ray tube (CRT) to provide a specific line built into the photographs of the stereopair against which subsequent orientation (or measurement) of the stereopair may be made. It is advisable at this stage to check the alignment of the tilt axis by translating the sample with the stage X or Y controls and simultaneously observing the movement of the sample as its image traverses the CRT (Fig. 2a). Either the working distance or the raster rotation must be altered so that the tilt axis does lie parallel to one of the sides of the CRT.

Step 2. The first micrograph of the stereopair can now be recorded (Fig. 2b). The subsequent relocation of the sample after an additional tilt has been applied is facilitated by marking the position of some prominent feature close to the center of the visual CRT with a wax crayon. Where

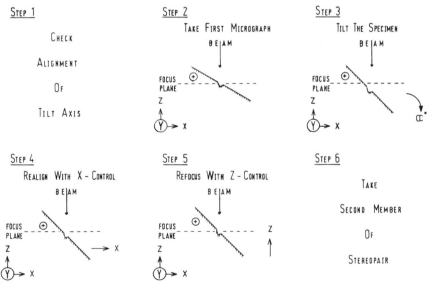

Figure 2. The sequence of events necessary to record a stereopair using the "tilt-only" method. Step 1: First check the alignment of the tilt axis to ensure it coincides with one of the sides of the CRT of the SEM ($\oplus$ = tilt axis). Step 2: Record the first member of the stereopair, marking some prominent feature to enable relocation of the specimen after tilting the sample. Step 3: Tilt the sample through the required tilt-angle difference. Step 4: Realign the sample with the X-stage control. Step 5: Refocus the image using the stage Z control. Step 6: Now record the second member of the stereopair.

surfaces have a repetitive or intricate structure, this procedure may have to be repeated at two or more decreasing magnification steps.

Step 3. The sample is now slowly tilted (Fig. 2c) and the direction of movement of the features across the CRT corrected by adjusting the appropriate stage movement so as to retain their image within the area of the screen. Unless the area of interest on the specimen surface lies exactly on the tilt axis, the sample moves across the screen in a direction perpendicular to the tilt axis when the tilt control is adjusted. At the same time it passes out of focus.

Step 4. It is necessary therefore to return the area of the sample into location under the electron beam with the stage X control (Fig. 2d).

Step 5. The sample is brought back into focus with the stage height or Z control (Fig. 2e). This latter is used in preference to refocusing with the final lens as this will alter the magnification between the pair of photographs.

Step 6. With the area of the sample relocated under the crayon marks on the CRT, the second member of the stereopair can be recorded (Fig. 2f).

Specially designed eucentric stages are now available (Houghton et al., 1971) that often provide a tilt in the "side-to-side" direction, θY, for use in x-ray microanalysis. Here the ability to position the mean plane of the sample surface toward the collimator of the x-ray detector greatly increases the x-ray yield. A sufficient degree of rotation about the Y-axis is usually provided to enable stereopair photographs to be recorded using this tilt or rotation. The benefits are that the gross inclination of the sample with respect to the electron detector remains constant, eliminating any alterations in brightness and contrast that may be observed while recording stereopairs with the "fore–aft" tilt stage. The orientation of the stereopair is the same as that of the screen when recorded from the SEM. The third axis about which it is possible to rotate the sample is θZ, that is, around the incident electron beam. Whereas this may, and will, generate stereopairs if there is a gross forward inclination of the sample, the subsequent alignment of the photographs thus created is tedious and is not recommended by the authors. Also the subsequent progression to stereophotogrammetry is difficult, if not impossible, under these recording conditions.

Low Magnification Shift-Only Case

In conventional aerial photogrammetry, stereoscopic images are obtained from aircraft by flying along a straight flight path during which photographs of the earth below are recorded sequentially. Adjacent photographs of the series include areas of the ground common to both images—the area of overlap. The analogous situation in the scanning electron microscope is to translate the sample across the field of view between sequential exposures (Fig. 3). This technique can be used only at relatively low instrumental magnifications (up to 50×), as above this the translation necessary to provide reasonable stereoscopic imaging would remove that area completely from the field of view.

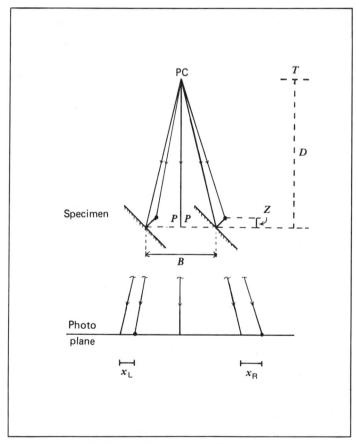

Figure 3. The low magnification "shift-only" case. The stereopair is generated at very low magnification by imaging the sample in two different areas of the divergent raster. The perspective center (PC) is the origin of the raster and lies a distance D from the principal point in the sample plane. The sample is moved by a distance B, the base shift. Corresponding distances between pairs of points in the two separate images recorded as the stereopair (the image or photo plane) are x_L and x_R and have a height difference Z between them.

Electronic Beam Tilting

The third method by which stereoscopic images may be made in the SEM is using the electronic beam tilting system (real time stereo). The two views that are required for stereoscopic images are produced by deflecting the beam with an accessory set of scan coils to produce a raster that approaches the sample from two slightly differing directions about the mean incident of the normal static electron beam. The deflection coils may be located either within the bore of the final lens (Fig. 4*b*), overwinding the normal scan coils (Boyde, 1974), or placed as a separate entity within the specimen chamber (Fig. 4*c*) (Chatfield et al., 1974). The raster is driven at normal television rates to produce a flicker-free image or reduced to a slow scan rate with a fixed offset to enable photographs to be recorded. The degree of convergence of the rasters depends on the accelerating voltage and working distance and may range from as little as 4° at

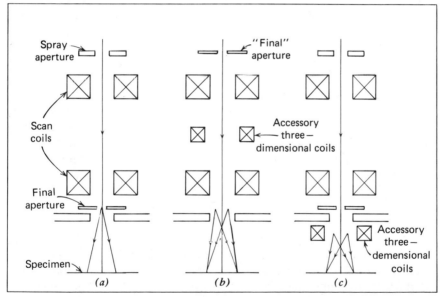

Figure 4. The normal arrangement of the scan coils and final aperture for the SEM (*a*). In the real time stereo SEM an accessory set of scan coils is used to provide an extra deflection so that two rasters are produced to give different imaging points necessary for stereoviewing. The coils may be situated either within the final lens assembly (*b*), when the "final" aperture is positioned above the scan coils, or as a separate item inside the specimen chamber of the SEM (*c*—after Chatfield et al. 1974), when the normal configuration of apertures is used. This latter position does reduce the available space if a micromanipulator is to be used in conjunction with the stereo SEM.

18 mm working distance and 10 kV up to 8° at 6 mm working distance and 5 kV (Boyde, 1974). The instantaneous three-dimensional appreciation of the surfaces within the microscope gives the ability to interact with a sample inside the specimen chamber of the SEM, enabling dynamic experiments (e.g., microdissection, micromilling) to be properly controlled.

How Much Tilt? Any value of tilt angle difference between 6 and 15° may be used. A value of 10° tilt angle difference was adopted in our laboratory because of the limitations imposed by the mechanical specimen stage for our SEM, which, when calibrated, gave the most accurate estimate of tilt angle difference for 10° intervals. It is also easy to remember! This value produces a stereoscopic effect when viewing stereopair micrographs of a surface subjectively similar to that which would be expected if the surface were viewed outside the microscope. As a rule of thumb it may be suggested that if surfaces are very rough (i.e., the height difference between a pair of points greatly exceeds the distance between them), then the value of 10° should be reduced substantially. This is because the eye–brain complex is unable to fuse the two images if there is too great a disparity between them, and results from the large parallax values obtained (greater than 15–20 mm). Conversely, where the surface to be examined is very flat (i.e., the distance between two points greatly exceeds the corresponding height difference) as, for example, with polished samples with shallow depressions, it is better to use a greatly increased value of tilt angle difference. Where the value of 10° is greatly exceeded, it is suggested that symmetrical tilts about the normal incidence of the electron beam be employed. Then the relative positions of a feature within the two members of the stereopair will lie within the same area of the micrograph and assist the brain in its interpretation.

14.2.2 Methods for Viewing Stereopairs

For stereoscopic imaging the photographs should be mounted side by side with the tilt axis lying perpendicular to the interocular line. Some time ago in our laboratory, a convention was adopted by which stereopair micrographs were to be mounted for stereoscopic viewing: The photographs are placed as for normal viewing with the direction of the secondary electron collector providing the apparent illumination from the top of the micrographs, the tilt axis lying parallel to the interocular line (if recorded with the stage fore–aft tilt mechanism); the photographs are then rotated to the left turning them through 90° so that the tilt axis now lies perpendicular to the interocular line. The direction of the apparent illumination of the sample (or the electron collector) lies to the left in the photographs. The photograph of the sample with the lesser forward tilt angle setting is placed for viewing by the *left* eye. The micrograph recorded with the greater tilt angle setting is placed for viewing by the *right* eye.

Single Viewer Systems

Unaided Vision. The simplest method for stereoscopic vision is that involving only the unaided eyes, yet for the initiate this is the most difficult. However, the time and effort required pays dividends subsequently. The eyes are trained (or deceived) into observing the two separate photographs of the stereopair by relaxing the muscles of the eye so that they are focused at infinity and with their optic axes parallel. The photographs, which are placed side by side with the tilt axis perpendicular to the interocular line, are introduced into the field of vision at arm's length. The eyes, initially, see three images—two of which are the separate images of the stereopair itself, while the third central image is a composite that when brought into focus will provide the stereoscopic model. The use of Polaroid film in the SEM camera allows the stereopair to be viewed immediately gaining a rapid insight into the surface structure of the sample.

Simple Two-Lens Viewer. Most readers will be familiar with the simple two-lens viewer system for studying stereopair photographs published in journals. The frame of the viewer holds the two lenses at their focal length (about 75–100 mm) above the micrographs, which are mounted side by side with the tilt axis perpendicular to the interocular line. The maximum width of the micrographs is limited to the interocular width itself of between 60 and 70 mm. The observer looking through these lenses is only able to see one micrograph with each eye which the "eye–brain" complex then fuses to produce a three-dimensional interpretation of the surface.

System Nesh. To overcome the limitation in the size of the micrograph imposed by the preceding method the System Nesh was developed (Neubauer and Schnitger, 1970). The photographs, which may be of any size, are mounted one above the other with the tilt axis parallel to the interocular line. The viewer consists of a mount in Plexiglas having two prisms of small angle that are placed in front of each eye. The prisms are of similar size but of opposite sense, such that when an observer stands at the appropriate distance from the stereopair two sets of micrographs are seen. By adjusting the observer–photograph distance the two sets of images are brought into register when three images of the photographs are seen. The middle image is the stereoscopic image. The limitation of this system is that the prism viewers are not currently as common as the two lens viewers.

Mirror Stereoscopes.   For a more detailed examination of stereopairs it is more normal to use some type of stereoscope. The original was described by Wheatstone in 1838; in this setup the photographs are placed either side of the observer, who looks into a pair of mirrors set at right angles to each other. Each eye then can see only one photograph which, when correctly aligned, permits stereoscopic imaging. However, the image of the stereopair is mirror inverted. More modern equivalent stereoscopes (Martin, 1966) have two pairs of mirrors (Fig. 5) and the mirror inversion of the first pair is corrected by the second pair so that the images are returned to their correct orientation. These stereoscopes effectively expand the interocular width, allowing large-format photographs to be examined comfortably.

Multiple Viewer Systems

Polarizing Filter System.   Where it is necessary to display stereoscopic images to a large audience, the use of the mirror stereoscope is not applicable. The two separate images of the stereopair must be projected so that each image can be seen by only one eye. This may be achieved by projecting the two slides through two identical projector systems into which are fitted polarizing filters (Wergin and Pawley, 1980). The filters are oriented such that their plane of polarization is at right angles to each other and at 45° to the horizontal. The images are projected onto a silvered screen that maintains the plane of orientation of the polarized light falling upon it and reflects it back to the audience, which wears spectacle frames holding polarizing filter whose orientation is similar to those in the projector. Each eye can then only see one image, and the two together are seen stereoscopically. The advantage of this system is the immediate visual impact of the stereoscopic model hanging in the plane of the silvered screen. There is a drawback for the lecturer who has to provide a polarizing light projector, a silvered screen, and a large number of polarizing filter spectacles for the audience. The loss of light from the passage through two sets of filters and the reflection from the silvered screen mean that the images are of relatively low intensity despite the use of high intensity projectors.

Anaglyph System.   The anaglyph method of stereoprojection has been used in an attempt to overcome some of the limitations of the polarizing filter system (Nemanic, 1974). The two members of the stereopair are

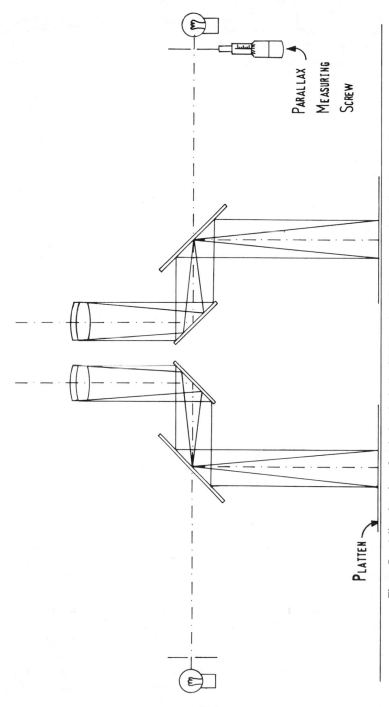

Figure 5. A line drawing of optical paths in the SB180 series of stereometers including the parallax-measuring facility. (After Martin, 1966.)

PARALLAX MEASURING SCREW

PLATEN

separately color coded in red and green (or red and blue). The slides can be recorded from the SEM directly using color film in the SEM camera and placing the appropriate color filter in front of the record CRT when taking the two exposures (Boyde, 1971). The single slide is projected through a normal projector onto a normal screen. The audience wears anaglyph spectacles with a green filter in front of the right eye and a red filter in front of the left eye. Alternatively, the normal black and white negatives from the SEM may be printed onto color film or diazo-type film and the two positives sandwiched together to form a single slide that is projected in the normal manner.

The anaglypyh method is especially suited for the display of images from the real time stereo SEM (Figs. 4b, 4c) (Boyde 1974; Chatfield et al., 1974). The two different rasters produced by electronically deflecting the beam to give the two views of the sample surface are displayed by directing the output from the detector to modulate the brightness of the red and green guns of a color television monitor. The observers wearing red/green spectacles see the sample directly at the SEM, permitting interaction between the operator, the sample, and a micromanipulator and avoiding unintentional damage of either the sample or the micromanipulator, because the operator can "feel" the third dimension.

14.3   LEVEL 2: PHOTOGRAMMETRY

Photogrammetry literally means measurement of a photograph and does not necessarily imply the aid of stereoscopic vision. The objective is to recover the real three-dimensional (X, Y, Z) coordinates of features on the surface of the sample represented in both images of the stereopair.

14.3.1   Frames of Reference

What can be measured from the stereopair to provide more than a purely subjective impression of the surface of the sample? All that it is possible to measure are the (x, y) coordinates of a feature in each of the separate members of the stereopair. Other than these relatively simple measurements, certain other information must be available to let the image geometry be reconstructed and thus derive a mathematical relationship to relate these measurements from the micrographs to the real three-dimensional coordinates of the sample surface. How are the results to be presented? To what reference plane are the calculated values to be referred—since within the SEM there is no *absolute* zero datum plane to enable a Cartesian coordinate system to be defined? A zero datum point could arbitrarily be defined but there are certain planes that present as obvious choices. The electron optic axis is an obvious choice to define the height or Z-axis in a proposed coordinate system. The X-axis and the Y-axis may be defined relative to the specimen stage movements or to any other arbitrary line that may or may not possess any significance at the time of recording the stereopair. Most obviously the tilt axis of the stage itself is the choice and this is used to define the Y-axis of the coordinate system; the X-axis then lies perpendicular to both these defined axes and is the direction in which the differences in the two images of the stereo pair represent themselves. Either of the micrographs may be used as the basis of the map for a coordinate system; then the other will be inclined to it by an angle α, the tilt angle difference. The frame of reference based on the position of the sample recorded with the lower tilt angle setting, and in accordance with the preceding convention, viewed by the *left* eye, is suffixed with the letter "L" (Fig. 6). Similarly the frame of reference based on the sample recorded with the greater tilt, viewed by the *right* eye is suffixed with the letter "R." When one views these two images, neither of them is observed separately but rather as a composite three-dimensional model based on a sample position intermediate to the two extremes recorded as

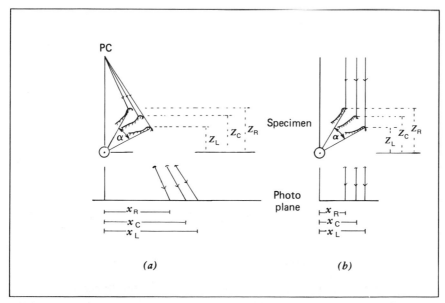

Figure 6.   Relations between the sample positions and the X and Z coordinates, for (*a*) the low magnification (perspective geometry) and (*b*) the high magnification (parallel projection) tilt-only cases.

the left and right images. This observed image *could* have been recorded as a third member of the stereo pair and could be used as the basis of a coordinate system to which the results may be referred. This frame of reference is suffixed with the letter "C" to signify its central position. In some instances, notably in materials science, the sample does indeed possess a specific orientation with which it is placed on the specimen stub. Under these conditions it may be desirable to relate the results to a frame of reference in which the X and Y coordinates are referred to a plane parallel and the Z-axis perpendicular to the stub face. The electron optic axis is now *not* the Z-axis as defined above but is inclined to it by the forward tilt angle setting (Fig. 7). This frame of reference is suffixed with the letter "S" to denote the use of the "stub-face" convention (Lane, 1969; Hilliard, 1972).

14.3.2   Requirements for Three-Dimensional Reconstruction

Low Magnification Shift-Only Method

This requires a knowledge of the following: the feature's (x, y) coordinate in the left and right photographs; the magnification, M; the principal image distance, D; the base shift, B; the distance moved by the sample between successive recordings; and the parallax, p (Fig. 3). With these values determined, a relatively simple mathematical relationship enables the calculation of the real (X, Y, Z) three-dimensional coordinates of those features measured.

They are as follows:

$$MX = x_{\mathrm{L}} \frac{(D - Z)}{Z} \tag{1}$$

$$MY = y_{\mathrm{L}} \frac{(D - Z)}{Z} \tag{2}$$

$$MZ = p \frac{D}{B} \tag{3}$$

Low Magnification Tilt-Only Method

Again using low magnification (i.e., operating with relatively short working distances below 500×) but with the sample tilted to generate the stereo pair, it is necessary to make a correction for the finite angle of scan of the raster across the sample surface (Fig. 1). A knowledge of the following is

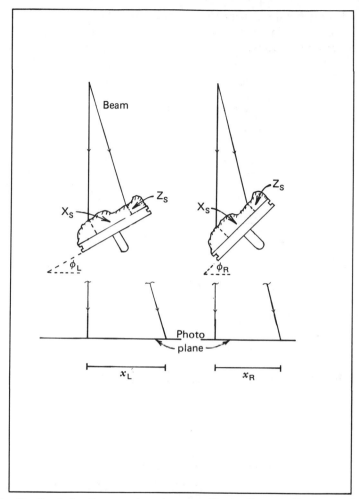

Figure 7. The "stub-face" convention—low magnification (perspective geometry) tilt-only case—relates measurements from the two photo planes to the real (X, Y, Z) coordinates at zero forward tilt—making the X-axis parallel and the Z-axis perpendicular to the plane of the stub face.

required: the feature's (x, y) coordinates; the magnification, M; the principal image distance, D: and, depending on the chosen frame of reference, either the forward tilt angle settings of the sample at which the two images were recorded θ_L and θ_R or the tilt angle difference α where

$$\alpha = \theta_L - \theta_R \tag{4}$$

All the distances taken from the micrographs should be measured from the principal point in the two images. Because of the perspective image geometry at low instrumental magnifications, complex formulas arise that are only justified when using a programmable calculator or minicomputer to reduce the raw numerical data to the real three-dimensional coordinates. They are shown here to indicate their complexity (Howell, 1975; Boyde and Howell, 1977; Howell, 1978).

For the left frame of reference:

$$MX_L = x_L \frac{(D - Z_L)}{Z_L} \tag{5}$$

$$MY_L = y_L \frac{(D - Z_L)}{Z_L} \tag{6}$$

$$MZ_L = \frac{(MD)^2[x_L \cos(\alpha) - x_R] + MD x_L x_R \sin(\alpha)}{[(MD)^2 + x_L x_R] \sin(\alpha) + MD(x_L - x_R) \cos(\alpha)} \tag{7}$$

For the "stub-face" convention:

$$MX_S = \frac{MD\{x_R[x_L \cos(\theta_L) - MD \sin(\theta_L)] - x_L[x_R \cos(\theta_R) - MD \sin(\theta_R)]\}}{[(MD)^2 + x_L x_R] \sin(\alpha) + MD(x_L - x_R) \cos(\alpha)} \tag{8}$$

$$MY_S = \frac{y_L[MD - MX_S \sin(\theta_L) - MZ_S \cos(\theta_L)]}{MD} \tag{9}$$

$$MZ_S = \frac{MD\{x_L[x_R \sin(\theta_R) + MD \cos(\theta_R)] - x_R[x_L \sin(\theta_L) + MD \cos(\theta_L)]\}}{[(MD)^2 + x_L x_R] \sin(\alpha) + MD(x_L - x_R) \cos(\alpha)} \tag{10}$$

High Magnification Tilt-Only Method

At high magnification (or lower magnifications with very long working distances) the perspective image geometry that held true above may be simplified as the divergence of the raster under these operating conditions is severely reduced. A parallel projection may then be considered to represent the image formation within the SEM (Fig. 1b). In consequence, relatively simple mathematical relationships result to relate the measurements from the photographs to the height measurements in the chosen frame of reference (Howell, 1975).

For the left frame of reference:

$$MX_L = x_L \tag{11}$$

$$MY_L = y_L \tag{12}$$

$$MZ_L = \frac{x_L \cos \alpha - x_R}{\sin \alpha} \tag{13}$$

or

$$MZ_L = \frac{p - x_L(1 - \cos \alpha)}{\sin \alpha} \tag{14}$$

where

$$p = x_L - x_R = \text{parallax} \tag{15}$$

For the right frame of reference:

$$MX_R = x_R \tag{16}$$

$$MY_R = y_R \tag{17}$$

$$MZ_R = \frac{x_L - x_R \cos \alpha}{\sin \alpha} \tag{18}$$

For the central frame of reference:

$$MX_C = x_L - \frac{p}{2} = x_R + \frac{p}{2} \tag{19}$$

$$MY_C = y_L = y_R \tag{20}$$

$$MZ_C = \frac{p}{2 \sin(\alpha/2)} \tag{21}$$

For the "stub-face" convention:

$$MX_S = \frac{x_L \sin \theta_R - x_R \sin \theta_L}{\sin \alpha} \tag{22}$$

$$MY_S = y_L = y_R \tag{23}$$

$$MZ_S = \frac{x_L \cos \theta_R - x_R \cos \theta_L}{\sin \alpha} \tag{24}$$

Full analyses of the possible combinations of rotations and tilts that may be applied to a sample when recording stereopairs have been given by Tovey (1973) and Hallert (1970), the latter for the analogous case of x-ray photogrammetry.

14.4   LEVEL 3: STEREOPHOTOGRAMMETRY

Stereophotogrammetry implies the use of stereoscopic imaging techniques while making measurements from stereopairs and represents by far the most accurate approach to the solution of three-dimensional analysis. The stereometer used for imaging the stereopair is provided with a parallax measuring facility (Fig. 5), which is an optical marker introduced into the viewing system and observed as a light (or dark) spot in both members of the stereopair. The markers are initially adjusted to overlay the same feature in the two images, at which time they unite to become a single point that appears to "float" in the optical model close to the surface of the sample. With fine adjustment of the parallax micrometer the floating marker is made to lie in the plane of the surface: the parallax value is then read from the vernier. A series of parallax measurements may be used to build up a section of profile (Fig. 8b) through the sample from the calculated (X, Y, Z) coordinates. Alternatively, lines of equal parallax or iso-height lines may be drawn by keeping the optical marker in contact with the ground and tracing round the surface of the optical model (Fig. 8c).

The measuring instruments used in SEM stereophotogrammetry may be roughly divided into three: (1) those instruments that permit measurement directly from the optical model, (2) those that have been modified from existing stereometers, and (3) those instruments that were designed specifically to solve the particular problem of measurement from scanning electron micrographs.

Direct Measurement from the Optical Model. The Stereosketch* may be used to make direct measurements in the stereoscopic optical model, or indeed, to model this model. The instrument is configured such that the optical model hangs in a physical space below and between the two photographs of the stereo pair, being viewed by a four-mirror stereoscope system with the first pair of mirrors semi-silvered. A modeling material can be molded to fit the detail of the optical model, or features

---

* Originally manufactured by Hilger and Watts and now constructed and sold by Cartographic Engineering, Ltd., Landford Manor, Landford, Near Salisbury, Wiltshire, England.

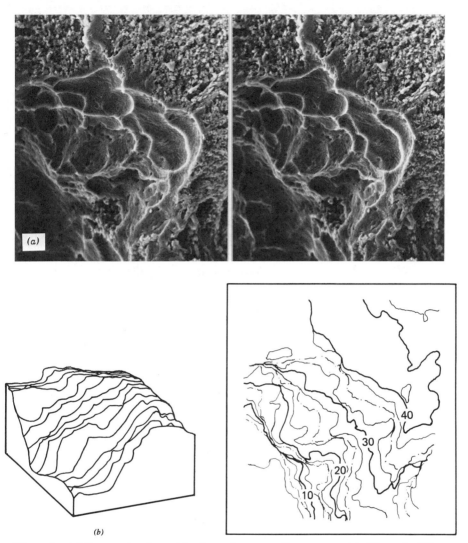

Figure 8. (*a*) Stereopair of resorbing bone (human malleus, anorganic). Tilt-angle difference 10°. Field width = 65 μm. (*b*) A series of profiles across the surface of the sample shown in part *a*. (*c*) Contour map of sample with contour lines at $2\frac{1}{2}$ μm intervals. These contour maps and profiles were drawn with the RS3 stereometer using rotary encoders to digitize the data for the X, Y coordinates and the parallax. The height differences were calculated by the on-line microcomputer, which output the data to a graph plotter.

343

can be abstracted. For example, twisted filamentous structures can be "copied" in flexible wire, cut to length: this can then be straightened to obtain true lengths. Distances and angles can be measured directly. We would not like to imply that the precision of this method would be adequate for rigorous analyses. It does, however, provide an opportunity for close scrutiny of three-dimensional relationships, in one way combining the senses of touch and vision.

Instruments Suitable for Measuring Parallaxes and Plotting "Contour" Maps. A simple mirror stereoscope with a parallax measuring facility of this type is exemplified by the SB180 series of stereometers (Martin, 1966). Other manufacturers make a wide variety of simple stereometers (e.g., Zeiss and Wild). The more expensive stereocomparators, used by conventional photogrammetrists in the field of aerial surveying, have been modified to take measurements from SEM stereopairs (Ghosh, 1971; Wood, 1972) but their high capital cost excludes their use on other than an occasional basis.

Stereometers Designed for SEM. Three stereometers have been specially constructed for our laboratory: EMPD1 (Boyde, 1974a), EMPD2 (Boyde and Ross, 1975) and RS3 (Boyde, 1981). They were made to accommodate negatives, diapositives, or prints derived from the electron microscope, and to enable profiles and contour maps to be made directly from the stereopair without resorting to an intermediate stage of calculation. The first two were an attempt to find a mechanical solution to the measurement from high magnification tilt-only generated stereopairs. Looking at Eq. (13) it is seen that the height difference Z is not directly related to the parallax, which is readily derived from any stereometer, but requires a correction factor—cos α—to the left coordinate measurement. Whereas this is necessarily small (cos $10° = 0.9848$), it represents a 2% error across the field of view. To compensate for this a cylindrical lens was inserted into the optical path of one of the images so that this correction was made automatically and enabled a modified parallax p' to be defined where

$$p' = x_L \cos \alpha - x_R \tag{25}$$

The height difference of Eq. (13) may now be rewritten:

$$MZ_L = \frac{p'}{\sin \alpha} \tag{26}$$

where sin α is a constant for a particular tilt angle difference. This then permitted the solution of plotting and profiling from these stereopairs.

With the latest instrument (RS3) and advantages of the minicomputer have been used to enable a software solution to profiling and plotting to be made. Data from two rotary encoders give the (x, y) coordinate of a feature while a third encoder provides the parallax value. The data is output directly to a microcomputer that calculates the real three-dimensional coordinates (X, Y, Z), which may be displayed on the VDU of the computer or output to a graph plotter (Figs. 8b, 8c).

14.4.1   How Are the Parameters Measured?

Image Feature Coordinate (x, y)

The Cartesian coordinate system is defined so that the Z-axis is coincident with the electron optic axis in the photograph of the stereopair used as the map. The Y-axis lies parallel to the tilt axis so the X-axis *must* be perpendicular to this. The (x, y) values are required to enable plotting of the calculated height difference in any frame of reference, and to provide a correction factor for the slope of the sample during the recording of the stereopair. The separate left and right x coordinates may be used to derive the parallax, but it is preferable to derive this parameter directly from the stereometer as this is more accurate.

The (x, y) coordinate of a feature within the image may be measured with:

1. A ruler, but this introduces considerable errors due to the round-off or end reading errors.

2. Graph paper attached to the base of the stereometer. This gives a cheap method by which the coordinate values may be obtained: With a vernier attached to one of the photo platens an accuracy of ± 0.1 mm is possible.

3. A digitizing tablet may be used as an alternative to graph paper but is expensive unless attached to a minicomputer for the direct calculation of (X, Y, Z).

Any errors in the (x, y) coordinate measurement will give a corresponding error in the real (X, Y) coordinate at high magnifications resulting in a degree of uncertainty as to the position of a feature in the profile or contour map. The error in the calculated height difference may be found by differentiating Eq. (14) with respect to x. This yields

$$dZ = dx * \text{const} \tag{27}$$

where the constant $= (1 - \cos \alpha)/\sin \alpha$. Thus the error in the height difference is directly proportional to the error in the coordinate value (Howell, 1978a).

Parallax p

The parallax is defined as the difference in distance between corresponding pairs of points in the two separate images of the stereopair, and *may* be calculated by subtracting the individual x coordinate values from the left and right images from (Eq. 15):

$$p = x_L - x_R$$

The values calculated in this manner are liable to serious errors due to the large round-off errors involved in measuring the individual (x, y) coordinates. It is better for the measurement of parallax to be made with the parallax facility on the stereometer as this is made with the adjunct of stereoscopic vision, which means that at the very worst the accuracy is three to four times that of the micrometer itself (± 0.01 mm).

Looking again at Eq. (14) but this time differentiating it with respect to the parallax yields

$$dZ_L = dp * \text{const} \tag{28}$$

where the const $= 1/\sin \alpha$. Thus the error in the calculated height difference is directly proportional to the error in the parallax.

Magnification M

The initial indication of the magnification is obtained from the meter or dial on the console of the SEM. This may, however, be in error by as much as 10%. It is necessary then to calibrate the SEM using either a cross-ruled grating or etched silicon wafer as a standard. With regular checks the magnification is likely to be within 1–2% of its correct value. Although this may seem to be an excessive error, it should be remembered that variation within a biological sample would completely outweigh this error.

Using the low magnification "shift-only" or the high magnification "tilt-only" methods for generating the stereopair, the magnification term only appears at the final stage of the calculation to provide a scaling factor between the micrograph and the sample in the SEM. As a result any error in the estimate of the magnification will result in a corresponding error in the calculated (X, Y, Z) coordinate values. With the low magnification "tilt-only" case the effect of an error in the magnification is more complex [see Eq. (7)], but the predominant effect is again one of an error in the scaling factor.

Principal Image Distance D

This measurement is required for both the low magnification "shift-only" and "tilt-only" methods for generating stereoscopic images. D is the distance between the perspective center (PC), or origin of the raster, and the principal point (PC) on the sample, or position on the sample at which the undeflected, central ray of the raster would strike a flat sample held at normal incidence. To assess this distance a measurement of the scanned width of some feature at various working distances is made and these values plotted against working distance. The curve is then extrapolated back to zero scanned width on the electron optic axis to give the position of the perspective center. If the lens currents are left static during the measurement procedure, the value obtained for the principal image distance places the perspective center within the plane of the final aperture. However, had the final lens current erroneously been altered to provide optimum focus, then the apparent position of the perspective center would lie several millimeters above the plane of the final aperture (10–12 mm for the Cambridge SII series SEMs).

In the low magnification "shift-only" case, with a working distance of 10 mm and assuming the false position of the perspective center high up in the final lens, Z calculations could be in error by 100%. Where the real principal image distance was 10 mm, an error of 1 mm in this parameter would give Z errors of about 1%. For similar conditions (10 mm working distance with a 1 mm error) using the low magnification "tilt-only" case the error in the estimate of the height difference would be again about 1% (Boyde, 1975).

Tilt Angle Difference α

It is necessary to estimate the forward tilt angle setting of the SEM stage by some calibration procedure—or accept the manufacturer's value! Calibration may be performed by constructing an arc of known radius and projecting a light from its periphery onto a surface mirror held in the specimen stub holder aligned to lie on the plane of the tilt axis of the stage. The angle of the reflected beam is read off from the outer rim of the previously calibrated arc (Boyde, 1970b). Alternatively, an autocollimator may be used as the light source and the angle of reflection measured with a theodolite (Houghton et al., 1971). These methods have been used to calibrate the specimen stages in our laboratory and have led to the choice of a 10° tilt angle difference for recording the stereopairs due to the hysteresis and backlash in the drive mechanisms for this particular design of stage. More accurate stages are now available that often have eucentric movements so that it is possible to adjust the sample to lie in the plane of the tilt axis, which increases the ease and rapidity of recording the stereopair. Errors in the measurement of tilt angle difference will result in an error in the calculated value of the height for all methods of generating stereopairs where the stage tilt movement is employed. Differentiating Eq. (27), the high magnification "tilt-only" case for the central frame of reference yields

$$dZ_C = \frac{d\alpha Z_C}{2 \tan(\alpha/2)} \tag{29}$$

with small values of α, which is usually the case; this may be simplified to

$$dZ_C = d\alpha \left(\frac{Z_C}{\alpha}\right) \tag{30}$$

so that the error in the height difference due to an error in the tilt angle difference is directly proportional to the height difference itself and inversely proportional to the tilt angle. An error of 1° in a tilt angle difference of 10° will result in a 7–10% error in the calculated height difference. This source of error is by far the largest in SEM photogrammetry, which is usually based on stereopairs generated using the high magnification "tilt-only" method.

REFERENCES

Ardenne, von, M. (1940). Z. Phys. *115*, 339–368.

Boyde, A. (1967). J. Microsc. *86*(4), 359–380.

Boyde, A. (1970a). Beitr. Elektronmikrosk. Direktabbild. Oberflächen, **3**, 403–410.

Boyde, A. (1970b). *Scanning Electron Microscopy/1970*, O. Johari, Ed., IIT Research Institute, pp. 105–112.

Boyde, A. (1971). Beitr. Elektronmikrosk. Direktabbild. Oberflächen. (Münster) *4*(2), 443–452.

Boyde, A. (1974). *Scanning Electron Microscopy/1974*, O. Johari, Ed., IIT Research Institute, pp. 93–100.

Boyde, A. (1975). J. Microsc. *105*(1), 97–105.

Boyde, A. (1981). *Scanning Electron Microscopy/1981*, O. Johari, Ed., SEM Inc., AMF, O'Hare, Chicago, pp. 91–95.

Boyde, A., and P. G. T. Howell (1977). *Scanning Electron Microscopy/1977*, O. Johari, Ed., IIT Research Institute, pp. 571–579.

Boyde, A., and H. F. Ross (1975). Photogram. Rec. *8*(46), 408–457.

Chatfield, E. J., More, J., and V. H. Nielsen (1974). *Scanning Electron Microscopy/1974*, O. Johari, Ed., IIT Research Institute, pp. 117–124.

Ghosh, K. (1971). Photogram. Eng. *37*(2), 187–191.

Hallert, B. (1970). Elsevier, Amsterdam.

Hilliard, J. E. (1972). J. Microsc. *95*(1), 45–58.

Houghton, A. H., D. Kynaston, P. G. T. Howell, and A. Boyde (1971). Beitr. Elektronmikrosk. Direktabbild. Oberflächen. (Münster) *4*(2), 429–441.

Howell, P. G. T. (1975). *Scanning Electron Microscopy/1975*, O. Johari, Ed., IIT Research Institute, pp. 697–706.

Howell, P. G. T. (1978a). Scanning *1*(2), 118–124.

Howell, P. G. T. (1978b). Scanning *1*(4), 230–232.

Lane, G. S. (1969). J. Sci. Instrum. (J. Phys., E.) *2*(2), 565–569.

Martin, D. B. (1966). Int. Arch. Photogram. *16*(III), 37–44.

Nemanic, M. (1974). *Principles and Techniques of Scanning Electron Microscopy*, Vol. I, M. A. Hayat, Ed., Van Nostrand, New York, pp. 135–148.

Neubauer, G., and A. Schnitger (1970). Beitr. Elektronmikrosk. Direktabbild. Oberflächen. (Münster) *3*, 411–414.

Tovey, N. K. (1973). *Scanning Electron Microscopy: Systems and Applications*, Institute of Physics, London and Bristol, pp. 82–87.

Wells, O. C. (1957). The construction of a scanning electron microscope and its application to the study of fibres. Ph.D. thesis, University of Cambridge.

Wergin, W. P., and J. B. Pawley (1980). *Scanning Electron Microscopy/1980*, O. Johari, Ed., SEM Inc., AMF, O'Hare, Chicago, pp. 239–249.

Wood, R. (1972). *Photogram. Rec. 7*(40), 454–465.

CHAPTER

15

HIGH VOLTAGE
ELECTRON MICROSCOPY

NINA FAVARD and PIERRE FAVARD

*Centre de Cytologie experimentale CNRS
et Laboratoire de Biologie cellulaire,
Université Pierre et Marie Curie,
67 Rue Maurice Gunsbourg
94200 Ivry sur Seine,
France*

15.1 INTRODUCTION

After the first electron microscope was built in 1931 by Ruska and Knoll
[1], it rapidly became apparent that the most useful and convenient oper-

351

ating voltage for applications in various fields was between 50 and 100 kV (conventional transmission electron microscope or CTEM). The use of such microscopes in biology during the past 20 years has provided a great deal of information on cell structure and function.

Very soon, however, instruments at higher voltages were developed [2]. Today the operating voltage of the most widely available high voltage electron microscopes (HVEM), some of them built in factories, is about 1 MV. Two 3-MV microscopes are in operation: one in the Laboratoire d'Optique Electronique, Toulouse, France, the other at the University of Osaka, Japan (Fig. 1). For many years HVEM were used mainly for metallurgy and materials science research. The application of HVEM to

Figure 1. Models of high voltage electron microscopes. (*a*) 1-MV electron microscope JEOL-1000C at the University of Boulder, Colorado. The pressurized tank enclosing high voltage generator and accelerator is at the top. (From Ref. 10.) (*b*) 3-MV electron microscope at the Laboratoire d'optique electronique du CNRS, Toulouse, France. Only the column and column console are included. (From Ref. 45.)

biomedical research is less than 15 years old. The number of users, very low a few years ago, is increasing.

The main argument against the high voltage electron microscope is that the instrument is very expensive; its cost is much higher than that of the CTEM or the scanning electron microscope (SEM); furthermore, its installation is also costly: This huge instrument must be set up in spacious rooms in buildings sometimes specially constructed for that purpose. Limitations of performance occur if mechanical, electrical, and magnetic factors are not carefully controlled.

In fact, most of the existing HVEM are installed in metallurgy or material sciences laboratories and are available only on a limited basis to biologists. Very few instruments, including three in the United States, are installed in biological laboratories, for use exclusively for biological purposes. However, over the past decade the use of HVEM in biological and medical sciences has steadily increased. In spite of the merits of other imaging modes, sometimes less expensive (some of which are investigated in other chapters of this book), certain specific features of the HVEM appeared to be useful for biological applications.

Unquestionably, the main application of HVEM in biology has been to provide three-dimensional information on thick biological specimens. This results from the ability of high energy electrons to penetrate thick specimens (much thicker than the specimens penetrated by 100-kV electrons) and to produce good resolution images of these specimens.

Improved resolution can also be expected in thin specimens by increasing the accelerating voltage; for example, for a mineral specimen 0.2 nm has been obtained at 600 KV [3]. It is difficult nonetheless to take advantage of this feature to observe biological specimens since damage is a limiting factor. The resolving ability of HVEM, however, is used to study biological molecules, especially at low temperatures. In this chapter a description is given of the main techniques used to prepare and observe biological specimens in HVEM. Special attention is paid to the artifacts or damage that may occur. Some examples of results are described and discussed in order to elucidate the advantages and indications for use of the HVEM in biology. Finally, the prospects for HVEM as compared to other imaging techniques are considered.

15.2 PHYSICAL PRINCIPLES OF HIGH VOLTAGE ELECTRON MICROSCOPY OF INTEREST FOR BIOLOGICAL STUDIES

The main advantages of high voltage electron microscopes for biological studies, as compared to the conventional 100-kV microscope, are the greater penetration of the high energy electron beam and the improved resolution obtained by increasing the accelerating voltage. The physical principles that explain these properties are described in a number of reviews [2,4–9]; only a brief comment will be given here.

15.2.1   Penetration Power and Electron Energy Loss

Interactions of Electrons with Matter

When incident electrons of a beam interact with matter, they are scattered. If we consider an electron incident on a single atom, it may be scattered in different manners: *Inelastic scattering* occurs when the interaction provokes a change in the state of the atom, which is ionized or left in an excited state. The incident electron loses energy. *Elastic scattering* occurs when the state of the atom is not changed by interaction with the incident electron. Energy loss is very small in this case [2,4]. Inelastically and elastically, scattered electrons contribute to image formation; in fact, inelastic scattering has deleterious effects on image quality, which depends in part on the relative contribution of elastic and inelastic scattering to its formation. Also, plural scatterings are involved in image formation.

Energy Loss and Accelerating Voltage

The amount of energy loss of electrons interacting with a specimen depends on many parameters, such as the chemical composition and the thickness of the specimen and the accelerating voltage of the electron beam (or the kinetic energy of the electrons). High energy electrons have an increased velocity. When they pass through a specimen of a given thickness, the energy loss decreases as the electron energy increases. Interactions of such electrons with the specimen can occur over longer trajectories before the electron energy is reduced to a certain extent, though still compatible with imaging conditions.

Calculations, for several substances (carbon, muscle, etc.), of the energy loss values as a function of the accelerating voltage show a minimum around 1 MV [2]. Consequently, the HVEM allows an appreciable increase in the usual maximum specimen thickness. Penetration is approximately three times greater at 1 MV than at 100 kV for thick specimens [4]. This increased penetrating power of the electron beam at high accelerating voltages is most significant for biological applications, as described in Section 15.3 as it allows the study of thick specimens.

15.2.2 Resolution in Thin and Thick Specimens

Resolving Power and Electron Wavelength

The theoretical resolving power of an electron microscope (δ_m) is a function of the electron associated wavelength λ. It is, however, limited by the spherical aberration of the objective lens, characterized by a constant C_s and by diffraction. It is usually given by the expression

$$\delta_m = A\lambda^{3/4}C_s^{1/4} \tag{1}$$

δ_m denotes the least resolvable distance between two points, and A is a factor dependent on the manner in which spherical aberration and diffraction combine in the image. The value of λ must take into account the important relativistic effects that occur due to the fact that the incident electrons travel with a very high velocity in HVEM; for 3-MV electrons, the velocity is 0.99 times the velocity of light. The value of λ decreases when the accelerating voltage increases. The wavelength associated with a 3-MV electron is 10 times smaller than that of a 100-kV electron [10,11]. Thus the resolution that may be achieved in HVEM is theoretically improved compared to that of CTEM.

Resolving Power and Chromatic Aberration

In expression (1) the presence of a specimen is not taken into account. Indeed, in the presence of a specimen and especially a thick specimen, an important factor limits resolution: chromatic aberration, which is due to energy loss by electrons of the incident beam within the specimen. Briefly, (see Ref. 4 for equations and details) this can be explained as follows: The incident electron beam is generally composed of monokinetic electrons; in traversing the specimen, the electrons lose different amounts of energy, depending on the nature of the specimen, its thickness, and its own electron energy. Thus the beam that emerges from the specimen is composed of electrons of different energies, i.e., of different wavelengths.

As the focal length of the objective lens of the microscope is a function of the electron wavelength, the emerging electrons of different energies are focused in different positions. If a theoretical image plan of no-loss electrons is taken as a reference, a point of the object is imaged as a disk at the level of this image plan, the chromatic aberration disk. Energy loss decreases as the voltage rises, and the radius of the chromatic aberration disk also decreases. This is shown in Table 1 for a carbon film, 100 nm

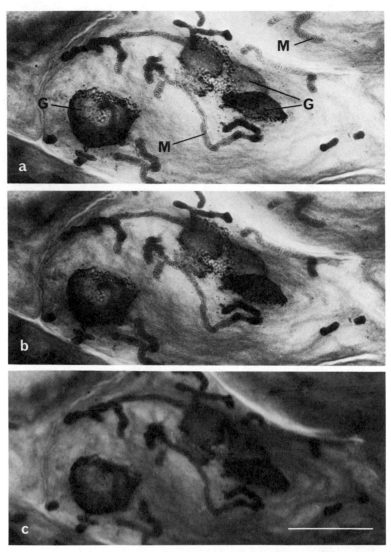

Figure 2. Resolving power and accelerating voltage. 5-μm-thick araldite section of an osmium-impregnated snail mucous gland recorded at various accelerating voltages: (*a*) 2 MV, (*b*) 1 MV, and (*c*) 0.5 MV. Image resolution is improved as the voltage is raised, especially from 0.5 to 1 MV. M, mitochondria; G, Golgi apparatus, forming face selectively stained. Bar = 2 μm.

TABLE 1[a]

V (kV)	Energy loss (MV/cm)	Radius chromatic aberration disk (rc/Å)
100	7.27	109
500	3.56	12
1000	3.22	4.8
3000	3.23	1.6

[a] From Ref. 11.

thick. Theoretically, for a thick biological specimen, chromatic aberration is reduced by at least 20 times in going from 100 kV to 1 MV [4].

Since chromatic aberration is a limiting factor of resolution, especially as the specimen thickness increases, the resolving power is considerably improved at high voltages [12]: sharp images of thicker specimens can be obtained. This feature, very important for biological studies, is illustrated in Fig. 2. Details of the effects of different parameters on resolution, such as elastic and inelastic scattering and energy loss, chromatic aberration, damage, and so on, are found in Refs. 2,4,6–8. This short overview shows that the main advantages expected by biologists using HVEM are (1) sharp images of thick specimens as a result of the increased penetrating power of high energy electrons and reduced chromatic aberration; and (2) very high resolution on thin specimens, in some cases, approaching atomic resolution.

The techniques used to study thick specimens as well as their limitations are now described.

15.3 THICK SPECIMENS AND THREE-DIMENSIONAL RECONSTRUCTIONS

Since thick specimens are directly imaged in HVEM with improved resolution, an important contribution to the understanding of the three-dimensional arrangement of the structures present within the specimen is obtained by stereoviewing single thick sections or, at times, by reconstruction from thick sections.

Appropriate techniques of preparation, examination, and imaging have been developed. The interpretation of information obtained using HVEM techniques must take into consideration artifacts resulting from the preparation and staining process, errors resulting from overlapping of all structural details within the thickness of the preparation, and damage arising from interactions between the high voltage electron beam and the specimen.

15.3.1 Preparation of Thick Specimens for HVEM

Special preparatory techniques have been developed for HVEM studies of thick biological materials, some of which are modifications of well-known CTEM techniques described in detail by several authors (see, e.g., Refs. 13–15). We will consider here only aspects of interest for HVEM. Two main categories of techniques are used in most HVEM studies:

1. Preparation of thick sections of embedded material.
2. Preparation of intact unembedded whole cells and organelles.

Frozen sections obtained by cryoultramicrotomy are also observed in HVEM in a dried or hydrated stage, but these interesting techniques are not widely employed.

The observation of thick wet specimens in electron microscopes is an important subject and is treated in a separate chapter of this book.

Thick Sections of Embedded Material

Fixation and Embedding. In general, the same techniques of fixation are employed for CTEM and HVEM work [14]; glutaraldehyde and osmium tetroxide are currently used. Embedding is also performed in a standard manner, and epoxy resins such as araldite and epon are often used. As these resins are not miscible with water, dehydration of the specimen occurs before embedding [13,14]. In thick as well as in thin sections, the resins usually remain around the embedded structures and constitute an unstained background that, however, scatters electrons. Removal of epoxy resins from the sections is difficult [15].

Some water-miscible embedding media such as polyethylene-glycol (PEG) are successfully used in HVEM [16–18]. This latter medium is subsequently removed with water; the absence of embedding medium when the section is viewed with a HVEM is an important contrast factor, since these media scatter electrons enough to obscure the relative contrast of the structures, as we will see later.

Sectioning and Section Thickness. Sections several micrometers thick are made with conventional microtomes [15]. When araldite is sectioned with glass knives, damage occurs to the surface of the thick sections. This can be avoided by alternating several thin sections with the thick sections [19].

Another problem lies in the choice of the appropriate section thickness, which depends on several factors: the nature of the biological structures under investigation and the use of simple thick sections or of serial sections for three-dimensional reconstitution [20], for example. Most frequently, sections in the range of 0.5–5 μm are observed. Several techniques can be used to measure the section thickness: stereotechnique [21], resection of thick reembedded sections perpendicular to their surface, and measurement with a CTEM [19], among others.

Adhesion of thick epoxy resin sections onto grids is obtained by various processes, for example, heating [22] or pressing between the two parts of folding grids [19].

Staining. Contrast in thin sections observed in CTEM is commonly obtained by general cytoplasmic stains, most of which are heavy metal salts such as uranyl acetate, lead citrate, and so on. In such stained sections, atoms of high atomic weight attach to some of the structures of the biological specimen. The selective electron scattering that results gives rise to the image contrast.

General Staining Methods. The general staining methods employed in CTEM require certain limitations when used in high voltage electron microscopy. The main reason is that the electron scattering of the background increases with section thickness; the comparative contrast of the stained structures simultaneously weakens. Furthermore, image contrast decreases as voltage increases (see Section 15.3.1, "Whole Unembedded Cells and Organelles"). Such general staining methods are, in fact, convenient for sections approximately 1 μm or less. Details on the practice of these various methods are found in Refs. 8,19,22–24 and are briefly summarized here. Most are adapted, with minor adjustments, from conventional techniques used in CTEM. Stains like uranyl acetate, lead citrate, and phosphotungstic acid have been commonly used. Specimens are sometimes stained before embedding or after sectioning. When sections are stained, they are immersed in the staining solution in the usual way, but it must be ensured that the stain has penetrated through the entire thickness of the section. The rate of penetration depends both on the nature of the staining solution and on the embedding medium. Most stains show poor penetration into epoxy resins, and it must be borne in mind that only the stained structures included within the thickness of the section are imaged [26].

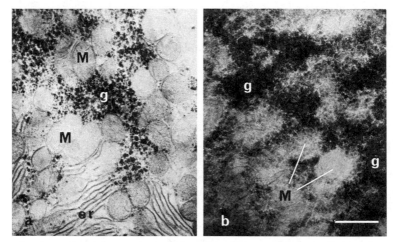

Figure 3. General staining methods and section thickness (*a*) 1-μm-thick section of rat hepatocyte stained before embedding by uranyl acetate and lead citrate and examined at 1 MV. (*b*) 10-μm section from the same material examined at 2.5 MV. The general staining method employed here is convenient to observe details in the 1-μm section; however, in spite of the increased accelerating voltage, no sharp details appear in the 10-μm section where important electron scattering by the surrounding embedding medium occurs. (*a*) Mitochondria M, endoplasmic reticulum ER, and glycogen particles G are visible. (*b*) Mitochondria M appear as empty areas among masses of glycogen particles g. Bar = 2 μm.

These techniques are the first to have been used for observing thick epoxy resins sections in HVEM [27–29; see reviews 22–24]. General staining methods are not convenient for sections over 1 μm thick (Fig. 3). As mentioned earlier, scattering by the embedding medium surrounding the contrasted structures increases dramatically in sections several microns thick; furthermore, as numerous cell components are stained by these methods, the overlapping in the final image is confusing. Indeed, all the stained structures, which are mainly membranous organelles included in the depth of the section, overlap in a two-dimensional, unclear picture.

Selective Staining Techniques. To effectively observe thick sections (several micrometers thick) in HVEM, suitable techniques have been developed. The principle consists of selectively staining a restricted number of cell components by deposits of heavy metals, so that they exhibit high contrast as compared to the background. In spite of the increased thickness of the section, and the resulting increased scattering of the unstained background that somewhat obscures the contrast of the stained parts, they remain clearly visible and interpretable. Only some of the most commonly used selective staining methods are briefly summarized here.

Heavy Metal Impregnation Techniques. Osmium tetroxide, the well-known fixative in electron microscopy, is also used to impregnate membranous surfaces of various organelles: Golgi apparatus, endoplasmic reticulum, and so on (see Section 15.3.5).

The impregnation method commonly used is that described by Friend and Murray [30]; fixed tissues are impregnated by immersion in an unbuffered aqueous solution of osmium tetroxide at 40°C for several days. The block-impregnated tissues are then embedded, and thick sections are observed in HVEM. Impregnation is also carried out with association of osmium to other metallic salts: osmium tetroxide and zinc iodine (ZIO) [31,32], osmium tetroxide, iron and cyanide (Os, Fe, CN) [33] (see details in Ref. 15). Impregnation occurs in blocks of previously fixed tissues, and (when sections are observed) reveals parts of the Golgi apparatus and endoplasmic reticulum (see Section 15.3.5). Glutaraldehyde-fixed tissues can be stained successively by various solutions: uranyl acetate, copper and lead citrate, and osmium (Fig. 4). The surfaces of most of the mem-

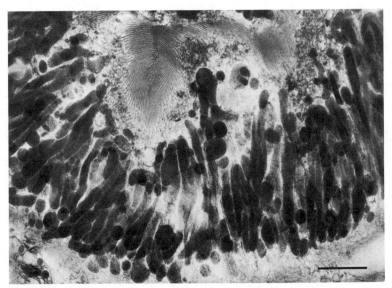

Figure 4. Heavy metal stained kidney cells. 1-μm-thick section of proximal tubule epithelial cell of rat stained by the Thiery Rambourg technique [34] and examined at 2.5 MV. Elongated mitochondria are well contrasted. Bar = 2 μm.

branous components of the cell are stained and can be observed in HVEM in thick sections [34–36].

Lanthanum stain has been used to study the T system in muscle [37]. Various techniques using silver impregnation, for example, the Ramon y Cajal technique (Fig. 5) for neurofibrils [38] and Golgi techniques [39,41], reveal components of nervous tissues.

Cytochemical Techniques. These techniques provide information on the specific localization of chemical components of the cells when heavy metals can be associated with these components. For example, Gomori's lead method can be used to detect the localization of phosphatase activity, since the product of the enzymatic reaction is lead phosphate [19]; the nature of the phosphatase revealed by this technique depends on various conditions, such as pH, which is an important factor. Some cellular membranous components are revealed by these methods: glucose-6-phospha-

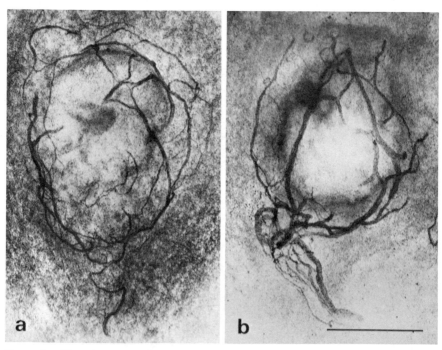

Figure 5. Silver-impregnated small neurons. 4-μm-thick section of a leech ganglion silver-impregnated by Ramon y Cajal technique and examined at 2.5 MV. The perninuclear neurofibrillar network is contrasted. Bar = 5 μm. (From Ref. 38.)

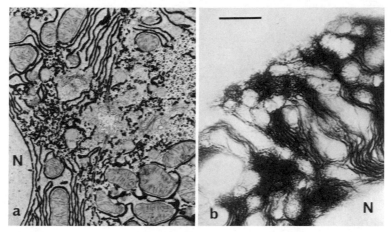

Figure 6. Gomori cytochemical lead method. Rat liver cells treated to reveal glucose-6-phosphatase activity. Cisternae of the ER contain lead phosphate precipitated: (*a*) thin section examined at 100 kV, (*b*) 2-μm-thick section examined at 2.5 MV. N, nucleus. Bar = 2 μm.

tase activity is associated to ER (Fig. 6), and acid phosphatase activity to Golgi apparatus [19].

A method for demonstrating endogenous peroxidase activity has been modified to reveal the distribution of the T system in mouse skeletal muscle [40,41].

Autoradiography on Thick Sections. A few investigations have explored the advantages of HVEM in autoradiography. Using thick sections, the exposure time of the cells to the emulsion is reduced, since a thick specimen contains more material and is more radioactive than a thin one (Fig. 7); for example, 0.3-μm-thick sections of [H^3]uridine-labeled amoeba give autoradiograms after only 5 days of exposure to the emulsion, while at least 28 days are necessary for 80-nm sections to obtain this result [42]. A reduction in the intensity of labeling also becomes possible when thick sections are exposed to the emulsion.

Whole Unembedded Cells and Organelles

The preparative methods for suitable thick sections for HVEM viewing involve chemical and mechanical treatments of very fragile biological materials. Some of these can be avoided as whole unembedded specimens are observed. Current methods have been developed to prepare these specimens: whole cells or isolated organelles, flat enough to be penetrated

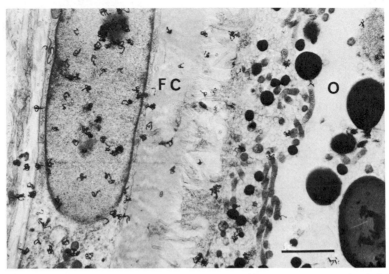

Figure 7. Autoradiography. HVEM autoradiography of [³H]lysine incorporation in follicular cells FC and oocyte O of newt. 1-μm-thick section examined at 1 MV. The exposure time is 7 days. Bar = 2 μm. (Courtesy P. Mentré and F. Jais.)

by the high energy beam of the HVEM and imaged at higher resolution are studied in HVEM.

Observation of whole cells is made on thinly spread cultured cells. The first attempt to observe such cells in CTEM was as early as 1945 by Porter et al. [43]. Methods have been developed to obtain cultured cells of a wide variety [18,44–46], plated on grids, the thickness of which is approximately 1 μm and even less at the periphery (Fig. 8). Suitable techniques have also been developed to prepare isolated organelles such as chromatin or chromosomes [47–49].

To avoid structural distortions that occur when specimens are air-dried, drying is carried out by the critical point method. The principles of this technique have been described by Anderson since 1951 [50], and it is currently employed to prepare whole cells and organelles for HVEM. Briefly, fixed cells are dehydrated through increasing concentrations of ethanol. From anhydrous alcohol, they are transferred to liquid anhydrous CO_2 in the pressure chamber. Liquid CO_2 replaces the alcohol and is taken through the critical point (34°C) for drying. The specimen dried by this technique is not exposed to the surface tension of an air–water interface and three-dimensional configuration of the structures is preserved.

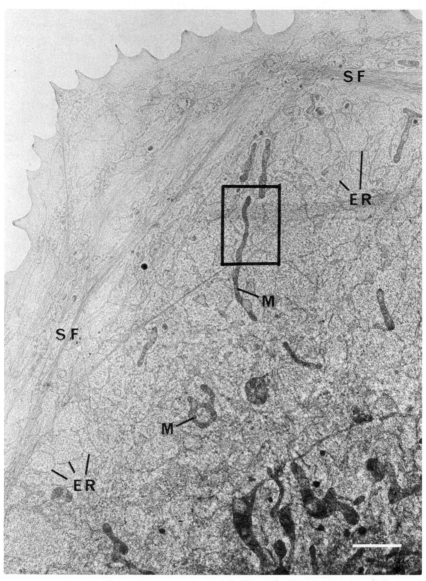

Figure 8. Whole unembedded cultured cell. Part of a BSC cell grown in culture, glutaraldehyde-fixed, OsO_4 postfixed, and critical-point-dried. M, mitochondria; ER, endoplasmic reticulum. Stress fibers SF, which are bundles of actin filaments, are visible in this whole cell. The are outlined is shown at higher magnification in Fig. 17. Examination at 1 MV. Bar = 2 μm. (From Ref. 83.)

365

The observation of a whole unembedded specimen does not necessarily require staining to contrast the structures. Most of them appear contrasted against a transparent background that does not scatter the electrons. This technique has numerous fields of application that have not as yet been exploited: whole cell autoradiography [51], immunolocalization of specific proteins in whole cells, and so on. [Frozen whole cells can be observed in cryomicroscopy, using high voltage electron microscopes fitted with cold stages (Fig. 9). The frozen cells are critical point dried or remain hydrated during examination. This interesting and promising method is still in its infancy.]

Cryosections

Only relatively thin, whole, unembedded specimens can be directly observed in HVEM, as seen earlier. To investigate thicker biological specimens without embedding, cryosections have been cut from fixed frozen tissue blocks. These sections have been critical point dried and examined with a 1-MV microscope [52].

Techniques of preparation of 1–2-μm-thick hydrated frozen sections of untreated biological material have been recently developed, mainly in view of scanning electron microscopy observations [53]. Such sections could also be viewed by transmission electron microscopes fitted with cold stages. However, these techniques have not been developed as yet for HVEM observations. As they avoid various chemical treatments and particularly the removal of water, they allow interesting comparisons with results obtained on dehydrated samples.

15.3.2 Examination Conditions of Thick Specimens

Various interacting parameters, relevant to the properties of the specimen itself and to the operating conditions of the high voltage microscope, must be taken into consideration to obtain optimum resolution images from thick biological specimens. Some of them are briefly summarized here (see details in Refs. 2,4,12,55). Under fixed instrumental conditions, the thickness of the specimen and its orientation in the beam affect the resolution. Elastically and inelastically scattered electrons contribute to image formation (see Section 15.2.1.). As thickness increases, more inelastic scatterings occur and have a deleterious effect on image quality. Under standard conditions, specimens of suitable thickness are approximately 0.5–5 μm thick (see also Section 15.3.1, "Sectioning and Section Thickness"). The position in the beam of the structures of interest within a thick specimen is also a factor affecting resolution. When these struc-

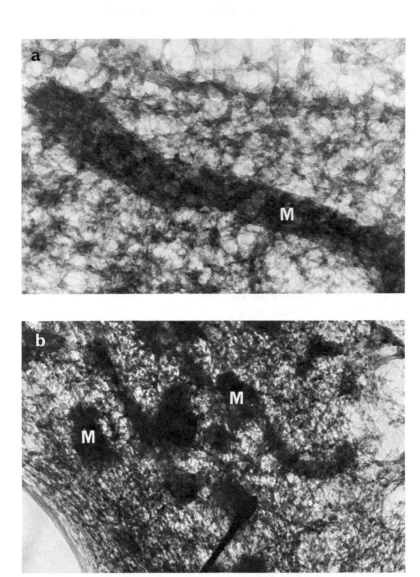

Figure 9. Whole unembedded cells examined at low temperature. (*a*) Part of cytoplasm of a fibroblast critical-point-dried and examined at 1 MV, 6 K (courtesy of M. Fotino). (*b*) NRK cell (newborn rat kidney) washed 2 s in 0.16 M NH$_4$ acetate at pH 7.2, frozen in propane, and transferred in liquid N$_2$. The specimen has been inserted into the column on a stage at 78 K, and temperature raised to 180 K to etch; examination at 1 MV (courtesy of M. Stearns). The microtrabecular lattice is visible in both micrograhs. M, mitochondria. Bar = 0.5 μm.

tures are located on the lower side of the specimen, resolution is improved; this effect is known as the "top–bottom" effect.

For a given thickness of the specimen (in the range mentioned earlier) and for a given objective lens focal length, resolution also depends on the diameter of the objective aperture. As already mentioned (see Section 15.2.2), the resolution is limited by two physical phenomena, the spherical aberration and the diffraction. The variations of these effects as a function of the acceptance angle of the objective lens (half-angle of the cone subtended by the objective lens aperture) are inverse. Consequently, the best attainable instrumental resolution corresponds to an optimum value of the acceptance angle determined by the diameter of the objective aperture. But this acceptance angle is also a limiting factor for the contrast—the smaller it is, the better the contrast. Furthermore, as thick specimens are used in HVEM for three-dimensional reconstruction purposes, the choice of acceptance angle and lens aperture must take into account the question of depth of field, for only when the thickness of the specimen is compatible with the depth of field can all the contrasted structures within the specimen be viewed in focus and imaged. This condition is implicitly fulfilled for thin sections in CTEM.

The depth of field D is related to the resolving power δ and to the acceptance angle α by the relation $D = \delta/\alpha$. Thus the value of an adapted aperture must be compatible with specimen thickness and expected resolution, and vice versa. The usual apertures set up in the high voltage electron microscope have a diameter varying from 10 to 50 μm.

15.3.3 Stereoscopic Imaging for Three-Dimensional Reconstruction

Thick biological specimens, prepared for examination in high voltage electron microscopy, contain various structures positioned above one another. When viewed and imaged in HVEM, these structures are projected onto a two-dimensional image and overlap. Their real positions within the specimen depth cannot be ascertained from the confusing image that results from this projection, even if only a selected number of details of interest have been contrasted by adapted preparatory methods (see Section 15.3.1, "Thick Sections of Embedded Material").

The simplest way to ameliorate this problem is by stereoscopic imaging and viewing, a method increasingly used to resolve problems requiring three-dimensional information. Details on this technique are given elsewhere in this book and its applications are found in a number of reviews (see Refs. 4,8,56–60, for example). Some of them are summarized here.

Stereoscopic imaging is performed as follows: Two pictures (stereopair) are recorded from the same area of the specimen, tilted in relation to the beam axis by an equal amount in two opposite directions. Each of the pictures of the stereopair records one of the two tilted positions of the specimen.

When these two micrographs are set side by side in such a way that they are viewed simultaneously with a stereoscope, the stereo effect is achieved; each eye of the observer views only one micrograph, and the two retinal images fuse into a single perceived image where the spatial relationships of the structures appear.

To obtain a stereoscopic view of two points A and C positioned one above the other in the specimen depth, tilting must actually produce a parallax between these object points. Parallax P exists when the linear distances of the image points in each of the two tilted micrographs are different, and its value corresponds to the difference between these linear distances, as shown in Fig. 10 according to Turner [56].

When $P = 0$, the stereoscopic effect cannot be obtained. The value of the parallax depends, of course, on the relative positions of the details of interest in the specimen depth.

Other parameters must be taken into consideration in stereo imaging and viewing: stability of goniometric stages, individual ability of depth perception, mounting of the stereopairs in good conditions, etc. In fact, even when stereopairs are imaged under optimum conditions, the stereo effect is lost if their side by side mounting is inaccurate. Some rules have been defined (references above) to obtain the best depth perception: Trim-

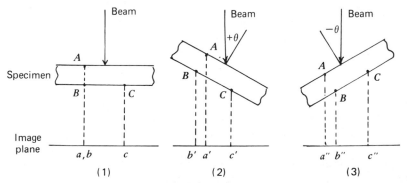

Figure 10. Schematic representation of the relative positions of a specimen to the incident beam in (1) untilted position and in tilted (2) $+\theta$ and (3) $-\theta$ orientation. Three points A, B, and C of the specimen are viewed as a, b, c image points in parallel projection. Parallax corresponding to the specimen A and C points is $P = a'c'-a''c''$.

ming must be done in such a way that only the same details appear in both images of the stereopair (this defines a "stereo window"); the deleterious effect of blurred edges is avoided in this case; horizontal and especially vertical alignment must be accurate, as it is difficult for the eye to correct a vertical misalignment; the respective positions of the two images mounted side by side also have an influence on stereo effect, which is better when the detail of interest seems to come up toward the observer.

Optimum spacing between the two micrographs of the stereopair depends on several factors and, in part, on the way the prints are to be viewed; for example, for a pocket viewer, prints must be mounted with their centers $\simeq$65 mm apart.

15.3.4   Beam Damage in Thick Biological Material

When biological specimens are irradiated by an electron beam, damage occurs. This results from interactions between the specimen atoms and the electrons, and causes a number of effects, limiting the expected performance of the microscope. The nature of such interactions under various physical conditions, and the techniques proposed to minimize their deleterious effects, are analyzed in exhaustive surveys in which a detailed bibliography is found [4,8,61–65]. Only some general data and facts of interest in high voltage microscopy research are treated here, especially as far as thick specimens are concerned.

General Data.   Irradiation effects are usually classified into primary and secondary effects [61]. Primary effects consist mainly of excitation and ionization of atoms of the sample and atom displacement; such secondary effects as temperature rise, sublimation of part of the specimen material, and formation of carbonaceous coating and bound breakages are those that provoke the most damage when thick specimens are observed in HVEM. The extent of radiation damage depends, of course, on both the nature and mode of preparation of the specimens, and on electron beam characteristics: energy, electron dose or electron fluence (with these terms describing the quantity of incident electrons per unit area) [8,61], and electron exposure during the period of time needed to record a picture.

The critical electron dose is the parameter used to evaluate the electron exposure and below which value no significant damage can be viewed in the specimen. It varies according to the nature of the details of interest that are under study and is in some cases difficult to evaluate. However, most biological specimens can tolerate an exposure of no more than 100 to 500 electrons/nm^2 [63] and even less for crystal proteins [65], without significant damage. Higher doses are invariably used in CTEM and HVEM.

Generally speaking, however, irradiation damage is reduced in high voltage microscopy as compared to CTEM. This is an important advantage for biological application. The reason is that electron velocity increases by raising the accelerating voltage and faster electrons deposit less energy in the specimen. The theoretical probability for a single electron excitation of producing ionization damage would be reduced by a factor of about 3, going from 100 kV to 1 MV [4,8,61]. However, in practice, one must keep in mind that a greater number of electrons is required for recording in HVEM conditions, as available photographic emulsions are optimized for CTEM and do not have a high enough sensitivity for high energy electron microscopy [4,8,61].

Damage in Thick Sections of Embedded Material

Much of the information obtained from high voltage electron microscopy is derived from thick-section stereoviewing. Radiation damage and overheating effects occurring in such specimens when examined in the high energy beam must therefore be evaluated with care; results should be interpreted taking into account various factors such as the nature of the embedding medium of the structures of interest, the contrasting agents, the thickness of the section, and the electron exposure (see Refs. 2,4,8,19,24 for details).

The main damage occurring to thick sections affects the embedment media, and consists of a mass loss (thinning) and shrinkage which in turn affects the distribution and configuration of the embedded biological structures. Few quantitative evaluations are available. Some results, however, have been obtained in HVEM for epon sections [8]; these preliminary results are of great interest, since epoxy resins are commonly used as embedding media. Thinning is observed in a 0.6-μm section of epon-embedded tissue, carbon coated on one side and irradiated in a 1-MV beam under fixed conditions, together with a lateral shrinkage of approximately 3%. An overall thinning is also observed in araldite sections, since the thickness is approximately less than 10 μm. Specimens thicker than 10 μm are more severely affected, with craters appearing in the araldite (Fig. 11), the number and deepness of which vary with section thickness and, consequently, with the electron dose necessary to penetrate the section [19].

To a lesser degree, heavy metals used to contrast the structures [see Section 15.3.1, "Thick Sections of Embedded Material"] are modified by high electron doses and prolonged irradiation; for example, lead citrate,

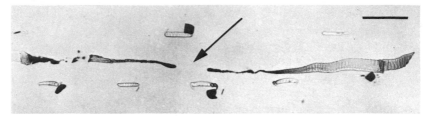

Figure 11. Beam damage in an araldite section thicker than 10 μm. Semithin section resectioned from a 25-μm araldite thick section, previously irradiated at 2.5 MV and then reembedded into araldite and sectioned perpendicularly to its surface. The semithin section is examined with a light microscope; irradiated areas appear dark; important thinning has occurred and a hole can be observed (arrow). Black spots on both parts of the irradiated section are copper bars of the folding grid onto which the 25-μm-thick section was trapped. Bar = 100 μm.

commonly used to contrast membranous structures coagulates locally to form lead particles 3–30 nm in diameter [66]; silver nitrate used for contrasting neurofibrils in nerve cells aggregates, forming enlarged silver grains [38]; these phenomena cause a decrease in resolution and furnish a coarse, sometimes uninterpretable image of the structures. Migration or volatilization of stains under the beam have also been described [8].

This short overview underlines the necessity for an accurate evaluation of the nature and extent of specimen damage when thick sections are examined in HVEM and interpreted for three-dimensional reconstruction. The types of damage described above also make it difficult to obtain multiple successive pictures of the same specimen area without alterations or shifts from picture to picture. Indeed, this is an essential requirement for obtaining tilted stereopairs. The same problem arises when successive serial thick sections are needed for three-dimensional reconstruction of some thick specimens [e.g., 20,49,81].

Various techniques are proposed to minimize radiation damage, such as coating the specimen with carbon, irradiating with a low dose, stabilizing the specimen at a low temperature, or use of more suitable resins [8,61,63].

Damage in Whole Unembedded Specimens

Little precise information exists on the effects resulting from critical point drying prepared specimen–high voltage beam interactions. It seems that no noticeable damage occurs; this is deducted by comparing pictures of cells prepared by various methods, some of them examined in the scanning electron microscopy, with pictures of CPD prepared cells viewed in HVEM [18,46,67].

15.3.5 Results: Some Examples

Over the past 10 years, the number of publications by biologists using high voltage electron microscopes has increased noticeably [18,22–24,41]. An exhaustive review is beyond the scope of this article and only a few examples, relating important results mainly due to the use of HVEM for examining thick specimens, are summarized here.

Results Obtained by Examination of Thick Sections

As is well known, stereoviews of thick sections of embedded tissue contrasted by selective staining techniques have provided important information on the three-dimensional configuration of membranous components of the cell and, particularly, the Golgi apparatus and endoplasmic reticulum.

Golgi Apparatus. In two-dimensional pictures recorded from thin sections prepared for CTEM, the Golgi apparatus appears to be composed of separate units called dictyosomes widely scattered throughout the cytoplasm of the cell. Each dictyosome consists of a stack of closely parallel saccules; one part of each stack faces a reticulum endoplasmic sheet and appears to arise by fusion of vesicles originating from that sheet: This is the proximal, or forming, face. The opposite side of the stack, which in grandular cells gives rise to secretion granules, is the distal or mature face. Thus dictyosomes display a morphological polarity.

Osmium impregnation (see Section 15.3.1, "Thick Sections of Embedded Material") in animal cells usually reveals the forming faces of the Golgi apparatus; thus their three-dimensional configuration can be studied on stereoviews of thick sections of osmium-impregnated material examined at high accelerating voltage (1–2.5 MV). New information has been obtained on the saccule ultrastructure and on the relationships between the dictyosomes of the cell. An examination of various cell types shows that the first saccules of the forming face are perforated sheets (Figs. 12 and 13) and have a network structure (snail mucous gland [19,68–70], nerve ganglion cells, Sertoli cells [71–73]). The polygonal network described in nerve cells [71–73] has been termed the primary network; stereoscopic examination reveals that the elements making up the meshes run at various angles to the general plan of the sheet.

Another important face emerged from these studies: The osmium-impregnated portions of the dictyosomes are interconnected and constitute a continuous network throughout the cytoplasm of the cells [69–73], termed the secondary network [71–73].

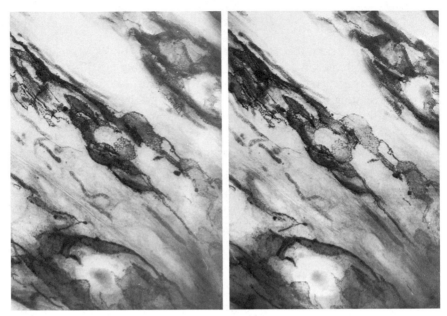

Figure 12. Osmium-impregnated Golgi apparatus of snail mucous gland. Stereopair show-ing the fenestrated structure of the forming face; tubes interconnect the various dictyo-somes. 1.5-μm-thick section examined at 2.5 MV. Tilt: 20° × 6.500.

Thus the scattered stacks observed in thin sections are, in fact, por-tions of the same functional system and correspond to the "internal retic-ular apparatus" discovered by Golgi with the light microscope; its signifi-cance has long been poorly understood and much discussed.

Endoplasmic Reticulum.   The three-dimensional configurations of the endoplasmic reticulum have been studied in HVEM, mainly in selectively stained plant cells and muscle. In plant cells, osmium impregnation and zinc–iodine impregnation reveal both the ER and the Golgi apparatus. ER ultrastructure and three-dimensional configurations in various cell types have been studied on impregnated tissue thick sections [22,70,74–76]. Changes in ER ultrastructure have been examined in HVEM during ger-mination of storage parenchyma of cotyledons [70,75,76] and, quantita-tively, the extent of interconnected tubular ER and cisternal ER have been followed during the protein deposition phase [75,76].

Changes occurring during mitosis have been studied in sections of osmium-impregnated meristematic cells. In interphase cells, ER is abun-dant, mainly in the vicinity of the nuclear envelope and near the periphery

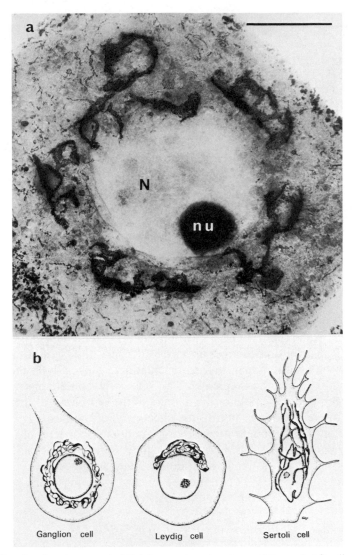

Figure 13. Osmium-impregnated Golgi apparatus of various cell types. (*a*) Small nerve cell of a rat trigeminal ganglion. The impregnated portions of the primary network are interconnected and form the secondary network, all around the nucleus; N, nucleus; nu, nucleolus; 3-μm-thick section examined at 1 MV. Bar = 5 μm. (*b*) Illustration of the organization of the secondary network of three different cell types. Secondary network is perinuclear in ganglion cell, caps one side of the nucleus in a Leydig cell, and extends throughout the cytoplasm in Sertoli cell. (From Ref. 73.)

375

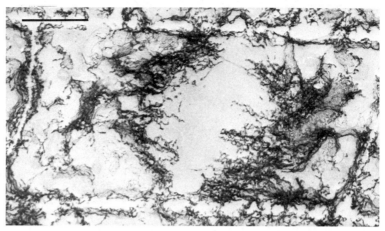

Figure 14. Osmium-impregnated plant cell ER. Root meristematic cell of a castor oil plant in metaphase. ER elements surround the spindle and form two caps at the poles; some elements penetrate into the spindle matrix. 2-μm-thick section examined at 2.5 MV. Bar = 2 μm.

of the cell. Some tubules and perforated sheets extend between these two zones. ER distribution changes as mitosis occurs. Almost all the cisternae surround the mitotic apparatus, and accumulate especially in the regions of the spindle poles. Some elements go across the spindle matrix; others are oriented from one pole of the spindle to the periphery of the cell. ER consists of continuous or perforated sheets and of tubules (Fig. 14). This three-dimensional configuration is poorly understood when only thin sections are examined in CTEM. A possible involvement of ER in the regulation of Ca^{2+} concentration in the spindle region, suggested by other methods of investigation [77] is supported by these findings; this regulation could be related to chromosome movements.

Sarcoplasmic Reticulum and T System in Striated Muscle.   In striated muscle cells, the sarcoplasmic reticulum (smooth ER) that surrounds the myofibrils, and the T system, constituted of sarcolemma invaginations extending transversally into the cell, are two interconnected systems, both involved in the contraction mechanism. Information on the extent and distribution of the T system (Fig. 15) and on its relationships to the sarcomere bands is derived from HVEM studies of selectively stained thick sections (see Section 15.3.1, "Thick Sections of Embedded Material") of various muscles. Sarcoplasmic reticulum distribution has also

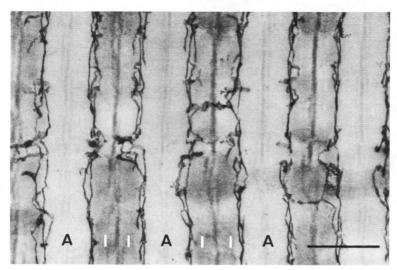

Figure 15. Distribution of the T system in skeletal muscle. Part of a longitudinally sectioned muscle fiber of mouse diaphragm. The T system is selectively stained. At the level of A-I junction, the regular pattern of the T-system distribution is observed. A, A band; I, I band. 1-μm-thick section examined at 1 MV. Bar = 2 μm. (From Ref. 78.)

been investigated by these techniques [37,40,78,79; see references in Ref. 41].

Vacuoles in Plant Cells. Vacuolar apparatus development has been studied in HVEM on meristematic cell thick sections stained by Gomori's lead method. Initially constituted of anastomosing tubes, the vacuolar apparatus form changes; tubes in older cells appear as chains of swellings which later on form individual globular vacuoles [80] (Fig. 16).

Serial Thick Sections and Three-Dimensional Reconstructions. Three-dimensional information cannot always be obtained by stereoviewing single thick sections. It can sometimes be obtained through reconstruction from serial sections. The use of thick sections facilitates the reconstruction as fewer sections are needed to reconstitute the structure of interest (see Ref. 20 for details and references). This technique has been used successfully by several authors (see Section 15.3.4, "Damage in Thick Sections," for damage limitations). For example arrangement of microtubules in kinetochore fibers of the mitotic apparatus has been studied using 0.25-μm serial sections examined in HVEM [81]: three-dimen-

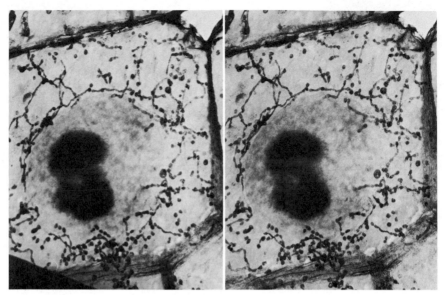

Figure 16. Vacuolar compartment in pea root meristematic cell. Stereopair in a young meristematic cell, showing the vacuolar apparatus contrasted by lead precipitate deposits (Gomori method, see Section 15.3.1). The vacuolar compartment consists of a tubular network and some isolated vesicles. 2-μm-thick section examined at 1 MV; 6000×. Tilt: 15°.

sional models of trypanosomatids mitochondria have been constructed from HVEM recorded micrographs of serially thick-sectioned *Trypanosoma* (e.g., Ref. 82; see Ref. 20).

Results Obtained by Examination of Unembedded Whole Specimens

Whole Cells Grown in Tissue Culture. Using the technique based on critical point drying and HVEM stereoscopic viewing, Porter and his co-workers obtained important new findings on the cytoplasmic ground substance (CGS) organization, in whole unembedded cells grown in tissue culture [18,45,46,83]. CGS consists of slender strands 4–10 nm thick, termed "microtrabeculae" by the authors (Figs. 17 and 18). Microtrabeculae are interconnected and form an irregular three-dimensional lattice, the microtrabecular lattice (MTL), which is continuous throughout the cell. The lattice is associated with membranous (endoplasmic reticulum, mitochondria) and nonmembranous (microtubules, filaments) structures. It has been suggested [46,83] that microtrabeculae constitute a protein-rich phase in the cytoplasm, including structures such as polysomes, microfilaments, microtubules, and cisternae of the ER: intratra-

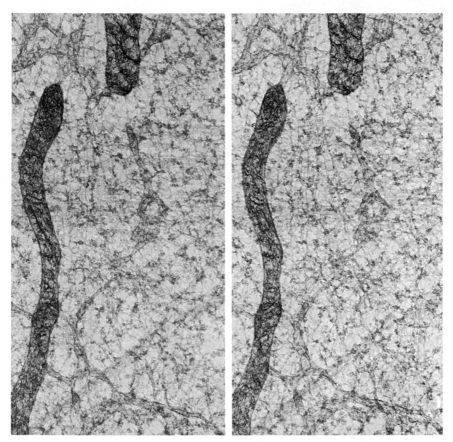

Figure 17. Microtrabecular lattice in a whole unembedded cultured cell. Stereopair showing relationships between microtrabecular lattice (MTL) and other cytoplasmic components: numerous fine strands (microtrabeculae) link microtubules to one another; some strands join polysomes. Examinations at 1 MV; 40,000×. (From Ref. 83.)

becular spaces, outlined by the trabeculae of the lattice, should contain water-rich solutions of low molecular weight metabolites.

The consequences of these findings are important. The microtrabecular lattice is thought to represent the nonrandom organization of the CGS. It probably restrains the Brownian motion of organelles in living cells and provides a distribution of the non-membrane-associated proteins in the CGS in such a way that it permits efficient interactions: between proteins as related enzymes or between polymers as microfilaments and their subunits pools [17]. These possibilities are analyzed in Refs. 17, 18 and 83.

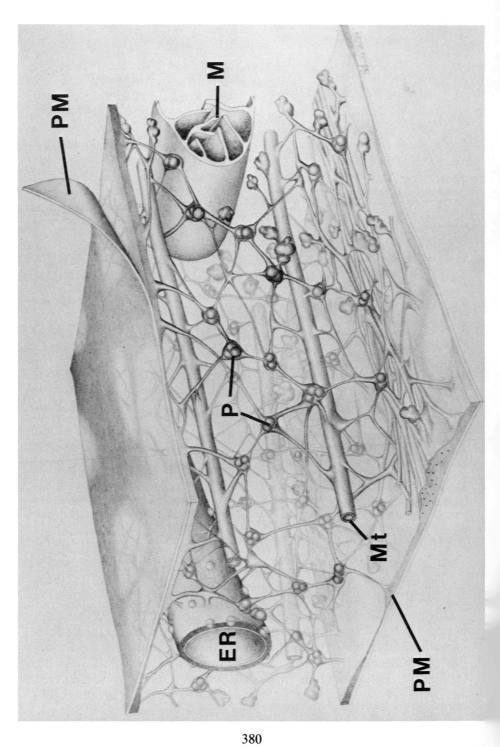

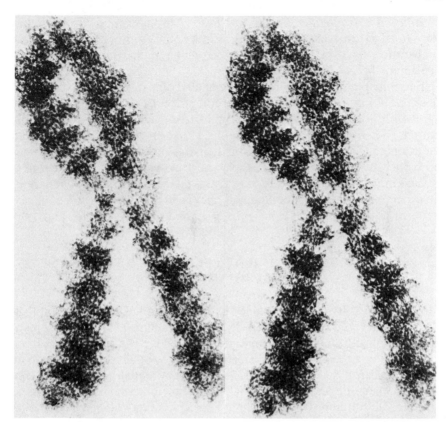

Figure 19. Isolated unembedded metaphase chromosome. Stereopair showing an isolated metaphase chromosome from a Chinese hamster ovary (CHO cultured cell). The component fiber is 50-nm thick and gyres of the chromosome helix are preserved. Whole organelle examined at 1 MV. Tilt: 10°. 20,000×. (From Ref. 49.)

The hypothesis that the CGS lattice structure is an artifact derived from specimen preparation techniques has been discussed. Arguments for and against its reality in living cells have been developed. However, various experiments using different preparation techniques [18,46] have concluded in favor of MTL existence; the form observed in HVEM would be similar or equivalent to the form existing in living cells [83].

Figure 18. Interpretative diagram of the MTL and associated structures of the cytoplasm. Microtrabeculae interconnect cell components—plasma membrane, PM; mitochondria, M; microtubules, Mt; polysomes, P; ER, endoplasmic reticulum. (From Ref. 83.)

Isolated Unembedded Cell Organelles. Chromatin and chromosomes have been extensively studied in HVEM by Ris. Appropriate methods of preparation have been developed to examine such whole unembedded specimens [47,49,84].

For example, nucleosomes have been observed in the 20-nm chromatin fiber, resulting from the helical supercoil of the 10-nm fiber, isolated and spread on a water surface. During mitosis, a high condensation of the chromosomes occurs. Isolated metaphasic rod-shaped chromosomes can be examined in HVEM, in spite of their thickness; these chromosomes are generally so compact that their substructure is not visible, unless treated in such a way that they swell and their details become apparent. One can observe that the component fiber is 50 nm thick; in some places, gyres of chromonema helix are preserved and directly visible (Fig. 19).

15.4   THIN BIOLOGICAL SPECIMENS AND HIGH RESOLUTION

Most of the results thus far obtained by introduction of HVEM for biological research have contributed to the understanding of the three-dimensional organization of biological material, because sharp images of thick specimens can be obtained and stereoviewed. Another slowly developing field of research derives from the theoretical potentiality of high voltage electron microscopes to provide high resolution images of thin specimens, since resolution improves by raising the accelerating voltage (see Section 15.2.2) and radiation damage is minimized. To carry out high resolution performances, the high voltage microscope must be constructed in such a way that electron optical, electrical, mechanical, and thermal factors are carefully controlled and are of sufficiently high standard [85]. For example, resolution approaching 0.2 nm has been obtained using a 600-kV microscope, especially constructed for achieving high resolution [3]. Therefore it should be theoretically possible to image individual atoms in their position in biological crystals in various biochemical compounds or in membranes [86]. Atomic resolution, if attainable, would provide important contributions to molecular biology.

In practice, this attractive aim is not easily achieved in organic systems mainly due to radiation damage introduced by the beam during examination. In addition to this major factor limiting resolution, methods of preparation can introduce some deformation in these fragile biological specimens. A number of reviews have considered the various problems concerning the high resolution electron microscopy of biological material [61,63–65] and provide numerous details and references on this subject.

Some preparation and examination procedures developed to reduce damage in biological thin specimen molecules and macromolecules are briefly considered here in relation to HVEM, also because they raise new prospects for its use in high resolution research.

15.4.1 Preparation of Thin Biological Specimens for High Resolution Electron Microscopy

Various preparation techniques of thin biological specimens, especially molecules and macromolecules, have been developed in view of conventional or high voltage high resolution microscopy. These techniques must preserve the specimen ultrastructure from deformations; they must also increase its radiation resistance, in order to reduce the damage resulting from electron irradiation.

This is done in different manners [61], for example, by surrounding the specimen with appropriate media. Mounting in the presence of glucose has been carried out to study purple membranes of *Halobacterium halobium* and catalase [87], mounting in the presence of methalcrylate to study catalase [61], freezing the specimen in ice [88], and so on. The mechanical containment of the specimen should prevent the displacement of molecular fragments formed by irradiation. Other methods involve conductive coating (aluminum, gold), chemical combinations of the specimens such as halogenation, and so on. All have stabilizing and protective effects.

15.4.2 Examination Conditions of Thin Biological Specimens for High Resolution Electron Microscopy

Since thin biological specimens are much more radiation sensitive, they must be examined with the minimum electron dose compatible with detectable contrast and image recording. This dose is often above the critical dose (maximum value of electron exposure above which significant damage occurs), which can be determined for crystals by fading of the diffraction pattern during irradiation; for example, in the case of protein crystals, the critical dose varies between 50 and 150 electrons/nm^2 [65].

The gap between the critical dose and the dose required for viewing and imaging small details increases when the magnification is raised. High magnifications are sometimes necessary to visualize small details at high resolution. As magnification M increases, electron flux through the specimen must be increased in proportion to M^2 [61] for imaging, as the recording emulsion has the same sensitivity; consequently, magnification must be adjusted taking into account the radiation sensitivity of the specimen.

TABLE 2   Values of Experimental Critical Exposures (Fluxes Necessary to Destroy Crystallinity of L-valine) as a Function of Electron Energy[a]

Electron Energy (MeV)	Main Values of Measured Critical Exposure. ($\times 10^{-2}$ coulombs/cm^2)
0.05	0.11
0.1	0.30
0.5	1.2
1	1.5
2.5	1.1
3	1.3

[a] From Ref. 89.

Operating at high voltage reduces the damage rate and increases the value of the critical dose. This appears in Table 2, showing the energy dependence of the critical exposure in the case of aliphatic amino acid L-valine [89].

The reasons for this improvement are discussed in general reviews [8,61,63]. In particular, the inelastic scattering cross section declines by a factor of 2.5–3 over the energy range from 100 kV to 1 MV; elastic scattering also declines, but the proportion of elastically scattered electrons collected in the objective aperture increases at high voltage. Damage is therefore reduced. In addition, faster electrons lose less energy in a specimen. As the velocity of the electron is increased at high voltage, damage is, in theory, reduced (see Refs. 8 and 61 for discussion). Operating at low temperature also reduces radiation damage; many organic specimens have an increased radiation resistance when kept at low temperature in the electron microscope. This has been observed in CTEM. Promising results have been obtained when examinations are done at low temperature using high voltage electron microscopes with cold stages or supraconducting lenses [91].

15.4.3   Results: Some Examples

An improvement in resolution of a factor of 2 can be expected at 1 MV as compared to 100 kV when aliphatic amino acid L-valine is studied over the range of electron energy mentioned above [89].

Also, the radiation sensitivity of organic compounds such as anthracene and coronene examined at 500 kV is reduced by a factor of 3 to 4 at liquid helium temperature as compared to room temperature [90]. A factor of 5 to 8 can be gained at −120°C in the critical doses of catalase or purple membrane crystals [65].

Interesting results are obtained using supraconducting lenses with the electron microscope operating at 220 kV and 4 K. This microscope re-

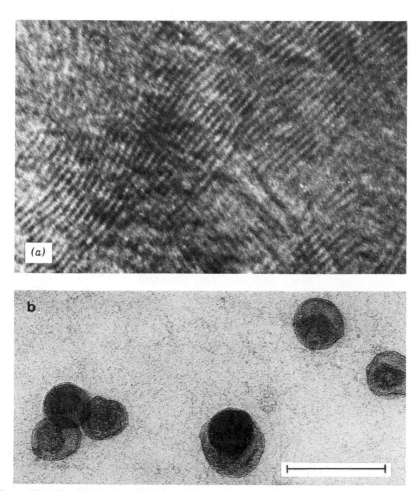

Figure 20. Examination of thin specimens at low temperature. (*a*) Graphitized carbon examined at 1 MV, 125 K; 0.34-nm spacings are resolved. 5.5.10^6× (courtesy of M. Fotino). (*b*) Murine RNA tumor virus examined at 1 MV, 9 K. Bar = 0.2 μm. (Courtesy of E. de Harven.)

solves 0.17-nm object details in an amorphous carbon film ([91]; see review in Ref. 65). An important gain in critical dose has been found in various organic crystals [92]. Using that microscope a gain of about 10 is found, on crotoxin complex crystals [65].

These examples show that HVEM operating at low temperatures offers some interesting prospects. The construction of cold stages adapted to commercial high voltage microscopes is being developed (Mircea Fotino, personal communication). Sharp images of various biological materials have been obtained at −120 to −150°C (Figs. 20a,b).

15.5   CONCLUSIONS AND PROSPECTS

Some of the more important results obtained by high voltage electron microscopy in the biological field have been summarized in this review, taking into account the characteristics of the specimens, the potential of the instrument, and the degree of reliability of the results. Many questions must now be debated including the importance of the HVEM's contribution to biology, based on results thus far obtained, future developments and trends, and the merits of HVEM compared to other techniques able to give similar information.

15.5.1   Three-Dimensional Information

HVEM has been widely used to obtain information on the three-dimensional organization of various cell components and has provided valuable answers in this field. This is an important development. An increasing number of problems require three-dimensional morphological information, since distribution of cell structures and interrelationships between cell components are closely related to cell function.

Among the results thus far obtained, some concern *structures previously and extensively studied by conventional electron microscopy* of thin sections. Three-dimensional morphology has not always been derived from the two-dimensional images obtained in CTEM. Examples were described in Section 15.3.5 of new information obtained by HVEM resulting from the ability of this instrument to provide sharp images and depth perspective from thick biological specimens.

The future prospects for HVEM in this field are vast. HVEM stereoviewing of thick specimens may become a standard method to complement information obtained by other techniques and to reconstitute the three-dimensional appearance of specimens, which is not provided by thin-section examination in CTEM.

The capacity of HVEM to provide resolution images of *whole thick specimens* and to allow the visualization of minute elements in the depth dimension has permitted the introduction of *new concepts* of cell ultra-structure. This is the case, for example, for the microtrabecular lattice (MTL) observed in the cytoplasmic ground substance (CGS) of whole unembedded cells (see Section 15.3.5). The demonstration of this component and of its interrelationships with minute filamentous structures of cells cannot be ascertained from thin sections, since they do not prominently stain and since portions viewed on thin sections are too small to allow valuable interpretation.

New achievements can be expected in the future from HVEM investigations on whole unembedded material as adapted preparation techniques develop. As an example, specific proteins characteristic of filamentous components of cells could be localized in whole cells by immunocyto-chemical methods; using this means, specific three-dimensional information might be gained as labeling allows the differentiation between various categories of filamentous components [18].

New findings may also be expected from *HV cryomicroscopy of frozen thick specimens,* since tissues and cells may be reasonably well preserved by freezing and many chemical preparative treatments are avoided. Results obtained with such methods could be correlated to those of such alternative CTEM techniques as freeze-etching, freeze-fracturing, or the recently developed technique of Heuser [93] that includes quick-freeze, deep-etch, and rotary replication and provides sharp CTEM images of a stereoviewed replica of frozen structures as thin as cytoplasmic filaments or extracellular coat.

However, contrary to what occurs when whole specimens are penetrated in depth by a high energy electron beam, opaque metal replicas are viewed in these techniques; damage is probably reduced, as the replicas are composed of radiation-resistant materials and made on a well-preserved tissue surface. However, differential characterization of the structures, using specific staining methods, for example, is of course impossible.

Prospects in HVEM examination of frozen specimens are evidently linked to progress in preparative techniques and to progress in the construction of cold stages fitted to HV cryomicroscopes.

Techniques allowing the examination of *thick wet specimens* in HVEM fitted with the appropriate environmental chambers have been carried out mainly by Parsons and his colleagues [8]. Prospects and attempts to resolve this difficult problem are developed in King et al. [8] and further details are given in another chapter in this book.

Interesting prospects for obtaining accurate quantitative three-dimensional information from HVEM examined thick specimens are offered by *computer quantitative reconstruction techniques* [94,95]. The development of accurate quantitative models, rendering a point-by-point account of the specimen three-dimensional organization, has been attempted using multiple-tilt-angle series of HVEM images. Such precise information cannot be derived from stereoviewing or from serial-section reconstruction methods. Methods for this application must be developed in the future.

15.5.2 High Resolution High Voltage Microscopy

In recent years the problem of high resolution in electron microscopy of biological material has been actively studied. Resolution at atomic dimensions is of course very desirable, extending the contribution of electron microscopy in cell biology to the molecular level.

Beam damage is the major factor limiting resolution when biological specimens are irradiated in the electron microscope. Hence progress in this field depends on instrumentation refinement, allowing examination at high voltage and low temperature where damage is reduced (see Section 15.4.2). To take full advantage of the increased accelerating voltage, mechanical and electrical stabilities of the high voltage electron microscope have to be of a very high standard.

The favorable effect of operating at low temperature on the radiation sensitivity of biological specimens can be exploited only with a cold stage fitted to the commercially available high voltage electron microscope. Such stages have been constructed in recent years. Some are under construction. Alternatively, a cryomicroscope, equipped with short-focal-length supraconducting lenses might be used for high resolution work and would offer interesting prospects.

15.5.3 High Voltage Electron Microscopy and Scanning Electron Microscopy

The scanning electron microscope (SEM) provides three-dimensional information on metal-coated specimen surfaces. Thus the instrument lends itself readily to correlatives studies using HVEM. For example, the internal organization of whole cells can be analyzed by HVEM, and then the metal-coated external surface of the same cell can be viewed in stereo SEM [18]. Information on correlative changes of internal architecture and surface morphology under various experimental conditions is becoming available.

The relative merits of scanning transmission electron microscope (STEM) and high voltage electron microscope are subject to discussion. Conventional-voltage high resolution STEM equipped with high brightness sources, that is, a field emission gun, can provide advantages such as great penetration and detection of low contrast.

It is generally agreed that a high voltage version STEM would provide new potential; greater facility in manipulating the signal image and ease of energy filtration and of elemental analysis by electron energy loss methods are the main expected improvements. Projects to construct high voltage STEM have been undertaken at the University of Chicago [96] and at the Laboratoire d'Optique Electronique of Toulouse, France (B. Jouffrey, personal communication) and are now under development.

ACKNOWLEDGMENTS

The authors wish to thank Professor Mircea Fotino for his helpful advice and discussion. We are also grateful to E. de Harven, M. Fotino, H. Hishikawa, F. Jaïs, P. Mentré, K. R. Porter, A. Rambourg, H. Ris, and M. Stearns, for supplying micrographs as cited in the text and to A. Marraud for reading parts of the text. We would also like to thank Danièle Coulon and Nicole Gatellier for typing the manuscript.

REFERENCES

1. M. Knoll and E. Ruska, Z. Phys. *78*, 318 (1932).

2. M. Fotino, in *Electron Microscopy in Biology,* Vol. 1, J. D. Griffith, Ed., Wiley, New York, 1981, p. 89.

3. V. E. Cosslett, R. A. Camps, W. O. Saxton, D. J. Smith, W. C. Nixon, H. Ahmed, C. J. D. Catto, J. R. A. Claever, K. C. A. Smith, A. E. Timbs, P. W. Turner, and P. M. Ross, Nature *281*, 49 (1979).

4. C. Humphreys, *Principles and Techniques of Electron Microscopy: Biological Applications,* Vol. 6, in M. A. Hayat, Ed., Van Nostrand Reinhold Co., New York, 1976, p. 1.

5. V. E. Cosslett, Proc. Roy. Soc. Lond. *338*, 1 (1974).

6. G. Dupouy, Ultramicroscopy *2*, 199 (1977).

7. J. M. Cowley, in *Principles and Techniques of Electron Microscopy: Biological Applications,* Vol. 6, M. A. Hayat, Ed., Van Nostrand Reinhold Co., New York, 1976, p. 40.

8. K. V. King, D. F. Parsons, J. N. Turner, B. B. Chang, and A. J. Ratkowski, *Cell Biophys.* 2, 1 (1980).
9. K. Hama, in *Advanced Techniques in Biological Electron Microscopy*, J. K. Koehler, Ed., Springer Verlag, Berlin, 1973, p. 275.
10. G. Dupouy, F. Perrier, and L. Durrieu, J. Microsc. *9*, 575 (1970).
11. G. Dupouy, Jermkont Ann. *155*, 393 (1971).
12. M. Fotino, in *Microscopie Electronique à Haute Tension, 4ème Congrès International*, B. Jouffrey and P. Favard, Ed., SFME, Paris, 1975, p. 361.
13. A. M. Glauert, in *Practical Methods in Electron Microscopy*, Vol. 3, A. M. Glauert, Ed., North-Holland/American Elsevier, Amsterdam and New York, 1974, pp. 1–207.
14. M. A. Hayat, *Fixation for Electron Microscopy*, Academic Press, New York, 1981.
15. M. A. Hayat, *Principles and Techniques of Electron Microscopy: Biological Applications*, Arnold, London, 1981.
16. J. J. Wolosewick, J. Cell Biol. *86*, 675 (1980).
17. J. C. Guatelli, K. R. Porter, K. L. Anderson, and D. P. Boggs, Biol. Cell. *43*, 69 (1982).
18. K. R. Porter and M. E. Sterns, in *Methods in Cell Biology*, Vol. 22, J. N. Turner, Ed., Academic Press, New York, 1981, p. 53.
19. P. Favard and N. Carasso, J. Microsc. (Oxf.) *97*, 59 (1973).
20. C. L. Rieder, in *Methods in Cell Biology*, Vol. 22, J. N. Turner, Ed., Academic Press, New York, 1981, p. 215.
21. H. F. Premsela, Micron *7*, 171 (1976).
22. C. R. Hawes, Micron *12*, 227 (1981).
23. A. M. Glauert, J. Cell Biol. *63*, 717 (1974).
24. A. M. Glauert, J. Microsc. (Oxf.) *117*, 93 (1979).
25. T. A. Shalla, T. W. Carroll, and G. A. de Zoeten, Stain Technol. *39*, 257 (1964).
26. N. Carasso, M. C. Delaunay, P. Favard, and J. P. Lechaire, J. Microsc. *16*, 275 (1973).
27. K. Hama and K. R. Porter, J. Microsc. *8*, 149 (1969).
28. H. Hama and F. Nagate, Micron *1*, 229 (1969).
29. H. Ris, J. Microsc. *8*, 761 (1969).
30. D. S. Friend and M. J. Murray, Am. J. Anat. *117*, 135 (1965).
31. M. Maillet, Z. Mikr. Anat. Forsch. *70*, 397 (1963).
32. M. Maillet, Bull. Assoc. Anat. Paris *1*, 233 (1968).
33. M. S. Forbes, B. A. Plantholt, and N. Sperelakis, J. Ultrastruct. Res. *60*, 306 (1977).
34. G. Thiéry and A. Rambourg, J. Microsc. Biol. Cell. *26*, 103 (1976).
35. Y. Clermont and A. Rambourg, Am. J. Anat. *151*, 191 (1978).

36. A. Rambourg, Y. Clermont and L. Hermo, Am. J. Anat. *154,* 455 (1979).

37. L. D. Peachey and C. Franzini-Armstrong, *Proc. 35th EMSA Meeting,* Boston 1977, p. 570.

38. R. Couteaux, N. Carasso, and P. Favard, J. Microsc. *24,* 283 (1975).

39. K. Hama, K. Hirosawa, and T. Kosada, in *Proc. 5th Int. Conf. High Voltage Electron Microscopy,* T. Imuna and H. Hashimoto, Ed., Kyoto, 1977, p. 333.

40. L. D. Peachey, J. Cell Biol. *70,* 357a (1976).

41. E. Yamada and H. Ishikawa, in *Methods in Cell Biology* Vol. 22, J. N. Turner, Ed., Academic Press, New York, 1981, p. 123.

42. L. Goldstein, G. E. Wise, and C. Ko, J. Cell Biol. *73,* 322 (1977).

43. K. R. Porter, A. Claude, and E. F. Fullam, J. Exp. Med. *81,* 233 (1945).

44. I. K. Buckley and K. R. Porter, J. Microsc. (Oxf.) *104,* 107 (1975).

45. J. J. Wolosewick and K. R. Porter, in *Practical Tissue Culture Applications,* K. Maramorosch and H. Hiromi, Ed., Academic Press, New York, 1979, p. 59.

46. J. J. Wolosewick and K. R. Porter, J. Cell Biol. *82,* 114 (1979).

47. H. Ris, in *Proc. 9th Int. Congr. Electron Microscopy,* Vol. 3, J. M. Sturgess, Ed., Toronto, Microscopical Society of Canada, 1978, p. 545.

48. H. Ris, in *Methods in Cell Biology,* Vol. 18, G. Stein and L. J. Kleinsmith, Eds., Academic Press, New York, 1978, p. 229.

49. H. Ris, in *Methods in Cell Biology,* Vol. 22, J. N. Turner, Ed., Academic Press, New York, 1981, p. 77.

50. T. F. Anderson, Trans. N.Y. Acad. Sci. *13,* 130 (1951).

51. T. A. Barber, T. W. Rademacker, and W. H. Orme-Johnson, *Proc. 37th EMSA Meeting,* San Antonio 1979, p. 164.

52. E. Yamada and H. Watanabe, in T. Imura and H. Hashimoto, Ed., *Proc. 5th Int. Conf. High Voltage Electron Microscopy, Kyoto,* 1977, p. 339.

53. A. J. Saubermann and P. Echlin, J. Microsc. (Oxf.) *105,* 155 (1975).

54. E. Varriano-Marston, J. Gordon, E. A. Davis, and T. E. Hutchinson, J. Microsc. (Oxf.) *109,* 193 (1977).

55. M. Fotino, in *Proc. 6th Eur. Cong. Electron Microscopy,* D. G. Brandon, Ed., Vol. 1, Tal International, Jerusalem, 1976, p. 277.

56. J. N. Turner, in *Methods in Cell Biology,* Vol. 22, J. N. Turner, Ed., Academic Press, New York, 1981, p. 1.

57. J. N. Turner, in *Methods in Cell Biology,* Vol. 22, J. N. Turner, Ed., Academic Press, New York, 1981, p. 33.

58. M. V. King, in *Methods in Cell Biology,* Vol. 22, J. N. Turner, Ed., Academic Press, New York, 1981, p. 13.

59. M. V. King, in *Methods in Cell Biology,* Vol. 22, J. N. Turner, Ed., Academic Press, New York, 1981, p. 147.

60. F. N. Low, G. E. Olson, B. Persky, and J. J. Van Rybroek, in *Three Dimensional Microanatomy of Cells and Tissue Surfaces*, D. J. Allen, P. M. Motta, and L. J. A. Didio, Eds., Elsevier/North-Holland, New York, 1981, p. 1.

61. V. E. Cosslett, J. Microsc. (Oxf.) *113*, 113 (1978).

62. R. M. Glaeser, in *Physical Aspects of Electron Microscopy and Microbeam Analysis*, B. M. Siegel and D. R. Beaman, Eds., Wiley, New York, 1975, p. 205.

63. R. M. Glaeser and K. A. Taylor, J. Microsc. (Oxf.) *112*, 127 (1978).

64. M. Isaacson, in *Principles and Techniques of Electron Microscopy*, Vol. 7, H. Hayat, Ed., Van Nostrand Reinhold, New York, 1977, p. 1.

65. W. Chiu, in *Electron Microscopy of Proteins*, Vol. 2, J. R. Harris, Ed., Academic Press, New York, 1982, p. 233.

66. F. Nagata, T. Matruda, and K. Hama, J. Electron Microsc. *23*, 219 (1974).

67. J. J. Wolosewick and K. R. Porter, Am. J. Anat. *147*, 303 (1976).

68. P. Favard, L. Ovtracht, and N. Carasso, J. Microsc. *12*, 301 (1971).

69. N. Carasso, L. Ovtracht, and P. Favard, C. R. Acad. Sci. Paris, *273D*, 876 (1971).

70. N. Carasso, P. Favard, P. Mentré, and N. Poux, in *High Voltage Electron Microscopy, Proc. 3rd International Conference*, P. R. Swann, C. J. Humphreys, and M. J. Goringe, Eds., Academic Press, New York, 1974, p. 414.

71. A. Rambourg, A. Marraud, and M. Chrétien, J. Microsc. (Oxf.) *97*, 49 (1973).

72. A. Rambourg, A. Marraud, and Y. Clermont, in *High Voltage Electron Microscopy, Proc. 3rd International Conference*, P. R. Swann, C. J. Humphreys, and M. J. Goringe, Eds., Academic Press, New York, 1974, p. 419.

73. A. Rambourg, Y. Clermont, and A. Marraud, Am. J. Anat. *140*, 27 (1974).

74. F. Marty, C. R. Acad. Sci. Paris *277D*, 2681 (1973).

75. N. Harris, Planta *146*, 63 (1979).

76. N. Harris, Planta *148*, 293 (1980).

77. P. K. Hepler, J. Cell Biol. *86*, 490 (1980).

78. H. Ishikawa and S. Tsukita, in *Proc. 5th Int. Conf. High Voltage Electron Microscpy*, T. Imura and H. Hashimoto, Eds., Kyoto, 1977, p. 359.

79. L. D. Peachey and B. R. Eisenberg, Biophys. J. *22*, 145 (1978).

80. N. Poux, P. Favard, and N. Carasso, J. Microsc. *21*, 173 (1974).

81. P. L. Witt, H. Ris, and G. G. Borisky, Chromosoma *81*, 483 (1980).

82. J. J. Paulin, J. Cell Biol. *66*, 404 (1975).

83. K. R. Porter, M. C. Beckerle, and M. A. McNiven, in *Modern Cell Biology*, Vol. 2, *Spatial Organization of Eucaryotic Cells: A Symposium in Honor of K. R. Porter*, J. R. McIntosh Ed., Alan Liss Inc., New York, 1983, p. 259.

84. H. Ris and J. Korenberg, in *Cell Biology: A Comprehensive Treatise*, Vol. 2,

D. M. Prescott and L. Goldstein, Eds., Academic Press, New York, 1979, p. 267.

85. D. J. Smith, *Proc. 38th EMSA Meeting,* San Francisco, 1980, p. 822.

86. W. Baumeister and M. Hahn, in *Progress in Surface and Membrane Science,* Vol. 11, D. A. Cadenhead and J. F. Danielli, Eds., Academic Press, New York, 1976, p. 227.

87. P. N. T. Unwin and R. Henderson, J. Mol. Biol. *94,* 425 (1975).

88. K. Taylor and R. M. Glaeser, J. Ultrastruct. Res. *55,* 448 (1976).

89. D. G. Howitt, R. M. Glaeser, and G. Thomas, J. Ultrastruct. Res. *55,* 457 (1976).

90. S. M. Salih and V. E. Cosslett, J. Microsc. (Oxf.) *105,* 269 (1975).

91. I. Dietrich, F. Fox, E. Knapek, G. LeFranc, K. Nachtrieb, R. Weyland, and H. Zerbst, Ultramicroscopy *2,* 241 (1977).

92. E. Knapek and J. Dubochet, J. Mol. Biol. *141,* 147 (1980).

93. J. Heuser, in *Methods in Cell Biology,* Vol. 22, J. N. Turner, Ed., Academic Press, New York, 1981, p. 97.

94. J. Frank, in *Methods in Cell Biology,* Vol. 22, J. N. Turner, Ed., Academic Press, New York, 1981, p. 199.

95. G. F. Bahr, J. A. Boccia, W. F. Engler, R. A. Robb, I. Jurkevich, and A. F. Petty, Ultramicroscopy, *4,* 45 (1979).

96. P. S. Lin, J. J. Shuler, and A. W. Crewe, *Proc. 38th EMSA Meeting,* San Francisco, 1980, p. 6.

Errata for:

Analysis of Organic and Biological Surfaces
Edited by Patrick Echlin

These are the correct captions for the figures in Chapter 16.

Figure 1. Plot of the Fabre characteristic cross section for a beam energy of 20 keV as a function of atomic number. Note the rapid decrease in ionization probability with increase in atomic number.

Figure 2. Plot of the Bethe-Fermi relativistic cross section for a beam energy of 100 keV as a function of atomic number. Note the rapid decrease in ionization probability with increase in atomic number.

Figure 3. Plot of continuum distribution demonstrating anisotropic property. Beam direction is left to right. Continuum anisotropy is a complicated function of not only beam energy but also photon energy and target atomic number.

Figure 4. Plot of emitted continuum energy as function of photon energy for two elements and three observation angles. Note essentially uniform distribution up to beam energy V_0. (From Ref. 15).

Figure 5. Plot of the quantities Z^2/A and $Z/2$.

Figure 6. Plot of the characteristic (Fabre) cross section for the element silicon as a function of beam energy.

Figure 7. Dual plot consisting of distribution of energy loss in silicon (*energy*) for an initial electron energy of 20 keV (left vertical axis) and ionization probability Q along this path (right vertical axis).

Figure 8. Variation of the backscatter coefficient as a function of average atomic number for beam energies of 10 and 49 keV. (Data from Ref. 18. From Ref. 25.)

Figure 9. Energy distribution of backscattered electrons for several elements. The measurements were made at a take-off angle of 45° above the surface. $V_0 = 30$ keV. (Data from Ref. 31. From Ref. 25.)

Figure 10. Dual curve showing mass-absorption coefficient of element nickel for its own Kα radiation (top curve, left axis), and fraction transmitted through 1-μm Ni foil (right axis).

Figure 11. Depth distribution of the production of Mg K alpha x-rays in aluminum. Experiment: Castaing and Henoc, solid curve; Monte Carlo calculations, Newbury and Yakowitz, dots. Note that vertical scale is logarithmic. (From Ref. 25.)

X-RAY MICROANALYSIS
BY MEANS OF FOCUSED
ELECTRON PROBES

C. E. FIORI and C. R. SWYT

Division of Research Services
Biomedical Engineering and Instrumenation Branch
National Institutes of Health
Bethesda, Maryland

16.1 INTRODUCTION

In electron beam x-ray microanalysis one focuses a beam of fast electrons onto a small region of a specimen surface and examines the x-ray signals that emanate because of the electron beam–specimen interaction. The diameter of the electron probe is usually in the range of 10 nm to 1 μm. Depending on the volume of interaction, the total mass analyzed can be anywhere from 10^{-11} to 10^{-16} g. However, the limits of detection vary from 0.001 to 1%. The technique is thus microanalysis rather than trace analysis. Instruments on which electron beam microanalysis can be performed are the scanning electron microscope equipped with an energy dispersive detector, the analytical electron microscope, and the electron beam x-ray microanalyzer. Present commercial instruments can form a probe less than 10 nm in diameter. Electron beam microanalysis is thus an important tool in biology because it provides the capability to measure the concentration and distribution of a large number of elements at the subcellular level. However, small, intense electron probes can be extremely destructive to a biological specimen. The energy input to the analyzed region can break chemical bonds and cause selective mass loss and translocation of mobile ions out of the analyzed volume [1,3]. Successful application of the technique in biology has usually occurred because the investigator has developed an appropriate analytical strategy to reduce the degree of damage to an acceptable level. It must be recognized, however, that the analytical results are frequently affected as much by preparation of the biological specimen as by the amount of beam damage. For example, most of the preparative techniques for conventional electron microscopy are not suitable for analytical microscopy because they cause major changes in the chemistry of the specimen by either replacement or depletion of elemental constituents by interaction with the dehydrating agents, stains, or fixatives. Early attempts by many biologists to use electron column instruments to analyze their specimens often led to failure because of just such specimen alteration due to preparation. However, considerable progress has been made in the last several years in reducing the effects of both radiation damage and specimen preparation so that important biological problems are now being solved by the technique. This chapter proceeds from a recognition of the importance of the preceding problems and sets as its goal a description of the physics germane to the analysis of "thin" (<200 nm), and "opaque" (>10 μm) biological samples. We do not have space in this chapter to cover the considerably more complicated situation that results when a specimen has an intermediate thickness.

We assume a successful and correct preparation procedure. An understanding of the physical processes is required of the analyst if he is to choose the optimum analytical conditions for a particular problem or specimen from the wide range of possibilities. The quality of a chemical assay depends on this choice. We will begin our discussion with the "thin" specimen case, since this, in most respects, is the easiest to understand, occurs most often in biological work, and provides a framework in which to examine the more complicated case of thick specimens. Since the rate of x-ray production from any type of biological sample is low, the x-ray detector of choice is the Si(Li) solid state detector [4,5]. This lithium-diffused silicon device has high geometric and quantum efficiency but poor spectral resolution. Because of the wide use of the solid state detector rather than the alternative crystal spectrometer detector for most biological applications, we assume its use in our discussions. This is not a "how to" chapter; there are no equations into which one can stick numbers from a specimen, turn the crank, and obtain a chemical answer. Instead, we hope to convey an appreciation of the underlying physical principles of analysis in a sufficiently, but not overly, rigorous manner. The treatment is certainly not exhaustive; we generally discuss only one approach to any given aspect of the overall problem of quantitation. The interested reader should assume that other approaches exist and can be found in the reference list. By its nature, the subject requires a mathematical treatment and there is no way to avoid this fact. The authors appreciate that the readers of this book have varying levels of mathematical and physical expertise. We have attempted to surround the required equations with sufficient definitions and explanations to make them at least understandable. In this chapter we will use the words *specimen* and *standard* in their self-explanatory context. We will use the word *target* when the subject under discussion is applicable to both. Parts of the section on thin targets in this chapter were presented at the joint meeting of the Microbeam Analysis Society and the Electron Microscopy Society of America in Washington, D.C., and appear in the MAS proceedings [20].

16.2   X-RAY PRODUCTION

X rays observed by the energy dispersive detector in an analytical electron column instrument arise from two types of inelastic or energy loss interactions between fast beam electrons and target atoms. In one case a beam electron interacts strongly with a core electron and imparts sufficient energy to remove it from the atom. The ejected core electron can have any energy up to the beam energy less the characteristic shell energy. The beam electron is depreciated in energy by whatever kinetic energy the ejected electron has acquired plus the characteristic energy required to remove it. A characteristic x-ray is occasionally emitted when the ionized atom relaxes to a lower energy state by a transition of an outer shell electron to the vacancy in the core shell. The x-ray is called characteristic because its energy equals the energy difference between the two levels involved in the transition and this difference is characteristic of the element. Since the energies of these levels are well known, the photon energy distribution that results from a statistically meaningful quantity of x-rays, from the same transition and from an ensemble of atoms of the same atomic number, generally suffices to identify the atom species. This x-ray "line" is characterized by a "natural width," normally specified by the full width at half the peak maximum (FWHM). Because there is usually instrumental broadening when an x-ray line is measured, we must make the distinction between the measured distribution and the line. Consequently, we call the measured distribution a "peak" and specify its width, also, as FWHM. The difference between these distributions is very large when a solid state energy dispersive detector is used. For example, the natural width of the Mn Kα line is 1.50 eV. This line is broadened to a peak of typically 155 eV by a 30-mm^2 lithium-drifted silicon detector and its associated electronics.

The second type of inelastic interaction we must consider occurs between a fast beam electron and the nucleus of a target atom. A beam electron can decelerate in the Coulomb field of an atom, which consists of the net field due to the nucleus and core electrons. Depending on the deceleration, a photon is emitted that can have an energy ranging from near zero up to the energy of the beam electron. X-rays that emanate due to this interaction process are referred to in this chapter as continuum x-rays. They are called alternatively "background," "white," or *Bremsstrahlung* x-rays.

By counting characteristic x-rays we obtain a measure of the number of analyte atoms present in the volume of target excited by the electron beam. By counting continuum x-rays we obtain a measure approximately proportional to the product of the average atomic number and the mass thickness of the excited volume. If the average atomic number of the target is known, the continuum signal gives a measure of target mass thickness. In the ratio of a characteristic signal to a continuum signal, the factors that affect both signals equally cancel. Since in the usual energy dispersive x-ray analysis system the two signals are recorded simultaneously, the effects of the recording time and incident probe current cancel. Marshall and Hall [6] developed a highly successful analytical procedure, based on such a ratio, which is widely used in biological applications for thin targets in the scanning electron microscope and microprobe and analytical electron microscope.

16.3   THE THIN BIOLOGICAL TARGET

Because relatively few inelastic collisions suffered by the incident beam electrons in any target result in x-ray production, most inelastic collisions result in the loss of only very small amounts of energy through mechanisms such as ionization of an atom by the ejection of an outer shell electron or collective excitation of outer shell electrons. Also, only a few electrons are multiply scattered in a thin target. Therefore, most scattered beam electrons lose very little energy in total in traversing a thin target and almost none are absorbed. That this is the case in fact defines a target as thin. We may therefore conveniently assume that the energy of the electrons involved in collisions producing x-rays from a thin target is the beam energy.

Let us first consider in some detail the equation describing the generation of characteristic x-rays.

16.3.1 The Characteristic Signal

We can predict the number of characteristic x rays generated into 4π steradians from the following relation:

$$I_{ch} = \left(\frac{N_0 \rho C_A}{A_A}\right) Q_A \omega_A F_A N_e \, dz \tag{1}$$

where N_0 is Avogadro's number, ρ is the density of the target in the analyzed volume, C_A is the weight fraction of the analyte in the volume, A_A is the atomic weight of element "a," Q_A is the ionization cross section for the shell of interest and has dimensions of area; ω_A, the fluorescent yield, is the probability that an x-ray will be emitted due to the ionization of a given shell; F_A is the probability of emission of the x-ray line of interest relative to all the lines that can be emitted due to ionization of the same shell; N_e is the number of electrons that have irradiated the target during the measurement time; and dz is the target thickness in the same units of length as used in Q and ρ.

The quantity in parentheses in Eq. (1) is the number of atoms of analyte per unit volume of target and is obtained by the following rationale. By definition the weight fraction of an element "a" is the mass of "a" divided by the total mass. The mass of "a" per cm^3 is the weight fraction of "a" times the density ρ in the analytical volume given in g/cm^3. We emphasize that density is a measured quantity and refers to the mass of all the atoms per unit of volume. To convert mass to number of atoms we use Avogadro's number, $N_0 = 6.02 \times 10^{23}$: the number of atoms in a mole of an element. Therefore, the number of atoms of element "a" per cm^3 is $(C_A \cdot \rho) \cdot (N_0/A_A)$ where A_A is the gram atomic weight of a mole of element "a." The total number of atoms in a volume of 1 cm^3 is then the sum over all the elements i in the volume:

$$N = N_0 \rho \sum_i \left(\frac{C_i}{A_i}\right) \tag{2}$$

The basic functional form of the characteristic cross section is due to Bethe [7]:

$$Q = 6.51 \times 10^{-20} \frac{n_s b_s}{UV_c^2} \ln(C_s U) \tag{3}$$

where the constant is the product πe^4 (in keV^2 − cm^2; e is the charge of an electron), n_s is the number of electrons which populate the sth shell or subshell of interest ($n_k = 2$), V_c is the energy required to remove an electron from a given shell or subshell (the critical excitation energy), U is the overvoltage ratio V_0/V_c (where V_0 is the energy of the impinging beam electron) and b_s and C_s are constants for the sth shell or subshell. For the remainder of this chapter we use the symbol V to denote electron kinetic energy and the symbol E to denote photon energy. The "area" in cm^2 of Q is essentially the size of the K, L, or M shell "target" that a beam electron must "hit" to produce an ionization of that shell. In general this area is about 100 times smaller than the area of the entire atom. Powell [8,9] reviewed a number of semi-empirical cross sections for the beam energy range commonly found in the electron microprobe (<40 keV). He recommended, for the K shell, that $b_k = 0.9$ and $C_k = 0.65$ when the energies are expressed in keV. Powell favorably notes a modification to the Bethe formula by Fabre:

$$Q_F = \frac{6.51 \times 10^{-20} n_k \ln(U)}{V_c^2 \cdot a(U + b)} \tag{4}$$

where $a = 1.18$ and $b = 1.32$ are constants for the K shell. The range of the overvoltage ratio is recommended to be $1.5 < U < 25$. The other terms are as above. It was not intended that these cross sections should be applied above about 40 keV, and it should cause no great surprise if they fail in the range of beam energies found in the analytical electron microscope. There are several forms that have been suggested for the ionization cross section for the energy range 70–200 keV. This is an energy region where the effects of relativity should not be ignored and a generally useful cross section should accommodate the effect explicitly or at least provide empirical adjustment. Zaluzec [10,11], for example, has recommended the formulation, given in Mott and Massey [12], of the relativistic cross section derived by Bethe and Fermi [13] and Williams [14]:

$$Q = \frac{6.51 \times 10^{-20} n_k a_k}{V_c V_0} \ln[b_k U - \ln(1 - \beta^2) - \beta^2] \tag{5}$$

where all energies are given in keV, $a_k = 0.35$ (K shell),

$$b_k = \frac{0.8U}{(1 - e^{-\gamma})(1 - e^{-\delta})}, \qquad \gamma = \frac{1250}{V_c U^2}, \qquad \delta = \frac{1}{2V_c^2}$$

and

$$\beta = \frac{v}{c} = \sqrt{1 - \left[1 + \left(\frac{V_0}{511}\right)\right]^{-2}}$$

is the relativistic correction factor. Figure 1 shows the Fabre cross section plotted as a function of atomic number for a 20 keV electron beam (a practical beam energy for the scanning electron microscope and microprobe). In Fig. 2 is plotted the Bethe-Fermi cross section (with Zaluzec's coefficients) for a 100-keV electron beam (a typical working voltage in the analytical electron microscope) as a function of atomic number. Note the change in the vertical scale between the two plots and the rapid decrease in ionization probability as the atomic number increases.

16.3.2 The Continuum Signal

The interaction of a large number of beam electrons with a thin foil of a given element will produce an emitted continuum spectrum having a distribution approximately proportional to $1/E$, where E is the photon en-

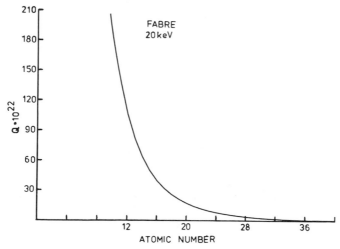

Figure 1. Energy distribution of backscattered electrons for several elements. The measurements were made at a take-off angle of 45° above the surface. $V_0 = 30$ keV. (Data from Ref. 31. From Ref. 25.)

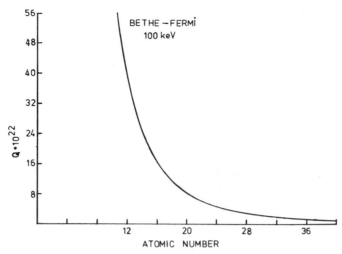

Figure 2. Variation of the backscatter coefficient as a function of average atomic number for beam energies of 10 and 49 keV. (Data from Ref. 18. From Ref. 25.)

ergy. The magnitude of this distribution is proportional to the number of atoms comprising the foil. Other factors, such as beam current and measurement time, scale the magnitude linearly. The beam energy, V_0, and observation angle, θ, defined later, affect the overall shape of the distribution in a complex manner. We can predict the number of continuum x-rays generated into a unit steradian from the following relation:

$$I_{co} = \sum_i (N_i Q_i) N_e \qquad (6)$$

where N_i is the number of atoms of element i and Q_i is the continuum cross section for that element and has dimensions of area and N_e is the number of electrons that have irradiated the N_i target atoms during the measurement time. Q_i is usually assumed to be differential in photon energy and observation angle. Consequently, it is not necessary to include dE and $d\theta$ terms in Eq. (6). Using the fact that the total number of atoms in 1 cm^3 is:

$$N_0 \rho \sum_i \left(\frac{C_i}{A_i} \right) \qquad (7)$$

Equation (6) can be rewritten for a thin film as:

$$I_{co} = N_0\rho \sum_i \left(\frac{C_iQ_i}{A_i}\right) N_e \, dz \tag{8}$$

where dz is the target thickness in the same units of length as used in Q_i and ρ. We note the following differences between the generation of characteristic and the continuum radiation. For our purposes, the ejection of a core electron by a fast beam electron and occasional subsequent emission of a characteristic x-ray photon are independent events. A characteristic photon has equal probability of being emitted in any direction (isotropy) after the ionization. Relativistic considerations apply only to the probability of ionization. The probability of continuum emission, on the other hand, is intimately related to the probability that a fast beam electron will decelerate in the Coulomb field of an atom. Indeed, the emission is a direct consequence of the deceleration. Furthermore, the probability of continuum emission is directionally dependent and peaked toward the forward direction as defined by the direction of travel of the beam electrons (Fig. 3). We define the observation angle to be the angle between the

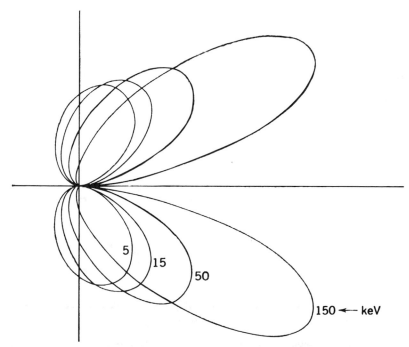

Figure 3. Dual curve showing mass-absorption coefficient of element nickel for its own Kα radiation (top curve, left axis), and fraction transmitted through 1-μm Ni foil (right axis).

incident beam direction and the x-ray detector. The "angle" and magnitude of the maximum of each of the lobes shown in the figure depends on the photon energy being observed, beam energy, and the average atomic number of the target. Target tilt has no effect (other than to change the apparent thickness of the target). Because of this anisotropy, continuum radiation is usually expressed as emission into a unit steradian at a specified angle, rather than uniformly into 4π steradians, as is the characteristic cross section.

We continue our discussion of the continuum cross section by deriving the expression used in the Marshall–Hall procedure for biological microanalysis. We first note two properties of continuum radiation.

1.   The maximum energy that can be given up by an incident beam electron is its kinetic energy which is numerically equal to the beam voltage V_0. Consequently, there is a highest photon energy, E_0, in a continuum spectrum due to beam electrons of energy V_0. This highest energy is the so-called high energy limit, or the Duane–Hunt limit.

2.   In a thin target both theory and experiment indicate that the amount of emitted energy in an energy interval dE is approximately uniformly distributed from near zero up to the high energy limit [15], (Fig. 4).

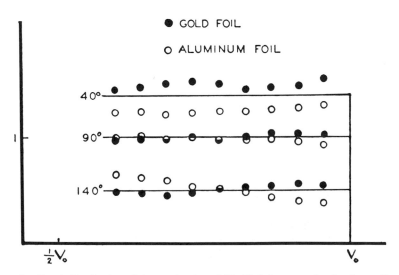

Figure 4.   Depth distribution of the production of Mg K alpha x-rays in aluminum. Experiment: Castaing and Henoc, solid curve; Monte Carlo calculations, Newbury and Yakowitz, dots. Note that vertical scale is logarithmic. (From Ref. 25.)

From item 2 we can define the fraction of the total emitted continuum energy E_t, in the interval from E to $E + dE$, as $dE/(E_0 - 0)$. The efficiency of the generation of continuum in a thin film is defined to be the total continuum energy (from near zero to E_0) generated by electrons that lose an amount dV of their energy divided by the energy lost. Kirkpatrick and Wiedmann [16] have determined from the theory of Sommerfeld that this efficiency of production is $2.8 \cdot 10^{-9} ZV_0$, where V_0 is the beam voltage in kV and Z is the atomic number. We can thus express the total radiated continuum energy as:

$$E_t = 2.8 \times 10^{-9} ZV_0 \, dV \qquad (9)$$

The fraction of this quantity in the energy interval dE provides us with the number of photons I_{co} of energy E. The number of photons in the vanishingly small interval dE is obtained by dividing the amount of energy in that interval by the photon energy E. Consequently, the number of photons in the energy interval dE is given by the quantity $(E_t \, dE/E_0)/E$ or:

$$I_{co} = 2.8 \times 10^{-9} \cdot (Z/E) \, dE \, dV \qquad (10)$$

The energy loss dV of electrons traversing a thin film of a single element of thickness dz can be given by the Bethe equation for the slowing down of an electron [17]:

$$-dV = \frac{2\pi e^4 \, N_0 Z\rho}{V_0 A} \ln \left[\frac{1.166 V_0}{J} \right] \, dz \qquad (11)$$

where J is the mean excitation energy of an atom [18,19], 1.166 is the square root of half the base of the natural logrithms, dz is the target thickness in cm if ρ is given in g/cm^3 and the other terms are as defined earlier. The derivation of the equation is based on the assumption that the energy transferred to the relatively massive nucleus when an electron interacts with an atom is negligible so that the interaction can be assumed to involve only the atomic electrons. (For a more detailed discussion of the assumptions used by Bethe in deriving the equation and of potential errors from its application see Heinrich [18], pp. 226–232.) We note that, although the continuum results from interactions of beam electrons and atomic nuclei and Eq. (11) applies to energy assumed lost only in interactions between the beam electrons and atomic electrons, the energy loss in Eq. (11) can be used in Eq. (9). This is because Eq. (9) describes only the "efficiency of production"; no connection between the energy loss and the continuum production is implied. Combining these two equations, we obtain the number of continuum photons, I_{co}, with energy E, generated in the interval dE by a beam electron with energy V_0, traversing a thin film of thickness dz of one element:

$$I_{co} = \frac{220Z^2\rho \, \ln(1.166V_0/J) \, dz \, dE}{(AV_0E)} \qquad (12)$$

The energy terms are in electron volts. Equation (12) can be expressed in units of a cross section, differential in photon energy, by multiplying by the quantity $A/(N_0\rho \, dz)$, giving:

$$Q_{iu} = \frac{3.65 \times 10^{-22}Z^2 \, \ln(1.166V_0/J) \, dE}{V_0E} \qquad (13)$$

The subscripts iu on Q distinguish this cross section by its principal characteristics: isotropy and uniformity in energy distribution from the preceding property 2. The resulting cross section now has the required dimensions of area in cm^2 per beam electron. There are several versions of the J factor available in the literature. We present as an example the Sternheimer formulation (given in 19):

$$J = Z(9.76 + 58.82Z^{-1.19}) \cdot 0.001 \quad \text{(keV)} \qquad (14)$$

The use of the J factor in Eq. (11) permits the application of the equation, originally derived from a quantum mechanical treatment of the hydrogen atom, to higher atomic number elements. Equation (13) is an apparently simple expression of the continuum cross section. The simplicity is, however, a result of the approximations and assumptions used in the derivation. The cross section takes no account of the strong relativistic effects for beam energies above several keV. Furthermore, the equation poorly predicts the continuum distribution as a function of photon energy and, as mentioned earlier, it takes no account of the strong anisotropy of the continuum. The degree of anisotropy is a complicated function of beam energy, photon energy, and target atomic number. However, despite these considerable shortcomings, the cross section, as used by Marshall and Hall, is remarkably effective in many biological applications. Care must be exercised, however, not to use the equation outside the regime they carefully specified. For a discussion of cross sections that take more exact account of the physical processes involved in continuum generation see [20–22].

At this point we make a short digression and note the following features concerning continuum cross sections: The generalized formulation of the continuous radiation cannot be solved in closed form. As a result, all current cross sections derive from approximations of varying accuracy. However, with all continuum cross sections it is possible to factor out the square of the atomic number. Generally, the factored cross section will retain additional terms involving Z. In the more complete cross sections the residual terms include a photon energy dependence, such as in the Kirkpatrick–Wiedmann algebraic fit to the Sommerfeld theory discussed in Ref. 20. However, in Eq. (8), which predicts the number of continuum photons in a given material for a given electron flux, we note that the square of the atomic number is divided by the atomic weight. This quantity, Z^2/A, is approximately equal to $Z/2$ throughout the periodic table (i.e., $A \approx 2Z$). We stress the word *approximately*. As can be seen in Fig. 5, the function does not plot as a straight line and, indeed, is not even monotonic, that is, the same value of Z^2/A occurs at more than one Z. Consequently, according to Eq. (8), the continuum radiation as a function of atomic number has the undesirable shape as plotted in Fig. 5. The argument can be put forward that by stating the problem in terms of Eq. (6) we will get a smoothly varying continuum distribution as a function of the square of the atomic number. However, real world specimens do not come such that they will all have a known number of atoms at all points. Consequently, for most practical applications we use Eq. (8).

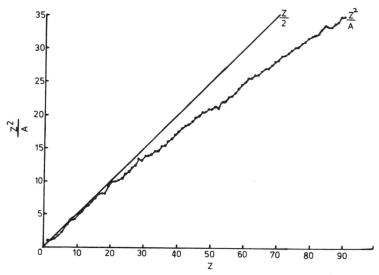

Figure 5. Plot of the characteristic (Fabre) cross section for the element silicon as a function of beam energy.

16.4 DERIVATION OF THE HALL PROCEDURE IN TERMS OF X-RAY CROSS SECTIONS

The density of the target within the analytical volume of the electron probe is generally not known unless the material being irradiated is a standard reference material. We also note that the thickness of a target at the site of impact of the electron beam is rarely a known quantity. However, it is obvious that a change in either local density or thickness causes a proportional change in the number of atoms with which a beam electron is likely to interact in passing through a thin target. Characteristic and continuum generation are equally affected. In order to treat more easily these changes, density and thickness are frequently combined into a single quantity, *mass thickness*, denoted by the notation $(\rho \, dz)$, which has the somewhat confusing dimensions of g/cm^2 (i.e., g/cm$^3 \cdot$ cm = g/cm^2). We must stress that the concept of mass thickness is useful only if the beam energy does not change by more than a very small amount in traversing the target thickness, dz. A typical biological target should be less than several thousand angstroms in thickness for the preceding assumption to be sufficiently accurate. With this caveat we form the ratio of Eqs. (1) and (8).

$$\frac{I_{ch}}{I_{co}} = \frac{(N_0 C_A/A_A) Q_A \omega_A F_A N_e(\rho \, dz)}{N_0 \sum_i (C_i Q_i/A_i) N_e(\rho \, dz)} \tag{15}$$

We note that $(\rho \, dz)$, the mass thickness, Avogadro's number, and the number of electrons incident on the film during the measurement period all cancel in the ratio. To simplify the formula further we take advantage of the fact that, for the analysis of a given element, the fluorescent yield and associated relative transition probability are constants that can be gathered into one grand constant, k. Thus we have:

$$\frac{I_{ch}}{I_{co}} = k \frac{(C_A Q_A/A_A)}{\sum_i (C_i Q_i/A_i)} \tag{16}$$

Where Q_A is the characteristic cross section for a particular element "a" and Q_i is the continuum cross section for the ith element; the subscript i must span all the elements present in the electron beam–target interaction volume. Equation (16) is the most general formula for the method proposed by Hall and his co-workers.

We now put into Eq. (16) the cross sections used by Marshall and Hall, Eqs. (3) and (13), and collect all numerical constants into k to obtain:

$$\frac{I_{ch}}{I_{co}} = k \frac{C_A/(A_A U V_c^2)\ln(C_s U)}{\sum_i [(C_i Z_i^2/A_i)/(V_0 E) \cdot \ln(1.166 \ V_0/J_i)]} dE \qquad (17)$$

We can further simplify this formula by taking advantage of the fact that the operating voltage V_0 is held constant during the analysis and the energy band dE of continuum measurement, centered at the energy E, is usually not changed during the analysis. Consequently, dE, E, V_0 (and hence U) can be absorbed into k, resulting in the simpler form:

$$\frac{I_{ch}}{I_{co}} = k \frac{C_A/A_A}{\sum_i [(C_i Z_i^2/A_i)\ln(1.166 \ V_0/J_i)]} \qquad (18)$$

where J_i is a function of atomic number such as that given in Eq. (14). This is the usual representation of the Marshall–Hall [6] equation, which is a more rigorous expression of the continuum correction concept proposed by Hall et al. earlier [23] and often referred to as the Hall method or correction. Usually, the equation is used in the following manner: A

TABLE 1

	Number of Atoms						
	H	C	N	O	P	S	Z^2/A
Water	2.0			1.0			3.67
Fatty acid (oleic)	34.0	18.0		2.0			2.87
Triglyceride	107.0	60.0		12.0			2.98
Glucose	12.0	6.0		6.0			3.40
Deoxyribose	10.0	5.0		4.0			3.33
Nucleic acid cytidine monophosphate	14.0	9.0	3.0	8.0	1.0		3.78
same without P							3.41
Protein (with S)	112.0	66.7	18.3	25.0		1.0	3.28
Protein (no S)	112.0	66.7	18.3	25.0			3.20
Araldite	8.4	5.8	0.02	1.19		0.08	3.15
Nylon	11.0	6.0	1.0	1.0			3.01
Polycarbonate	14.0	16.0		3.0			3.08

measured characteristic to continuum ratio from a characterized material is set equal to the right side of Eq. (18). Since the target being irradiated is a reference standard, we presumably know the atomic numbers Z_i, atomic weights A_i, and weight fractions C_i. Since all other terms except the constant k are known, k can be calculated for the given set of experimental conditions. Next, we hold these conditions constant and measure the characteristic and continuum intensities from the specimen. To calculate a weight fraction of analyte, C_A, from the specimen, we must know the weight fraction of each and every one of the elements that comprise that part of the specimen irradiated by the electron beam. Crucial to the Hall procedure is the following assumption: The quantity $\Sigma(C_i Z_i^2/A_i)$ for the biological specimen to be analyzed is dominated by the matrix so that the unknown contribution of the analyte, C_A, to the sum may be neglected. Furthermore, the value of the sum is known or can be estimated from other information about the detail being analyzed. A table of $\Sigma(C_i Z_i^2/A_i)$ values for a number of typical biological materials is presented in Table 1.

16.5 BULK TARGETS

For the purposes of simplifying the explanation of the germane physics we begin this section with an admittedly arbitrary definition of a "bulk" target. We require that the target has one planar surface that is oriented at a right angle to the impinging electron beam. All the other surfaces are sufficiently distant that no incident electron or generated x-ray exits through any of them. The required distance from the point of impact may be as small as several micrometers to as large as several hundred, depending on the nature of the target and operating conditions.

We will not discuss in this chapter the case of bulk specimens and standards that are inclined with respect to the electron beam because the additional complication obscures the underlying principles we wish to emphasize. We simply caution that, in a practical analytical situation with such a configuration, errors in results may occur because of approximations in the quantitation scheme presented. We refer the reader to several publications which discuss inclined targets in great detail [18, 25, 26]. The field of electron beam x-ray microanalysis of bulk targets is a mature one and has had an immense impact on a number of disciplines, including biology. The basic concepts of microprobe analysis of bulk specimens were originally presented in 1951 by R. Castaing in his Ph.D. thesis [27]. A number of workers have added substantially to Castaing's work and a large body of literature, including several textbooks, exists for the field.

Our approach differs from the standard treatment to some degree in that we separate the major ideas along the lines of inelastic and elastic scattering of electrons. However, a mature field can only be written about in so many ways and little is to be gained by trying to be too different. The present authors have leaned heavily on the development of the major ideas laid out in an early but superb book by Reed [26], which is now, unfortunately, out of print.

Within the relatively extensive bulk target volume, beam electrons undergo multiple scattering, both elastic and inelastic. Elastic scattering causes the beam electrons to deviate from their initial direction but their speed (and, therefore, kinetic energy) remains constant. If perfect elastic scattering were the only interaction process, the beam of electrons would permeate the entire target and exit from all the surfaces with the incident beam energy. However, inelastic scattering causes the beam electrons to lose energy, mostly with little change in direction. Indeed, if inelastic scattering were the only interaction between the beam and target, most of the volume of interaction would lie within a cone having a top diameter equal to that of the electron beam and a bottom diameter only somewhat larger. The height of the cone would depend on, among other things, the initial energy of the electrons. Since any given beam electron is scattered many times in a purely random manner by both types of interactions, the shape and extent of the volume of interaction for a large number of electrons are determined by the balance between the two.

An "analysis" of a bulk specimen consists of counting x-ray photons characteristic of the analyte element from the specimen and comparing the measured intensity to that obtained from a standard material also containing the analyte. Except in those cases where the composition of the specimen is very close to that of the standard, there is not a simple linear relation between the x-ray intensity ratio and the composition ratio of specimen to standard. Several "corrections" must be applied to both of the measured intensities. These are the "stopping power" correction, F_s, a "backscatter" correction, F_b, an x-ray absorption correction, F_a, a secondary characteristic x-ray fluorescence correction, F_f, and the secondary fluorescence correction due to the continuum radiation, F_c. They arise from multiple scattering of the electrons and the effect of the target on the intensity of the emitted x rays. Often the F_s and F_b corrections are combined into a single correction, F_z, which is called an "atomic number" correction, since both are functions of the average atomic number of the target being irradiated. The correction for fluorescence due to the continuum is frequently neglected. However, in biological applications it is sometimes the most important correction and must not be neglected. If the corrections F_z, F_a, and F_f are formulated such that their composite effect, F, can be obtained by multiplication of the individual corrections, the resulting procedure is called a "ZAF" correction. The form of the correction is:

$$C_A = C_A^* \frac{I_A F}{I_A^* F^*} \tag{19}$$

where I_A/I_A^* is the ratio of the characteristic x-ray signals of element "a" from the specimen and standard. The symbol "*" denotes a quantity that is related to, or obtained from, a standard. We will see that each of the various terms in F and F^* requires a knowledge of the chemical composition of the target. The composition of the standard, of course, we know. The composition of the specimen is the purpose of the exercise in the first place. We avoid this seemingly hopeless situation by the procedure of iteration. We estimate a weight fraction for each element present. These values are usually chosen to be the measured intensity ratios of the specimen to standard. We then solve Eq. (19) and obtain new estimates. By repeated application of this procedure we quickly converge on a result. Depending on the iteration algorithm, convergence usually occurs in a few steps.

Let us first consider the physics that leads to the stopping power correction. From an examination of the equation of the characteristic x-ray cross section, Eq. (3), we see that the probability of characteristic x-ray generation for a given characteristic x-ray line is roughly proportional to $\ln(U)/U$. U is the instantaneous "overvoltage", V/V_c, where V, the energy of the incident electron, depends on the depth in the target. The cross section for inner shell ionization (Q) is plotted in Fig. 6 (the required first step for the generation of a characteristic x-ray photon), as a function of electron energy for the element silicon. We note in particular the following characteristics of the curve: The silicon K shell cannot be ionized by electrons lower in energy than V_c, the critical excitation energy of the K shell (1.84 keV). The probability of ionization is greatest for electron energies of $e \cdot V_c$, where e is the base of the natural logarithms (2.714 . . .). The curve is moderately constant from about $1.5 \cdot V_c$ to over 20 keV. Characteristic cross section plots for other elements exhibit the same properties and general form. The shape of this curve is important because understanding it is the first step in acquiring a qualitative appreciation for the distribution of the generation of characteristic x-rays as a function of depth below the surface. We make the assumption that there are only pure inelastic scattering processes and ask the question, "what is the average energy of the electrons as a function of depth in the narrow cone, hypothesized above?" An approximate answer is obtained by re-

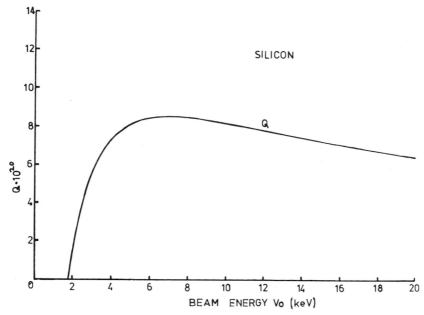

Figure 6. Dual plot consisting of distribution of energy loss in silicon (*energy*) for an initial electron energy of 20 keV (left vertical axis) and ionization probability Q along this path (right vertical axis).

peated evaluation of the Bethe equation [Eq (11)]. As an example we again use the element silicon, put in the appropriate numbers, and collect the constants to give:

$$-\frac{dV}{dz} = \frac{91452}{V} \ln(6.769 \cdot V) \quad (\text{keV/cm}) \tag{20}$$

where V and dV are in keV and dz is in cm. (We assume the density is constant and use depth rather than mass thickness.) We want to determine the average energy loss of a beam electron as it passes through the first layer of the target. This layer must be sufficiently thin (say 100 nm) that the differential equation [Eq. (20)] can be reasonably approximated by a difference equation. Evaluating Eq. (20) to find the energy loss for a 20-keV beam electron gives 224 eV. Consequently, the average energy of the beam electrons entering a second 100-nm layer, immediately beneath the top layer, is 19.7759 keV. Continuing this process and plotting the results (Fig. 7) reveals the average electron energy as a function of distance traveled in silicon for a beam of electrons having an initial energy of

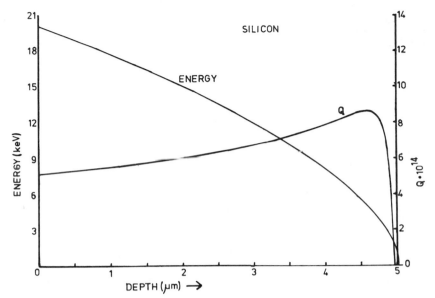

Figure 7. Plot of emitted continuum energy as function of photon energy for two elements and three observation angles. Note essentially uniform distribution up to beam energy V_0. (From Ref. 15).

20 keV. The curve labeled *energy* in the figure is this distribution and the left vertical axis gives the scale. We emphasize the fact that this curve approximates the average energy as a function of depth. In reality, because of the statistical nature of the scattering process, there is a distribution of energies at any given depth; this effect is called *straggling*.

We observe that for the first 2.5 μm of penetration the average beam electron loses only about 5 keV of energy. In the next 2.5 μm, however, it loses virtually all its remaining energy. The curve denoted Q is the cross section for K shell ionization as a function of depth below the surface and reveals the distribution of silicon K radiation for our idealized situation of pure inelastic scatter. The right vertical axis gives the scale for this distribution. We note that, as the average beam electron penetrates into the target and loses energy, the likelihood of causing a K shell ionization increases until its kinetic energy becomes less than V_c and it is incapable of removing a K shell electron from a silicon atom. It is extremely important to keep in mind both here and for the remainder of the chapter the difference between generated and emitted x-rays. Here we refer strictly to the generation process within the target. Indeed, what we have done by neglecting elastic scattering is "straighten out" the electron trajectories.

This has permitted us to obtain more simply the energy loss, and hence the distribution of both energy and the probability of ionization along the path, as well as to determine the total distance (range) that an average electron travels. We will consider the effect of elastic scattering only when it is required to understand the difference between generated and emitted x-ray intensities.

At this point we define the concept of "stopping power" S as:

$$S = -\frac{dV}{(\rho\ dz)} \quad (\text{keV} \cdot \text{cm}^2/\text{g}) \tag{21}$$

The stopping power is the measure of the energy lost by an electron in traversing a small mass thickness (i.e., interacting with a given number of atoms). As was shown for the example of silicon, the energy lost depends on the energy of the electron and increases as the energy of the electron decreases. The most often used formulation of stopping power derives from the Bethe equation [Eq. (11)]. To accommodate multielement targets requires that we extend its applicability to more than one atomic number. We can do this by assuming that the stopping power of the atomic electrons of various atoms in a multielement target combine additively. Consequently, the stopping power equation becomes:

$$\bar{S} = 7.84 \times 10^4 \cdot \frac{N_0}{E} \sum_i \frac{C_i Z_i \ln(1.166 E/J_i)}{A_i} \tag{22}$$

where the terms are as explained earlier. We note that since J increases with increasing atomic number, the logarithmic term decreases with Z. Similarly, as Z increases, the quantity Z/A tends to decrease. The net effect is that the stopping power decreases with an increase in average atomic number. A main reason for defining "stopping power" in terms of mass thickness is that it is then dependent on the type and number of atoms rather than on the density with which they are packed. Fortunately, it has been possible to develop an entire theory of microprobe quantitation for bulk targets without the need to know the density at each region of analysis. The use of the variable $(\rho\ dz)$ permits this development.

Let us now consider how the stopping power effect is quantitatively accounted for in the prediction of x-ray intensity. The number of ionizations, dn, of a given shell of an element along some path depends on the number of atoms encountered, or the mass thickness, and the ionization cross section for the given shell of the element, as follows:

$$dn = \frac{N_0 Q}{A} \rho \, dz \tag{23}$$

The total number of ionizations caused by a single electron as it decelerates along our hypothetical straight line path in the pure element is obtained by integrating Eq. (23). However, Q is a function of energy and this changes as the electron traverses the target. We could parametrically make the connection between energy and mass thickness, but it is more straightforward to change the variable of integration to energy to give:

$$n = \int_{V_0}^{V_c} \frac{N_0 Q}{A} \frac{d(\rho z)}{dV} \, dV \tag{24}$$

The integration proceeds from the surface, where the energy is V_0, to a depth where the electron has lost sufficient energy to be unable to ionize the given shell of the element and $V < V_c$. But since stopping power, S, is defined as $-dV/(\rho \, dz)$, we write:

$$n = \frac{-N_0}{A} \int_{V_0}^{V_c} \frac{Q}{S} \, dV = \frac{N_0}{A} \int_{V_c}^{V_0} \frac{Q}{S} \, dV \tag{25}$$

where the integration limits are reversed to take care of the minus sign.

For a multielement target Eq. (25) becomes:

$$n = \frac{C_A N_0}{A_A} \int_{V_c}^{V_0} \frac{Q_A}{\bar{S}} \, dV \tag{26}$$

where $\bar{S}$ is defined by Eq. (22) and Q_A is the ionization cross section for the element "a" in the multielement target. Even though Q_A is not a function of the other elements, the stopping power, and therefore energy of the electron, clearly is.

It is useful to pause here and note an implication of Eq. (26). A decrease in stopping power implies an increase in the number of generated photons from element "a" per beam electron. Since a decrease in stopping power comes about by an increase in the average atomic number, it follows that a concentration, C_A, of element "a" generates more intensity in a higher atomic number matrix than in a lower one.

If we form the ratio of the number of ionizations, n, due to one electron that has decelerated in a specimen containing C_A weight fraction of analyte "a" to the number of ionizations, n_0, due to one electron which has decelerated in a standard containing C_A weight fraction of the analyte, we get:

$$\frac{n}{n_0} = \frac{C_A(N_0/A_A) \int_V^{V_0} (Q_A/\bar{S}) \, dV}{C_A(N_0/A_A) \int_V^{V_0} (Q_A/\bar{S}^*) \, dV} = \frac{C_A \int_V^{V_0} (Q_A/\bar{S}) \, dV}{C_A \int_V^{V_0} (Q_A/\bar{S}^*) \, dV} \qquad (27)$$

As explained earlier, the superscript * denotes a quantity associated with the "standard"; the remaining terms are as previously defined. The characteristic ionization cross section for element "a," Q, is identically the same inside both integrals. The stopping power, however, is not and this difference must be taken into account. We define the stopping power corrections as:

$$F_s = \frac{1}{\int_{V_C}^{V_0} (Q_A/\bar{S}) \, dV} \quad \text{and} \quad F_s^* = \frac{1}{\int_{V_C}^{V_0} (Q_A/\bar{S}^*) \, dV} \qquad (28)$$

for the element "a" in specimen and standard, respectively.

Before the advent of the fast laboratory digital computer the preceding integrations were, for practical purposes, not feasible. Consequently, in many of the computer programs that have been written to perform the conversion of x-ray intensity ratios to elemental concentrations the approximation $F_s = \bar{S}$ is often made (Poole and Thomas [28]). The particular energy at which to evaluate $\bar{S}$ is usually chosen to be $V = (2V_0 + V_c)/3$. Recently, however, sufficiently fast computers have become commonplace in the laboratory and the more accurate programs that do this integration should be considered [29].

Our discussion so far has not required any consideration of elastic scattering of the incident electrons, but significant corrections for quantitation in a bulk specimen arise from it. There are two major effects that modify generated x-ray intensities in a bulk specimen. The first is that elastic scattering considerably reduces penetration of the electron beam into the target. The total average distance traveled by a beam electron, its "range," is unchanged, as is the probability of causing ionization along the path; but the distribution of the characteristic x-rays as a function of depth is dramatically altered from that shown in the example of Fig. 7. This will be discussed in the section on x-ray absorption.

The second effect of elastic scatter, the phenomena known as *backscattering,* is the scattering of electrons out of the target. Many of these electrons have sufficient energy to have caused inner shell ionizations had they remained in the target. We will see that electron backscattering increases with increasing average atomic number. This "loss" of potentially ionizing electrons as a function of increasing average atomic number, very approximately, offsets the increase in x-ray generation efficiency in a higher Z matrix.

Because of the statistical nature of elastic scattering and the complicated nature of the backscattering phenomena, it is difficult to predict the effect on the generation of x-rays from first principles. Most of the information we have about backscattering comes from experimental measurement. The equations commonly used are the results of exercises in curve fitting and show little evidence of the underlying physical processes. Nevertheless, the algebraic fits have been made to a large body of data from a number of workers so that the empirical nature of the equations should not be a cause for concern, but only caution.

There are two aspects of backscatter that we need to consider: the fraction of incident electrons that are backscattered and their energy distribution. We examine first the distribution, as a function of atomic number, of the total fraction of electrons from the beam backscattered at all energies. The traditional symbol for the backscatter fraction is η; the dependence on average atomic number is shown in Fig. 8 for two beam energies. We note that only a few electrons incident on a low atomic number target are backscattered but nearly half are backscattered from a high atomic number target. The fraction is essentially independent of the energy of the electron beam for energies between 10 and 40 keV. Outside of this range, especially at low beam energies, this is not at all the case.

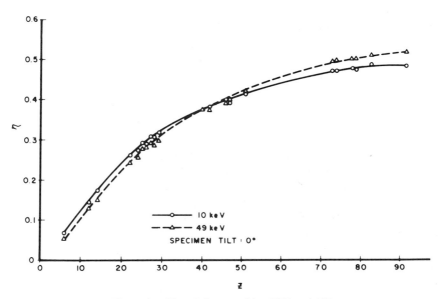

Figure 8.   Plot of the quantities Z^2/A and $Z/2$.

Darlington [30] has shown that η varies strongly with the beam energy between 500 eV and 10 keV. This is one reason among several why low beam energy ($<$10 keV) quantitative analysis should be undertaken with great care.

Knowledge of only the fraction of electrons that have backscattered is not sufficient to correct for the difference in x-ray production efficiency between a specimen and standard resulting from backscatter loss. The electrons that are scattered by an atom near the specimen surface into an angle exceeding 90° can leave the target with essentially the incident beam energy, but backscattered electrons have all energies from near zero to the beam energy. If we knew the energy of each backscattered electron as it emerged from the target, we could use Eq. (25) to calculate the number of additional x-rays that would have been produced by the backscattered electron. This value would be the basis of a "backscatter" correction. Let us proceed along this line.

As noted, both the absolute number and the energy distribution of the backscattered electrons as a function of atomic number are difficult to predict from first principles; we must rely on experimental measurements to obtain them. Bishop [31] made measurements of the number of electrons $d\eta$, backscattered with energies in the small interval of energy dV, as a function of energy V, in a number of pure elements. The data are usually displayed as a differential energy distribution, $d\eta/dW$, plotted as a function of W, where $W = V/V_0$, and V_0 is, as usual, the beam energy (W is not to be confused with $U = V_0/V_c$, the overvoltage). Figure 9 is a plot of some of Bishop's results obtained with an electron energy spectrometer viewing the target surface at 45°. From the definition of W, the value 1 on the horizontal axis corresponds to a backscattered electron energy equal to the beam energy; 0.5 is half that, etc. We observe that more than half the electrons that backscatter from a gold target, atomic number 79, have energies greater than 0.8 of the incident energy, the largest number having just less than V_0. In a low atomic number target, however, not only are there many fewer electrons backscattered, but the reduction is at the higher energies. This is because the electrons penetrate deeper into the low atomic number material, undergoing a greater number of inelastic (energy loss) events before they are backscattered. For example, fewer than a third of the electrons backscattered from the element copper exit with greater than 0.8 of the incident energy, the largest number with about $0.85 \cdot V_0$ (from Fig. 9). For very low atomic number elements, such as carbon, there are relatively few electrons backscattered at any energy and the differential distribution in energy is quite flat.

We note the difference between Figs. 8 and 9: Figure 8 predicts the fraction of electrons backscattered with all energies from targets of vari-

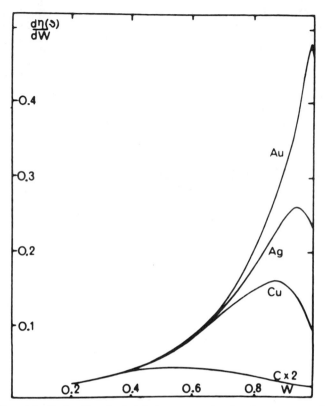

Figure 9. Plot of the Bethe–Fermi relativistic cross section for a beam energy of 100 keV as a function of atomic number. Note the rapid decrease in ionization probability with increase in atomic number.

ous atomic numbers. Figure 9 predicts the fraction of electrons backscattered with a particular energy, from a target of a particular atomic number. The total area under each of the curves in this figure contains the same information about the given element as that in Fig. 8.

The fraction of x-ray intensity lost by backscattering, n_x, is defined as the number of x-rays that backscattered electrons would produce if they remained in the specimen, divided by the number of x-rays that electrons with the incident energy V_0 would produce if there were no backscatter. The latter quantity is obtained by multiplying Eq. (26) by the number of incident electrons. Similarly, the x-ray intensity backscattered electrons with energy V would produce, if they remained in the specimen, is also given by Eq. (26), multiplied by the number of electrons backscattered with energies in the small energy interval from V to $V + dV$. This number is just equal to the product of the differential energy distribution function at energy V and the number of incident electrons. The total x-ray intensity lost to backscatter is thus the sum of x-ray intensities for all the backscattered electron energies from V_c to V_0. *The backscatter fraction,* n_x for a pure element, can therefore be written

$$n_x = \frac{\int_{W_0}^{1} (d\eta/dW) \int_{V_c}^{V} (Q/S) \, dV \, dW}{\int_{V_c}^{V_0} (Q/S) \, dV} \tag{29}$$

where $W_0 = V_c/V_0$.

The "backscatter correction," F_b, is defined as:

$$F_b = \frac{1}{1 - n_x} \tag{30}$$

However, the correction is usually given in terms of the quantity R, which is a measure of how much the backscattering effect reduces the x-ray intensity and is given as $R = 1 - n_x$. Therefore, $F_b = 1/R$. For composite materials it is common practice to use an empirically justified average [32]:

$$R = \sum_i C_i \cdot R_i \tag{31}$$

where C_i, as earlier, is the weight fraction of the element i and i spans all the elements in the composite material.

There are several formulas for R in the literature. We present, as an example, the algebraic fit due to Yakowitz et al. [33] to the R values calculated by Duncumb and Reed [32] from the experimental n_0 data of Bishop [31]:

$$R_{ij} = r_1 - r_2\ln(r_3Z_j + 25) \tag{32}$$

where

$$r_1 = 8.73 \times 10^{-3}U^3 - 0.1669U^2 + 0.9662U + 0.4523$$

$$r_2 = 2.703 \times 10^{-3}U^3 - 5.182 \times 10^{-2}U^2 + 0.302U - 0.1836$$

$$r_3 = \frac{0.887U^3 - 3.44U^2 + 9.33U - 6.43}{U^3}$$

and U is the overvoltage ratio V_0/V_c. The subscript i denotes the analyte element and the subscript j spans all the elements in the target, including the analyte. Consequently, from Eq. (31), the backscatter correction for the analyte element i is

$$F_b = \frac{1}{R_i} = \frac{1}{\sum_j c_j R_{ij}} \tag{33}$$

and

$$F_b^* = \frac{1}{R_i^*} = \frac{1}{\sum_j c_j^* R_{ij}}$$

for specimen and standard, respectively. In general, as the average atomic number of the target increases, the efficiency for the generation of characteristic x-rays increases. However, as the average atomic number increases there is an accompanying loss of generation efficiency caused by backscatter loss. The two effects tend to cancel but only approximately. A correction must be applied to account for any difference in average atomic number between specimen and standard.

16.6 X-RAY ABSORPTION

When an x-ray impinges on a thin film of matter, it can pass through the film without change in energy or direction, be scattered with or without energy loss, or be annihilated, giving up all its energy by the ejection of a core electron of an atom in the film. For photons in the energy range of interest to us (1–10 keV), only the first and last possibilities are germane to our needs. The annihilation process is called photoelectric absorption. It is the cause of the next correction to the measured x-ray intensity that we consider.

Assume a collimated beam of I_0 monochromatic (one energy) x-rays per unit time, incident on a single element thin film. The number of x-rays, I, that pass through a mass thickness (ρz) is given by Beer's law:

$$I = I_0 \exp\left[-\left(\frac{\mu}{\rho}\right)\rho z\right] \tag{34}$$

As the beam energy is increased, the number of x rays transmitted, I, increases but at a decreasing rate. An abrupt decrease in I occurs at the critical excitation energy of each shell of the target material; the number again increases as the energy of the incident x rays is increased further. The relatively large decreases in the number transmitted occur at energies equal to the critical excitation energies because the x rays then have sufficient energy to interact with the electrons of an additional shell through the process of photoelectric absorption. Although the quantity one measures is I, the quantity more often plotted as a function of photon energy is (μ/ρ), the mass absorption coefficient (dimensions of area/mass, usually given in cm^2/g). The mass absorption coefficient provides us with a measure of the absorbing power of a target of a given atomic number for photons of given energy. It is independent of the physical state of the material. Figure 10 presents plots of (μ/ρ) and I as functions of photon energy. The upper plot is of the mass absorption coefficient of nickel in the energy range 3–13 keV. It decreases exponentially with increasing photon energy except at the critical excitation energy, $V_c = 8.3$ keV, of nickel K electrons. The lower plot gives, from Eq. (34), the fractional transmission, I/I_0, of x rays through 1 μm of nickel (density 8.9 g/cm^3).

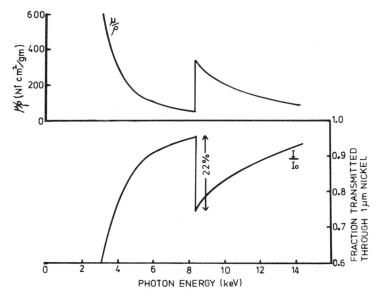

Figure 10. Plot of continuum distribution demonstrating anisotropic property. Beam direction is left to right. Continuum anisotropy is a complicated function of not only beam energy but also photon energy and target atomic number.

We note the abrupt 22% drop in fractional transmission at the K shell binding energy. Beer's law can be applied to composite foils by summing the individual contributions of the constituent elements:

$$\frac{\mu}{\rho} = \sum_i C_i \left(\frac{\mu}{\rho}\right)_i \qquad (35)$$

where C_i is the weight fraction of element i.

Beer's law predicts the fraction of incident monochromatic x-rays that emerge from a particular thickness of a material. We can use the equation to determine from the number of x-rays measured, I, the number of x-rays generated, I_0. The number of x-rays generated at any point in the target and the mass thickness they must traverse depend on the depth beneath the surface. Because of the nonlinearity of Eq. (34), it is not possible to assign an "average" depth to calculate the total number I_0 generated in a particular matrix from the number detected. Beer's law must be integrated over the distribution in depth for generation of x-rays of a particular energy to determine the total number of generated x-rays from the number that emerge from the specimen.

Before proceeding let us review the effects of elastic scatter on the incident electron beam and on the required distribution with depth of x-ray generation. Because of the frequency of large angle elastic scattering events, on the average, a large number of beam electrons with relatively high energy are directed back toward the surface at the full range of angles. This phenomenon results in the bombardment of the surface region of the target from below as well as above and thus an increase in x-ray generation near the surface. This increase occurs, of course, at the expense of the deeper regions of the target that the relatively energetic backscattered electrons would have reached had there been no elastic scatter. Consequently, the distribution in depth of the characteristic x-rays is considerably altered from that shown for the example of Fig. 7, where the effects of elastic scattering are neglected.

16.7 THE $\phi(\rho z)$ CURVE

The distribution of the generated characteristic x-rays as a function of depth is given by a function traditionally known as $\phi(\rho z)$. For many of the same reasons as in the case of backscattered electrons, it is not possible to derive the function from first principles. Again, the distribution is determined from experimental measurements. A number of methods are used to make the measurements, but we describe only one to communicate the essence of most of the techniques. This is the "tracer method" devised by Castaing and used by him and other workers to determine $\phi(\rho z)$ curves. In this method very thin layers of an element of a known thickness are vacuum-evaporated as free-standing films and onto thick substrates of a neighboring atomic number material. Increasing thicknesses of this substrate material are then deposited into the thin layers so that they effectively lie at varying depths (ρz) in the substrate material. The intensities of a characteristic line are measured from the embedded thin layer and the corresponding free-standing film. The generated intensity of the line from the thin film at the depth (ρz) in the matrix material is calculated from the measured intensity by applying Beer's law. Plotting the calculated generated characteristic x-ray intensities for the embedded thin layers, which have been "normalized" by dividing by the measured characteristic intensity from a free-standing film of the same thickness, gives the $\phi(\rho z)$ curve. Such a curve for an aluminum matrix with a magnesium tracer is shown in Fig. 11. Note that the vertical scale is logarithmic

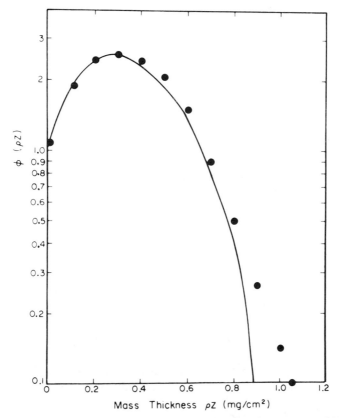

Figure 11.   Plot of the Fabre characteristics cross section for a beam energy of 20 keV as a function of atomic number. Note the rapid decrease in ionization probability with increase in atomic number.

and the horizontal scale is in units of mass thickness. We see that at a "depth" of 0.3 mg/cm^2 the intensity from a thin film of Mg in an Al matrix is almost 2.5 times that from an equally thick but free-standing film. The general shape of the $\phi(\rho z)$ curve is the same for all elements in any matrix. Note that the integral of $\phi(\rho z)$, from $(\rho z) = 0$ to (ρz) at which $\phi(\rho z)$ falls to zero, is the total normalized generated characteristic intensity from the material.

Despite the fact that we do not have a theoretically derived form for $\phi(\rho z)$, we can formulate the x-ray absorption correction in terms of this function. The next section outlines the procedure.

16.8 THE ABSORPTION CORRECTION, F_a

The increment of the total intensity generated in a thin layer of target of mass thickness $d(\rho z)$ at a "depth" (ρz), relative to that generated in a free-standing film of mass thickness $d(\rho z)$ is:

$$dI_0 = \phi(\rho z)d(\rho z) \tag{36}$$

However, as described earlier, absorption occurs as the x-rays travel toward the detector through the target. We account for the x-ray absorption using Beer's law in the following manner. The x-ray detector is usually located above the target at some angle ψ, called the "take-off" angle, specified with respect to the plane of the target surface. X-rays generated in the layer $d(\rho z)$ must traverse a path through the target, longer than (ρz), given by csc $\psi \cdot \rho z$, to reach the detector. Clearly, from Beer's law, the amount of dI_0 reaching the detector is then

$$dI = \phi(\rho z) \exp\left[-\left(\frac{\mu}{\rho}\right)(\rho z)\text{ csc }\psi\right]d(\rho z) \tag{37}$$

To simplify notation we define a quantity χ equal to $(\mu/\rho)\text{csc }\psi$. Equation (37) can then be written:

$$dI = \phi(\rho z) \exp[-\chi \cdot (\rho z)]\,d(\rho z) \tag{38}$$

The total normalized intensity of radiation that actually reaches the detector is the sum from all layers of the target from the surface down:

$$I = \int_0^\infty \phi(\rho z) \exp[-\chi \cdot (\rho z)]\,d(\rho z) \tag{39}$$

This is given as a fraction of the total normalized *generated* intensity I_0 by

$$f(\chi) = \frac{I}{I_0} = \frac{\int_0^\infty \phi(\rho z) \exp[-\chi(\rho z)] \, d(\rho z)}{\int_0^\infty \phi(\rho z) \, d(\rho z)} \tag{40}$$

The fraction $f(\chi)$ is referred to as the absorption factor. The range of $f(\chi)$ is 0 to 1 and can be considered as the probability that a characteristic x ray, generated at an angle toward the detector, will escape the target. We strive to maximize $f(\chi)$ in a real analytical situation. More will be said about this shortly. The absorption correction, F_a, can now be defined to be $F_a = 1/f(\chi)$.

Scattering is a statistical process and the balance between elastic and inelastic scattering of electrons is not amenable to an analytical treatment. Consequently, any attempt to model the depth distribution of the characteristic x-rays forces the modeler into making certain substantial assumptions, omissions, and approximations. A number of authors have engaged in this exercise and a detailed account and derivation of several of the most successful models can be found in the recent book by Heinrich [18] (Chapter 10). All of the models have one thing in common: fudge factors to force the model to agree as much as possible with experimental measurements. We mention here one of the models, from Heinrich and Yakowitz [34], which makes a modest attempt at physical representation but treats the problem more as a mathematical fitting exercise. The accuracy of their model results from the large number of accurate experimental measurements to which the fits were made. Their equation for $f(\chi)$ is

$$f(\chi) = (1 + 1.2 \times 10^{-6} \, \gamma\chi)^{-2} \tag{41}$$

where γ is a parameter that predicts the electron energy dependence of the absorption factor and is given by

$$\gamma = V_0^{1.65} - V_c^{1.65} \tag{42}$$

and V_0 and V_c are as defined earlier. The model ignores the expected dependence of the absorption factor on the target composition, and, indeed, the atomic weights and numbers of the target constituents do not appear in the equation. The only parameters are γ, which is the voltage term, and χ, which for a given instrumental configuration involves only the mass absorption coefficients of the elements of the target.

Given the accuracy of any of the current expressions for $f(\chi)$ the analyst should attempt to maintain the condition $f(\chi) > 0.7$ for all elements being measured. If it is not possible to achieve this value the accuracy of the analysis will deteriorate rapidly as the value of $f(\chi)$ diminishes below 0.7. To maximize $f(\chi)$ the use of low overvoltage ratios should be considered and the take-off angle should be as high as possible.

In general, the absorption correction is by far the most significant of the corrections that must be made.

16.9   THE CHARACTERISTIC FLUORESCENCE CORRECTION, F_f

In the section on x-ray absorption it was pointed out that photoelectric absorption of an x-ray consists of an x-ray photon giving up all of its energy to cause the ejection of an inner shell electron of a target atom. In the process the x-ray is annihilated and the atom is ionized. As noted at the beginning of this chapter, as a first step to returning to the ground state the ionized atom emits either an Auger electron or a characteristic x-ray. The probability of emission of a particular characteristic x-ray, such as Fe $K\alpha_1$, is the product of the fluorescent yield and the relative transition probability defined earlier in the chapter.

It should be clear that if we measure, for example, the $K\alpha_1$ line from element A of the target and the K shell of that element significantly absorbs the radiation of an x-ray line of element B, the measured x-ray intensity from element A will be enhanced. This enhancement, characteristic fluorescence, can be considerable in materials such as steel but is almost nonexistent in biological applications. We refer the interested reader to the literature and recommend Reed [26], pp. 261–282; Heinrich [18], pp. 303–328; and Goldstein et al. [25], pp. 322–325.

16.9   THE CONTINUUM FLUORESCENCE CORRECTION, F_c

In a manner similar to characteristic fluorescence, the continuum radiation also ionizes analyte atoms in the target: continuum photons that have an energy greater than the critical excitation energy of the shell responsible for the emission of the measured line can cause the ejection of electrons from that shell. The detailed physics of continuum fluorescence is considerably beyond the scope of this chapter; we again refer the reader to the literature. An excellent and complete description of the process can be found in Heinrich [18], pp. 328–338. The continuum fluorescence cannot be neglected in a biological application if an x-ray line greater in energy than about 4 keV is used to measure an element that occurs at several weight percent or less. An example of this analytical situation is any low concentration assay using the K alpha line of any element from vanadium through zinc. (For elements beyond zinc in the periodic table the analyst can utilize L or M characteristic x-ray lines that have energies less than 4 keV). In this case, continuum fluorescence produces at least as much of the measured signal as direct excitation by the beam electrons. In such situations it is important to use standards that are as close as possible in composition to that of the specimen. A particularly useful class of standards for biological microanalysis of bulk specimens is doped lithium borate glasses [35].

16.11 CONCLUSIONS

We have attempted in this chapter to provide an overview of the essential concepts of biological x-ray microanalysis by means of focused electron probes. It is hoped that the material was presented in a sufficiently, but not overly, rigorous manner. The interested reader is encouraged to go to the large body of literature that now exists for more detailed treatments of specific subjects. To those workers planning to do analysis in the analytical electron microscope we recommend caution. Elemental analysis on this machine is not a pushbutton operation; a commitment must be made to master the subtleties of the method.

ACKNOWLEDGMENTS

The authors thank their colleague Richard Leapman for his helpful comments.

REFERENCES

1. R. M. Glaser, Radiation damage with biological specimens and organic materials, in *Introduction to Analytical Electron Microscopy,* J. J. Hren, J. I. Goldstein, and D. C. Joy, Eds., Plenum, New York, 1979, pp. 423–436.
2. T. Barnard and L. Seveus, Preparation of biological material for x-ray microanalysis of diffusible elements, J. Microsc. *112,* 281 (1977).
3. A. V. Somlyo, H. Shuman, and A. P. Somlyo, Elemental distribution in striated muscle and effect of hypertonicity: electron probe analysis of cryosections, J. Cell Biol. *74,* 828 (1977).
4. C. E. Fiori and D. E. Newbury, Artifacts observed in energy-dispersive x-

ray spectrometry in the scanning electron microscope, Scanning Electron Microsc. *1*, 401–422 (1978).

5. C. E. Fiori and D. E. Newbury, Artifacts in energy dispersive x-ray spectrometry in the scanning electron microscope (II), Scanning Electron Microsc. *2*, 250–258 (1980).

6. D. J. Marshall and T. Hall, in *X-ray Optics and Microanalysis,* Castaing, Descamps, and Philibert, Eds., Hermann, Paris, 1966, p. 374.

7. H. Bethe, Ann. Phys. (Leipzig) *5*, 325 (1930).

8. C. J. Powell, Rev. Mod. Phys. *48*, 33 (1976).

9. C. J. Powell, in *Use of Monte Carlo Calculations in Electron Probe Microanalysis and Scanning Electron Microscopy,* K. F. J. Heinrich, D. E. Newbury, and H. Yakowitz, NBS Spec. Pub. 460, Washington, D.C., 1976, pp. 97–104.

10. N. J. Zaluzec, Quantitative x-ray microanalysis: instrumental considerations and applications to materials science, in *Introduction to Analytical Electron Microscopy,* J. J. Hren, J. I. Goldstein, and D. C. Joy, Eds. Plenum, New York, 1979, p. 121.

11. N. J. Zaluzec, An analytical electron microscope study of the omega phase transformation in a zirconium–niobium alloy, PhD. Thesis, University of Illinois, Urbana-Champaign, 1978.

12. N. F. Mott and H. S. W. Massey, *The Theory of Atomic Collisions,* Oxford University Press, 3rd ed., New York, 1965.

13. H. Bethe and E. Fermi, Zeits. Phys. *77*, 296 (1932).

14. E. J. Williams, Proc. Roy. Soc. *139*, 163 (1933).

15. A. H. Compton and S. K. Allison, *X-Rays in Theory and Experiment,* 2nd ed., Van Nostrand, New York, 1935.

16. P. Kirkpatrick and L. Wiedmann, Phys. Rev. *67* (11, 12), 321–339 (1945).

17. H. A. Bethe, in *Handbook of Physics,* Vol. 24, Springer, Berlin, 1933, p. 273.

18. K. F. J. Heinrich, *Electron Beam X-Ray Microanalysis,* Van Nostrand Reinhold, New York, 1981.

19. M. J. Berger, and S. M. Seltzer, National Research Council Publ. 1133, National Academy of Sciences, Washington, D.C., 1964, p. 205.

20. C. E. Fiori and C. R. Swyt, The theoretical characteristic to continuum ratio in the analytical electron microscope, in *Microbeam Analysis—1982,* K. F. J. Heinrich, Ed., San Francisco Press, San Francisco, 1982.

21. H. W. Koch and J. W. Motz, Rev. Mod. Phys. *31*(4), 920–955 (1959).

22. W. Heitler, *The Quantum Theory of Radiation,* 3rd ed., Oxford University Press, New York, 1954.

23. T. A. Hall et al., in *The Electron Microprobe,* T. D. McKinley, K. F. J. Heinrich, and D. Wittry, Eds., Wiley, New York, 1966, p. 805.

24. T. A. Hall, Problems of the continuum-normalization method for the quanti-

tative analysis of sections of soft tissue, in *Microbeam Analysis in Biology,* C. Lechene and R. Warner, Eds., Academic, New York, 1979.

25. J. I. Goldstein et al., *Scanning Electron Microscopy and X-ray Microanalysis,* Plenum, New York, 1981.

26. S. Reed, *Electron Microprobe Analysis,* Cambridge University Press, New York, 1975.

27. R. Castaing, Thesis, Univ. Paris, Paris, France, 1951.

28. D. M. Poole and P. M. Thomas, J. Inst. Metals 90, 228 (1961–62).

29. J. Henoc, K. F. J. Heinrich, and R. L. Myklebust, NBS Technical Note 769, National Bureau of Standards, Washington, D.C., 1973.

30. E. H. Darlington, J. Phys. D., Appl. Phys. *8,* 85 (1975).

31. H. E. Bishop, *Proc. 4th Int. Cong. on X-ray Optics and Microanalysis,* R. Castaing, P. Deschamps, and J. Philibert, Eds., Hermann, Paris, 1966, p. 153.

32. P. Duncumb and S. Reed, in *Quantitative Electron Probe Microanalysis,* K. F. J. Heinrich, Ed., NBS Special Publication 298, National Bureau of Standards, Washington, D.C., 1968, p. 133.

33. H. Yakowitz, R. Myklebust, and K. F. J. Heinrich, National Bureau of Standards Technical Note 796, Washington, D.C., 1973.

34. K. F. J. Heinrich and H. Yakowitz, Washington, D.C., Anal. Chem. *47,* 2408 (1975).

35. C. E. Fiori and D. L. Blackburn, "Low $\bar{Z}$ Glass Standards for Biological X-Ray Microanalysis," J. Microsc. Vol. 127, Pt. 2, 1982, pp. 223–226.

HIGH RESOLUTION LIGHT MICROSCOPE PHOTOMETRY

ANDREW W. WAYNE

Department of Haematology
King's College Hospital
London, England

17.1 INTRODUCTION

Photometry is readily utilized to analyze the image produced with a conventional optical microscope. *Microscope photometry* can be applied to most organic and biological specimens to obtain information on the distribution of the composite morphological elements or the amount of material localized at a given site. Photometry offers particular convenience in its use as it can be employed with most modern research microscopes while retaining the availability of their conventional observational facilities. The heart of the photometer is a photoelectric device, normally a photomultiplier tube, that converts light energy into electric current. The system provides a highly sensitive measure of comparative density, usually represented by the percentage transmission or relative absorption.

Absorbance photometry was first applied in a quantitative evaluation of the nuclear DNA content of the cell [1]. The measurement of cellular DNA by cytophotometry or fluorometry remains one of the most frequently employed quantitative investigations performed on biological specimens. Photometric measurement may be applied in the determination of absolute values, as in the measurement of DNA, or on a relative densitometric basis, to study the distribution of morphological features. In the former, a dye that specifically stains the material to be analyzed is used and the integrated optical density is a measure of the total mass of chromophore present.

High resolution light microscope photometry is essentially a mode of area scanning (or scanned image technique) that can be applied to any small field of measurement. Photometric measuring fields equivalent to less than 0.4 μm diameter can be employed with this technique, provided that the geometrical pattern of the object alone is not required. The technique normally requires computer control of the scanning program and output of data. The relative transmission or absorbance values yielded by these measurements can be graphically represented as a simple comparative density profile or more sophisticated picture processing techniques can be applied in the analysis of the data. The method provides qualitative or comparative quantitative data on the distribution and size relationships between objects close to the limit of resolution of the optical microscope. As such it is readily applicable to the analysis of many of the subcellular components and has been particularly developed to help study the fine "banding" structure of chromosomes [2].

This chapter outlines the principles and practice of high resolution photometry as utilized primarily with the conventional optical microscope in the transmitted light mode. The fundamental details of microscope photometry have already been propounded by several authors, see for example, [3,4].

17.2 PRINCIPLES OF PHOTOMETRIC MEASUREMENT

The photoelectric sensor transforms the impinging light energy, *radiant flux,* into an electric current; this measured quantity, usually a voltage, can then be displayed by the instrumentation. The factors affecting the total radiant flux measured by the system are essentially determined by the light source and the optical pathway of the microscope. A detailed account of the transfer of light energy in the optical system is given by [4]. In practice, it is only necessary to compare the radiant flux resulting from the interaction with the specimen and with the reference material. The radiant flux should be directly proportional to the actual measured value with a given set of equipment. The *transmittance*

$$T = \frac{I_s}{I_{ref}}$$

where I_s is the measured value of the specimen and I_{ref} is the measured value of the reference material.

In the simplest case, an empty area of the specimen may be set to give a reading of 100% and subsequent measurements can be directly displayed as a percentage transmission.

The absorbance (also sometimes called extinction or optical density) of the material under investigation is proportional to the light path through the substance and its concentration. Absorbance can be expressed directly in terms of the measured transmittance:

$$A = -\log_{10} T$$

Lambert-Beer's law expresses the relationship between absorbance A, light path or thickness of specimen l (cm), concentration c (mmol cm^{-3}), and a proportionality factor ε, known as the molar absorptivity (cm^2 mmol^{-3}):

$$A = lc\varepsilon$$

This rule holds true for clear nonscattering, nonreactive substances, preferably present in small amounts or high concentrations at a small light-path length.

As a practical example, the mass of DNA (m) in a Feulgen-stained (ideally flattened) nucleus (diameter $2R$) could be approximated by the relationship:

$$m = \pi\varepsilon^{-1}R^2A$$

17.3 ELEMENTS OF THE SYSTEM

The basic photometer ray path is shown in Fig. 1 and an example of a complete microscope photometer system is illustrated in Fig. 2. Some of

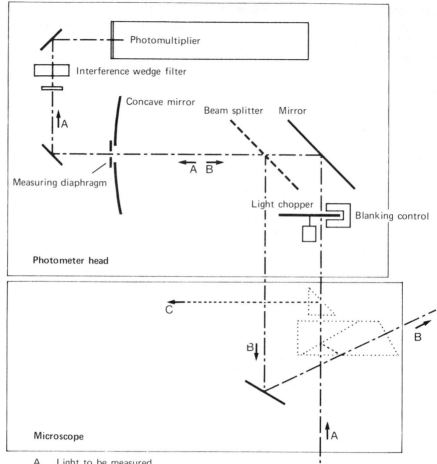

A   Light to be measured
B   Imaging of measuring diaphragm and object plane
C   Photomicrography

Figure 1. Photometer (light) ray path.

the basic elements of the system are detailed below (see [3], for a more detailed discussion).

17.3.1 Microscope

Although a photometer potentially can be used with any microscope, for high resolution studies, a high quality research instrument is essential for optimal performance. This should provide a solid framework to help mini-

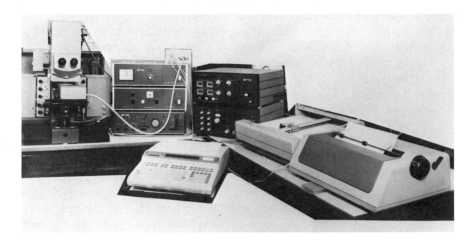

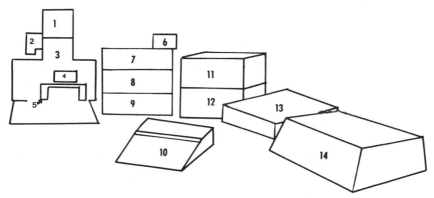

Figure 2. Reichert "Univar" microscope photometry system together with Hewlett Packard '9825' computer system. 1, Photometer; 2, 35-mm camera body; 3, microscope body; 4, scanning lens; 5, scanning stage; 6, scanning stage/lens switch; 7, filament lamp power supply; 8, arc lamp power supply; 9, automatic camera control; 10, desktop computer; 11, scanning routine control; 12, photometer control; 13, graphics plotter; 14, printer.

mize vibration, but the use of additional damping measures may be required, depending on the location of the instrument. The optics should be of the highest quality and of a type suitable for the work to be performed.

17.3.2 Light Source

An incandescent filament lamp, commonly a quartz halogen, or a gas discharge arc lamp, preferably a xenon arc lamp, can be used for photometry. In the case of fluorescence excitation, a mercury vapor or xenon arc lamp can be employed. It is essential for the lamp to have a DC-stabilized power supply, voltage stabilization being more effective for filament lamps and current stabilization more efficient for arc lamps.

The light reaching the photosensor can either be continuous or intermittent (modulated). In the former, stray light and photosensor dark current will produce a background "noise" or measuring value that is superimposed on any actual reading, whereas when the light is modulated (e.g., by the use of a mechanical chopper and phase synchronization of the amplifier), only the light from the specimen is measured.

The wavelength of light used for each measurement or series of measurements is selected in accordance with the nature of the investigation to be performed. Some studies may require the use of a single wavelength for all readings, two-wavelength scanning [5], or multiple wavelengths. The wavelength can be controlled by the use of broad-pass band or narrow-pass band filters. The latter are necessary, for example, in the measurement of the mass or concentration of a specific substance. The molar absorptivity factor ε (Lambert-Beer's law) is dependent on the wavelength at which the absorption is measured; the wavelength at which maximum absorption is obtained is normally used in its determination. A narrow-pass band filter can be provided either by the use of a monochromatic interference filter, the simplest being a color filter, or by a monochomatizing device. The continuous interference wedge filter is a commonly used example of the latter that supplies a monochromatic light of which the central wavelength can be varied, usually between 400 and 700 nm. The prism and grating monochromator are alternative types of continuous spectrum device operating in the range 250–1100 nm.

17.3.3 The Photometer

The photometer head and associated control modules are often supplied by the microscope manufacturer so as to be compatible with the rest of the microscope system. The positioning of the photoelectric sensor is determined by the design of photometer housing but is usually close to the exit pupil of the microscope or alternatively close to the primary image plane (or conjugate planes). The area and profile of the photometric measuring field is delimited by the insertion of a measuring diaphragm in the real image plane. Several different-sized diaphragms may be conveniently mounted in a turret to enable easy switching between alternative sizes of measuring field. The measuring diaphragm may also be motorized to enable area scanning. The shape of the measuring diaphragm can be selected according to the needs of the particular study. In general, for small apertures (0.1–0.4 μm in the object plane) circular diaphragms are technically easier to manufacture; square diaphragms provide more radiant flux for a given diameter, but it is difficult to produce the necessary sharp corners at this exiguous level. The measuring field is often conveniently projected onto the primary image plane and hence can be viewed via the main ocular. The photoelectric device most commonly and successfully employed for microscope photometry is the photomultiplier tube. There is a wide choice of different tube types available and the selection of a suitable one will depend, for example, on the range of measurable wavelengths, overall sensitivity, and concomitant "noise" characteristics.

17.3.4 Photometer Controls

The amplification factor applied to the photomultiplier tube is usually selected stepwise, a continuously variable voltage selector operating within each range to allow the indicated measuring value to be set at the calibration point (e.g., at 100% transmittance). The range of amplification available should obviously be sufficient to compensate for a wide range of variation in the radiant flux incident on the photosensor. However, where the radiant flux is small, the greater the amplification needed and the higher the proportion of background "noise."

The integration or sampling time of the instrument represents the delay between the moment the photosensor is illuminated to the time of display (or recording) of the full measured value. Most modern instruments incorporate a digital voltmeter with a light-emitting diode display. The integration time is selected with respect to the amplification factor applied to the photomultiplier and also the level of electronic damping that may be applied to the displayed measurement. For more accurate determinations it is also possible to average a series of sampling periods to provide a mean measuring value for a given photometric field.

17.3.5 Scanning System

High resolution microscope photometry is basically an area scanning (or scanned image) technique. The size of the photometric measuring field and the interval between measurements will determine the precise mode of scanning and the mechanics necessary for its attainment. Where a step size of greater than 0.5 μm is required between measurements, a mechanical stage can be employed to move the microscope slide. For step sizes below 0.5 μm and particularly less than 0.25 μm it is difficult to generate discrete steps of this magnitude directly in the object stage. Consequently, scanning may be more effectively performed in the plane of the magnified image (or one of the conjugate planes) or alternatively a contin-

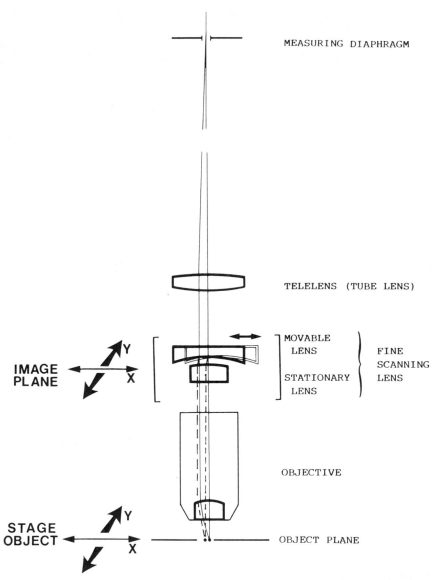

Figure 3. The use of a scanning lens in high resolution microscope photometry. The figure shows an objective (infinity-corrected) and the appertaining tube lens, in the focal plane of which the primary intermediate image is formed (not shown). The measuring diaphragm is mounted in this plane or a conjugate plane. The scanning lens (known as a "fine scanning lens") is introduced between the objective and the tube lens and consists of a pair of lenses with equal radii. The lateral displacement of one of these lenses causes the image to shift across the measuring diaphragm so that another point in the specimen can be measured. When working with a 100× objective, an optical reduction of the step size equivalent to ~300× can be obtained. A scanning object stage is also employed for course positioning of the specimen and rapid scanning procedures.

440

uously moving stage (with a defined sampling interval) could be utilized. Image scanning techniques operate on the principle of optical reduction of step size; for a given lateral displacement in the object stage, a proportionally smaller movement is required in the image plane. This reduction may be typically in the order of 300× (when working with a 100× objective) (see Fig. 3). The specimen can be scanned by movement of a lens, the lateral displacement of which brings successive points of the specimen to impinge on the measuring diaphragm, or alternatively the measuring diaphragm itself can be shifted in relation to the specimen image. One example of a scanning lens incorporated into a microscope photometry system is illustrated in Figs. 2 and 3; this particular motorized lens (known as a *fine scanning lens*) has been shown to be effective in producing step sizes equivalent to 0.05–0.5 μm in the object plane [2]. In order to obtain an equivalent step size using a continuously moving stage, the sampling frequency must be synchronized with the stage movement and the sampling time adjusted accordingly. In practice, a sufficiently brief sampling time can only be obtained using direct light (without modulation) with its inherently higher level of background "noise."

The size of step and scanning program sequence is usually effected by a computer via an electronic control module. The scanning program consists of a series of parallel line scans (which may overlap), most usually in the form of a comb or meander pattern.

17.4 PHOTOMETRY PROCEDURES

17.4.1 Specimen Preparation

Photometry can be applied in the study of virtually any biological specimen previously prepared for microscopic examination. However, for optimal performance of high-resolution studies, special care should be taken to ensure that the preparation is of the highest achievable quality; in particular, homogeneously thin sections (preferably less than 1 μm) should be employed. The dye used for staining should be selected according to the nature of the intended studies. In the qualitative investigation of morphological features, most stains are suitable for photometric evaluation, provided that a suitable light wavelength is selected for the readings. On the other hand, quantitative studies, especially the determination of the concentration or mass of a substance, require the use of a chromophore that specifically binds to the substance to be measured and should be stained under precisely controlled conditions.

17.4.2 Preliminary Adjustments and Precautions

The first and probably one of the most important procedures in photometry is the initial adjustment of the microscope part of the system. This requires the proper centralization and focusing of the light source and microscope optics to produce an acceptable image of the specimen. High resolution studies invariably require the use of an oil-immersed high-power objective (100× planapochromatic) and also an oil-covered condenser for maximum light transmission through the system.

Prior to the initialization of the photometry part of the system, some simple precautions should be observed to prevent external electrical interference affecting the sensitive circuitry of the equipment. In particular, times at which large fluctuations in supply voltage might occur should be avoided, as they may be outside the tolerance of even a voltage-stabilized power module. Also arc lamps should not be ignited when the photometer unit is in operation. Lamps should be run for a reasonable time prior to the measurement sequence to allow them to stabilize at their running voltage.

17.4.3 Photometric Calibration and Measurement

Once the specimen has been examined under high power and a suitable area for study has been selected, consideration can be given to the photometric measurement procedure. A luminous field stop or diaphragm should be employed with the condenser to limit the amount of peripheral stray light, the diameter of which should be equivalent to two or three times the photometric field. The size and shape of the field of measurement and the interval between measurements (sampling frequency) must be determined in relation to the size of the individual composite elements of the specimen. Circular or square measuring diaphragms with diameters down to 0.1 μm (in the object plane) can be employed and correspondingly small step sizes can be generated between consecutive readings by scanning in the plane of the magnified image. The use of overlapping fields of measurement helps to minimize random sampling errors when using small aperture diaphragms and, in the case of circular diaphragms, effectively gives a 100% coverage of the object area to be measured. The effect of increasing the sampling frequency (by utilizing decreasing step sizes) together with an increasing percentage of field overlap using both circular and square diaphragms can be visualized in Fig. 4. The light wavelength is set at an appropriate value for the type of specimen staining using a fixed homogeneous interference filter or continuous-spectrum interference wedge. Spectral scanning can be performed to help determine the wave-

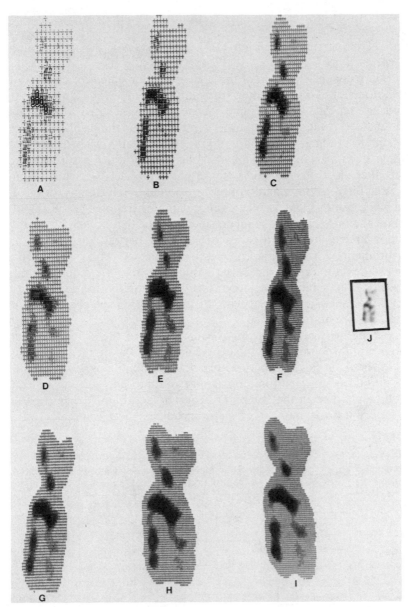

Figure 4. Comparison of a photomicrograph of chromosome No. 10 (inset J) together with photometric scans of the same member performed with different sampling frequencies. Square and circular measuring diaphragms were employed with a reducing step size to obtain a greater number of measurements in a given area and also to vary the percentage of field overlap between consecutive readings. A, 0.2 μm square diaphragm: 0% overlap; B, 0.2 μm square diaphragm: 25% overlap; C, 0.2 μm square diaphragm: 50% overlap; D, 0.1 μm square diaphragm: 0% overlap; E, 0.1 μm square diaphragm: 25% overlap; F, 0.1 μm square diaphragm: 50% overlap; G, 0.16 μm round diaphragm: 50% overlap; H, 0.12 μm round diaphragm: 50% overlap; I, 0.1 μm round diaphragm: 50% overlap.

length of maximum absorption for a given stain if this is not already known.

A relatively clear portion of the specimen is selected for calibration of the photometer at the maximum transmittance value. The amplification is increased stepwise until the appropriate range is located; the voltage is then adjusted using the continuous regulator until the display reads 100. Subsequent readings then will be expressed directly in terms of the percentage transmission. Alternatively, the display can be set to indicate absorbance electronically and is set at 0 in this case. Electronic damping can be applied to the digital voltmeter to enable the display to be stabilized while setting. The correct level of damping (relative to the amplification) for the measurement sequence itself can be determined by averaging the measured values for a series of sampling periods (usually 20 or 30 measurements). A damping range is then selected where the mean and standard deviation of the measured value are within the required limits of the study. The sampling or integration time required for each measurement is then set in accordance with the amplification and electronic damping factors previously applied.

A computer is necessary for overall control of the scanning program sequence and recording the measured values at the defined sampling interval. The main physical parameters of the scanning program can be set with the aid of an electronic (scanning routine) control unit or can be directly embodied in the computer program (see Fig. 5). These parameters determine the number of steps (readings) necessary to cover the

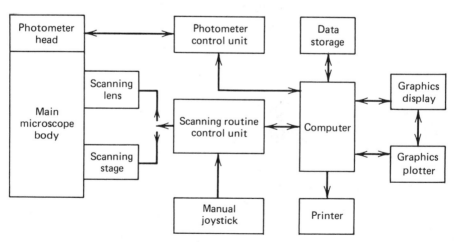

Figure 5. Equipment configuration for control of the scanning program and data presentation modes in high resolution microscope photometry.

object area and the pattern of scan (e.g., meander or comb sequence). The speed of the scanning device should also be adjusted to maintain a smooth motion between all readings, thus avoiding defocusing the specimen. For long scanning sequences it may be necessary to correct the focus between measurements (usually at the beginning or end of a line scan), or alternatively, an autofocus device can be incorporated into the design of the system.

Once all these parameters have been defined, the measuring diaphragm is positioned at the start point of the measurement sequence and the scanning program is initiated under computer control. The measurement sequence itself is usually automatic, the measured values for each successive coordinate being stored by the computer for subsequent display and analysis.

17.4.4 Data Presentation

The measured values (representing percentage transmission or relative absorbance) of each consecutive point in the measuring sequence can be conveniently stored in a numerical or string matrix of the computer memory. The stored data can be retrieved for immediate display (Fig. 6) or can be transferred to a permanent data store for subsequent handling (see Fig. 5). A simple comparative density profile can be provided by substitution of the appropriate symbols for each discrete range of measured values (Fig. 7). This form of representation will suffice in the study of most morphological features and also the calculation of the dimensions of their composite elements. More sophisticated picture processing techniques are also readily applicable in the analysis of densitometric data (see, for example, [6], [7], [8] and [9]).

17.5 PERFORMANCE CHARACTERISTICS

A set of typical performance parameters for one system of high resolution light microscope photometry is given in Table 1.

17.5.1 Spatial Resolution

The spatial resolution of the optical system is limited by the physical constraints of conventional light microscopy. Objects close to the limit of resolution of the microscope have relatively large diffraction fringes; the geometrical pattern of the object is only visible as such where the smallest elements are at least 0.4 μm in size [4]. The accurate determination

Figure 6. Display of numerical transmission values for a simple area scan. The values above 79 (darker figures) represent the background area and the values below this number (lighter figures) constitute the object of interest. The measured values for the object area could be further divided into discrete ranges and displayed, for example, in different colors or as a comparative density profile (see Fig. 7).

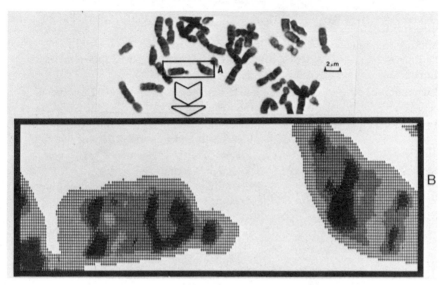

Figure 7. Part of a metaphase spread showing (*A*) two No. 14 chromosomes and (*B*) a densitometric representation of the same members using high resolution light microscope photometry.

of absolute photometric values requires both the geometrical pattern of the object and sufficient radiant flux to maintain the reliability of the individual reading. Consequently, the smallest diameter of measuring field suitable for this type of study is 0.4–0.5 μm. However, where photometric measurements are performed on a comparative point-to-point basis (area scanning), an accurate reflection of the underlying structure apparently can be produced, although the measured values may represent a mixture of geometrical and diffraction images [2].

The sampling theorem has established the basis of a theoretical minimum sampling rate for the reconstruction of a signal from a given set of sampled data [10]. The principles of sampled-data reconstruction are applicable to the analysis of specimens by scanned image methods. Therefore it intuitively follows that the adequate reconstruction of an object whose smallest visually perceivable elements are of 0.2 μm size (limit of resolution using oil-immersed objective of numerical aperture 1.3 at wavelength 550 nm) requires that the specimen be sampled at intervals of less than 0.1 μm apart. The use of photometric measuring fields of this order of size (0.1 μm diameter) is possible if the geometrical pattern alone is not required, but the individual readings become increasingly unreliable with the concomitant reduction in radiant flux. However, by using an overlapping measuring field for consecutive readings, a number of mea-

TABLE 1   High Resolution Microscope Photometry Performance Parameters[a]

Microscope optics	Infinity-corrected
Objective	Planapochromatic N.A. 1.32 (oil-immersed)
Condenser	N.A. 1.3 (oil-covered)
Light source	Quartz halogen filament lamp 12 V: 100 W operating at 11.5 V
Light wavelength selection	Interference wedge filter 60 mm length, 20 nm half-width and reproducibility of within ±3 nm
Scanning stage	Basic step: 0.5 μm Speed range: 4–4000 steps/s (2 μm–2 mm/s) Total range of motion: 25 × 50 mm
Scanning lens	Basic step: 1.25 μm (24 steps = 0.1 μm in the object plane using 100× objective) Total range of motion: 6 × 9.75 mm (20 × 32 μm in the object plane at 100× magnification)
Photometer:	
Measuring diaphragms	Minimum size 0.1 μm (in the object plane) (circular or square profiles)
Photomultiplier tube	EMI 9789B (S11 Bialkali cathode)
Amplification	10^4 maximum (in 12 steps)
Maximum sensitivity	5×10^{-11} lumen (indicator reading of 100)
Linearity	0.1% or better
Integration time	40 ms, 150 ms, 1 s, or 4 s
Zero consistency	0.5% (amplification step 9; integration time 1 s)
Signal-to-noise performance (for typical set of measurement conditions)	(As represented by the standard deviation in percentage of the measured value) 0.78% (at 100% transmission) 1.52% (at 15% transmission)

[a] Based on the Reichert "Univar" microscope photometry system, Fig. 2.

surements are effectively averaged together to provide the relative density determination for a given area and hence compensate for the higher level of sampling errors inherent with the technique. For example, for a step size of 0.05 μm and a circular measuring diaphragm of 0.16 μm, a 61% area of overlap is produced and the equivalent of eight measurements contribute to the relative density determination of each point in the object.

17.5.2 Sensitivity and Practical Limits of Detectability

The sensitivity of the system as a whole is determined by factors such as the effective optical flux and the response of the particular photoelectric device employed. (See the manufacturer's manuals for the response characteristics of individual instruments.) The minimum detectable radiant flux has been estimated by [3] to be equivalent to 2.95×10^{-12} W. Similarly, the minimum detectable mass of a substance has been estimated at 6×10^{-16} g.

17.6 SOURCES OF ERROR

The acuity of photometric measurement is limited by the accuracy of the measuring instrument and the precision achievable with the sampling technique. Systematic errors may arise because of inherent instrumental defects or as a result of improper adjustment or operation of the equipment. These are usually nonrandom and particular to the instrument. Sampling errors are random statistical variations that result from the particular sampling procedure employed and may be due to the effects of, for example, electronic "noise" or voltage fluctuations. Some examples of these sources of error follow.

17.6.1 Focus Dependence

The photometric measuring value is focus dependent; the displayed value will increase or decrease with a change in focus. Equivalent measurement values will only be obtained if the same focal plane is maintained for the whole measuring sequence.

17.6.2 Glare and Stray Light

Unwanted light or glare may arise from reflection or scatter of light at optical interfaces; in transmission mode in particular, multiple reflections often occur between condenser and specimen slide. The use of an oil-immersion objective and oiled condenser top lens eliminates this particular source of glare. Stray light may enter the microscope light path by way of any space or opening in the optical train. The effect of stray light can be minimized by enclosing the optical pathway wherever possible and by darkening the room in which the instrument is housed.

17.6.3   Distributional Error

The distribution of light-absorbing material in a biological specimen is usually nonuniform. Thus a determination of the mass of material (by absorption measurements) will be subject to a distributional error. The application of area scanning techniques and measurements performed with the use of two different wavelengths have been effective in reducing this source of error. The area scanning method was pioneered by [11] and entails the division of the total area to be measured into a series of spot measurements of smaller areas in which the material is homogeneously distributed. The total mass across the whole area can then be obtained by summation of the individual spot area measurements. A two-wavelength scanning method has also been developed to combine the advantages of both of these techniques [5].

17.7   ARTIFACTS AND DAMAGE

Microscope photometry does not have any significant deleterious effects on the specimen as a result of its application and, like conventional microscopic analysis, can be used repeatedly on the same sample. If the chromophore employed for the staining procedure is relatively unstable, perhaps as a result of inadequate control of the staining process, there may be a tendency for the stain to fade with prolonged light exposure. However, this is a rare occurrence except in the case of fluorochrome-stained specimens, which are naturally subject to fading and photo-oxidation when illuminated.

17.8   APPLICATIONS OF THE TECHNIQUE

Microscope photometry provides a measure of the radiant flux resultant from the interaction of light with the stage object. The technique can be applied to virtually any biological specimen for spot measurements or area scanning; the general applications have already been reviewed by several authors (e.g., [5], [3] and [4]).

High resolution photometry is basically an application of an area scanning (or scanned image) procedure for the analysis of objects at the limit of resolution of the conventional light microscope. It is thus of utility in the study of the distribution and size relationships between many of the subcellular components and of potential value in the examination of some unicellular organisms. The technique was particularly developed to aid in the analysis of the fine "banding" structure of chromosomes (see Fig. 7), both for the qualitative distribution of the bands and the quantitative evaluation of the amount of material exchanged between chromosomes [2], [12]. Other applications might include the study of cell and membrane boundaries, for example, measurement of the relative thickness of the basement membrane in blood capillaries (see Fig. 8).

17.9 SCANNED IMAGE ANALYTICAL TECHNIQUES

The relative performance of high resolution light microscope photometry may be most appropriately evaluated in comparison to other area scanning methods. The common form of all scanned-image techniques is the analysis of the specimen on a point-to-point basis. The analytical instrument can be based on the conventional light microscope or on a scanning microscope system [13].

The photomultiplier tube, TV camera, and CCD (charge-coupled-device) linear array are all examples of sensors that can be easily attached to the ordinary optical microscope. The photomultiplier tube forms the heart of most photometry systems and provides a highly reliable and sensitive measure of a wide range of radiant flux. The TV-camera-based instrument has the advantage of being readily available and of simplicity of operation. TV-based tools have been extensively used for image analysis procedures [14] and provide for rapid execution in routine applications. However, compared to the photomultiplier tube, the TV camera has a very limited dynamic range, probably less than one-tenth the sensitivity, and a restricted resolution (scanning source of the TV camera typically 100 μm). Image drift and distortion are also problems encountered in the use of the TV camera. Thus for high resolution studies, both in terms of spatial delineation and density discrimination, the photomultiplier tube provides a significantly superior performance. The CCD linear array has also been applied for scanned-image analysis and offers many advantages in its overall operation compared to the TV camera, particularly in terms of the stability and signal-to-noise ratio. The CCD array scanning system has been successfully operated at a resolution equivalent to 1 μm, but further developments may yield a pixel size approaching 0.125 μm, making it suitable for higher resolution studies [15].

Scanning microscope systems offer a range of important alternative methods to those based on the conventional optical microscope. The performance of the scanning microscope is not critically related to the coherence of the illumination and the resolution is not necessarily limited in the same manner as with conventional optical microscopy. Scanning

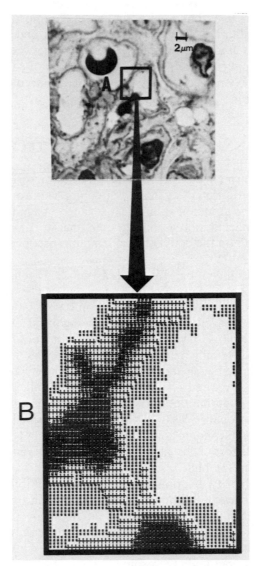

Figure 8.   Part of a section through a kidney glomerulus showing (*A*) the boundary between two adjacent blood capillaries and (*B*) a densitometric representation of the same area using high resolution light microscope photometry. (Specimen kindly provided courtesy of Dr. F. E. Dische, Department of Histopathology, Dulwich Hospital, London.)

systems are capable of recording both the amplitude and phase of the input signal. Consequently, picture processing techniques are readily applicable in the analysis of the recorded data [16]. Scanning instruments take a wide variety of forms: optical [17], acoustic [18], photoacoustic [19], and x-ray [20]. The main disadvantage of all these systems is in their highly specialized nature. Many of these instruments are custom-built assemblies and require considerable technical expertise in their operation and also in the interpretation of the results.

REFERENCES

1. Caspersson, T. (1936). Über den chemischen Aufbau der Strukturen des Zellkernes, Skand. Arch. Physiol. *73*, 1–151.

2. Wayne, A. W., and J. C. Sharp (1981). The use of high resolution microscope photometry in the discrimination of chromosome bands, J. Microsc. *124*, 163–167.

3. Piller, H. (1977). *Microscope Photometry,* Springer, Berlin.

4. Zimmer, H.-G. (1973). Microphotometry, in *Molecular Biology, Biochemistry and Biophysics,* Vol. 14, V. Neuhoff, Ed., Springer, Berlin, pp. 297–328.

5. Fukuda, M., N. Böhm, and S. Fujita (1978). Cytophotometry and its biological application, *Progress in Histochemistry and Cytochemistry,* Vol. 11 No. 1, Fischer, Stuttgart.

6. Duda, R. O., and P. E. Hart (1973). *Pattern Classification and Scene Analysis,* Wiley, New York.

7. Huang, T. S., Ed. (1975). What is picture processing? in *Topics in Applied Physics, Vol. 6, Picture Processing and Digital Filtering,* Springer, Berlin, pp. 1–20.

8. Bradbury, S. (1978). Microscopical image analysis: problems and approaches, J. Microsc. *115*, 137–150.

9. Zimmer, H.-G. (1979). Digital picture analysis in microphotometry of biological materials, J. Microsc. *116*, 365–372.

10. Kuo, B. C. (1963). The sampling process, in *Analysis and Synthesis of Sampled-Data Control Systems,* Prentice-Hall, Englewood Cliffs, N.J., pp. 18–38.

11. Caspersson, T., F. Jacobsson, and G. Lomakka (1951). An automatic scanning device for ultramicrospectrography, Exp. Cell. Res. *2*, 301–303.

12. Wayne, A. W., and J. C. Sharp (1982). A photometric study of the standard Philadelphia (Ph[1]) translocation of chronic myeloid leukemia (CML), Cancer Genetics and Cytogenetics *5*, 253–256.

13. Sheppard, C. J. R., and A. Choudhury (1977). Image formation in the scanning microscope, Optica Acta *24*, 1051–1073.

14. Caspersson, T. (1976). TV-based tools for rapid analysis of chromosome banding patterns, in *Automation of Cytogenetics: Asilomar Workshop*, M. L. Mendelsohn, Ed., Conf. 751158, U.S. Dept. of Commerce, Springfield, pp. 122–130.

15. Farrow, A. S. J. (1980). A CCD linear array based scanning system for rapid analysis of biomedical material on standard microscope slides, in *Scanned Image Microscopy*, E. A. Ash, Ed., Academic Press, London, pp. 241–245.

16. Kino, G. S. (1980). Fundamentals of scanning systems, in *Scanned Image Microscopy*, E. A. Ash, Ed., Academic Press, London, pp. 1–21.

17. Welford, W. T. (1980). Theory and principles of optical scanning microscopy, in *Scanned Image Microscopy*, E. A. Ash, Ed., Academic Press, London, pp. 165–182.

18. Quate, C. F. (1980). Microwaves, acoustics and scanning microscopy, in *Scanned Image Microscopy*, E. A. Ash, Ed., Academic Press, London, pp. 23–55.

19. Wong, Y. H. (1980). Scanning photo-acoustic microscopy (SPAM), in *Scanned Image Microscopy*, E. A. Ash, Ed., Academic Press, London, pp. 247–271.

20. Spiller, E. (1980). The scanning X-ray microscope—potential realizations and applications, in *Scanned Image Microscopy*, E. A. Ash, Ed., Academic Press, London, pp. 365–391.

PHOTOELECTRON MICROSCOPY

O. HAYES GRIFFITH

Institute of Molecular Biology and Department of Chemistry
University of Oregon
Eugene, Oregon

18.1   INTRODUCTION

Photoelectron microscopy (photoemission electron microscopy, or PEM) was one of the earliest electron optical techniques used to study surfaces, predating transmission electron microscopy (TEM) and scanning electron microscopy (SEM). In the early 1930s various types of emission microscopes, including photoemission [1–2], were used to investigate hot and cold cathodes. Most of the applications over the intervening decades have been in metallurgy and related fields [3]. The first organic and biological samples were examined by photoelectron microscopy in 1972 [4], and the first International Conference on Emission Electron Microscopy was held in Tübingen in 1979. The proceedings of this conference provide an overview of the range of applications and an extensive bibliography [5]. A recent review by Schwarzer emphasizes basic concepts and applications in physics [6].

Photoelectron microscopy is a relatively new technique in biology and is not yet widely employed. The work performed to date has established that PEM has unique advantages based on a new source of contrast, the photoelectric effect, and on the high sensitivity to fine relief. High quality images of cell surfaces and other biological specimens can be obtained with this technique. The purpose of this chapter is to explore the advantages and limitations of photoelectron microscopy and to discuss how this technique may be successfully used in studies of organic and biological surfaces.

18.2 PHYSICAL PRINCIPLES

The basic idea is illustrated in Fig. 1. UV light is focused on the specimen. Molecules may be ionized anywhere in the specimen, but only the photoelectrons that are produced at or very near the surface can escape into the vacuum. PEM is thus a surface technique. The escaping electrons are accelerated toward the anode, are passed through a conventional electron optics system, and are imaged on a phosphor screen or photographic plate. In contrast to TEM or SEM, there is no electron gun in PEM. The specimen itself is the source of electrons.

A more detailed diagram of the imaging process is given in Fig. 2. The specimen in PEM is in an electric field, unlike TEM or SEM. The electrons leaving the specimen in PEM have a very low kinetic energy, on the order of 1 eV, and the electric field is needed to accelerate these electrons quickly to 30–50 kV in order to maximize the resolution and produce a bright image on the screen. All of the electrons do not of course leave in a

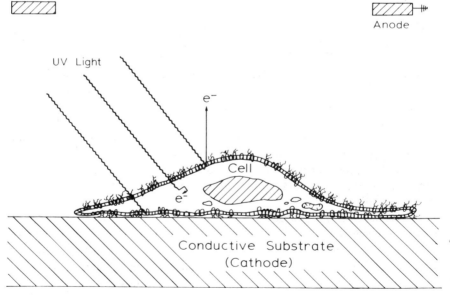

Figure 1.   Sketch of the photoelectron microscope experiment. The cathode–anode spacing is much greater than drawn here.

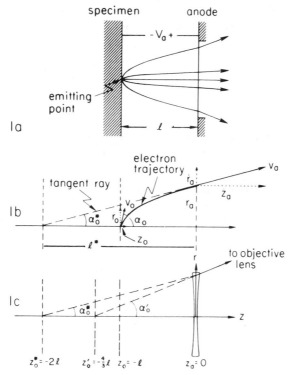

Figure 2. (*a*) Electron trajectories for a point on the axis showing the curved paths in the accelerating region and the diverging action of the anode lens. Radial distances have been exaggerated. (*b*) Detail of the accelerating region showing the trajectory and the tangent ray defining the position of the virtual object at a distance l^* from the anode, $\dot{r}_o$ and $\dot{z}_o$ are the components of the initial velocity v_o, and $\dot{r}_a$ and $\dot{z}_a$ are the components of the final velocity v_a. α_o and α_o^* are the angles made by the initial and final tangents, respectively. (*c*) Electron optical equivalent for the special case of a uniform accelerating field combined with the diverging anode lens. α_o' is the angle of the ray after the diverging effect of the anode lens. z_o^* is the location of the virtual object for the anode lens; z_o' is the location of the virtual image formed by the anode lens; z_o is the location of the emitting point and z_a is the location of the anode lens. (From Ref. 7 with permission.)

direction perpendicular to the specimen surface. They leave at various angles and the result is a family of approximately parabolic electron trajectories [6,7]. The tangents to these parabolas intersect the axis at nearly the same location (l^*), the position of the virtual object. That is, the electrons appear to be coming from a virtual specimen at a distance $l^* \sim 2l$ from the anode, as shown in Fig. 2*b*. The anode aperture acts as a weak diverging lens, producing a slightly demagnified virtual object at a distance of $-(4/3)l$, as shown in Fig. 2*c*. The electron lens system then magnifies the image in the same way as in TEM.

A useful comparison is with fluorescence microscopy. In fluorescence microscopy, the optical lens system images patterns of emitted fluorescent light, instead of photoelectrons, using conventional light optics. Thus photoelectron microscopy can be considered as an electron optical analog of fluorescence microscopy that is especially well suited for the study of surfaces.

18.3   THE PHOTOELECTRON MICROSCOPE AND SAMPLE PREPARATION

A diagram of a photoelectron microscope [8] is shown in Fig. 3 and an enlargement of the specimen area is given in Fig. 4. UV light from mercury short arc lamps is focused and reflected off the polished anode

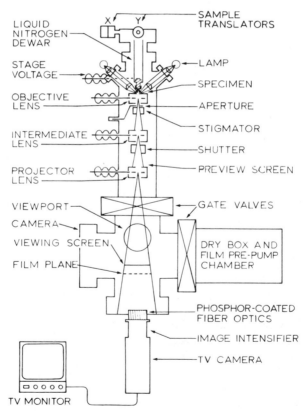

Figure 3. Simplified diagram of the University of Oregon photoelectron microscope.

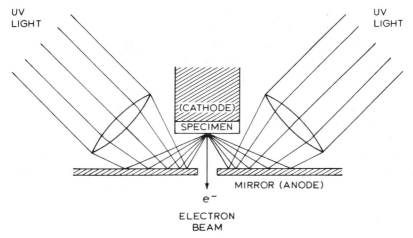

UV
LIGHT

UV
LIGHT

(CATHODE)

SPECIMEN

MIRROR (ANODE)

e⁻

ELECTRON
BEAM

Figure 4. Enlarged diagram of the sample area of the photoelectron microscope showing the UV light optics system. The design of the UV optics is adapted from W. Engel [9,10]. Not shown is the cathode cup that surrounds the specimen except for a central aperture 3 mm in diameter.

mirror onto the specimen. The specimen is held at cathode potential and the emitted electrons are accelerated, passed through the anode, and focused by the electron optics system. In this microscope the objective, intermediate, and projector lenses are electrostatic lenses, but in several other microscopes the lenses are magnetic. Either system may be used. The final image is projected either on a phosphor-coated fiber optics output window in contact with an image intensifier video system or on a photographic film within the vacuum chamber. Different areas of the specimen are viewed by moving the specimen stage. The sample support is connected to a dewar so that the specimen can be cooled when desired to near liquid nitrogen temperatures. The microscope is of ultrahigh vacuum design and there are no oil forepumps, diffusion pumps, or other sources of oil contamination. A clean vacuum is required in order to avoid a contamination layer that would mask the photoelectric properties of the biological surface. The microscope is described in somewhat more detail in Ref. 8.

Unfortunately, there are currently no commercial photoelectron microscopes designed for organic and biological specimens. Since the 1930s several emission microscopes have been built in various institutions, mostly in Germany. There have been occasional commercial developments, the most successful of which is the Balzers KE-3 emission microscope built under the direction of Dr. L. Wegmann at Balzers in Liechtenstein [3]. This microscope incorporates some of the important developments of Dr. W. Engel at the Fritz Haber Institute in Berlin [9,10]. It is not of ultrahigh vacuum design but is equipped with a variety of excitation sources and a hot stage useful in metallurgy and material science. Approximately one dozen of the Balzers KE-3 microscopes were built in the early 1970s and most are located in Europe. It is possible that these instruments could be modified by adding anticontamination devices and cold stages for the study of organic and biological specimens.

Sample preparation in the photoelectron microscope is relatively straightforward. In general, no staining, shadowing, or metal coating steps are used. Organic samples can be examined directly, providing they are relatively flat. PEM is not an appropriate technique for very rough samples such as fractured plastics. For biological samples the usual precautions to avoid drying artifacts must be taken as in TEM and SEM. Alternatively, frozen samples can be examined. The specimen must be on a smooth conducting substrate so that the emitted photoelectrons can be continuously replaced. Substrates that have been used successfully include polished stainless steel, carbon and other coatings on stainless steel, metal-coated glass discs, and tin-oxide-coated glass discs. The tin-oxide-coated glass discs are optically transparent and permit viewing of the biological specimens in an optical microscope prior to the PEM experiments. This is useful in cell culture and chromosomal preparations, for example.

18.4   CONTRAST, RESOLUTION, DEPTH OF INFORMATION, AND DEPTH OF FIELD

The two main contrast mechanisms in photoelectron microscopy are topographical contrast and photoelectron quantum yield contrast (material contrast). PEM is one of the most sensitive surface techniques for detecting fine topographical detail. This sensitivity arises because the electrons emerging from the specimen have very low kinetic energies and are easily deflected by small disturbances in the electric field produced by variations in sample topography. Reimer et al. [11] reported that steps as small as 30 Å are detectable. Topographical (relief) contrast effects in PEM have recently been calculated analytically for simplified models of negative and positive topography [7]. Electron cones emitted from sloping surfaces are deflected by the disturbed field, and when the angle of tilt of the cone axis exceeds the angular aperture accepted for untilted cones, pronounced contrast effects occur in the image. The effect is larger when smaller limiting apertures are used. The theory and experiments are in agreement, and topographical contrast is important, particularly in biological specimens where fine topographical detail can occur without significant material contrast. Where there is material contrast the topographical contrast provides the necessary information to determine the location of the emitting surface detail.

Photoelectron quantum yield contrast arises whenever molecules are present that have different ionization potentials. The standard method of determining these differences is by measuring photoelectron quantum yield curves, that is, plots of the number of electrons per incident photon vs. the wavelength of the exciting light. Photoelectron quantum yield curves of the common types of molecules found on biological surfaces have been reported [12–18]. However, by convention, all photoelectron quantum yield data are reported for low photon fluxes whereas the PEM utilizes intense UV illumination. It is commonly found that the quantum yield curves shift upward with UV exposure and therefore the quantum yield curve data are at best only an indication as to which molecules might be the best photoemitters in PEM [19]. Photoelectron quantum yield contrast in organic and biological specimens has been clearly demonstrated (see Section 18.7).

The lateral resolution, defined as the smallest point-to-point separation between two specimen details which can still be resolved in the image, depends on three factors. The first two factors are familiar in all microscopes: the diffraction limit imposed by the wavelength of the emitted electrons and the aberrations of the objective lens. The third factor in PEM is the aberration of the accelerating field. Referring back to Fig. 2b, the tangents to the family of parabolas do not intersect at exactly the same point on the axis and the virtual object point becomes a circle of confusion. Estimates of the theoretical resolution of PEM must also take into account properties of the sample such as the distribution in kinetic energies and angles of emission of the electrons, quantities that are not well established for most organic and biological specimens. Theoretical estimates range down to 40 Å [4,6]. To date, the experimental resolution achieved approaches 100 Å [8,9]. As a practical matter, the resolution is often limited by image brightness (inability to focus) rather than by the theoretical limits. There are several reasons to be optimistic about future increases in resolution including improved image intensification, improved stage design, and the eventual use of corrected electron optics that could largely remove the limitations imposed by the aberrations of the accelerating field.

The depth of information is also an important parameter, since all surface techniques probe a finite volume of the sample [6]. The depth of information is defined as the distance below the surface from which information is contributed at a specified resolution [20]. It is also referred to as depth resolution, that is, resolution in the direction normal to the specimen surface, as distinct from lateral resolution in the plane of the specimen [21]. The depth resolution is equal to the electron escape depth when the escape depth is small relative to the instrument resolution. This is expected to be the case for organic and biological specimens, since most of the escape depths reported to date for organic materials are in the range of 1–10 nm [20]. Because of the short escape depths, photoelectric contrast provides information from only an extremely shallow surface layer. As for topograhical contrast, this is purely a surface effect and fine topographical detail is always imaged at instrument resolution. There are a number of secondary mechanisms for propagating information to the surface, and these can lead to a diffuse image superimposed on the sharp image of photoelectric and topographic detail. However, these are generally considered to be artifactual in the typical biological experiment and can be recognized and minimized (see Section 18.6).

The depth of information should not be confused with the depth of field. The depth of field is the range over which the object distance can vary while the image appears to remain in focus. The calculation of depth of field in PEM differs from that for TEM or SEM since the object for the electron lens system is a virtual image of the specimen created by the accelerating field and the anode aperture lens [6,22]. Calculations show that the depth of field in PEM for substantially planar areas is comparable with that of TEM and SEM at all magnifications. However, for specimens with significant relief the working depth of field in PEM is substantially reduced. The reason for this is that the depths or heights in the virtual specimen formed by the accelerating field are exaggerated by the microfields near the specimen surface. The depth magnification does not distort the image but can readily cause the topography to exceed the depth of field of the electron optical system. The net result is a reduction in the effective or usable depth of field. The effect is greatest at high magnifications. However, we have not found this to be a practical limitation. The depth of field in PEM is sufficient to image fine detail encountered in biological specimens (e.g., cell surface detail) even at high magnifications.

18.5 COMPARISON OF PEM WITH CLOSELY ALLIED METHODS

Photoelectron micrographs have a close resemblance to scanning electron micrographs. However, the mechanisms for the production of the electron beam and image formation are fundamentally different. A comparison of PEM and SEM using metallurgical and inorganic samples was made by Pfefferkorn et al. [23]. The two techniques provide different kinds of information and are complementary. SEM is the recognized and established technique and is the method of choice where elemental analysis is desired or where a very broad range of sample morphologies is involved. PEM probes the valence state electrons rather than core electrons. It is possible therefore to have large differences in contrast between molecules (e.g., aromatic carcinogens and lipids) that have very similar elemental compositions and would be difficult to detect by SEM or electron microprobe techniques.

PEM is more sensitive to fine surface relief and has a shorter depth of information (i.e., higher depth resolution) than does SEM. A comparison of the depths of information in SEM and PEM using a flat aluminum bronze sample is shown in Fig. 5 and explained in Fig. 6 [23,24]. The SEM sees further into the specimen at the expense of surface resolution. In the PEM, information carried by electrons from a sharply defined surface detail is distinguishable from the superimposed circle of confusion of diffuse and less intense information from detail below the surface because the electrons are emitted from different locations on the surface and are imaged separately. In SEM the sharply defined area under the probe is surrounded by a circle of confusion defined by the bloom area and the signals from these two areas are not imaged separately. The resolution is therefore limited by the size of the bloom area [20].

It appears that the sample conductivity requirement in PEM is less than that in SEM or in electron microprobe techniques. One reason for

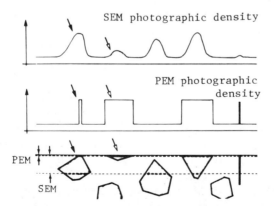

Figure 5. Comparison of the relative depths of information in SEM and PEM using an aluminum bronze sample. (*a*) SEM (30-kV primary electron energy); (*b–d*) PEM; (*b*) surface cleaned by ion etching; (*c*) after prolonged ion etching; (*d*) after more prolonged etching. (Reproduced from Ref. 24 with permission.)

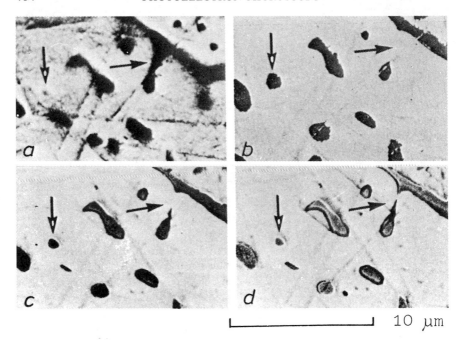

L_____J 10 μm

Figure 6. Cross-sectional schematic diagram illustrating the differences in depth of information in SEM and PEM for a two-phase mixture as in Fig. 5. The smaller depth of information in PEM produces a sharper density profile in the micrograph. (Reproduced from Ref. 24 with permission.)

this may be photoconduction induced by the high intensity light in PEM. Another factor is that the electrons are simultaneously collected over a wide area in PEM, whereas in the scanning techniques there is a high current density at one spot on the sample. It is possible to examine uncoated specimens in PEM that would be difficult to observe by SEM. Since metal coating tends to reduce the effective resolution and contrast by obscuring surface detail, it is an advantage to use uncoated specimens.

Other factors include sample damage and image distortion. PEM probably causes less sample damage than SEM or TEM because light causes less gross damage than a beam of ionizing radiation (see Section 18.6 for the related matter of image statistics). There is no foreshortening of the PEM image in one dimension as there is in SEM. This is because the specimen surface in PEM is always perpendicular to the optical axis, whereas in SEM the specimen surface is tilted with respect to the primary electron beam in order to enhance topographic contrast.

Another technique closely related to PEM is fluorescence microscopy. As mentioned earlier PEM is the electron optical analog of fluorescence microscopy. Both techniques use UV light as the excitation source, although PEM utilizes shorter wavelengths. Both techniques utilize an optical system to project an enlarged image with bright regions corresponding to sites of the specimen having greater photon emission (in fluorescence) or electron emission (in PEM). One of the major applications of fluorescence microscopy is in immunology [25]. Fluorescent molecules (markers) are covalently linked to site-directed labels such as antibodies or plant lectins, proteins that have high specificity for the recognition of a single class of sites on cell surfaces. By this means the distribution of specific proteins and lipopolysaccharides are mapped on normal and transformed cells. The advantages of fluorescence microscopy are the high contrast obtained and the fact that cells can be observed in aqueous solution. Two limitations are the resolution of the optical microscope and the fact that it is often difficult to distinguish fluorescence emitted at the surface from that coming from other parts of the cell. One goal in PEM research is to develop photoemissive markers that can extend the immunological experiments to high resolution. TEM and SEM are of course also being used in cell surface labeling experiments [26]. The advantage of PEM is that it is essentially a dark field experiment, as is fluorescence microscopy. It should be easier to recognize a photoemissive marker against a darker background than it is to recognize a small marker solely by its size and shape as in SEM.

Photoelectron microscopy is only one of several ways the photoelectric effect can be exploited in studies of organic and biological surfaces. Other examples are photoelectron spectroscopy using UV light (UPS) or an x-ray source (XPS or ESCA; see Chapter 2). As the word *spectroscopy* implies, UPS and ESCA utilize the kinetic energy distribution of the emitted electrons and discard the positional information (i.e., where the electrons leave the sample). PEM relies primarily on positional information, although the kinetic energy distribution has an effect on the resolution. One or more photoelectron microscopes have been equipped with energy analyzers [6]. In principle it is possible to obtain a PEM image as a function of the electron kinetic energy. This can be done by a modification of a standard PEM. In another approach, an electron beam is focused on the sample or sample support and the x rays emerging from the opposite side of the sample generate photoelectrons that, with energy analysis, produce an ESCA-type image [27].

Point projection microscopes have been built in which molecules on a field emission-type tip are irradiated with UV light. The voltage is reduced so that field emission does not occur. As in field emission microscopy and field ion microscopy (see Chapter 7) the diverging electric field provides the magnification without electron lenses. We were unable to locate any published micrographs using this technique, although some may well exist. Recently, a magnetic analog of the photoemission point projection microscope has been built and used to obtain micrographs [28]. The microscope utilizes a shorter wavelength UV lamp, the He(I) line (21.1 eV). This technique has been named photoelectron spectromicroscopy to emphasize that the image depends on both the positional information and kinetic energies of the photoelectrons [28]. The ESCA-type microscope and these latter microscopes are still in an early stage of development.

18.6 ADVANTAGES AND LIMITATIONS

The main advantages of PEM are the high sensitivity to fine topographical detail, the presence of a new source of contrast based on the photoelectric effect, high surface selectivity (i.e., short depth of information) and relatively low sample damage. There are, of course, also limitations. The same factors that lead to the sensitivity to fine topographical detail also make it difficult to image specimens with large surface relief. A second disadvantage is that PEM does not provide quantitative analytical information regarding elemental composition of surface layers. In this sense, PEM resembles fluorescence microscopy.

As with any new technique there are a number of artifacts or potential artifacts to be aware of and some of these have only recently been understood. The absorption coefficients and specimen thicknesses of most biological and organic samples are such that a fraction of the incident UV light can pass through the sample and be reflected off the cathode support. The reflected UV beam can cause interference effects and contribute to photoemission at the surface [20,29]. These effects can be minimized by using nonreflective substrates. Another artifact occurs when topography in the substrate induces topographical relief at the specimen surface [20]. This can be avoided by the use of smooth substrates. It is not always easy to distinguish topographical contrast and material contrast. For example, contrast in thin sections of specimens embedded in plastics was originally thought to be due to material contrast or charge injection [30,31]. However, these sections turned out not to be flat. Perhaps due to differences in specimen hardness or shrinkage of the plastic, the surface is a contour of the specimen. It has recently been shown that the contrast effects observed in the PEM images of these biological thin sections are largely topographical in origin [20]. Topographical and material contrast effects can be distinguished by observing the sample before and after depositing a thin layer of gold, which suppresses the material contrast.

In view of the high intensity UV illumination there is undoubtedly some photochemistry occurring during the PEM experiment and this probably accounts for the observed increased image brightness with time. However, there is no evidence that this causes gross molecular rearrangements during the course of the experiment. There is usually no observable degradation in image quality versus time, even after several hours of illumination (the typical PEM micrograph requires an exposure time of 1 s to 2 min). Thermal damage is always a possibility but is controllable by using the cold stage and employing IR filters with the UV optics as is routinely done.

A central question when considering sample damage is whether a given site on the sample surface can photoemit a sufficient number of electrons to be detected above the noise level before something happens to the site. This is a question of image statistics. The signal-to-noise ratio S/N is given by the formula [32]

$$\frac{S}{N} = \frac{n_2 - n_1}{(n_2 + n_1)^{1/2}}$$

where n_1 is the number of electrons emitted from a background resolution element and n_2 is the number of electrons emitted from the resolution element of interest. The relative values of n_1 and n_2 can be determined from the photoelectron quantum yield data or intensity ratios for uniform samples in the PEM. The calculations and measurements have been reported for a representative case, the aromatic carcinogen benzo[a]pyrene against a background of lipid, as could occur in a biological membrane [33]. It was found that these aromatic molecules can photoemit repeatedly and a sufficient number of times so that in principle a single molecule could be detected in the PEM. There was no evidence that the photoemitting sites were being destroyed, since the beam currents measured from a uniform layer of the carcinogen were stable over a period of hours [33].

18.7 APPLICATIONS TO ORGANIC AND BIOLOGICAL SPECIMENS

There are three conditions for obtaining useful photoelectron micrographs of organic and biological specimens: The samples must photoemit, they must have sufficient conductivity, and the topography must be limited. All substances emit electrons when subjected to sufficiently short wavelength light, so there is no theoretical limitation imposed by the photoemissivity. It has been our experience that every biological specimen produces an image in the photoelectron microscope, although some images are of course brighter than others. By this criterion the photoemissivity and conductivity of a wide variety of biological specimens are adequate. Charging due to insufficient conductivity can occur in some biological specimens and in sections of organic polymers such as Teflon, polystyrene, and polycarbonates unless the sections are extremely thin.

The third condition, the limitation on topography, arises because of the high sensitivity to topographical detail. We are philosophical about this problem. In order to achieve the extremely high sensitivity to fine topographic detail it is understandable that gross topography exceeds the range of the instrument. It would hardly be reasonable to expect otherwise. Types of samples that are likely to exceed the topography range of the instrument include parts of rounded up cells, surface ruffles, and microvilli. Cones of electrons from the steep slopes of these objects may be deflected at such large angles that they are stopped out before reaching the final image, leaving corresponding dark areas in the image. This effect is well known in the metallurgical PEM studies [3,5,6]. This has led physicists and metallurgists to cut and polish sections before observation in the PEM. However, this is not an appropriate procedure in biology, since the specimens are delicate and the topographic contrast is one of the most useful sources of information. The appropriate way to use PEM in biology is to make sure that the samples are well spread. This can be done with a surprising number of biological specimens, including membranes, chromosomes, viruses, DNA, and cytoskeletal and other components of cells. Many cultured cells attached to a substrate are appropriate samples as well. A few examples to illustrate the range of possibilities follow.

An illustration of the photoelectron quantum yield contrast is shown in Fig. 7. This is a photoelectron micrograph of one monolayer of bacteriochlorophyll a_{Gg} (Bchl a) from *Rhodospirillum rubrum* laid over a less photoemissive substrate (calcium arachidate on stainless steel) by classical monolayer techniques [34]. The Bchl a monolayer covers only the left half of the sample support, whereas the calcium arachidate is a continuous layer underneath. The uneven border toward the center is the torn edge of the Bchl a monolayer. Contrast between the left and right halves is very high, especially considering that the photoemitting sample is only one molecule thick. PEM detects imperfections in the monolayer (holes in the bright area of Fig. 7) and is sufficiently sensitive to distinguish between heads up and heads down configurations [34]. Figure 7 also illustrates a general principle in photoelectron microscopy. Large conjugated π-electron systems such as the tetrapyrrole head group of Bchl a have high photoelectron quantum yields. The hydrocarbon tails and related molecules such as fatty acids (e.g., calcium arachidate) and phospholipids are much less photoemissive. The plant chlorophylls a and b exhibit a

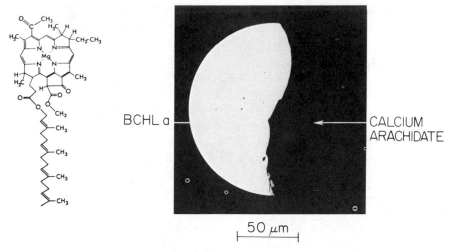

Figure 7. Structure of bacteriochlorophyll a_{Gg} (Bchl a) and photoelectron micrograph of one monolayer of bacteriochlorophyll (left side) deposited over two monolayers of calcium arachidate. (From Ref. 34 with permission.)

similar high brightness in PEM [14]. This suggests one possible type of study with PEM, the mapping of the photosynthetic pigment distributions in photosynthetic membranes. Initial images of chloroplasts and of photosynthetic bacteria have been reported [35,36].

A second example showing photoelectron quantum yield contrast is given in Fig. 8. This photoelectron micrograph is of the carcinogen benzo[a]pyrene sublimed onto a phospholipid background [33]. The bright objects are microcrystallites of the carcinogen. Benzo[a]pyrene, like Bchl a, is a large π-electron conjugated system that readily photoejects electrons into the vacuum upon excitation by a UV arc lamp. The process for benzo[a]pyrene can be diagrammed as follows:

$$h\nu + \text{(structure)} \rightarrow \left[\text{(structure)} \right]^{+} + e^{-}$$

The benzo[a]pyrene positive ion on the right then picks up another electron from the cathode support and the process can be repeated. Figure 8 demonstrates that high contrast is achieved and that there are no serious problems with sample charging or sample damage at this resolution. Car-

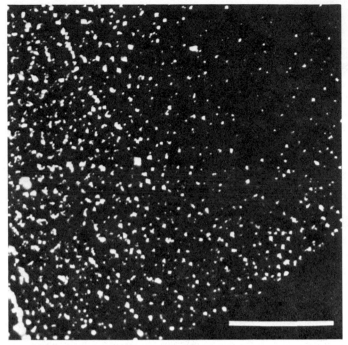

Figure 8. Photoelectron micrograph of the carcinogen benzo[*a*]pyrene sublimed onto a continuous thin layer of the phospholipid dimyristoyl phosphatidylcholine. Bar, 5 μm. (Micrograph by W. A. Houle.)

cinogens such as benzo[*a*]pyrene are accumulated by membranes and fatty tissues, then are selectively oxidized to a reactive form by enzymes of the endoplasmic reticulum and some of these molecules eventually react with DNA, altering the cell's genetic regulation mechanisms [37,38]. PEM may prove useful in clarifying some of these steps, in particular how aromatic carcinogens distribute in membranes and the distribution of reactive sites along DNA.

The remaining two examples illustrate topographical contrast. Figure 9 is a photoelectron micrograph of a rat lung fibroblast. Before fixation the living cell was growing, adhering to and traveling over the serum-coated conductive glass disc. The cell exhibits typical fibroblast morphology. The conclusion reached from viewing this and many other types of cells of known size and shape is that photoelectron microscopy provides a faithful image of the cells. Identifiable in Fig. 9 are the nucleus with nucleoli, many vacuoles, fibrous elements of the cytoskeleton, and the thin leading edge that has just come in contact with a neighboring cell (lower right).

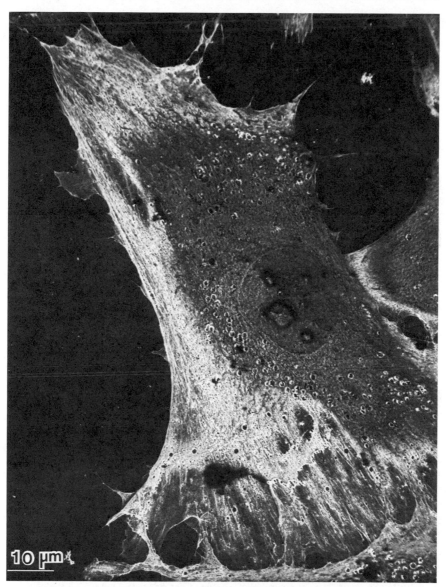

Figure 9. Photoelectron micrograph of an uncoated rat lung fibroblast. The cells were grown on glass disks coated with a conductive tin oxide layer and subsequently treated with calf serum. The cells were fixed in 2.5% glutaraldehyde, 0.05 M sodium cacodylate, 0.15 M sucrose, pH 7.4. They were dehydrated through a graded series of ethanol in water followed by 1:1 ethanol:amyl acetate and pure amyl acetate and dried in a stream of warm air. (Micrograph by K. K. Nadakavukaren and D. L. Habliston.)

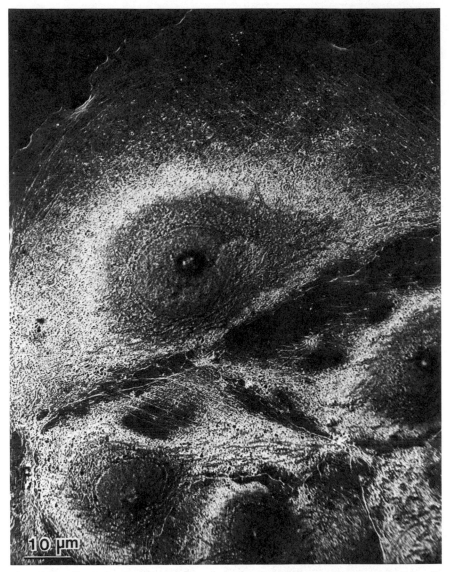

Figure 10. Photoelectron micrograph of uncoated rat kangaroo kidney epithelial cells (PTK-2 cell line) prepared as in Fig. 9. (Micrograph by K. K. Nadakavukaren and D. L. Habliston.)

Portions of other cells are visible in the right and bottom parts of the micrograph. The cells have not been coated as is the common practice in SEM. When the sample is coated with a thin layer of gold, the prominent features of cells can still be seen in the PEM but with a noticeable decrease in surface detail. This is evidence that the main contrast mechanism is topographical rather than photoelectron quantum yield contrast [18,39]. However, because of the loss in resolution the gold coating is normally avoided in PEM experiments.

Figure 10 is a photoelectron micrograph of an epithelial cell line. It is included here because of the unusually well-resolved cytoskeletal elements in the outer areas of the cells. One might reasonably ask why the structures are so well resolved if the depth of information is small. This question has not been examined in detail but it is likely that the thin membrane is partially collapsed over the cytoskeletal framework and/or these elements of the cytoskeleton may lie in close apposition to the plasma membrane. Some of the grainy surface detail may be a drying artifact and the variations in brightness are not completely understood (optical interference effects may contribute). One interesting possibility is to combine topographic contrast and photoelectron quantum yield contrast in labeling studies of cell surfaces. Labeling techniques are under development. The long-range goal is to correlate the topographic information present (e.g., Figs. 9 and 10) with the distribution of specific functional sites on the cell surfaces by using photoelectron markers attached to antibodies or plant lectins.

ACKNOWLEDGMENTS

It is a pleasure to thank the current and past members of the Oregon photoelectron microscopy group for their contributions, in particular Drs. Gertrude F. Rempfer, Karen K. Nadakavukaren, and G. Bruce Birrell, and George H. Lesch, Douglas L. Habliston, and William A. Houle. We also thank Dr. Lan Bo Chen for kindly supplying the cell lines and for useful discussions and Dr. Alex Fabergé for suggesting and working out the technique of using tin oxide coated glass discs as substrates. This work was supported by PHS Grant No. CA 11695 from the National Cancer Institute.

REFERENCES

1. E. Brüche, Elektronenmikroskopische Abbildung mit lichtelektrischen Elektronen, A. Phys. 86, 448–450 (1933).

2. H. Mahl and J. Pohl, Elektronenoptische Abbildung mit lichtelektrisch ausgelösten Elektronen, Z. Tech. Phys. *16*, 219–221 (1935).

3. L. Wegmann, The photoemission electron microscope: Its technique and applications, J. Microsc. *96*, 1–23 (1972).

4. O. H. Griffith, G. H. Lesch, G. F. Rempfer, G. B. Birrell, C. A. Burke, D. W. Schlosser, M. H. Mallon, G. B. Lee, R. G. Stafford, P. C. Jost, and T. B. Marriott, Photoelectron microscopy: A new approach to mapping organic and biological surfaces, Proc. Natl. Acad. Sci. USA *69*, 561–565 (1972).

5. G. Pfefferkorn and K. Schur, Eds., First International Conference on Emission Electron Microscopy, Beitr. Elektronenmikroskop. Direktabb. Oberfl. *12*(2), 1–234 (1979).

6. R. A. Schwarzer, Emission electron microscopy. A review. Part 1: Basic concepts and applications in physics, Microsc. Acta *84*, 51–86 (1981).

7. G. F. Rempfer, K. K. Nadakavukaren, and O. H. Griffith, Topographical effects in emission electron microscopy, Ultramicroscopy *5*, 437–448 (1980).

8. O. H. Griffith, G. F. Rempfer, and G. H. Lesch, A high vacuum photoelectron microscope for the study of biological specimens, SEM/1981/II, O. Johari, Ed., AMF O'Hare, Chicago, pp. 123–130.

9. W. Engel, Entwicklung eines Emissionsmikroskops hoher Auflösung mit photoelektrischer, kinetischer und thermischer Elektronenenauslösung. Dissertation, Freie Universität Berlin, 1968.

10. W. Engel, Emission microscope with different kinds of electron emission, Proc. 6th Int. Cong. Electron Microscopy, Kyoto, Vol. 1, 1966, pp. 217–218.

11. L. Reimer, E. Schulte, Chr. Schulte, and K. Schur, Abbildung von Oberflächenstufen im Photoemissions-Elektronenmikroskop, Beitr. Elektronenmikroskop. Direktabb. Oberfl. *5*, 987–1006 (1972).

12. R. J. Dam, C. A. Burke, and O. H. Griffith, Photoelectron quantum yields of the amino acids, Biophys. J. *14*, 467–472 (1974).

13. R. J. Dam, K. F. Kongslie, and O. H. Griffith, The photoelectron quantum yields of hemin, hemoglobin and apohemoglobin: Possible applications to photoelectron microscopy of heme proteins in biological membranes, Biophys. J. *14*, 933–939 (1974).

14. R. J. Dam, K. F. Kongslie, and O. H. Griffith, Photoelectron quantum yields and photoelectron microscopy of chlorophyll and chlorophyllin, Photochem. Photobiol. *22*, 265–268 (1975).

15. O. H. Griffith and R. J. Dam, Photoelectron microscopy and quantum yields of membrane phospholipids, Proc. Electron Microsc. Soc. Amer. *34*, 32–33 (1976).

16. W. Pong and C. S. Inouye, Vacuum ultraviolet photoemission studies of nucleic acid bases, J. Appl. Phys. *47*, 3444–3446 (1976).

17. H. M. Brown, P. C. Kingzett, and O. H. Griffith, Photoelectron quantum yield spectrum and photoelectron microscopy of β-carotene, Photochem. Photobiol. *27*, 445–449 (1978).

18. R. J. Dam, K. K. Nadakavukaren, and O. H. Griffith, Photoelectron microscopy of cell surfaces, J. Microsc. *111*, 211–217 (1977).

19. O. H. Griffith, D. L. Holmbo, D. L. Habliston, and K. K. Nadakavukaren, Contrast effects in photoelectron microscopy: UV dose-dependent quantum yields of biological surface components, Ultramicroscopy *6*, 149–156 (1981).

20. W. A. Houle, W. Engel, F. Willig, G. F. Rempfer, and O. H. Griffith, Depth of information in photoelectron microscopy, Ultramicroscopy *7*, 371–380 (1982).

21. C. A. Burke, G. B. Birrell, G. H. Lesch, and O. H. Griffith, Depth resolution in photoelectron microscopy of organic surfaces. The photoelectric effect of phthalocyanine thin films, Photochem. Photobiol. *19*, 29–34 (1974).

22. G. F. Rempfer, K. K. Nadakavukaren, and O. H. Griffith, Depth of field in emission microscopy, Ultramicroscopy *5*, 449–457 (1980).

23. G. Pfefferkorn, L. Weber, K. Schur, and H. R. Oswald, Comparison of photoemission electron microscopy and scanning electron microscopy, SEM/1976/I, O. Johari, Ed., ITT Research Institute, Chicago, pp. 129–142.

24. M. Bode, G. Pfefferkorn, K. Schur, and L. Wegmann, Influence of depth of information and of resolution in stereologic evaluation of surface electron micrographs, J. Microsc. *95*, 323–336 (1972).

25. R. C. Nairn, *Fluorescent Protein Tracing*, Churchill Livingstone, London, 1976.

26. J. K. Koehler, Ed. *Advanced Techniques in Biological Electron Microscopy Vol. II. Specific Ultrastructural Probes*, Springer-Verlag, Berlin, 1978.

27. J. Cazaux, Microscope photoélectronique pour l'analyse chimique des surfaces, Rev. Phys. Appl. *8*, 371–381 (1973); J. Cazaux, Microanalyse et microscopie photoélectroniques X, J. Microsc. Spectrosc. Electron. *1*, 73–80 (1976).

28. G. Beamson, H. Q. Porter, and D. W. Turner, Photoelectron spectromicroscopy, Nature *290*, 556–561 (1981).

29. R. J. Dam, O. H. Griffith, and G. F. Rempfer, Photoelectron microscopy of organic surfaces: The effect of substrate reflectivity, J. Appl. Phys. *47*, 861–865 (1976).

30. S. Grund, J. Eichberg, and W. Engel, Biologische Strukturen im Photoelektronenemissionsmikroskop, J. Ultrastructure Res. *64*, 191–203 (1978).

31. F. Willig, G. Frahm, W. Engel, N. Sato, and K. Seki. On the origin of photo-electron-microscope-pictures of organic samples, Z. Physikalische Chem. Neue Folge, *112S*, 59–68 (1979).

32. A. C. Van Dorsten, Statistical effects in electron microscope image recording in relation to optimal electron magnification and picture quality, in *Proc. Eur. Regional Conf. Electron Microscopy*, A. L. Houwink and B. J. Spit, Eds., Delft, Netherlands, Vol. 1, 1960, pp. 64–68.

33. W. A. Houle, H. M. Brown, and O. H. Griffith, Photoelectric properties and a method of detecting the aromatic carcinogens benzo(*a*)pyrene and dimethylbenzanthracene, Proc. Natl. Acad. Sci. USA *76*, 4180–4184 (1979).

34. R. B. Barnes, J. Amend, W. R. Sistrom, and O. H. Griffith, Quantum yield and image contrast of bacteriochlorophyll monolayers in photoelectron microscopy, Biophys. J. *21*, 195–202 (1978).

35. O. H. Griffith, H. M. Brown, and G. H. Lesch, Photoelectron microscopy of photosynthetic membranes, in *Light Transducing Membranes: Structure, Function and Evolution*, D. W. Deamer, Ed., Academic Press, New York, 1978, pp. 313–334.

36. G. B. Birrell, S. L. Moran, W. R. Sistrom, and O. H. Griffith, Photosynthetic bacteria. Photoelectron micrographs of pigment-containing biological specimens. Beitr. Elektronenmikroskop. Direktabb. Oberfl. *12*(2), 199–204 (1979).

37. K. M. Straub, T. Meehan, A. L. Burlingame, and M. Calvin, Major adducts formed by reaction of benzo(a)pyrenediolepoxide with DNA *in vitro*, Proc. Natl. Acad. Sci. USA *74*, 5285–5289 (1977).

38. A. M. Jeffrey, I. B. Weinstein, K. W. Jenette, K. Grzeskowiak, R. G. Harvey, H. Autrup, and C. Harris, Nucleic acid adducts formed in human and bovine bronchial explants, Nature (Lond.) *269*, 348–350 (1977).

39. K. K. Nadakavukaren, G. F. Rempfer, and O. H. Griffith, Photoelectron microscopy of cell surface topography, J. Microsc. *122*, 301–308 (1981).

STEREOLOGY AND SAMPLING
OF BIOLOGICAL SURFACES

HANS JØRGEN G. GUNDERSEN

Stereologic and Electronmicroscopic Laboratory for Diabetes Research
University Institute of Pathology
and Second University Clinic of Internal Medicine
University of Aarhus, Denmark

19.1   INTRODUCTION

Stereology is the branch of applied solid geometry that deals with methods for estimating structural quantities such as length, volume, curvature, thickness, diameter, surface, and so on, of a given specimen.

The fundamental feature of the methods used in stereological research is their ability to determine the size of such structures by measurements on the cut surface of the specimen. In stereology, two-dimensional figures provide information on three-dimensional structures. This principle was only realized after a century of developments in geometry, statistics, and mathematics. The basic principles and actual stereological methods were mainly due to petrographers and material scientists—a historical background that still shows up in current stereological practice [1–6]. The acknowledgement of the virtues of stereological methods in wider circles started with the formation of the International Society for Stereology in 1961. A large fraction of the exponential growth during recent decades in the application of stereological methods has been in biology rather than in material sciences, a development which to quite some extent is due to the work of Ewald R. Weibel and his group in Berne.

One important use of stereological methods in biological sciences is to determine the area of internal surfaces, that is, boundaries between recognizable compartments or spaces of an organ or a cell. This is often of considerable interest because the area and various *formal* quantities such as thickness of a surface often determine—wholly or partially—important *functional* events, for example, transport of ions or molecules, filtration or secretion, or synthesis.

The present chapter deals with stereological methods applied to the study of biological surfaces.

Sampling and statistics play a major role in stereology as in all natural sciences. The estimates of surface area are statistical in the usual sense that they pertain directly only to the sample of, say, seven tiny blocks of the organ. However, since we want to make inferences about the whole organ, we must—as always—take the sample in a way that allows us to make estimates for the entire organ. Moreover, since *propositions in biological sciences deal with species, not with individuals,* we need to calculate realistic confidence limits for the estimate of total area of the surface of interest in the group of individuals.

Stereological methods for estimating surface area are, however, statistical in yet another way: Only when a number of *random orientations* of the surface and the sections with respect to each other are encountered do they give precise estimates. Once we have realized that, we do not, in general, have much difficulty in implementing a proper sampling procedure for performing the estimation efficiently.

In the following discussion the underlying, extraordinarily simple geometrical–statistical principles of stereology of surfaces are outlined, followed by some remarks on practical sampling problems and a biological example to illustrate points of some importance.

19.2   PRINCIPLE FOR STEREOLOGICAL ESTIMATION OF SURFACE AREA

19.2.1   Hit and Miss

Our aim is to estimate the absolute area S of some specific, internal surface, which we shall denote α. Together with many other structures without particular interest right now, the surface is embedded in an organ, generally referred to as the reference space, the *known*, total volume of which we denote $V(\text{ref})$. The basic idea in the stereological estimation of the area of surface α is that lines or pins thrust into the organ will perforate the surface at various points of intersection, the number $I(\alpha)$ of which we can count. As demonstrated in the following, there is a constant relationship

$$\frac{I(\alpha)}{S(\alpha)} = \frac{1}{2} \frac{L(\text{ref})}{V(\text{ref})} \tag{1}$$

between the total length $L(\text{ref})$ of test lines penetrating the reference volume $V(\text{ref})$, on the one hand, and the total number of intersections $I(\alpha)$ between these test lines and the total surface $S(\alpha)$, on the other hand— provided we observe certain rules of the game. Intuitively, Eq. (1) seems to be satisfactory: Take a cheese and thrust some knitting pins completely inside it, take them out again, slice the cheese completely, examine the concave surface of all the holes (interior cheese surface), and count all the "points" of intersection $I(\alpha)$ between pins and hole surface. It is not necessary actually to repeat the experiment with twice as many pins in order to realize that this would give rise to twice as many intersections— as would twice the area of hole surface (more and/or larger holes) with an unchanged total pin length. Finally, if the same holes were dispersed in a much larger cheese penetrated by the same length of pins, these would hit many fewer holes. The surprising fact about Eq. (1) is that the factor of $\frac{1}{2}$ is exact, but its validity depends critically on whether we observe the rules or not. If so, measure the volume of the cheese, measure the length of the pins stuck into the cheese, and count the number of intersection points on the surface of the whole set of holes and

$$S(\text{holes}) = 2 \cdot I(\text{hole surface}) \, \frac{1}{L(\text{pins in cheese})/V(\text{cheese})}$$

We now need to derive the value of the constant factor of proportionality, that is, the integer 2, and the conditions for its validity to be specified.

19.2.2   Orientation of Surface

To construct valid estimators along the lines described above it is necessary first to consider in some detail the *orientation* of the target surface. The orientation of a vanishingly small element of surface is expressed by the direction of a normal erected on the element as shown in Fig. 1. By dividing our surface of interest into numerous small flat surfaces of con-

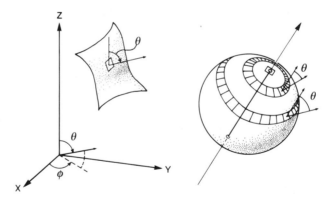

Figure 1.   Orientation of surface. To the left is a surface in three-dimensional space. The orientation of a small, flat element of the surface relative to a given direction (the Z-axis) is given by the angle from the Z-axis to the normal of the surface element. [The absolute orientation in three-dimensional space is given by the pair of angles (θ, ϕ)]. To the right is the ideal, uniform, three-dimensional orientation distribution: that of a sphere surface. As seen from the center of the sphere, all directions in three-dimensional space are equally represented. Relative to an arbitrary direction the surface normals do *not* show a uniform distribution of the angle θ: the greater this angle the larger is the number of normals. [The probability that the normal of a surface element selected at random is tilted at an angle θ is proportional to $\sin \theta$.] If the surface shown to the left is divided into many small elements which are transferred to the sphere without changing their direction, the normals on them will *not* be evenly distributed on the sphere surface.

stant area, we can consider the *distribution* of all normals, and thereby the orientation distribution of the surface α. There are two cases of interest for the estimation of $S(\alpha)$: (1) the *uniform* orientation distribution and (2) all others, that is, any nonuniform orientation distribution of α.

The uniform orientation distribution is that of a sphere surface, where all directions in space are equally represented. The surface α need not have the shape of a sphere (or of many spheres) to have a uniform orientation distribution, but if we take all surface normals and transfer them, without changing their direction, to the surface of a sphere, they must be distributed evenly on it. Similarly, all the surface elements of α must themselves form a sphere surface if transferred properly (see Fig. 1).

19.2.3 The Model

We now have a simple model for deriving a working principle for estimation of $S(\alpha)$ when it has a uniform orientation distribution (see Fig. 2). Since α is transformed to a sphere, its surface area is explicitly:

$$S(\alpha) = 4\pi r^2(\alpha) \text{ cm}^2 \qquad (2)$$

where $r(\alpha)$ is the (unknown) radius of the sphere. Let us imagine that the reference volume is penetrated by a bundle of test lines positioned randomly with respect to the reference space and of a density $1/a(t) \text{ cm}^{-2}$ This line density is the ratio between the total length of lines and the volume of three-dimensional space that contains them; that is, for the reference space

$$\frac{L(\text{ref})}{V(\text{ref})} = \frac{1}{a(t)} \text{ cm}^{-2} \qquad (3)$$

where $L(\text{ref})$ is the total, aggregate length of all parts of test lines *within* the reference space.

Some of the test lines within the reference space are likely to hit α. Evidently, any test line that aims at the shadow of α, projected onto a plane perpendicular to the line, hits α. The chance that a *randomly* positioned line hits α is therefore proportional to $A(\alpha)$, the constant area of the projection of α in any direction (this is where the uniformity of orientations on α really enters, only a sphere has this property). This area of the circular projection can be expressed in three ways: $A(\alpha) = \pi \cdot r^2(\alpha) = \frac{1}{4}S(\alpha) = n \cdot a(t)$ (see Fig. 2), where n is the dimensionless ratio (not necessarily integer-valued) between two areas

$$n = \frac{\pi \cdot r^2(\alpha)}{a(t)} = \frac{\frac{1}{4}S(\alpha)}{a(t)} \tag{4}$$

$S(\alpha)$ is fixed but unknown and $a(t)$ is random but observable (it has another value if we repeat the random positioning of the test lines). More explicitly, in statistical terms the average value of n is the expected or mean number of test lines that hit α when either $a(t)$ becomes small (the density of lines becomes high) or the experiment of throwing the bundle of

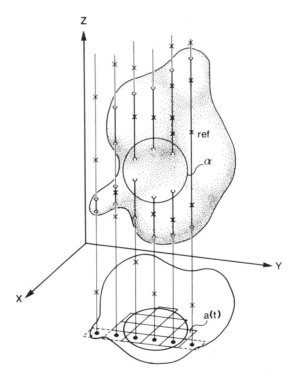

Figure 2.   Estimation of $S(\alpha)$. An arbitrary reference space within which the surface of interest, α, is embedded. If the orientation distribution of α is uniform, it is possible to transfer all surface elements of α in such a way that their aggregate is a sphere surface (or many sphere surfaces)—an imaginary situation that is depicted here. A bundle of test lines penetrates the reference space. Only one row of test lines is shown, drawn parallel to the arbitrary direction of the Z-axis but that is not a condition of any consequence. The exact position of the bundle is chosen *independently* of the projections of α and of the reference space. The area of the projection of α in any direction is $A(\alpha)$. The density of the lines of the bundle is $1/[a(t)]$ cm^{-2}; i.e., if the bundle is regular as shown, there is one line per element of area $a(t)$—but this regularity is of no consequence for the validity of the procedure. The crosses on test lines are markers with a density of $1/[l(t)]$ cm^{-1}.

lines into the reference space is repeated many times. A test line that hits α intersects it twice, and along the total length, $L(\text{ref})$, of test lines in the reference space we therefore count a total number of intersections that is $I(\alpha) = 2n$, or, by Eqs. (3) and (4)

$$I(\alpha) = \tfrac{1}{2}S(\alpha)\,\frac{L(\text{ref})}{V(\text{ref})} \tag{5}$$

It is necessary to measure the total length $L(\text{ref})$ of test lines within the reference space in order to use this expression, but if the lines carry notches with a known density $1/l(t)$ cm^{-1} we only have to count the total number $P(\text{ref})$ of notches *within* the reference space in order to estimate $L(\text{ref})$ (see Fig. 2). (Our ancestors used strings with knots for such tasks.) The density of points with respect to a set of lines is the ratio between the total number of points and the total line length:

$$\frac{P(\text{ref})}{L(\text{ref})} = \frac{1}{l(t)}\ \text{cm}^{-1} \tag{6}$$

Substituting into Eq. (5) and rearranging it, we get our final estimator

$$S(\alpha) = 2 \cdot I(\alpha)\,\frac{1}{P(\text{ref}) \cdot l(t)/V(\text{ref})}\ \text{cm}^2 \tag{7}$$

$$S(\alpha) = \left[\frac{I(\alpha)}{P(\text{ref})}\right]\frac{2 \cdot V(\text{ref})}{l(t)}\ \text{cm}^2 \tag{8}$$

where Eq. (8) emphasizes that to estimate $S(\alpha)$ we need to estimate (1) the ratio in brackets—by counting, solely, (2) the volume of the reference space, and (3) a measuring unit on our test lines, $l(t)$. That measuring unit is the length of test line associated with one marker or test point in the system. Its length is in centimeters on the true scale of the organ; that is, if we measure it after using a magnifying mechanism we must correct for the magnification.

19.2.4 The Reality

To make Eq. (8) an infallible workhorse that always does the job, we must get rid of two unrealistic assumptions concerning the surface α: its spherical shape and uniform orientation distribution. Let us first unscramble the surface elements of α and—without changing their orientation—move them back to their real position where they form the biologically interesting surface of, say, mitochondrial christae. Does that alter their chance of being hit by the test lines? {or can you save yourself by a change of position within a box with opaque walls into which someone is going to shoot, aiming at an *unknown* (random) point}. No, obviously not: the *random* position of the test lines means that the probability of hitting the surface elements is independent of their location within the reference space. The initial sphere model is just a shrewd way of getting you on the track—having done that we can dispose of all the geometric characteristics of the sphere and lose nothing. (The sphere "teaching model" was first used by Elias, Hennig, and Schwartz [7] in a still useful, condensed introduction to stereology as seen from a biologist's point of view.)

How do we dispose of the uniform orientation distribution of the surface? That particular distribution meant that the expected number of test lines hitting α was constant no matter how these lines were oriented: all parallel, just a few parallel, all converging toward an independent point in space, and so on; in short, the test lines were allowed to have an *arbitrary and unknown* orientation distribution. It did not matter at all because the uniform orientation of α ensured that with respect to any direction of a test line the angle θ to all surface normals had the same distribution: the larger θ was in the interval from 0 to 90°, the more normals were inclined at this angle to the arbitrary direction (look at Fig. 1 again). Since the position of test lines is independent of α, this angle θ is the one and only factor that determines the probability that a surface element is hit (see Fig. 3). By combining Eqs. (3) and (5) we can see that the chance of hitting a surface element of area $a(t)$ cm^2 by use of lines of density $1/a(t)$ cm^{-2} is $\frac{1}{2}$: $I(\alpha) = \frac{1}{2}[S(\alpha)/a(t)]$. This is where the factor $\frac{1}{2}$ in Eq. (1) came from}. It follows that when the distribution of θ for the aggregate of all normals on the surface of a sphere is *independent* of the chosen test line, then the overall chance of hitting an element is constant, irrespective of the direction of the test line(s). In other words, given that the normals

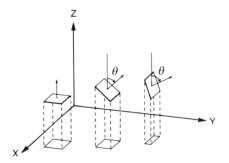

Figure 3. Probability of hitting surface. Small elements of surface (with area a) in a space inclined at various angles θ to an arbitrary direction (that of the Z-axis). For random positions of test lines in the chosen direction the chance that a surface element is hit is dependent on θ: the greater θ is the less is the projected area A of the element and hence its chance of being hit [it is proportional to $A = a \cos \theta$].

have a uniform orientation distribution, the distribution of the angle between them and any arbitrarily oriented test line is invariant and the overall chance of hitting a surface element is contant. Let us rephrase it again: Given that the test lines have uniform orientation distribution, the distribution of the angle between them and any arbitrarily oriented surface normal is invariant and so is the overall chance of hitting a surface element.

That is the "magician's trick": Select the test lines so that all positions on the silhouette of the organ *and* all directions in space are *equally likely,* forget about the orientation distribution of the surface α, and Eq. (8) is a valid estimator of $S(\alpha)$.

There are a few, very special cases of nonuniform orientation distributions of α that under certain restricting model assumptions may also be handled using special orientations of test lines. The reader is referred to the chapter on anisotropic structures and stereology by L.-M. Cruz-Orive [6], which, in the unsentimental language of geometrical probability and calculus, outlines the serious difficulties one faces when trying to use orientation-specific methods. Since nonuniform orientation distribution of surfaces is the rule rather than the exception in biology, we would be at a loss were it not for the fact, stated earlier, that a uniform orientation and a random location of *test lines* is all that is required for Eq. (8) to be valid. The following practical procedure is developed for this general type of problem; the exceptions where the surface of interest has a uniform orientation distribution are considered special cases, where certain, obvious shortcuts are permitted.

19.3 PRACTICAL PROCEDURE FOR STEREOLOGICAL ESTIMATION OF SURFACE AREA

19.3.1 Estimation of Total Organ Volume

In most cases weighing after fixation will be the easiest way, provided the specific gravity is close to 1.0. If not, the specific gravity may be estimated by the Archimedean principle. For in vivo studies noninvasive methods such as ultrasound provide convenient and safe methods for estimating total organ volume.

19.3.2 Sectioning of Tissue

When implementing the preceding principles in a method for estimation of $S(\alpha)$, it is obviously necessary to *see* the points of intersection $I(\alpha)$ between test lines and $S(\alpha)$ in order to count them. In practice, this means that we must cut or section the organ into slices or pieces first and then place our test lines on these, now exposed, slice surfaces.

The two-dimensional surface is now reduced to a one-dimensional curve in the section plane, the intersections $I(\alpha)$ to be counted are the zero-dimensional points of intersection between the surface trace and the test lines. In theory, we may *measure* the total length $B(\alpha)$ of the surface traces and the total areas $A(\text{ref})$ of the reference volume seen in the sections and estimate $S(\alpha)$ by another stereological relationship:

$$S(\alpha) = \left[\frac{B(\alpha)}{A(\text{ref})} \right] \frac{4}{\pi} \, V(\text{ref})$$

analogous to that of Eq. (8). In practice, this measuring procedure is considerably less efficient than the counting of $I(\alpha)$ and $P(\text{ref})$, as discussed later. The sectioning step does, however, raise another difficulty: The requirements of random position and uniform orientation are now to be fulfilled by the *sectioning* process.

The most convenient way of sectioning an organ—by a series of equidistantly spaced cuts, the first of which is randomly positioned within the first interval on the organ—obviously does *not* lead to fulfillment of the requirements. The sections obtained this way can only be used when it is known or can safely be assumed that the surface α has a uniform orientation distribution. For the general case it seems that the only reasonable way to sample properly is to cut the parallel slices into small pieces, select a sufficiently large number of these blocks, embed them with random orientation in plastic, and then cut *one* thin section from each block. A safe and very efficient way to make the selection of blocks is to take a transparent plate, drill a systematic array of holes in it, put all the organ slices underneath it (one by one or side by side), and then with a sharp biopsy needle punch out a block in every hole with tissue underneath it— or in every second hole in every second row, and so on, as necessary for getting a reasonable number of blocks. In this way a systematic sample of blocks is obtained: more blocks from large slices than from small ones, which exactly is one of the points. The conveniently small blocks (1–2 mm all over) can each be embedded in the sticky and viscous plastic medium in small vials with round bottoms, where for all practical purposes they can be considered to settle with a random orientation. The thin section is therefore simply taken perpendicular to the axis of the plastic vial.

19.3.3   Selection of Fields

In the microscope (light or electron) a predetermined number of systematically selected fields is studied. Evidently, it is best to space the fields as widely as possible, and, as always in systematic sampling, the first is taken randomly (e.g., the first upper left field within the section where some marks on the handles of the stage control both are pointing at 6 o'clock) and all the rest are taken systematically (after, say, $1\frac{1}{2}$ turn of one stage control).

19.3.4   Viewing of Fields

Finally, to get at the intersections between test lines and the surface α we must put the test lines and their notches on the field of vision. On each of the two types of microscopes there are two good ways of doing this. For EM work use either negatives projected onto ground glass with the test system in the image-forming plane [8] or use paper prints. In the latter case the transparent test system is fixed to cardboard along two adjacent edges and the print is put between them in a fixed, predetermined position against the two nonopen edges. For light microscopical work one must project the field of vision *out* of the microscope. This can be done either on a ground glass screen on top of the microscope or onto the table by an oblique mirror. (The projected field must be circular, not ellipsoidal!) Do *not* use ocular test systems; they will rapidly make the operator so tired that it is unreasonable to expect accurate counts. In any case it is most convenient to make a drawing of the test system and then photoreproduce it on transparent, durable film (e.g., Kodak Ortholith, type 3). The questions of how to make the test system, how many points, how long test lines, and so on, are dealt with below.

19.3.5   Tissue Inhomogeneity

If the surface α of interest can only be resolved at high magnification by EM, it is often more efficient to *redefine* the reference space and then use a two-step procedure. In a study of, for example, the outer mitochondrial membrane in Kupfer cells it is necessary to use such a EM magnification that, although some fields of vision only show a Kupfer cell, numerous fields do not contain Kupfer cells at all, they only show liver tissue, the primary reference space. In other words, at this magnification the liver is very inhomogeneous. Since every field of vision that shows anything of the *reference* space must be photographed, this is an inefficient sampling procedure. If we define the smallest secondary reference space that contains the entire surface α, we only have to sample a minimum of high magnification fields within that minimal reference space—but we must estimate its volume V(ref 2).

   In the preceding example, a minimal, well-defined secondary reference space is the Kupfer cell cytoplasm, that is, the whole cell exclusive of the nucleus, which obviously cannot contain mitochondria. This secondary reference space can be resolved at a much lower magnification than can the space containing the membranes of the mitochondria. At a magnification that is 100 times lower we observe on each field of vision an area of the section that is 10,000 times larger than before. In almost all the fields of vision a few percent of the whole field is now Kupfer cells. At this, lower magnification, the image is much more homogeneous, which is what makes the two-stage procedure more efficient. Notice that it is essential to have effectively the same definition of the secondary reference space at both magnifications. One cannot exclude, for example, the Golgi apparatus from the secondary reference space at high magnification—even though there are no mitochondria in the Golgi zone—because it is not possible to resolve the Golgi zone at the low magnification. The nucleus, on the other hand, is resolved equally well at both magnifications.

For estimating $V(\text{ref } 2)$ we have to use another stereological relationship:

$$\frac{P(\text{ref } 2)}{V(\text{ref } 2)} = \frac{P(\text{ref } 1)}{V(\text{ref } 1)} \tag{9}$$

entirely analogous to the one in Eq. (1), but now dealing with points hitting a volume and not with lines hitting a surface. Equation (9) expresses the constant relationship between the total number of points hitting the secondary reference space and its volume, on the one hand, and the total number of points hitting the primary reference space and its volume, on the other hand, when points with a density of

$$\frac{1}{v(t)} = \frac{P(\text{ref } 1)}{V(\text{ref } 1)} \text{ cm}^{-3}$$

are thrown into the primary reference space. As before we can put in marker points P' with a (dimensionless) density of

$$\frac{1}{p(t)} = \frac{P'(\text{ref } 1)}{P(\text{ref } 1)}$$

among the test points. Thereby the counting effort is reduced to counting the fewer marker points P' in the larger volume of the primary reference space, whereas we only have to count the more numerous points P in the small secondary reference space. Notice that by definition the secondary reference space is part of the primary one so that $P'(\text{ref } 1)$ includes points P' hitting the secondary reference volume. The estimate of $V(\text{ref } 2)$ is therefore

$$V(\text{ref } 2) = \left[\frac{P(\text{ref } 2)}{P'(\text{ref } 1)} \right] \frac{V(\text{ref } 1)}{p(t)} \qquad (10)$$

where, as before, $V(\text{ref } 1)$ is measured, $p(t)$ is a constant of the test system we generate ourselves, and the ratio in brackets is estimated from sections with a random position in the primary reference space. A big difference between the requirements for estimating a volume and the earlier situation, where a surface area was the target, is that the sections need only have a random position—their *orientation* does *not* matter. In many instances it is therefore possible to estimate $V(\text{ref } 2)$ at low magnification on the randomly positioned set of *parallel* sections, which one usually has to make, as described in Section 19.3.2.

Our "organ" is now the secondary reference space and the preceding estimate of $V(\text{ref } 2)$ is used in Eq. (8) for the determination of $S(\alpha)$. The ratio $I(\alpha)/P(\text{ref } 2)$ is estimated on all fields that contain anything of the secondary reference space. Those high magnification fields that show only the primary reference space are "empty" or outside the "organ" and we need not consider or photograph them at this magnification.

There are cases where it may be an advantage to have even more than two hierarchically defined reference spaces. In other organs the inhomogeneity is at a very high level of organization. The brain cortex, for example, may be defined as a secondary reference space for large neurons without magnification and the estimation of its volume $V(\text{ref } 2)$ may be carried out by the naked eye [7]. More details about multistage sampling are found in a recent paper [9] and in [5]. If the secondary reference space is very sparsely distributed in the organ, special sampling procedures have to be invented, examples of which are given by Pfeiffer [10] and in [11]. Obviously, there are also tissues so homogeneous that it is more efficient to take a few extra micrographs at high magnification than to get involved in a two-stage design. Finally, some organs have an easily identifiable axis and an internal structural organization which along that axis shows a simple pattern that can be modeled. In extensive studies of such organs it may be advantageous to take the systematic set of sections through the organ in a fixed relationship to the axis (e.g., always perpendicular to the axis), and then describe sampling variation in terms of variation of the model parameter(s) [12].

19.3.6 Effective Magnification—Magnification Effect

It is a rule without exceptions that the most efficient final magnification to work at is the lowest magnification at which the surface α still can be resolved with reasonable certainty. It follows from the preceding remarks that the organ inhomogeneity on the scale of fields of vision at this magnification determines whether it is worthwhile including an extra sampling level with a secondary reference space.

Besides these considerations, which are solely related to sampling efficiency, the magnification or resolution plays a direct role for the final estimate of $S(\alpha)$, an effect generally known as the "coast of Britain effect." Viewed at a range of improving resolutions or increasing magnifications, the final image will show finer and finer details of the surface trace and the estimate of $S(\alpha)$ will increase accordingly. (The total estimated length of the coastline of the British Isles in kilometers is increasing on maps drawn at a smaller and smaller scale, and measured directly on the beach it is even larger—exactly how much depends on the magnification used! The effect may be approximated within the mathematical concept of "fractal dimension" due to Mandelbrot; see Paumgartner and Weibel [13]). For irregular surfaces the effect can be appreciable; it is also present for estimates of volumes, but usually to a much smaller degree.

19.3.7   Section Thickness

It has been assumed so far that the section thickness t is virtually zero, that is, very much smaller than the smallest dimension of the structure elements, the surface of which is under study. If this is not the case, the estimator in Eq. (8) is biased (the estimates are too high), because of overprojection. The positive bias arises because in addition to surface α cut by the upper section face, the projected image seen in the microscope contains some "extra" part of surface α that is within the section and some part of surface α cut by the lower face of the section. If the structure is roughly spheroidal (i.e., all three dimensions are roughly equal), the bias is nearly proportional to t/diameter. For $t \sim$ diameter, the bias is therefore $\sim 100\%$ and corrections are obviously necessary, but they are quite elaborate and are only approximations [5]. Whenever possible one should use sections so thin that the remaining bias is acceptably small. For cylindrical structures (where two dimensions are small) it is quite easy to avoid the bias by an *indirect* estimation of surface based on the following relation: surface = length $\cdot \pi \cdot$ diameter, valid also for tortuous tubules, as described elsewhere [14]. For platelike structures (where only one dimension is small) the bias in surface estimates due to section thickness is usually negligible.

19.3.8   Recapitulation

The preceding description deliberately lacks a lot of details that are likely to be specific for the tissue under study. A number of useful, practical hints about sampling are found in several recent papers [9,12,15–17] and in Refs. 4 and 5. If one is not very familiar with the *Basic Ideas of Scientific Sampling,* the excellent booklet with this title by Alan Stuart [18] is a must; it requires absolutely no prior formal statistical knowledge.

Before embarking on specific sampling procedures or a given tissue, it is important to make sure that one *respects the principles* underlying the estimation. Although these principles are crucial, it is possible to make changes to fit in the details of the system being studied.

1. The volume of the reference space must be *known*.
2. The sections must have a *random position* in the tissue.
3. The surface α must have a *uniform orientation* distribution *or* the sections must be taken with *random orientations*.
4. The selection, location, and orientation of fields within sections and of test systems within fields must be *independent* of the surface α; that is, decisions are made before looking at the field.

The list calls for a few further comments:

ad 1. We have assumed nothing *whatsoever* about the reference space, except that we can define it and measure its volume.

ad 2. In the general case, where the tissue blocks are sectioned randomly, we have assumed nothing whatsoever about the surface α: Its orientation distribution does not matter and it may be located or concentrated anywhere inside the reference space, the estimate we obtain is *unbiased* anyhow. It is only when *efficiency* is considered (e.g., the coefficient of error of the estimate) that homogeneity becomes important: It is much easier to get a stable estimate in a very homogeneous organ.

ad 3. It is difficult to make a more rigorous procedure than simply letting the small tissue blocks settle at random on the curved bottom of the small embedding vials. However, if this requirement is *deliberately* broken (e.g., by taking sections perpendicular to a given direction), the estimate is in general left undefined.

ad 4. In practice, *independence* means systematic sampling where the position of the whole systematic set is determined by a mechanism or factor independent of the structure under study: the position of stage controls, grid bars, cardboard edges, and so on. The systematic set itself is predetermined: $\frac{3}{4}$ turn of handles, a test system where all lines and points are fixed in relation to each other, and so on. The requirement of *independent* selection is the most important, and if one selects the "good" fields in whatever sense except the purely technical one of discarding sections with uneven thickness, for example, the result may have nothing to do with the problem being studied.

19.4   EXPERIMENTAL DESIGN

19.4.1   Overhead

The most important aspect of a stereological–biological experiment is the content and formulation of the biological question that has to be answered. The problem is, however, too specific to be dealt with seriously in this condensed text. Here it is simply assumed that there does exist a well-defined and concisely formulated question that determines the species, the treatment of various groups, the organ, the mode of fixation of the tissue, and so on.

19.4.2   An Example

Fact: Patients with acute-onset diabetes mellitus have an enlarged surface of the glomerular capillaries in the kidney at the time of diagnosis (i.e., probably after only one or two weeks of the metabolic derangement).

Question: Is it possible for the glomerular capillary surface to become enlarged in such a short time?

Experiment: Rats are made diabetic by a chemical (streptozotocin). Four days later the left kidney is perfusion fixed in vivo. Control animals are included. Inferences about the disease in humans, diabetes mellitus, are drawn by *analogy* [19].

19.5   SAMPLING DESIGN

19.5.1   Overhead

Following the outline of the experiment, it is necessary to consider some of the details of the sampling design; for example: How many individuals, sections or blocks, fields of vision, test lines and points are enough?

The key to such questions is the meaning of *enough*. Concerning the first question, "enough individuals" to all scientists obviously means "sufficiently many biologically distinct and independent units to provide information." [The question is definitely one of the 20th century: it recognizes that the number of individuals is the critical factor. It could not have been formulated some centuries ago when it all started and when information—or knowledge, as it was called—was still *qualitative*: 'Is there continuity between the venous and arterial systems?']

Being of this century, we only need to find a factual yardstick for 'enough' or a measure of 'information' in the *quantitative* sense in order to begin deciding on the sampling scheme. It is in general an easy problem in biology where the only yardstick universally agreed on is the coefficient of variation among individuals, $CV(x) = [SD(x)/\bar{x}]$, where x is the surface α in one individual, $\bar{x}$ is the mean of the n individual values of x in the sample, and $SD(x)$ is the estimate of the spread or standard deviation among individuals with the same definition as those in the sample. The calculations are well known:

$$CV(x) = \frac{SD(x)}{\bar{x}} = \frac{\sqrt{S^2(x)}}{\bar{x}} = \frac{\sqrt{\left[\Sigma\, x_i^2 - \dfrac{1}{n}(\Sigma\, x_i)^2\right]\Big/(n-1)}}{\bar{x}} \tag{11}$$

$S^2(x)$ is the variance of x defined by the formula under the square root, in which $n - 1$ in the numerator signifies that it is never the spread of the sample that is of interest (it equals $\sqrt{[\Sigma\, x_i^2 - (1/n)(\Sigma\, x_i)^2]/n}$). What is of interest is an estimate of the spread in the population of individuals from which the sample was taken. We want to supply information about diabetes mellitus, namely, numbers related to the area of the surface of glomerular capillaries, a piece of information that may or may not be inferred by analogy from a study of a sample of *Rattus norvegicus*, some of which had diabetes.

19.5.2 How Many Individuals in an Experiment?

The point is to make sure that information obtained from the experiment is reliable and convincing. Conventionally, this means that the relative difference between the surface α in the two groups, must be "statistically significant at the 5% level," or equivalently:

$$\text{Rel. diff. } \bar{S}(\alpha) \geq \frac{2\sqrt{2} \cdot CV\{S(\alpha)\}}{\sqrt{n}}$$

The equation is a simple rearranging of Student's t pertaining to the comparison of the mean value in two groups with roughly the same size and roughly the same variance. (Notice that we are not performing any statistical test now; we just want to get a handle on the approximate number of individuals needed in a study that has not started yet. For that purpose we can safely forget about the rigorous statistical conditions for the validity of the t-test, like normal distribution and homogeneity of variance.) In terms of the total number of individuals studied, $N = n_1 + n_2$,

$$N \sim 16 \cdot \left[\frac{CV\{S(\alpha)\}}{\text{Rel. diff. } \bar{S}(\alpha)} \right]^2 \tag{12}$$

Therefore, the number of animals must be of a magnitude that depends on the unknown answers of the study, namely, the relative variance of animals *within* each group and the relative variation (i.e., the relative difference) *among* groups within the experiment. If, for example, a 10% increase in surface is to be considered significant in animals where the $CV\{S(\alpha)\}$ is 10%, the number of animals must be roughly 8 in each group, whereas if $CV\{S(\alpha)\}$ is 20%, it is necessary to estimate $S(\alpha)$ in *four* times as many animals.

At the beginning of a study Eq. (12) is of little use for deciding the number of individuals since the values of the two key terms—the total variation *within* groups (among animals) and the experimentally induced variation or difference *among* groups—are both unknown. A *pilot experiment* is then inevitable, and it is often a welcome opportunity to discover unforeseen problems at various stages of the experiment. The rule of thumb for a pilot experiment is to study about five individuals in each group: Less is not likely to be enough and more can always be included afterward.

In the *example* being discussed nothing was known beforehand about the variation of the glomerular capillary surface among animals, but the volume of the glomeruli was known to be increased by 30% after four days of diabetes [20]. Considering that plastic-embedded blocks from perfusion-fixed organs were already at hand from that study, it was decided to start out with as many as 9 + 9 animals for the EM-study.

19.5.3 How Many Blocks in an Individual?

The principle adopted for deciding on the number of blocks sampled in each individual is identical to that described earlier, since it depends equally on the organ inhomogeneity and the variation among individuals: The larger the variation of blocks *within* an individual compared to the variation *among* individuals within a group, the greater the number of blocks to be examined. Simple as it is, the principle is nevertheless based on two quite frequently overlooked facts. First, if animals vary a lot (i.e., the total area of surface α in one animal of the group is likely to be very different from that of the next animal), then one must study many animals to get a reliable answer. There is no possibility at all for getting that answer by looking at many blocks in fewer animals. Second, the number of blocks necessary does not depend on the primary outcome of the experiment [i.e., whether the observed difference between the two groups is large or not, it only depends on the CV$\{S(\alpha)\}$ *among* individuals within the groups and on the inhomogeneity of organs *within* the animals].

The details of how to make the decision on the number of blocks are, however, different, for now a block is not fixed, independent, information-bearing unit analogous to a biologically defined individual. As stated earlier, inhomogeneity depends on the scale at which the organ is viewed. Therefore, the size of the block and thereby the size of the section through it must be chosen before making decisions regarding the number of blocks. The previous remarks on the most efficient magnification carry over to the size of the section: It should be as large as possible, the limit usually being the purely technical problems involved in processing large blocks and making large sections of good quality.

The arbitrariness of the fixed scale at which block-to-block variation has to be expressed makes it unlikely that precise and useful information about organ inhomogeneity is at hand from previous studies. Two very practical considerations therefore determine the number of blocks to be sampled: (1) In the systematic sampling of tissue it is really just as simple to sample 20 blocks as 3 blocks. (2) When nothing is known about organ inhomogeneity with respect to the surface α the rule of thumb is to start by examining two or three blocks per animal in the pilot study. If in the end that turns out to be enough, no effort was wasted on many blocks, and if it is too few, more blocks can be studied afterward, *provided* they were sampled and embedded from the beginning. Obviously, when section-to-section variation is known or strongly suspected to be pronounced, one may as well make sections from more blocks right away, *provided* it is known or strongly suspected that animals do not vary a lot!

It is, however, generally not a good design to use more than one section per block (the next section from a block carries less new information than a section from the next block) or to use less than two blocks per animal, since that excludes the possibility of getting an idea of inhomogeneity by later analyzing the results of the pilot experiment. Notice that section-to-section variation really includes more than organ inhomogeneity at the scale of sections since we are dealing with the estimation of surface. The random orientation of sections means in the general case that some of the section-to-section variation of $I(\alpha)/P(\text{ref})$ is due to the nonuniform orientation distribution of $S(\alpha)$ in the organ, still to be considered at the scale of a section. This special aspect of organ inhomogeneity only becomes significant in the rare cases where the surface α has a markedly nonuniform orientation like that of plane lamellae with a size comparable to the section.

In the *example* under discussion a two-level sampling design is used, with the glomeruli as the secondary reference space. Using a set of fixed razor blades, the kidney is cut into a systematic series of parallel 0.8-mm slices with a 2-mm separation. The total glomerular volume, V(ref 2) is estimated by light microscopy on sections from the 2-mm-thick slices [20]. From 10 to 20 blocks are systematically punched out from the 0.8 mm slices through a plate with holes in a 6- by 6-mm square pattern. For analysis three randomly selected blocks with at least one glomerulus on a random section are taken. Only one glomerular profile from each block is selected for ultrathin sectioning at about 70 nm. The most central glomerulus, which has its center more than a maximal radius off the edge of the section is used irrespective of its size, see Ref. 11 for further details of the unbiased selection of such a secondary reference space. The reason for taking three rather than two blocks per animal was that the total filtration surface of the kidney, $S(\alpha)$, was expected to show rather little variation among animals of the same sex, age, and size, whereas the variation within animals (from glomerulus to glomerulus) was unknown.

19.5.4   How Many Fields of Vision on a Section?

When the most efficient magnification is chosen (the lowest at which the trace of the surface α can be reasonably resolved), the scale of the section inhomogeneity is fixed. Within systematically and independently selected sets of fields of vision or micrographs there is a certain variation of the ratio $I(\alpha)/P$(ref) or of the more "visible phenomenon" $B(\alpha)/A$(ref) from field to field. If that variation is large compared to the section-to-section variation, we must examine many fields. The arbitrariness of both sampling units and the fact that we are unlikely to have much idea about their variation beforehand means that quite simple considerations apply: (1) No less than two fields per section. (2) It is unlikely that section-to-section variation is so low that it is necessary to examine more than 5–10 fields per section. Moreover, the question concerns, for example, 10 animals and 3 blocks from each, so the extra effort and cost involved in studying 300 rather than 150 fields may well make it reasonable to start with five fields per section in the pilot study.

In the *example* the surface of the glomerular capillaries can be studied at a final magnification of about 2000× [21]. At this magnification the image of the largest glomerular diameter is roughly 25 cm. Most glomeruli can be photographed on one micrograph, and none of them take up more

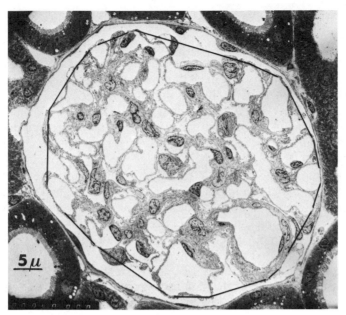

Figure 4. Secondary reference space. Glomerular profile from rat kidney reproduced at 1 : 3 of actual size of the micrograph. The black line is part of the convex string polygon, i.e., straight lines connecting the most extreme loops of basement membrane. The string polygon is used as the secondary reference space since it corresponds to the definition of glomeruli at the light microscopical level where the total glomerular volume was estimated.

than two, overlapping fields. Consequently, all glomeruli were photographed *in toto* on at most two micrographs, which were subsequently pasted together (see Fig. 4). This is a rather special case due to the small secondary reference space and the low magnification. Ordinarily a total photomontage is not an efficient way of sampling.

19.5.5 How Many Test Lines and Points per Field?

The previous paragraphs have dealt with the fact that with respect to the area of the surface α, individuals within groups vary, blocks or sections within individuals vary, and fields within sections vary. Furthermore, for obvious reasons it is only possible to study a quite limited number of sampling items. Individuals are generally so costly that the amount of information or confidence in the final answer is mainly a function of the low total number of individuals studied with a certain minor contribution to overall variation from the limited number of blocks. Fields of vision do show some but usually an even lesser contribution to the final variation due to their limited number, which nevertheless is often a total number of one or more hundreds.

The lesson is now clear: With all this variation among sampling items at higher levels there is virtually no possibility for improving the quality of the final answer by analyzing each field with a large number of test points and an increased length of test line. If the ratio $I(\alpha)/P(\text{ref})$ for each individual is approximately 200/200, there is no published stereological–biological study in which a higher number of counted intercepts and points per individual would have been of significance for the quality of the final result of the study.

The problem of constructing a test system that will provide intersection and point counts that run into a few hundred per individual is apparently one of the best kept secrets in stereology: Take a few animals from each experimental group and make the chosen number of sections. Use *any* available test system of sufficient size with some test lines and points and count intersections $I(\alpha)$ and $P(\text{ref})$ on the chosen number of fields. The average $I(\alpha)$ for the four animals is, for example, 76 and $P(\text{ref}) = 487$. Now draw a test system of the same size as before but with a total length of test lines which is roughly 200/76 times longer and a total number of test points that is roughly 200/487 times that used before. Draw the test lines and points in any pattern (the test system is unbiased anyhow), but it is easier and more efficient to use if the pattern is somewhat regular [16,17]. This is the test system that will provide an $I(\alpha)/P(\text{ref})$ ratio of about 200/ 200 per animal.

In the *example* the glomeruli from 4 animals were preliminarily analyzed by a 28.4 by 38.0 cm test system or grid with a total length for the test line of 722 cm and 130 test points arranged in the pattern shown in Fig. 5. On average, $I(\alpha) = 464$ and $P(\text{ref}) = 112$ per animal. We therefore need a test system with less than half the line length. In addition, the continuous lines in this test system are very inefficient. Since virtually all the surface traces of α in the micrograph are closed curves and intersections consequently come in pairs, we have to count two for every *independent* intersection. We therefore want a grid with short segments of test line. Their length should be roughly equal to or less than the diameter of a characteristic loop of the surface trace, that is, from 1 to 2 cm (see Fig. 8). A test system with about $722 \cdot (200/464)$ cm ~ 300 cm of test line in short segments would be appropriate. Since a grid of the previous size with 224.4 cm test line in segments of 1.1 cm is at hand, we will use that one (see Fig. 6).

When $P(\text{ref}) = 112$ we should perhaps increase the number of test points in the grid. However, these 112 points are counted on three convex

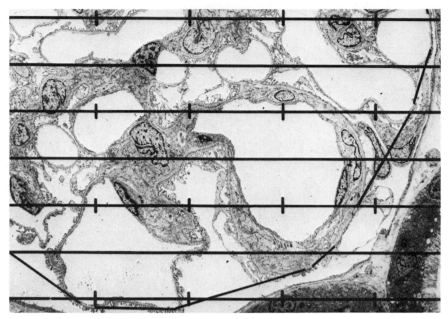

Figure 5. Pilot test system. Part of the glomerular profile shown in Fig. 4, but at the actual size of the micrograph and with pilot grid superposed. Intersections $I(\alpha)$ between all test lines and the glomerular surface are counted. The surface α is the outer (urinary) aspect of the basement membrane (which is covered everywhere with epithelial podocytes); see also Fig. 7. The uppermost border of the thick test lines are the true, one-dimensional test lines and the upper, right corner of the test points (crosses) are the true, zero-dimensional test points. The number of all test points within the string polygon (including its limiting line) is $P(\text{ref } 2)$.

and smoothly delineated areas: the glomerular string polygons, illustrated in Fig. 4. If we repeat the estimation, the variation of $P(\text{ref})$ in a given animal is very small. This is because for a new position of the test system only the few points close to the boundary of the string polygon may change from hits to misses or from misses to hits. Even though the grid at hand had less than half the number of points of the pilot grid (51 vs. 130), we shall use it as it is. Notice that the points need not actually be *on* the test lines; we may put them anywhere on the grid [17]. The grid constant or the measuring unit, $l(t)$, which we must know, of course, is simply the ratio between the total length of test line and the total number of points in the grid: 224.4/51 cm = 4.4 cm. At the scale of the tissue, viewed at a final magnification of 2063X, $l(t) = 4.4$ cm/2063 = 0.00213 cm.

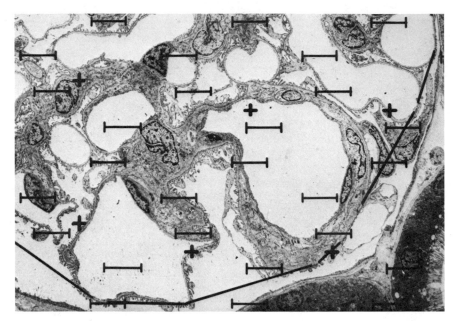

Figure 6. Final test system. The same part of the rat glomerular profile as shown in Fig. 5 but with the final test system superposed. The upper border of test lines within the end bars are true, one-dimensional test lines. The test points are the heavily drawn crosses, of which six are seen.

19.6   THE RESULT

In Eq. (8):

$$S(\alpha) = \left[\frac{I(\alpha)}{P(\text{ref } 2)} \right] \frac{2 \cdot V(\text{ref } 2)}{l(t)} \ \text{cm}^2$$

the two terms in the primary ratio, $I(\alpha)$ and $P(\text{ref } 2)$, are summed over all the fields and all the sections from one individual before the calculation of $S(\alpha)$ in that individual. The three variables pertaining to each individual 4-day-diabetic rat and the final estimates of glomerular surface are shown in Table 1. The mean surface area of the glomerular capillaries in a 4-day-diabetic rat is 165 cm² with a CV of 11%. Control rats (data not shown here; see Ref. 21) showed a glomerular surface of 120 cm², CV = 9.3%. The increase in glomerular surface of 45 cm² with a CE of 15% (CE = SE(x)/x) is highly significant as judged by Student's t-test: $2p = 8.0 \times 10^{-6}$. For a further discussion of the biological question to which this result is an answer, see Ref. 21.

TABLE 1 Estimates of Total Surface Area[a]

Animal	V (glom) cm^3	I (α)	P (glom)	S (α) cm^2
1	0.0670	171	58	185
2	0.0504	157	44	169
3	0.0578	115	40	156
4	0.0601	165	58	160
5	0.0582	203	63	176
6	0.0592	136	46	164
7	0.0584	136	39	191
8	0.0485	97	33	134
9	0.0469	115	34	149
Mean	0.0563	144	46.1	165
CV	0.11	0.23	0.24	0.11

[a] Individual values and mean and CV (coefficient of variation) of total volume of the secondary reference space V (glom) and total intersection count $I(\alpha)$ and P(glom) on three glomerular cross sections per animal. The last column shows the total glomerular surface in one kidney, calculated from Eq. 8, using $l(t) = 0.00213$ cm.

19.7 ANALYSIS OF SURFACE ESTIMATES

19.7.1 Outcome of Pilot Study

In contrast to the example used here, it may well be that the pilot study shows that the CV of $S(\alpha)$ among individuals is so great that the necessary number of them, calculated from Eq. (12), is prohibitively large. There are two possible reasons for this finding: (1) individuals do vary so much (then change it, i.e., make another experiment with more well-defined groups— or forget it altogether) or individuals do not truly vary so much; (2) *sections* vary greatly and so few are studied that the final CV among animals is increased too much. This reasoning continues through all levels of sampling: Maybe sections do not truly vary very much, but (3) *fields* vary a lot within sections and too few fields were studied. The way to tackle the problem is to analyze the design from the highest level (animals) and down to the level where more items are needed. The outline of such an analysis is given elsewhere [22]. The rather simple technique is well known in statistics and stereological examples are found in Refs. 12, 23– 25. One very useful result of such a complete analysis of sampling variation is that it leads to an *optimal* sampling design, that is, a sampling where the information gained per unit cost spent in the experiment is at optimum. Such a design may even be read off in a simple nomogram [26].

19.7.2 Measuring Precision

Usually, it is not necessary to analyze sampling at more than the first one or two levels. Concerning the lowest level (i.e., the test system), there is, as stated earlier, no known case where more than a few hundred counted intersections and points on the reference space were needed, *given* that sampling was efficient at all the important, higher levels.

Analysis of the design used in the example shows that true variation among animals and among glomerular cross sections each contributes 45% to the total variation, whereas the variation in the point and intersection count (i.e., the test system) contributes 10% to the $CV^2\{S(\alpha)\}$ of 0.11^2. Had we counted 10 times as many points and intersections by making a 10 times denser grid, we would have got a $CV\{S(\alpha)\}$ of at least

$$\sqrt{0.11^2 \left(100 - 10 + \frac{10}{10} \right) \bigg/ 100} = 0.105$$

instead of 0.11; that is, the effect of 10 times the counting effort would be nil! The only feasible way to improve the design would be to use more glomeruli. If, for example, five were used instead of three, we would expect a reduction in $CV\{S(\alpha)\}$ from 0.11 to

$$\sqrt{0.11^2 \left(100 - 45 + \frac{45 \cdot 3}{5} \right) \bigg/ 100} = 0.10$$

a very modest improvement that may not be worthwhile, considering the effort in sectioning and photographing 66% more glomeruli.

A final word about measuring precision: Since the crude grid that was used contributed 10% to the final variance among animals, we would at most gain an improvement in $CV\{S(\alpha)\}$ from 0.11 to $\sqrt{0.11^2(100 - 10)/100}$ = 0.104 after substituting the grid with a semiautomatic measuring device (no existing automatic systems can analyze biological EM-micrographs like Figs. 4 and 5). Moreover, to use the machines is much more time-consuming than to do point counting, as amply shown in the two available comparative studies [27,28]: Take a pencil and follow *precisely* the *complete* outline of $S(\alpha)$ on the "micrograph" in Fig. 7; it takes 10 times more time than counting the 15 intersections!

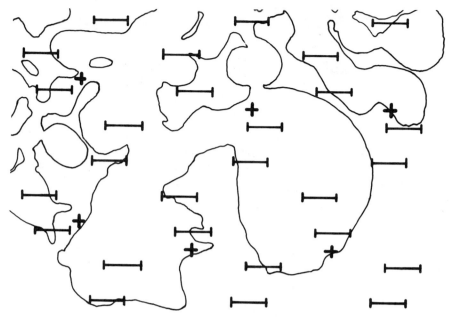

Figure 7. Surface α. Drawing of the trace of surface α from the same part of a glomerular profile as shown in Figs. 5 and 6 and with the test system in Fig. 6 superposed.

19.7.3 Absolute Versus Relative Measures

It is still quite common in biology to see stereological estimates reported in terms of a stereological ratio: $V(\alpha)/V(\text{ref})$, $S(\alpha)/V(\text{ref})$, and so on—a most unfortunate and ill-founded practice. Evidently, $S(\alpha)/V(\text{ref})$ and $S(\alpha)$ report exactly the same changes in surface if, and only if, the reference volume is unchanged, a quite unlikely situation in biology. To be certain, it is necessary to measure $V(\text{ref})$, and then there is no reason for not estimating $S(\alpha)$! These well known considerations obviously pertain to *all* ratio estimates, stereological or not. $S(\alpha)/V(\text{ref})$ is, however, a particularly uncertain basis for biological conclusions, as illustrated in Fig. 8 and Table 2. This uncertainty it shares with all fundamental stereological ratios except $V(\alpha)/V(\text{ref})$ because it is not dimensionless. $S(\alpha)/V(\text{ref})$ has dimension length^{-1} and represents the reciprocal mean thickness of the reference space if it were smeared onto the flat surface (α). Had the result of the example been given in terms of $S(\alpha)/V(\text{ref})$, the biological conclusion would have been quite different: "There is no in-

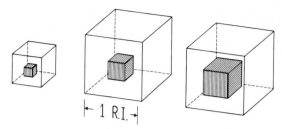

Figure 8. Varying reference space. Schematic organ or reference space containing some embedded surface α (hatched). In situations A and B the organs are isomorphous since all dimensions differ by a factor of 2. The organ in situation C is anisomorphous with both A (the dimensions of the reference spaces differ by a factor of 2, but those of surface α by a factor of $2\sqrt{2}$) and B (identical reference spaces, but dimensions of α differ by a factor of $\sqrt{2}$). The arbitrary unit of length, Ruritanian inch, is used in Table 2, where values of $S(\alpha)$, $V(\text{ref})$, and ratios are given.

crease in the glomerular surface in acute diabetes, the change of 7% with a CE of 58% is far from significant.''

The large number of biological reports in which random conclusions were drawn from stereological ratio estimates is a sad chapter in stereology. As outlined in the introduction, the main reason for this may be that biologists uncritically took over the stereological tools from material scientists, who use them in sciences in which $S(\alpha)/V(\text{ref})$ properly handled does report important physical properties of materials. The lesson learned is unambiguously: Never fail to estimate the total volume of the reference space. Even if the result then is given as a $S(\alpha)/V(\text{ref})$ ratio, there are few biologists who, when comparing, for example, situation A to situation B in Fig. 8 and Table 2 would conclude from the 50% fall in $S(\alpha)/V(\text{ref})$ that $S(\alpha)$ decreased, knowing that it *increased* by 300%.

TABLE 2   Varying Reference Space[a]

	$S(\alpha)$ RI^2	$V(\text{ref})$ RI^3	$\dfrac{S(\alpha)}{V(\text{ref})}$ RI^{-1}	$\dfrac{S^{1/2}(\alpha)}{V^{1/3}(\text{ref})}$ RI^0
A	0.157	0.157	1.00	0.735
B	0.630	1.26	0.50	0.735
C	1.26	1.26	1.00	1.04

[a] Total surface area, reference volume, the surface–volume ratio (arbitrary units) in the three situations shown in Fig. 7. The last column shows a dimensionless ratio, changes of which directly report linear magnitude and direction of anisomorphous changes in the organ; see also Ref. 21.

There is probably only one exception to the rule just mentioned: If the surface α is distributed on or within a number N(ref) of distinct and well-defined structural units, one may estimate directly the average area of α per structural unit, $\bar{s}(\alpha) = S(\alpha)/N(\text{ref})$, *without* estimating $S(\alpha)$ and N(ref) separately. This possibility for estimating $\bar{s}(\alpha)$ directly even when the N(ref) structural units do not have the same and simple shape has arisen from the recent development of a three-dimensional stereological probe [29]. Notice, however, that the difference in *interpretation* of $\bar{s}(\alpha)$ and $S(\alpha)/V(\text{ref})$ is only that we have substituted the assumption of a constant volume of the reference space by the assumption of a constant number of structural reference units. The example used in this text is one where such an assumption might well be acceptable—within the four-day-period of the experiment it is most unlikely that adult rats should develop new glomeruli or lose some of the existing ones without trace. It is important when dealing with human biopsies from macroscopically very inhomogeneous organs like the kidney, to be able to estimate the average area of surface α in an unchanged number of structural units.

ACKNOWLEDGMENTS

I am much indebted to Knud Lundbaek and Ruth Østerby for the privilege of countless fruitful discussions and substantial contributions to this text.

REFERENCES

1. R. T. De Hoff and F. N. Rhines, *Quantitative Microscopy*, McGraw-Hill, New York, 1968.

2. E. E. Underwood, *Quantitative Stereology*, Addison-Wesley, Reading, Mass., 1970.

3. S. A. Saltykov, *Stereometrische Metallographie*, VEB Deutscher Verlag für Grundstoffindustrie, Leipzig, 1974.

4. M. A. Williams, *Quantitative Methods in Biology*, Part II, North-Holland, Amsterdam, 1978.

5. E. R. Weibel, *Stereological Methods, Vol. 1, Practical Methods for Biological Morphometry*, Academic Press, London, 1980.

6. E. R. Weibel, *Stereological Methods, Vol. 2, Theoretical Foundations*, Academic Press, London, 1980.

7. H. Elias, A. Hennig, and D. E. Schwartz, *Stereology*, Physiol. Rev. *51*, 158 (1971).

8. E. R. Weibel, Stereological techniques for electron microscopic morphometry, in *Principles and Techniques of Electron Microscopy, Vol. 3*, M. A. Hayat, Ed., Van Nostrand Reinhold, New York, 1973, p. 237.

9. L.-M. Cruz-Orive and E. R. Weibel, Microsc. *122*, 235 (1981).

10. U. Pfeifer, J. Cell. Biol. *78*, 152 (1978).

11. R. Østerby and H. J. G. Gundersen, Sampling problems in the kidney, in *Lecture Notes in Biomathematics, Vol. 23*, R. E. Miles and J. Serra, Eds., Springer, Berlin, 1978, p. 185.

12. J. P. Kroustrup and H. J. G. Gundersen, J. Microsc. *132*, 43 (1983).

13. D. Paumgartner, G. Losa, and E. R. Weibel, J. Microsc. *121*, 51 (1981).

14. H. J. G. Gundersen, J. Microsc. *117*, 333 (1979).

15. R. Østerby and H. J. G. Gundersen, Mikroskopie *37*, suppl., 161 (1980).

16. L.-M. Cruz-Orive, J. Microsc. *125*, 89 (1982).

17. E. B. Jensen and H. J. G. Gundersen, J. Microsc. *125*, 51 (1982).

18. A. Stuart, *Basic Ideas of Scientific Sampling*, 2nd ed. Griffin, London, 1976.

19. K. Lundbaek, *Experimental Disease and Analogy* (unpublished data).

20. K. Seyer-Hansen, Joan Hansen, and H. J. G. Gundersen, Diabetologia *18*, 501 (1980).

21. R. Østerby and H. J. G. Gundersen, Diabetologia *18*, 493 (1980).

22. H. J. G. Gundersen, O. Götzsche, and R. Østerby, in Metab. Bone Dis. Rel. Res. *2*, 443 (1980).

23. H. J. G. Gundersen and R. Østerby, J. Microsc. *121*, 65 (1981).

24. J. Shay, Am. J. Pathol. *81*, 503 (1975).

25. W. L. Nicholson, J. Microsc. *113*, 223 (1978).

26. H. J. G. Gundersen, Microsc. Acta *83*, 409 (1980).

27. O. Mathieu, L.-M. Cruz-Orive, H. Hoppeler, and E. R. Weibel, J. Microsc. *121*, 75 (1981).

28. H. J. G. Gundersen, M. Boysen, and A. Reith, Virchows Arch. (Cell Pathol.) *37*, 317 (1981).

29. D. C. Sterio, J. Microsc. *134*, 127 (1984).

LASER DOPPLER MICROSCOPY AND FIBER OPTIC DOPPLER ANEMOMETRY

RICHARD P. C. JOHNSON and DOUGLAS A. ROSS

The Department of Botany
Aberdeen University
Aberdeen, Scotland
The Department of Electrical and Computer Engineering
University of Colorado at Denver
1100 Fourteenth Street
Denver, Colorado 80202, U.S.A.

20.1 INTRODUCTION

The advent of the laser and of fast digital electronics has provided the means to measure rapidly the small Doppler shifts given to the frequency of light when it is scattered from particles in motion. These Doppler shifts can provide information about the size and movement of microorganisms, cells, organelles, and molecules of interest to biologists and about particles of interest to industry (e.g., suspensions of pigments for paint or particles of silver halides for photographic emulsions).

When light is scattered from a moving particle the frequency of the light is Doppler-shifted by an amount proportional to the velocity of the particle. When coherent light, as from a laser, is scattered by many particles all flowing at the same speed in one direction, a single Doppler-shifted frequency can be detected and measured to give the velocity of flow [1,2]. If the light is scattered from many particles moving at random in thermal (Brownian) motion, then a spectrum of frequencies is produced. This spectrum can be analyzed to measure the diffusion constant of the particles [2–5]. The translational diffusion constant (D) of spherical, inelastic particles is related to their radii and to the temperature and the viscosity of the fluid around them, as shown in Eq. (1):

$$D = \frac{kT}{6\pi\eta a} \tag{1}$$

Where k = Boltzmann's constant (1.3806×10^{-16} ergs/K), T is the temperature in Kelvin degrees, a is the radius of the particles (cm), and η is the viscosity (Poises) of the medium around them. Thus if D and any two of the other three parameters are known, the third can be determined. Furthermore, the diffusion constants of molecules can be used to determine their molecular weights [6,7].

The Doppler shifts of light scattered from test tube quantities of specimens can be analyzed by the now well-developed method of photon correlation spectroscopy, sometimes called intensity fluctuation spectroscopy [1,3,53]. Velocities of particles in liquids or gases can also be measured by means of the related method, laser Doppler velocimetry [2,8]. However, the laser beams used in these methods are usually about 100 μm wide and cannot be placed accurately in, or detected precisely from, small specimens (e.g., plant or animal cells), or from small parts of specimens. Our chapter describes two ways to overcome this limitation—by applying and detecting light through microscopes or, alternatively, through fine fiber-optics light guides by means of the fiber-optic Doppler anemometer (FODA) [9–14].

20.2   THE BASIC PRINCIPLES OF PHOTON CORRELATION SPECTROSCOPY AND LASER-DOPPLER VELOCIMETRY

Particles with diameters of the order of 1 μm, as often found in living cells, have velocities of the order of 1 μm/s when undergoing Brownian motion in water. The Doppler shift in frequency given to light scattered from particles moving at this speed is only a few thousand cycles per second. Even if an object is moving away at 60 miles an hour, the maximum Doppler shift of light from it is only 8×10^7 Hz, depending on the angles at which the object is illuminated and viewed. These shifts in frequency are far too small to be perceived as a change in wavelength in conventional spectrometers. Even a Fabry-Perot interferometer cannot detect changes less than the equivalent of about one megacycle per second. Neither can the frequency of light be measured directly with photoelectronic detectors. This is because at frequencies above about 10^{11} Hz, corresponding to the border between microwave and infrared wavelengths, the energy of trains of electromagnetic waves is detected discontinuously as single electrons emitted when quanta are absorbed by atoms. Also, the frequency response of present photoelectric detectors is limited by two factors— carrier drift and capacitance; bandwidths of a few gigahertz (10^9 Hz) are the maximum obtainable. The frequency of light emitted from, for example, a helium–neon laser with a wavelength of 632.8 nm, is about 4.7×10^{14} cycles per second. Hence photoelectric detectors cannot directly follow the oscillations of light.

In spite of these limitations the Doppler shift in the frequency of light scattered from moving particles may be measured to within about one cycle per second if the particles are illuminated with coherent light, obtained from a laser. The scattered light is allowed to mix with a reference beam, of light from the same laser, on the photocathode of a photomultiplier or in the junction region of a semiconductor photodiode. The reference beam may be arranged to come directly from a laser or via scattering from the specimen, or both [see the discussion of Eqs. (5) and (6) and Figs. 6a, 6b and 7]. The number of electrons emitted by the photodetector varies at the beat frequency between the two beams, which is slow enough to be measured electronically. If enough electrons are produced, then the beat frequency can be measured with the aid of tuned circuits in a spectrum analyzer. If only a few electrons are being produced, they can be made to trigger pulses whose frequency of arrival can be analyzed statistically with a digital spectrum analyzer or with a signal correlator. This method, known as the heterodyne or optical beating method, was proposed by Yeh and Cummins in the early 1960s.

The Doppler frequency is proportional to the scattering wavenumber (Q, though some authors use K).

$$Q = \frac{4\pi n}{\lambda} \sin \left(\frac{\theta_s}{2}\right) \tag{2}$$

Where λ is the wavelength of the incident light in meters, n is the refractive index of the medium round the particles and θ_s is the scattering angle, the angle between the incident beam and a line drawn between specimen and detector (Fig. 1).

A practical photon correlation spectrometer is essentially as shown in

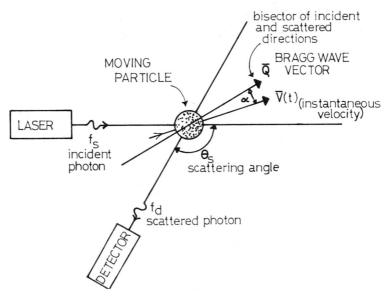

Figure 1.   The basic arrangement of laser and detector in a photon correlation spectrometer.

Fig. 1, except that usually many particles in suspension in a sample tube would be illuminated at once. A photomultiplier tube mounted on a rotatable support enables light scattered from a specimen to be collected at a known angle to the beam from the laser. Apparatus of this kind has been used to determine, for example, the diffusion constants and the electrophoretic behavior [5,50] and hence the molecular weights of protein molecules [6,7] and to measure the velocity of streaming cytoplasm in cells of the alga *Nitella* [15] and in plasmodia of the slime mold *Physarum* [16].

Various forms of photon correlation spectrometer are available commercially (e.g., from Malvern Instruments, Malvern WR14 1AL, U.K., or Langley-Ford Instruments, Amherst, Mass. 01002, U.S.A.).

The Doppler shift (Δf) given to light scattered by a moving particle (Fig. 1) is given by

$$\Delta f = f_s - f_d$$

$$= \frac{2n}{\lambda} \sin\left(\frac{\theta_s}{2}\right) V \cos(\alpha) = \frac{QV}{2\pi} \cos(\alpha) = \frac{\bar{Q} \cdot \bar{V}}{2\pi} \tag{3}$$

note that $\Delta\omega = 2\pi \, \Delta f = \dfrac{4\pi n}{\lambda} \sin(\theta_s/2) \, V\cos(\alpha)$

where f_s is the frequency in cycles per second of the light from the source, f_d is the frequency of light at the detector, V is the velocity of the particle (cm/s), θ_s is the scattering angle (degrees) and α is the angle between the direction of motion of the particle and the Bragg wave vector $\bar{Q}$ (Fig. 1).

However, if the light is scattered from particles undergoing random thermal (Brownian) motion, then a spectrum of Doppler-shifted frequencies will be produced. This will extend between zero frequency, where particles are moving at right angles to the Bragg wave vector, and a maximum frequency produced by the fastest particles moving parallel to the Bragg wave vector. This power spectrum, a measure of the intensity of light at the various frequencies, will have the form of a Lorentzian with a decrease in intensity of 6 db per octave (Fig. 2a). In effect, the linewidth of the laser light has been broadened (Fig. 2b), but the Doppler shifts of frequency are all recorded as positive.

A line drawn to the x (frequency) axis from a point where the curve has half its maximum height (Fig. 2a) gives Q^2D, where D is the diffusion constant of the particles as defined in Eq. (1) and Q is the scattering wave number as defined in Eq. (2). Thus if any two of the temperature, the viscosity, and the particle diameter are known, the remaining parameter may be determined.

If the diffusing particles are also all flowing together in one direction, then the frequency spectrum caused by the diffusion (Fig. 2a) will be shifted away from zero to appear as in Fig. 2b. From Eq. (3) the mean velocity along the Bragg wave vector of all the particles together will be given by

$$V = \frac{2\pi \, \Delta f}{Q \, \cos(\alpha)} \tag{4}$$

In practice it is now more usual to measure not the frequency spectrum with a spectrum analyzer, but the autocorrelation function of the intensity of the scattered light by means of a signal correlator. Descriptions of signal correlators may be found in Refs. 1–4 and 17. A main reason for using correlators is that they are faster than spectrum analyzers. Analogue spectrum analyzers are especially inefficient because they sample the signal for different frequencies in turn and thus waste much of it.

The autocorrelation function of a signal gives a curve composed of points that are correlation coefficients (Figs. 2c,d,e, and 8). Each coefficient shows how the intensity of the signal at a given instant compares with what it was at some specified time previously. At time zero the correlation coefficient of the signal intensity with itself will be 1 (Figs. 2c and 8). The autocorrelation function of a signal and its power spectrum

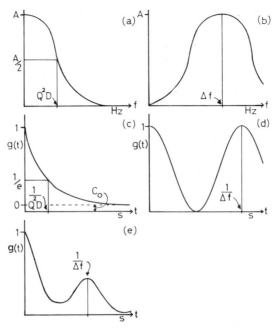

Figure 2. (*a*) The power spectrum of laser light scattered from a population of particles in Brownian motion. The line width of the light is broadened by Doppler shifts. A = amplitude; Q^2D is at the half-height at half-bandwidth. (*b*) The power spectrum of laser light scattered from particles that are in Brownian motion and that also have a common velocity in one direction; Δf is the Doppler shift due to their mean velocity. (*c*) The autocorrelation function of the intensity of laser light scattered from particles in Brownian motion; $g(t)$ is the normalized autocorrelation at time t; $1/Q^2D$ is the time constant of the exponential where Q is the scattering wavevector (Fig. 1) and D the translational diffusion constant of the particles; C_0 is a baseline due to the sum of random noise. (*d*) The autocorrelation function of laser light scattered from particles all moving at the same velocity in one direction is a cosine wave; $1/\Delta f$, the time per cycle, is inversely proportional to the velocity. (*e*) The autocorrelation function of the intensity of laser light scattered from particles in Brownian motion which also have a common velocity in one direction; the cosine wave dies away exponentially with a time constant of $1/Q^2D$.

are Fourier transforms of each other (the Weiner-Khintchine theorem). They bear a similar relation to each other as the image and the diffraction pattern of an object, which can be transformed one to the other by a lens. If necessary, power spectra and autocorrelation functions may be Fourier-transformed one to the other with the aid of a suitable computer program.

For a single frequency, representing a uniform motion in one direction, the autocorrelation function of the intensity of the scattered light would appear as a single cosine wave (Fig. 2*d*) having a frequency directly proportional to the velocity. For many spherical particles all of the same diameter and in Brownian motion, the autocorrelation function is an exponential (Fig. 2c). The exponential may be taken to represent the decay of a structure made of particles all moving at random. The time constant of the exponential is given by the position on the time axis at which the height of the curve has fallen to $1/e$ of its initial value (Fig. 2c). This time corresponds to

$$\frac{1}{Q^2 D}$$

and thus, as from a frequency spectrum, the diffusion constant of the particles may be calculated. In practice the curve may contain noise and, as discussed later, may contain the sum of several exponentials rather than one; curve-fitting by computer becomes necessary.

The autocorrelation function of particles undergoing Brownian motion and flow together consists of exponential and cosine terms multiplied together, as shown in Fig. 2e.

Where a reference beam is shone directly on the detector or where the reference light is light scattered from stationary parts of the specimen or instrument (such as the sides of test tubes, microscope slides, plant cell walls, or lens surfaces), the autocorrelation function G of the detected intensity of the light scattered from one or more particles, all of one size and in Brownian motion, can be represented for a time shift t as

$$G(t) = C_0 + C_1 \exp(-Q^2 \, Dt) + C_2 \exp(-2Q^2 \, Dt) \tag{5}$$

Note that in Figs. 2 and 8 $g(t)$ is the autocorrelation function $G(t)$ normalized to 1, that is, $g(t) = G(t)/G(t = 0)$.

The first term, the coefficient C_0, is a constant that represents a baseline due to the intensity of a background reference light scattered from stationary objects. The second term, with coefficient C_1, represents the scattered field autocorrelation that one would ideally like to measure, usually called the heterodyne term. The third term, with coefficient C_2, represents the intensity autocorrelation of scattered light from moving objects only, usually called the self-beating, or sometimes homodyne, term.

If a flow were present then a cosine term would also occur, the period of the cosine being directly proportional to the velocity of the component of flow along the Bragg wave vector.

$$G(t) = C_0 + C_1 \exp(-Q^2 \, Dt) \cos(2\pi \, \Delta ft) + C_2 \exp(-2Q^2 \, Dt) \tag{6}$$

If the intensity of light scattered from stationary objects is much greater than that from moving particles, only the first two terms are significant. Conversely, if the light scattered from stationary objects is weak, only the first and third terms are significant. A cosine, flow, term is absent in self-beating because the moving particles themselves act as local oscillators to provide a reference signal. The time constants of the self-beating (homodyne) terms are half those of the heterodyne terms because, on average at any instant, the diffusing particles are moving twice as fast relative to each other as they are moving relative to stationary scatterers. Terms due to secondary or higher order scattering, in other words due to light scattered from particles that has already been scattered once or more from other particles, are negligible and are left out of the equations.

Of course, if particles of more than one size are present then separate terms appear for each size of particle. Thus the fitting of model equations to correlograms obtained from suspensions containing particles of different sizes (polydisperse suspensions) becomes complex [1,3,10,18,40,48,52], especially since the intensity of light scattered from particles is a complex function of their radii varying with the sixth power of their radii for particles small compared to the wavelength. It is therefore an advantage to arrange the optical part of the apparatus so that either the heterodyne or the self-beating signal is small enough to be ignored. In a photon correlation spectrometer for test tube quantities of sample it is usually the heterodyne signal that is made negligible, by screening the photomultiplier so that it sees only light scattered from the sample and not that scattered from test tube walls or other stationary surfaces. Conversely, in a fiber optic Doppler anemometer (FODA) the end of the optic fiber in the specimen reflects enough light to swamp the self-beating signal, so that only the heterodyne signal need be considered. In a laser Doppler microscope too a predominantly heterodyne signal can be obtained by arranging for the intensity of the reference scattering, from stationary parts of the apparatus or specimen, to be about 30 times the intensity of the self-beating signal from moving particles.

An alternative way to measure the velocity of moving particles is to use two beams of laser light, derived from a single beam by means of a beamsplitter. The two beams are arranged to intersect in the specimen to produce a set of parallel interference fringes in the region of interest. Particles moving through the fringes cause flashes of light that may be timed with suitable apparatus, either a signal correlator or some form of frequency tracker (2,19–21) to provide their flow rate. Strictly speaking, the rate of flashing is the beat frequency between the Doppler shifts of light scattered from each beam [21]. The double beam, or fringe method has two advantages. It can work when only a few particles are present. Also, the rate of flashing is independent of the direction in which it is detected. However, the angle between the flow, the fringes and the plane of the fringe pattern must be known. If the frequency of one beam is shifted slightly, with the aid of a Bragg tank or a rotating diffraction grating, running fringes are produced and it is then possible to distinguish whether the particles are moving from left to right or from right to left in the fringes. Details of laser velocimetry with crossed beams on a nonmicroscopic scale may be found in Refs. 2 and 8. The velocity of flow of particles might also be measured from the autocorrelation function of the intensity of light scattered from two beams intersecting the flow a known distance apart; particles moving on to pass through the second beam would produce a delayed version of the signal they produced in the first beam, the delay being proportional to the velocity. Intensity fluctuations of light from fluorescent particles may also be used to study their motions [51].

20.3   COMPUTING

In order to obtain the required parameters for flow or Brownian motion a model equation, such as (5) or (6), may be fitted [52] to the sets of correlation coefficients, that is, to the autocorrelation function (Figs. 2 and 8), obtained from the signal correlator [1,3,4].

A method of curve-fitting based either on the Gauss-Newton method or on an exhaustive search may be used. The method minimises the sum of squares of the residuals which represent the differences between the model equation and the set of correlation coefficients. For example, for conditions where there is diffusion but no flow, Eq. (7) may be rewritten as:

$$R(t) = C_0 + C_1 \exp(-\Gamma t) + C_2 (\exp(-\Gamma t))^2 \tag{7}$$

where $\Gamma = Q^2 D$.

Suppose that D_n, where $n = 0, 1, 2, 3, \ldots N - 1$, are correlation coefficients measured each with a time shift $t_n = n\Delta t$, Δt being the sample time increment for which the correlator was set, that is, the interval of time represented by the distance between each data point (Fig. 8). Then the problem is to optimize

$$F = \sum_{n=0}^{N-1} [D_n - R(t_n)]^2 \tag{8}$$

by varying all relevant parameters until the minimum value of F is found. The computer search may be simplified by expressing all linear parameters (e.g., C_0, C_1, C_2) in terms of Γ. At the best fit between the data and the model equation, when F is minimum, the rate of change of F with C_0, C_1, C_2 is zero, that is,

$$\partial F/\partial C_0 = \partial F/\partial C_1 = \partial F/\partial C_2 = 0$$

This gives three equations in three unknowns that may be used to find C_0, C_1, C_2, and so on, for given values of the other parameters.

A large mainframe computer takes only a few seconds to do all these calculations, whereas a microcomputer, such as a Commodore PET, may take about half an hour. However, in our experience, the input and output processes of large time-sharing computers cause delays that limit progress. Because a microcomputer is an efficient and economic form of data-logger for recording the output of a signal correlator, it is convenient for one to be part of the laser-Doppler apparatus (Fig. 4). Because the microcomputer can be located with the rest of the apparatus, the correlation data may be fed to it and results extracted from it without delay. This immediate availability tends to offset the microcomputer's comparative slowness in the actual calculations of curve-fitting. The microcomputer may also be used to plot graphs of data as they are produced during experiments. We find that it is essential to be able to plot graphs and fit curves quickly if experiments are to be controlled as they proceed.

20.4 THE FIBER OPTIC DOPPLER ANEMOMETER

The fiber optic Doppler anemometer (FODA) (Fig. 3) was invented by R. B. Dyott [9] as an application of multimode optical communications fiber. The optical fiber of the FODA carries laser light with negligible attenuation to the point of measurement. This is particularly advantageous in the measurement of particle motion in dense opaque liquids or gases, since it is difficult to detect scattered light from within a dense medium. The fiber forms a window into the substance to be measured which may be some distance from the photodetector. Another advantage of the FODA is that it requires relatively small volumes of samples for measurement. Less than 1 ml of liquid in a 1-mm-diameter drinking straw is sufficient; indeed, a drop of liquid on the tip of the fiber may be adequate. A version of the FODA is marketed by SIRA Institute Ltd., South Hill, Chislehurst, BR7 5EH, U.K.

The arrangement of the FODA is shown in Fig. 3. A polarized laser beam passing through a 1-mm hole in a mirror is focused onto the end of an optical fiber, which carries the light to where it radiates from the tip of the fiber. Very little light is lost in the fiber, which may have an attenua-

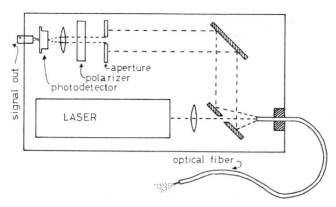

Figure 3. The arrangement of a fiber optic Doppler anemometer (FODA). The signal is passed to apparatus as in Fig. 4.

tion of less than 1 dB/km. Light scattered from moving particles near the tip reenters over the full numerical aperture of the fiber, within a cone of angles defined by a critical acceptance angle. Also, the outgoing light is partially reflected at the tip because of the difference in refractive index between the fiber's core and that of the surrounding medium (usually air or liquid). The amount of this reflection may be used to measure the refractive index of the specimen. The returned light, consisting of Doppler-shifted scattered light from moving particles and stationary reflected light from the fiber's tip, completely fills the numerical aperture of the fiber and therefore radiates from the launching end within a cone of angles θm. This light is reflected from the holed mirror, passed through a polarizer, and focused onto the surface of a photodetector. The purpose of the polarizer is to reject light back-reflected from the mirror surfaces and from the launching end of the fiber. It passes 50% of the light reflected from the tip of the fiber because the polarization of light is scrambled by propagation in multimode fiber. One may calculate that 90% of the returning light is reflected by the holed mirror, and therefore 45% is detected after passing through the polarizer. This efficiency of collection of returned light is far greater than that of other systems reported in the literature [12] and accounts for the success of the FODA over other systems.

The FODA gives a heterodyne detection of back-scattered light because, in most cases, the returned light reflected from the tip of the fiber is much more intense than the returned scattered light. For this reason a simple photodiode gives adequate means of detection of returned light. The resultant photocurrent is transformed to a low impedance voltage by a transimpedance preamplifier, providing a randomly fluctuating voltage whose net average value represents the intensity of light reflected from the tip of the fiber. The signal from the preamplifier is passed to additional apparatus for analysis (Fig. 4).

The FODA gives a very precise back-scatter detection. This may be seen from the scattering wavenumber [see Eq. (2)]. For scattered light within a cone of angles $\pi - \theta_m \leq \theta_s \leq \pi + \theta_m$, the scattering wavenumber Q is always greater than

$$\left(4\pi \frac{n}{\lambda}\right) \cos \left(\frac{\theta_m}{2}\right) \tag{9}$$

Thus for $\theta_m = 6°$, for example, Q deviates from what it would be for a precisely back-scattered direction by no more than 0.14%.

A limit to the length of optical fiber that may be used between the FODA and the measurement volume is set by the acoustic properties of the optical fiber. Random vibration of the fiber imparts a low frequency modulation of the returned light. This is because the multimode propagation is particularly sensitive to flexing of the fiber, which modulates the propagation path. In most cases this low frequency noise may be removed by filtering the voltage output of the FODA, provided that useful signal is not also removed. A fiber length of 1 or 2 m may be used without difficulty.

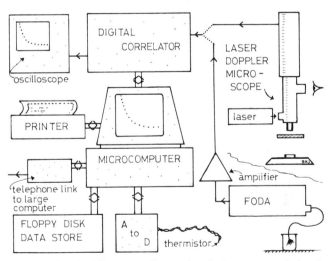

Figure 4. Apparatus for obtaining, recording, and analyzing autocorrelation data produced by the FODA or the laser Doppler microscope.

20.4.1 Accuracy, Sensitivity, and Spatial Resolution of the FODA

The accuracy of particle size distributions measured by the FODA has been tested by comparing them with size distributions measured from electron micrographs of silica spheres in Syton, a commercially available grinding compound. It was concluded from electron micrographs of Syton W-30 that the mean and standard deviation of particle radius were 26.9 and 11.2 nm, while results calculated from FODA data were 25.6 and 9.0 nm, respectively [10]. In measurements of velocity (e.g., sperm motility), it has not been possible to measure the velocity of particles by other methods and at the same time to assess the accuracy of results provided by the FODA. Also, for sperm at least, the FODA may be measuring rotational motion, which is related to but not identical with linear motion. However, as discussed earlier, the Doppler frequency is related to velocity by a scattering wavenumber that is very close to backscatter, to better than 0.14%. This means that the measured Doppler shifts will be directly proportional to the velocity toward the tip of the fiber.

The sensitivity of the FODA may be expressed by the minimum number of particles required to provide scattered light intensity above a lower limit set by shot noise in the photodetector. This minimum depends also on the particle size. For Syton W-30 the shot noise limit was equivalent to somewhere between 10^9 and 10^{11} particles per ml, depending on the photodetector preamplifier bandwidth. With 1-μm-diameter Dow latex beads the shot noise was equivalent to 10^7 particles per ml. These limits could be reduced by increasing the maximum power of the laser.

The scattering volume of the FODA is determined by the pattern of light radiated from the tip of the optical fiber. The one used so far has had a core diameter of about 60 μm, though smaller-diameter fibers could be used. This spatial resolution is determined also by the inverse variation of scattering intensity with the square of the distance between particles and the tip of the fiber. It may be estimated that the FODA detects particles within a region, near the tip of the fiber, which is roughly a cylinder 60 μm in diameter and less than 1 mm long, representing a scattering volume less than 0.06 mm^3. The use of an optical fiber with different characteristics would change the size of the scattering volume.

20.4.2 Applications of the FODA

Many applications of the FODA have been investigated, including the measurement of Brownian motion of particles in a colloid [10,11], in vivo blood flow [12], motility of sperm [22,23], and the instrumentation of manufacturing processes. The FODA may be used to measure Brownian motion to find the size distribution of a colloidal suspension of particles obeying Stokes–Einstein diffusion. It has been shown that the size distribution of particles obtained by the FODA agrees to within experimental accuracy with that obtained by electron microscopy [10]. Intravascular blood velocities of aortic and coronary artery flow of up to 20 cm/s, corresponding to Doppler frequencies of up to 1 MHz have been measured [12]. The motility of sperm swimming in raw diluted semen and in a mixture of semen and lactose/egg yolk has also been reported [23]. The motility is characterized by a mean swimming speed in the range 50 μm/s to 200 μm/s. The intensity of signal from motile sperm indicates the percentage of swimming sperm in a sample. The FODA has also been used to detect the rate of pulling of glass formers in the manufacture of optical fibers and has measured the acoustically driven motion of loudspeaker cones [9]. Figure 5 shows the result of an experiment to measure the change in viscosity of a solution of protein (20% w/v bovine serum albumen in water) while it was being treated with glutaraldehyde, a fixative used to prepare biological material for electron microscopy. Latex beads with a diameter of 0.264 μm were mixed with the protein. The tip of the optical fiber was placed in the solution, the glutaraldehyde was added to a final concentration of 0.33%, and the gradual restriction of motion of the beads by the protein molecules as they became cross-linked was followed with the FODA. Since the temperature (297 K) and the size of the beads were known, the viscosity could be calculated using Eq. (1).

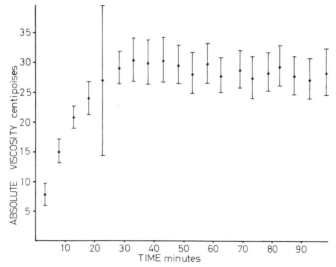

Figure 5. The change in viscosity, with time, of protein (bovine serum albumen) during fixation with glutaraldehyde, measured with the FODA (see text).

20.5 THE LASER DOPPLER MICROSCOPE

A laser Doppler microscope was first described by Maeda and Fujime in 1972 [24]. Since then details of various others have been published [25–35], (see Ref. 49 for further references). Excellent accounts of laser Doppler microscopy are provided by Earnshaw and Steer [36,53].

Three main optical arrangements have been used for laser Doppler microscopes. Single beams projected through an objective lens (Fig. 7) have been used to illuminate the specimen in some [26,34–36]. Two beams, one going directly to the detector to provide a reference signal and the other going via the objective lens to illuminate the specimen (Fig. 6a), have been used in others [26]. Two beams, projected onto the specimen and crossed to provide interference fringes in the plane of the specimen (Fig. 6b), have also been used [20,25–27,29,37], as have other arrangements [24,31].

An advantage of the crossed beam system is that, because the angle of detection does not matter so long as the direction of motion relative to the fringes is known, all of the light from an image of the specimen may be focused on the detector. Thus it can provide a strong signal. Also it requires simpler optics between objective lens and detector than single beam microscopes. However, a disadvantage is that errors may arise if the fringe pattern is distorted by variations of refractive index in the specimen and if the particle diameter approaches the distance between the fringes [38,39]. Also, usually only about 10 fringes can be obtained in

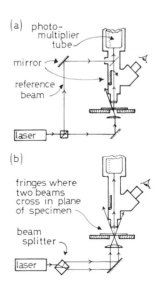

Figure 6. (a) A laser Doppler microscope with a separate reference beam provided via a beam splitter. (b) A laser Doppler microscope using crossed beams to produce a pattern of interference fringes in the specimen.

the scattering volume [38] so that sophisticated apparatus may be required to recognize select and record short trains of light flashes for measurement [19–21,25].

Microscopes that illuminate the specimen with a single beam are more suitable for measuring the diffusion of particles than microscopes that use crossed beams. But for measuring flow they have the disadvantage that any flow component in the signal appears multiplied by the diffusion component [Eqs. (5) and (6)]. Thus, in a correlogram, the flow component will fall to zero with the diffusion component (Fig. 2e). This means that, in order to measure flow, the diffusion constant of the particles scattering light in the flow and the component of flow along the Bragg wave vector must be arranged to match; the diffusion exponential must include enough cycles of the cosine wave caused by flow for the frequency of the wave and hence the flow rate to be measured. Also, in the single beam microscope, the angle at which the scattered light is being detected must be known. This angle may be defined in the microscope by projecting onto the detector not an image of the specimen but part of a diffraction pattern of the specimen image. This can be done with the aid of a Bertrand lens to project the back focal plane of the objective lens into the plane of the detector (Fig. 7). A movable aperture may then be positioned in the plane of this diffraction pattern to select a chosen angle of scatter. A small removable telescope may be used to view the aperture and diffraction pattern (Fig. 7). In our own microscope, which works in this way, the angle at which the aperture collects scattered light is calibrated by placing

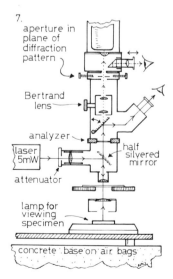

7.
aperture in plane of diffraction pattern

Bertrand lens

analyzer

laser 5mW

attenuator

half silvered mirror

lamp for viewing specimen

concrete base on air bags

Figure 7. The construction of our single-beam laser Doppler microscope [35,49]. The scattered light from the specimen is detected at an angle selected by an aperture in a diffraction pattern of the specimen, projected from the back focal plane of the objective lens by a Bertrand lens. The reference light is provided by scattering from the objective lens, microscope slides, and stationary parts of the specimen.

a replica of a diffraction grating under the objective lens instead of a specimen. The aperture is placed to accept the zero, first, second, or third order diffraction spots from a grating with 15,000 lines per inch. The centers of these spots define scattering angles of 180° (backscatter, like the FODA) and about 164, 146, and 123°, respectively, as described by Johnson [35]).

20.5.1   Accuracy, Sensitivity, and Spatial Resolution of the Laser Doppler Microscope

The accuracy of photon correlation spectrometry is discussed in Refs. 1, 3, 4, 5, and 36 and the problem of resolving size differences in suspensions of particles of different sizes is considered especially in Refs. 11, 18, 48, and 52. Within the limitations of this parent method the main special problems of laser Doppler microscopy are to define the scattering angle, to define the scattering volume, and to obtain a large enough signal-to-noise ratio in a short enough time without overirradiating the specimen. As with the FODA, the accuracy of resolution of particle sizes may be tested with particles that have been measured in the electron microscope. Light-scattering methods, however, may provide the hydrodynamic radii of particles. These may be greater than their radii measured on micrographs [18]. Measurements made with our laser Doppler microscope (Fig. 7) agree with electron microscopy to within about ±10% over a range of diameters between 0.088 μm and 1 μm. We are not yet able to be more precise about the components of this error.

The definition of the scattering angle in our kind of microscope (Fig. 7) depends on the size of the aperture placed in the diffraction pattern. We find that an aperture of about 25 μm diameter is necessary to give an adequate signal strength, but this size allows a spread in the scattering angle of collected light of, at worst, about ±3%. Laser Doppler micro-scopes that use crossed beams (Fig. 6b) do not require the scattering angle to be known and do not suffer from this problem. But, as mentioned earlier, distortion of the fringe pattern in them by the specimen may cause errors and further errors may arise if the diameters of the particles are great enough to fill the distance between fringes [38,39].

If particles in a single beam microscope (Fig. 7) are of the same order of diameter as the laser beam, then the intensity of the detected scattered light may fluctuate if particles move in and out of the beam and will fluctuate with their position in the beam if its intensity is not uniform across its diameter [31,47]. If the correlation time of these intensity fluctu-ations is comparable with that of the exponential decay representing Brownian motion, then the autocorrelation data may show decays that could be attributed mistakenly to diffusion. This fluctuation may be prevented by broadening the beam. However, it should be feasible to measure the size of a single large particle in Brownian motion if the beam illuminates it constantly; the statistics of the motion of a large particle depend on impacts from a large number of small molecules. The size of the sample of fluid containing particles may also be made small; drops with a diameter of 100 μm or less may be mounted to hang beneath a coverslip under the lens of the microscope.

The scattering volume may be defined if the laser beam is projected onto the specimen at an angle to the microscope axis [35]. An aperture in the image plane can then be arranged to define the length of beam, of known width, that contributes to the detected scattered light [31]. This, like an aperture in the diffraction plane (Fig. 7), will cut down the amount of light detected. If too little light reaches the detector, the random arrival of photons will cause the signal to be noisy. Thus there is a balance to be found, a trade-off, between loss of resolution of Q^2D due to noise and loss of resolution due to spread in scattering angle. However, if the condition of the specimen is not changing too quickly, noise may be averaged out by accumulating the autocorrelation data over a longer time.

An alternative way to lessen noise caused by random arrival of photons is to increase the intensity of the laser beam incident on the specimen. However, if the beam is too intense, it may damage the specimen. Sing [40] discusses effects of laser light on erythrocytes. We have found that the streaming of cytoplasm in cells of the leaves of the water plant *Elodea canadensis* is halted within about 1 min if the unattenuated beam of a 5 mW He–Ne laser is shone on it via a half-silvered mirror and a 100× microscope objective lens (Fig. 7). However, an adequate signal may be obtained with the beam attenuated to a small fraction of this, to an intensity that is barely visible on the specimen even against a dark background. This intensity does not seem to affect the streaming. Oliver (quoted by Earnshaw and Steer [36]) states that the detection of only 300 photons can give reasonable estimates of both Δf and D.

20.5.2   Applications of the Laser Doppler Microscope

A major advantage of photon correlation spectroscopy is that it may be used to measure the motions and sizes of particles below the limit of resolution of the light microscope. It can do this because it measures the frequency of light scattered from particles and does not require them to be resolved in an image. Photon correlation spectroscopy has been used to measure the motion or size of particles over a range of sizes between those of the zinc sulphate molecule and microorganisms, and under some

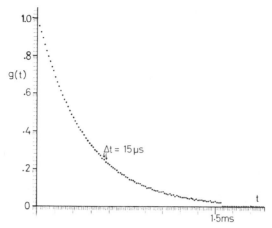

Figure 8. The normalized autocorrelation function of the intensity of laser light scattered from 0.234-μm-diameter (S.D. 0.003 μm) Dow latex beads in water under our microscope shown in Figs. 4 and 7. Δt is the sample time, set in the correlator. Each point represents a correlation coefficient showing how the intensity of the signal at that time compares with what it was $n \Delta t$ s ago, where n is the number of sample times between the point and $t = 0$ on the x-axis. The particle diameter calculated from this curve was 0.229 μm.

circumstances it may be able to measure movement as slow as a few nanometers per second [36]. These features and the ability to average the motion of many particles together (Fig. 8) give the FODA and the laser Doppler microscope an advantage over those methods that require microscopic particles to be resolved or followed one at a time. Also, some methods, such as frame-by-frame analysis of cine film [41] or even of enhanced-contrast video recordings [42,43], may be more time-consuming and tedious. Nevertheless, these and other methods (e.g., the use of the tracking microscope to follow the movements of individual bacteria [44] or the use of a rotating prism to arrest and measure the movement of an image of flowing particles in a microscope [45]) may be more appropriate, or cheaper, or at least complimentary for some kinds of specimens. Also, other methods may at first be easier for nonmathematicians to understand.

So far, the laser Doppler microscope has been applied mainly to study the flow and diffusion of blood in capillaries [20,27,28,32,33,37]. The motions of particles in some cells have also been examined [34,35]. However, most publications about laser Doppler microscopes have been about methods and instruments; the full potential of photon correlation spectroscopy applied via microscopes has yet to be exploited. The laser Doppler microscope may help to balance the current emphasis on biochemical aspects of cell biology; it may help to give greater meaning to such words as *transfer* [46].

REFERENCES

1. B. Chu, *Laser Light Scattering,* Academic Press, New York, 1974.
2. L. E. Drain, *The Laser Doppler Technique,* Wiley, Chichester, 1980.
3. H. Z. Cummins and E. R. Pike, Eds., *Photon Correlation and Light Beating Spectroscopy,* NATO Advanced Study Institute Series B: Physics, Plenum Press, New York, 1974.
4. P. N. Pusey and J. M. Vaughan, in *Dielectric and Related Molecular Processes,* Vol. 2, M. Davies, Ed. The Chemical Society, London, 1975, pp. 48–105.
5. B. R. Ware, *Chemical and Biochemical Applications of Lasers,* Vol. II, C. Bradley Moore, Ed., Academic Press, New York, 1977, pp. 199–239.
6. R. Foord, E. Jakeman, C. J. Oliver, E. R. Pike, R. J. Blagrove, E. Wood, and A. R. Peacocke, Nature (Lond.) *227,* 242–245 (1970).
7. J. S. Gethner, G. W. Flynn, B. J. Berne, and F. Gaskin, Biochemistry *16,* 5776–5781 (1977).
8. J. B. Abbiss, T. W. Chubb, and E. R. Pike, Opt. Laser Technol. 249–261 (December, 1974).
9. R. B. Dyott, Microwaves, Opt. Acoust. *2,* 13–17 (1978).
10. D. A. Ross, H. S. Dhadwal, and R. B. Dyott, J. Colloid Interface Sci. *64,* 533–541 (1978).
11. H. S. Dhadwal and D. A. Ross, J. Colloid Interface Sci. *76,* 478–489 (1980).
12. T. Tanaka and G. B. Benedek, Appl. Opt. *14,* 189–196 (1975).
13. D. Kilpatrick, J. V. Tyberg, and W. W. Parmley, IEEE JBME (in press).
14. C. J. Bates, J. Phys. E: Sci. Instrum. *10,* 669–675 (1977).
15. R. V. Mustacich and B. R. Ware, Biophys. J. *17,* 229–241 (1977).
16. R. V. Mustacich and B. R. Ware, Protoplasma *91,* 351–367 (1977).
17. S. Jen, J. Shook, and B. Chu, Rev. Sci. Instrum., *48,* 414–418 (1977).
18. Er. Gulari, Es. Gulari, Y. Tsunashima, and B. Chu, J. Chem. Phys. *70,* 3965–3972 (1979).
19. H. Mishina and T. Asakura, Appl. Phys. *5,* 351–359 (1975).
20. G. V. R. Born, A. Melling, and J. H. Whitelaw, Biorheology *15,* 163–172 (1978).
21. P. Le-Cong and R. H. Lovberg, Appl. Opt. *19,* 4222–4225 (1980).
22. D. A. Ross and J. G. Bullock, *Proc. Conf. Biomedical Applications of laser light scattering,* Cambridge, England, Sept. 8–10, 1981, B. R. Ware, W. L.

Lee, and D. B. Sattelle, Eds., Elsevier/North Holland, Amsterdam 1982, pp. 239–248.

23. D. A. Ross, H. S. Dhadwal, and J. A. Foulkes, J. Reprod. Fert. (in press).

24. T. Maeda and S. Fujime, Rev. Sci. Instrum. *43*, 566–567 (1972).

25. H. Mīshina, T. Ushizaka, and T. Asakura, Opt. Laser Technol. *8*, 121–127 (June, 1976).

26. T. Cochrane and J. C. Earnshaw, J. Phys. E: Sci. Instrum. *11*, 196–198 (1978).

27. P. R. DiGiovanni, B. Manoushagian, S. Einav, and H. J. Berman, *Proc. 6th New Engl. Bioengineering Conf.*, Pergamon Press, New York, 1978, pp. 113–116.

28. S. Rahat, D. C. Howard, S. Einav, and H. J. Berman, Fed. Proc. *37*, 214 (1978).

29. A. Koniuta, M. T. Dudermel, and P. M. Adler, J. Phys. E: Sci. Instrum. *12*, 918–920 (1979).

30. B. S. Rinkevichyus, A. V. Tolkachev, V. N. Sutoshin, and V. L. Chudov, Radio Eng. Electron Phys. (USA) *24*, 114–116 (1979).

31. T. J. Herbert and J. D. Acton, Appl. Opt. *18*, 588–590 (1979).

32. T. Koyama, M. Horimoto, H. Mishina, T. Asakura, M. Horimoto, and M. Murao, Experentia *35*, 65–67 (1979).

33. M. Horimoto and T. Koyama, Biorheology *18*, 77–78 (1981).

34. J. C. Earnshaw and M. W. Steer, Proc. Roy. Micr. Soc., *14*, 108–110 (1979).

35. R. P. C. Johnson, *Proc. Conf. Biomedical Applications of Laser Light Scattering*, Cambridge, Sept. 8–10, 1981, B. R. Ware, W. L. Lee, and D. B. Sattelle, Eds., Elsevier/North-Holland, Amsterdam 1982, pp. 391–402.

36. J. C. Earnshaw and M. W. Steer, Pestic. Sci. *10*, 358–368 (1979).

37. M. Horimoto, T. Koyama, Y. Kikuchi, Y. Kakiuchi, and M. Murao, Respir. Physiol., *43*, 31–41 (1981).

38. H. Mishina, Y. Kawase, T. Asakura, Japan. J. Appl. Phys. *15*, 633–640 (1976).

39. Y. Kawase, H. Mishina, and T. Asakura, J. Appl. Phys. *15*, 2173–2179 (1976).

40. M. Sing, Curr. Sci. *48*, 720–722 (1979).

41. G. F. Barclay and R. P. C. Johnson, *Plant, Cell and Environment 5*, 173–178 (1982).

42. R. D. Allen, N. S. Allen, and J. C. Travis, Cell Motility *1*, 291–302 (1981).

43. R. D. Allen, J. C. Travis, N. S. Allen, and H. Yilmaz, Cell Motility *1*, 275–289 (1981).

44. H. C. Berg and D. A. Brown, Nature (Lond.) *239*, 500–504 (1972).

45. T. J. Kotas and E. J. LeFevre, J. Mech. Eng. Sci. *17*, 65–70 (1975).

46. V. Ya Alexandrov, J. Theor. Biol. *35*, 1–26 (1972).

47. P. N. Pusey, Statistical properties of scattered radiation in *Photon Correlation Spectroscopy and Velocimetry,* H. Z. Cummins and E. R. Pike, Eds., NATO Advanced Study Institute Series B, Plenum Press, New York, 1977, pp. 45–133.

48. D. A. Ross, Laser sizing by orthogonal polynomial in *4th Int. Conf. Photon Correlation Techniques in Fluid Mechanics,* W. T. Mayo and A. E. Smart, Eds., Joint Institute for Aeronautics and Acoustics, Stanford University, 1980, pp. 1–15.

49. R. P. C. Johnson in *The Application of Laser Light Scattering to the Study of Biological Motion,* NATO Advanced Study Institute Series A: Life Sciences Vol. 59, J. C. Earnshaw and M. W. Steer, Eds., Plenum Press, New York, 1983, pp. 147–164.

50. B. R. Ware, Electrophoretic light scattering: Modern methods and recent applications to biological membranes and polyelectrolytes, in *The Application of Laser Light Scattering to the Study of Biological Motion,* J. C. Earnshaw and M. W. Steer, Eds., Plenum Press, New York, 1983, pp. 89–122.

51. D. E. Koppel, Fluorescence techniques for the study of biological motion, in *The Application of Laser Light Scattering to the Study of Biological Motion,* J. C. Earnshaw and M. W. Steer Eds., Plenum Press, New York, 1983, pp. 245–273.

52. B. Chu, Correlation function profile analysis in laser light scattering, in *The Application of Laser Light Scattering to the Study of Biological Motion,* J. C. Earnshaw and M. W. Steer Eds., Plenum Press, New York, 1983, pp. 53–76.

53. M. W. Steer, J. M. Picton and J. C. Earnshaw, Laser light scattering in biological research, in Advances in Botanical research, 11, Academic Press, London (in press).

CHAPTER

21

PROCEDURES FOR LOW TEMPERATURE SCANNING ELECTRON MICROSCOPY AND X-RAY MICROANALYSIS

PATRICK ECHLIN

The Botany School
University of Cambridge
Cambridge, England

21.1   INTRODUCTION

It is the purpose of this chapter to show how low temperatures may be used to advantage in the preparation, examination, and analysis of organic samples by simply converting the liquid phases of the specimen to metastable solids. In this context, the temperature range that will be considered is that at which water is in a solid state, that is, 0–273 K.

There appear to be four main advantages to using low temperatures in the study of organic material. (1) By lowering the temperature but still maintaining the liquid state, chemical reaction rates and transport process such as viscosity and diffusion are slowed down. This slowing down makes such processes more amenable to study in situ. (2) Water and many organic materials when in the solid state show an increase in mechanical strength as their temperature is lowered. This increase in mechanical strength then allows these materials to be sectioned, fractured, or dissected to reveal subsurface details. (3) There is a diminution in the amount of damage to the specimen that is an inevitable consequence of using most of the high energy beam imaging and analyzing systems. (4) There may be no need to use the deleterious chemical methods that are frequently used to stabilize and strengthen organic samples during specimen preparation. Leaving aside for the moment the vitrification of water, which is a special case, many nonaqueous organic materials, liquids, elastomers, plastics, can be converted to a solid phase in which only the physical properties (i.e., density, heat capacity, viscosity, tensile strength, hardness) have been changed. Their chemical composition remains unaltered because of the low rates of transformation processes at low temperatures and the strength of the bonds between the atoms. Nearly all biological material and a surprising amount of nonbiological organic material contain a high proportion of water, and it is the presence (and properties) of the water which gives rise to most of the problems in cryomicroscopy.

This chapter will not enter into a detailed discussion of the physicochemical properties of water and ice, as this has been elegantly and concisely achieved in a recent article [1]. It would seem more useful to consider some of the techniques that are being used to solidify liquid specimens, to discuss how such solid samples may be further processed for microscopy and analysis, and finally to consider the problems of damage that may occur to specimens during the process of preparation and data acquisition. Much of what will be discussed will be centered on biological material, for it is here that the problems associated with cryomicroscopy are most exacting. But the underlying principles will also apply to any system in which it is hoped to preserve and study the triphasic state of matter (i.e., soils, food products, oil and brine mixtures, etc.).

21.2 LOW TEMPERATURE STATES OF WATER

At atmospheric pressure, water exists entirely as a vapor above 373 K; as a solid, below 273 K; and as a liquid between these two extremes of temperature. It is this solid phase of water which we must now consider. Depending on temperature and pressure, the solid phase of water can exist in eight stable crystalline polymorphs, one metastable crystalline form and one metastable amorphous form [2][3]. Within this context it is convenient to consider the three main forms of ice with which most microscopists are familiar—cubic, hexagonal and amorphous, or vitreous ice.

The liquid-to-solid transformation is initiated by nucleation events of which there are two types. *Homogeneous nucleation* occurs when the random movement and clustering of water molecules produces a cluster of the right proportions onto which more water molecules can condense. There is a specified temperature range at which this probability event occurs—the homogeneous nucleation temperature—and it is possible to subcool water to this temperature (~233 K), at which point the water clusters are considered to be sufficiently icelike that they act as nuclei for spontaneous crystal growth. Thus, if one can carefully subcool water, the crystallization process results in many small ice crystals. It is more common for water, particularly in biological systems, to undergo *heterogeneous nucleation* where a foreign particle within the aqueous matrix acts as the catalyst on which ice nuclei growth is initiated. It is assumed that living cells would abound with structures of the correct dimensions (~10 nm) and correct physical state to provide nucleation sites. The transition of liquid water to ice is a first order phase change and the ice crystal grows by accreting pure water at its surface. This is the process by which hexagonal ice crystals are formed.

Amorphous ice is formed when water vapor condenses at temperatures below the glass transition point of ice (143 K). Falls et al. [4] point out that because there is always a partial pressure of water in an electron microscope, a uniform layer of water will invariably be deposited on substrates maintained at below 143 K. When amorphous ice is warmed above the glass-transition point, the water molecules undergo relaxation and reorientation and an irreversible process of crystallization occurs. In this instance the transition is a devitrification and the amorphous ice changes to cubic ice. As the temperature is increased, further recrystallization occurs that includes the resumption of growth of preexisting nuclei or crystallites, the formation of new crystals from amorphous solids, and the growth of larger crystals at the expense of smaller crystals. The actual kinetics of these processes are unclear, but they are certainly temperature dependent. A further transformation occurs at temperatures above 163 K and the cubic ice is slowly but irreversibly changed to hexagonal ice. At 273 K this hexagonal ice melts to form liquid water. These crystalline forms of ice can, like any crystalline material, contain matrix defects such as grain boundaries, dislocations, bend contours, and stacking faults (Falls et al., [4]). Such defects may affect the properties of ice during manipulations such as fracturing and sectioning.

The cryopragmatist may argue that this distinction between microcrystalline and amorphous or vitreous ice is a question of semantics. Not so, for accepting that microcrystalline ice is a practical possibility and that bulk vitrified water is a theoretical dream will allow us to understand better what is happening when water in biological systems is cooled. It is hoped that the foregoing discussion explains some of the problems associated with cryomicroscopy and can provide us with a better understanding of how to minimize the inevitable formation of ice crystals.

21.3   CONVERSION OF THE LIQUID TO THE SOLID STATE

Conversion of the liquid to the solid state can be considered under two general headings: (1) specimen treatment before cooling and (2) methods of rapid cooling.

21.3.1   Specimen Pretreatment

Chemical Fixation

On the assumption that chemical fixatives cross-link and stabilize macromolecules, some form of mild, nonoxidizing fixatives are being used at ambient temperatures in the preparation of specimens for morphological studies. Such studies fall into two main categories: freeze-etching and low temperature embedding procedures. Much of the earlier work on freeze-etching made use of mild fixatives such as some of the organic aldehydes that stabilized and strengthened the cell structure and caused sufficient changes in membrane permeability to permit the more easy access of cryoprotecting agents. These fixatives play no part in minimizing ice crystal formation and growth, but because they are considered to strengthen organic material, may help to minimize ice crystal damage.

Artificial Nucleators

As discussed earlier, one important factor determining the size of ice crystals is the rate of formation of water clusters of a critical size to initiate the nucleating event. The higher the nucleation rate, the smaller the ice crystals. As it is not possible to vitrify specimens of a size convenient and practical for most cryomicroscopy, it would be appropriate to settle for the next best thing in which the water had been converted to a large number of very small ice crystals. To this end a number of artificial nucleators have been proposed, the idea being that the introduction of such compounds into cells and tissues would promote nucleation, in the same way that silver iodide crystals are used to seed a rain cloud to form rain. Unfortunately, the likely candidates for biological artificial nucleators are either toxic (e.g., chloroform) or difficult to use (e.g. liquid CO_2). In any case it is doubtful whether artificial nucleators would be of much use in the cytoplasm of cells, because there is already a large number of hydrophilic macromolecules and inclusions, of the appropriate size and dimension, in the cytoplasm that would act as heterogeneous nucleators.

Cryoprotection

It has been known for some time that certain chemicals can protect cells and tissues from the disruptive action of freezing and thawing.

Penetrating Cryofixatives. Penetrating cryofixatives include methanol, ethanol, DMSO (dimethyl sulfoxide), glycerol, ethylene glycol, and some sugars.

All penetrating cryofixatives alter the permeability of the cell membrane, which invariably results in a redistribution of ions. Although penetrating cryofixatives should not be used in studies involving the analysis of diffusible ions, they have been used to great effect in freeze etch morphological studies. The most widely used penetrating cryofixative is glycerol. It penetrates most animal and some plant tissues quite readily, effectively lowers the freezing point of the cell contents, and retards ice crystallization.

Nonpenetrating cryofixatives. Nonpenetrating cryofixatives do not appear to penetrate cells to any appreciable extent and exert their effect from outside the cell. The molecules that have been used include serum albumin, dextran, polyethylene glycol, and more effectively, polyvinylpyrolidone (PVP) and hydroxyethyl starch (HES).

Studies by Franks and Skaer [5] showed that buffered solutions of PVP effectively diminished ice crystal damage in insect tissue. These studies were extended to a wide range of plant and animal tissues ([6],[7], and [8] et al., 1977; Echlin et al., 1977; Skaer et al., 1977) where it was shown that solutions of PVP and HES gave acceptable preservation as demonstrated in frozen sections, freeze-fracture images, and microanalytic studies. Because these polymers do not penetrate the tissues, they allow ice to be sublimed from the interior cells. This is particularly useful for examining the fracture faces of bulk frozen material, for although it is possible to see the cell outlines in unetched specimens, removal of a surface layer of ice facilitates visualization of subcellular detail. This is not to suggest that these polymers are ideal for all investigations, as there are indications that even after brief exposure some of these molecules can affect the natural physiological processes of some tissues. The polymers may also exert a noncolligative osmotic pressure, resulting in shrinkage, and adequate cryofixation has not been achieved in all tissues.

Mode of Action of Cryofixatives

The mode of action of cryofixatives is not clearly understood. They are all very soluble in water and their presence will perturb solutions to the extent that they can no longer be considered in an ideal state. The ability of penetrating cryofixatives such as glycerol to minimize ice crystal damage is believed to lie in their increasing the viscosity of solutions, thus slowing down the movement of water molecules to growing ice crystals and in their ability to act as a solvent, thereby keeping potentially harmful salts in solution as they undergo freeze concentration. The presence of penetrating cryofixatives inside cells also lowers the nucleation temperature and thus diminishes the size of ice crystals.

The nonpenetrating cryofixatives probably act in the same way as their penetrating counterparts. They too have high solution viscosities, which effectively retards crystallization.

Embedding Agents

It has long been known that it is useful to embed pieces of tissue in 10–20% gelatine prior to freezing. The gelatine provides additional support to the frozen sample and allows it to be more easily sectioned or fractured. Following these earlier studies, serum albumin, methyl cellulose, polyethylene glycol, and dextran have also been used to the same purpose, and effective encapsulation has also been achieved using polymeric cryofixatives such as PVP and HES. The use of PVP and HES has an added advantage: Solutions of these materials vitrify when frozen. Such vitrified solids are more readily sectioned and fractured than crystalline materials.

Nonchemical Pretreatment

The preparative procedures outlined in the preceding sections all suffer from the disadvantage that they involve the addition of foreign substances to the cells and tissues. It is possible, however, to carry out some manipulations that are compatible with life processes in biological systems. One of the greatest problems in freezing tissues is the extraction of heat from the sample. From this it follows that the size and shape of specimens are important parameters when it comes to optimizing cryofixation. It is no accident that the best preservation has been obtained in individual cells spray-frozen in microdroplets or in thin films of liquid suspensions placed between metal foils of high thermal conductivity. In many instances we have no opportunity of dictating the size and shape of the specimen, but we should always seek to use specimens as small as is compatible with the physiological and morphological aims of the experiment.

21.3.2 Methods of Rapid Cooling

The rate of cooling at the center of a specimen is dependent on the shape and size of the sample, the temperature difference between the sample and the cooling medium, as well as the density, specific heat, and thermal conductivity of both sample and cooling medium. The greatest problem in cooling samples that contain a substantial amount of water is the removal of latent heat, because the ice that is formed has a low thermal conductivity. This means that even with very rapid cooling, although the surface layers of the sample may have acceptably small ice crystals (i.e., ~10 nm), there will be a progressive increase in crystallite size the further one progresses into the sample; hence the necessity of removing the heat as rapidly as possible.

A number of ingenious methods have been devised to facilitate rapid cooling. These will be discussed in turn and a value given for the optimal cooling rate that has been achieved with each technique.

Plunge Cooling

As the name suggests, the sample is plunged, mechanically or by hand, into a liquid cryogen. A number of cryogens have been used including propane, fluorocarbons, liquid nitrogen, and liquid nitrogen slush. A recent paper by Elder et al. [9] summarizes clearly the different ways this rapid cooling may be achieved and the precautions it is necessary to take to achieve the optimal rate of cooling.

Spray Cooling

There is a rather specialized method in which specimen suspensions (cells, organelles, macromolecules) are sprayed as ~20-μm droplets, with a spray gun into a liquid cryogen (Bachmann and Schmitt [40]). The cells show good preservation but it is unclear whether this is due to the small size of the specimen or fast contact with the cryogen.

Jet Cooling

In jet cooling a jet of cryogen at several atmospheres pressure is sprayed onto a specimen from one or both sides simultaneously. This ensures a rapid renewal of the cryogen during cooling and, it is presumed, a rapid removal of heat from the sample. A recent paper by Knoll et al. [10] demonstrates how this method is carried out (see Fig. 1.A).

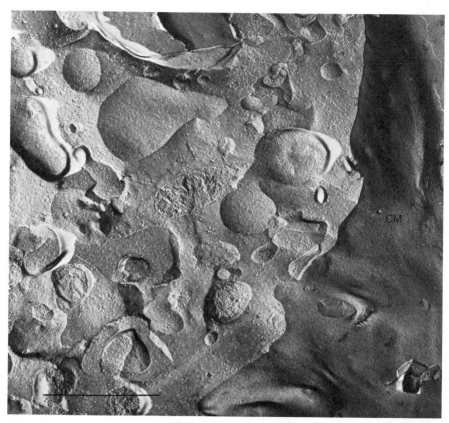

Figure 1.A. Fibroblast after jet-cooling. 3T3 mouse fibroblasts, cell line 101 (obtained, BCK Biocult-Chemie, Karlsruhe, FRG) were grown at 310K in Dulbecco's minimal essential medium with 5% calf serum added and buffered with $NaHCO_3$ (in exchange with a 10% CO_2 incubator atmosphere). Fibroblasts were grown as monolayers on Thermanox tissue culture sheets, and then used for the assembly of sandwich samples for cryofixation. CM, cell membrane (outer fracture face); M, mitochondria; and various other organelles are visible. There are no remarkable signs of intracellular ice crystal formation, membrane distortions or unusual organization of membranes 45,000×. Bar $\equiv$ 1 μm. (From Knoll et al., 1982 [10].)

Metal-Block Cooling

Specimens are pressed rapidly against a highly polished metal surface maintained at or near the temperature of liquid helium (4 K). The main advantage of this method is that although metals such as copper or silver have a thermal capacity about the same as organic liquids, they have a much higher thermal conductivity and consequently the same amount of heat is transferred 10,000 times faster through copper than an organic cryogen such as propane. The method that was originally devised by Van Harreveld and Crowell [11] has been improved by Heuser et al. [12] and Escaig [13].

Cooling at High Pressures

An examination of the phase diagram of water will show that vitreous water should form when liquid water is cooled under high pressure. Moor and his colleagues at Zurich have exploited this phenomenon and have built a high pressure cooling device. Specimens are subjected to pressures of 2000 bar a few milliseconds before being cooled by a high pressure jet of liquid nitrogen. This simple idea has been difficult to achieve in practice and at the time of writing only the prototype of this equipment exists in Moor's laboratory. One of the difficulties in assessing the relative usefulness of the different methods of cooling is to obtain some comparative measurement of cooling rates. These rates may be calculated from theory (Bald, [14]) deduced from variations in certain electrical coefficients during cooling (Heuser et al., [12]) or directly measured using thermocouples. Although thermocouples give consistent results when used to measure the cooling rate of a given cryogen, it is very difficult to use them to measure a cooling rate in the middle of an organic sample. The cooling rates given in Table 1 are average values for a thin (100–200 μm)

TABLE 1 Comparison of Cooling Rates

Method	Best Rate $K \times 10^3/sec^{-1}$	Artefact Free Zone (m)	Cryogen and Temperature (K)
Plunge	20–30	5–10	Propane (83)
Spray	100	20	Propane (83)
Jet	30–40	10	Propane (83)
Block	50–60	15–20	Helium (4)
Pressure	100	500	Nitrogen (77)
Vitreous	1000	1000	?
Micro-crystalline	25–50	50	?

sample. The values given are probably an underestimate of the cooling rate at the surface and a gross overestimate of the value at the center of the sample.

21.4 SAMPLE HANDLING AFTER RAPID COOLING

There are a number of options open to the experimenter regarding the further treatment of the sample after it has been rapidly cooled. These options include (1) sectioning, (2) fracturing, (3) etching and replication, (4) chemical substitution, and (5) freeze-drying.

21.4.1 Sectioning

A convenient way of exposing the internal features of a solid sample is to section it into slices thin enough to be examined by some form of transmitted illumination. Although there is still a great deal of uncertainty about the process of cryosectioning, it is convenient to consider thin (50–100 nm) and thick (0.2–2.0 μm) sections separately.

Thin Sections

The nature and success of cryosectioning is very dependent on the way the specimens have been prepared. Specimens that are prepared with a minimum phase separation of the aqueous and nonaqueous components either by infiltration with cryoprotectants or by rapid cooling, can be sectioned more thinly, more easily and at lower temperatures. In other words, the larger the ice crystals, the thicker the section and the warmer the sectioning temperature has to be in order to obtain smooth sections. Another important variable in the sectioning process is of course the nature of the specimen. The lower the amount of free or unbound water in the sample, the easier it is to section (Fig. 1.B). Thus while an elastomer such as rubber is easily sectioned at 143 K, it is virtually impossible to obtain consistently smooth sections of a mature vacuolate plant cell. A good general introduction to the subject can be found in the papers by Sitte [15a].

Thick Sections

The sectioning procedure is basically the same as for thin sections, except that warmer temperatures (193–233 K) appear to give better results (i.e., smooth sections). The thicker sections are cut using glass, diamond, and metal knives and it is interesting to note that the use of metal knives

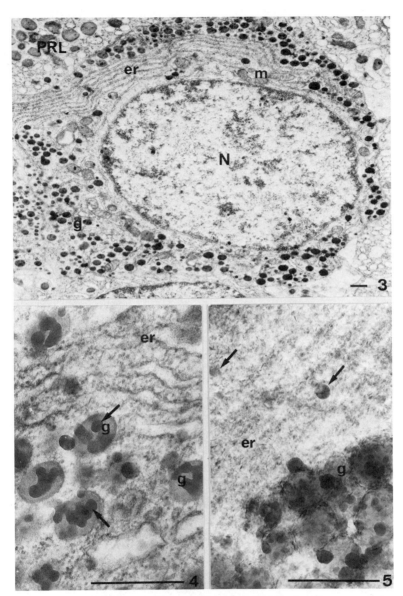

Figure 1.B. Anti-hGH antiserum. Glutaraldehyde-osmium fixation. Immunocytochemical reaction (→) was detected in cells characterized by regular medium-sized secretory granules, parallel arrays of endoplasmic reticulum, and large mitochrondria. After embedding (3 and 4, uranyl acetate and lead citrate counterstain), GH immunoreactivity was localized on secretory granules with antiserum diluted to 1/120 for 10 min. After freezing (5, uranyl acetate counterstain), immunoreactivity could be found on secretory granules and in endoplasmic reticulum with antiserum diluted to 1/20,000 for 10 min. (N, nucleus; PRL, prolactin cell; er, endoplasmic reticulum; g, secretory granule; m, mitochondrion). Scale bar, 0.5 μm. (From Hemming et al., 1983 [15].)

539

permits warmer sectioning temperatures to be employed. The papers by Beeuwkes et al. [16] and Biddlecombe et al. [17] review the current procedures used in cutting these thicker cryosections.

Although the success of cutting is a reflection of the quality of cooling, a firm theoretical and practical base for understanding cryomicrotomy has yet to be established. It will be necessary to cut thick and thin sections at a variety of different rates. Only then can we begin to understand what is happening during cryomicrotomy and in turn interpret the features seen and analyzed in sections such as shown in Fig. 2.

21.4.2 Fracturing

An alternative way of exposing the internal surfaces of a hydrated or frozen-dried sample is to fracture the sample at low temperature and

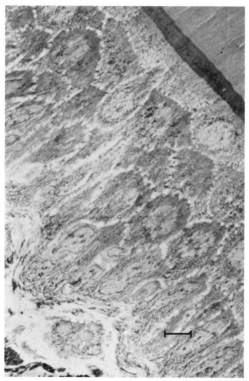

Figure 2. Scanning transmission electron micrographs of a frozen-dried 1-μm-thick cryosection of toad skin epithelium. Scale bar, 5 μm. (From Rick et al., [18].)

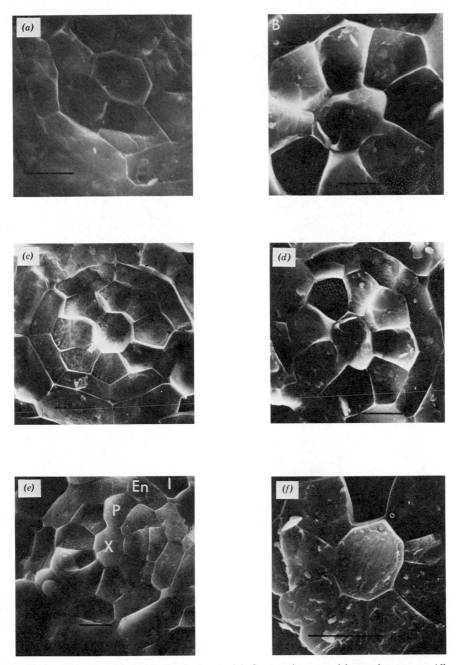

Figure 3. Frozen-hydrated bulk plant material, fractured to reveal internal contents. All samples from *Lemna minor* root tip and in a fully hydrated condition samples (C) and (E) etched to reveal more of internal contents. Carbon coated 15 kV, 100 K. Samples used for x-ray microanalysis. X, xylem; p, phloem; En, endodermis; I, inner cortex. Bar ≡ 10 μm.

541

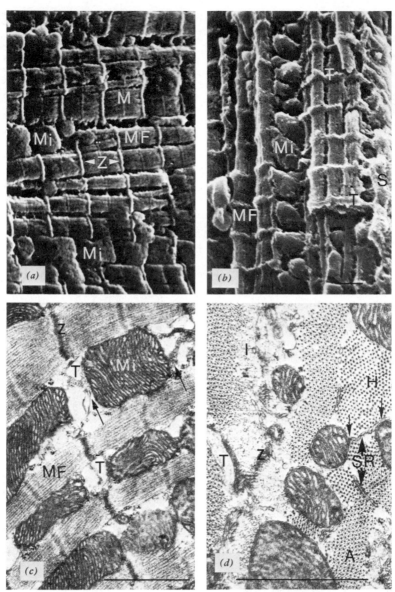

Figure 4. (*a*) Cryofracturing of ethanol-impregnated gerbil myocardial tissue quenched in liquid N_2 resulted in a smooth fracture plane, which frequently passed through the cell organelles. (*b*) However, the relief of the cracked surfaces was improved considerably if the cryofracturing had been carried out on critical-point-dried myocardium. In the latter specimens the fracture plane followed the surfaces of the cell organelles. Tissue shrinkage, manifested as separation of the cell organelles, was most pronounced in the critical-point-dried material. S, sarcolemma; Mi, mitochondria; T, T tubules; MF, myofibrils; M, M band; Z, elevations at the Z-band level. $10,000\times$. (*c* and *d*) TEM micrographs of (*c*) longitudinal and (*d*) transverse sections of gerbil myocardial tissue, which had been critical-point-dried

542

examine and analyze either the frozen-dried or frozen-hydrated surface. The fracturing procedure is a relatively straightforward process. Specimens are rapidly cooled and transferred under liquid nitrogen to the precooled stage of the fracturing device.

Recent papers by Marshall [19] and Echlin et al. [20] (see Fig. 3) show how the fracture technique is used in conjunction with x-ray microanalysis of biological material. The studies of Beckett and Porter [21] on plant material, by Carr et al. [22] on animal material by Schmidt and van Hooydonk [23] on dairy cream, and by Pesheck et al. [24] on porous oil bearing rock samples serve to illustrate how this technique can be used to expose the internal surface morphology of specimens.

An alternative approach to fracturing hydrated samples is chemically to fix the material and infiltrate it with an organic solvent prior to cooling and fracturing. Fracture faces of frozen ethanol (Fig. 4) and of frozen epoxy resin infiltrated materials reveal a wealth of morphological detail, but the use of chemicals during specimen preparation precludes the use of this technique in connection with analytical studies.

21.4.3 Etching and Replication

Freeze-fracturing and freeze-etching are now a standard primary preparative technique for ultrastructural investigations at the molecular level. A variety of different instruments and methods have been described since the technique was introduced 25 years ago by Russell Steere, but the basic procedure remains the same. Rapidly frozen specimens are fractured at low temperatures in a high vacuum and a heavy metal–carbon replica is made either of the fracture face or of the fracture face from which a surface layer of water has been removed by sublimation.

It should be stressed that the conventional freeze-fracturing technique only provides replicas of fractured surfaces and while they are providing a great deal of highly significant structural information at the molecular and macromolecular level (i.e., membranes, as shown in Fig. 5), such replicas are devoid of any chemical information about the specimen.

---

prior to resin (Epon) embedding. Although the cell constituents were well preserved, the cellular shrinkage due to critical point drying had frequently resulted in separation of cell organelles. As a consequence of this the close association between the sarcoplasmic reticulum (SR) and the outer mitochondrial membrane (indicated by arrows) had been better visualized. T, T tubules; Z, Z band; A, A band; H, H band; I, I band; Mi, mitochondrion. (a) 33,000×; (b) 50,000×. Bar ≡ 1 μm. (From Dalen [25].)

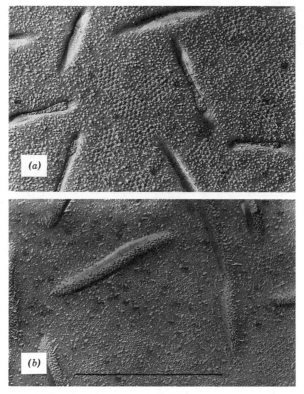

Figure 5. Yeast cell plasmalemma fractured at 12 K. (*a*) E-face; (*b*) P-face. The micrographs show defined arrays of ringlike depressions on the E-face. The hexagonal arrangement of the paracrystalline pattern can be seen directly. Some particles are plastic deformed; on the P-face the craterlike structures form a hexagonal array. However, deformed particles can be observed especially within the crystalline arrays 100,000× Bar ≡ 0.5 μm. (From Niedermyer, 1982 [26].)

21.4.4   Chemical Substitution

The rapidly cooled sample may be dehydrated by a process of low temperature chemical substitution using organic liquids. The dehydrated sample is subsequently infiltrated at low temperature with a liquid resin which is then polymerized to a solid form from which thin sections may be cut (Fig. 6). Although the main force of the investigations has been directed toward preserving the morphological integrity of the specimen, there is now sufficient evidence to suggest that freeze substitution might also be a useful preparative technique for microanalysis. The recent review by

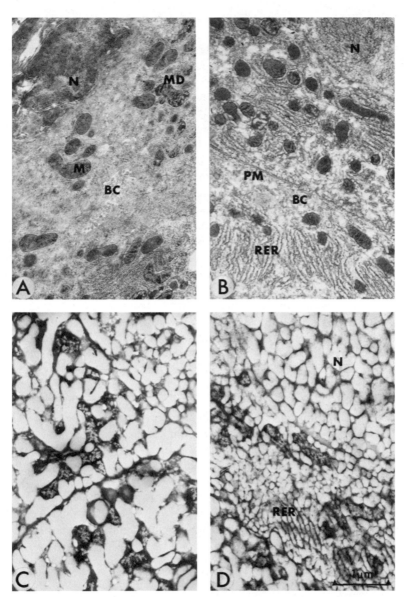

Figure 6. Electron micrographs of freeze-substituted rat liver taken at various depths from the tissue–quenchant interface. Tissue was quenched in propane at 83 K and an entry velocity of 1.5 m/s. (*A*) Shows zone 1, the surface layer. The free surface was just out of the frame at top left. (*B*) Shows zone 2, <15 μm from the free surface. (*C*) Shows zone 3, 45 μm from the free surface. (*D*) Shows zone 4, 85 μm from the free surface. BC, bile canaliculus; L, probable site of lipid droplet extraction; M, mitochondrion; MD, mitochondrion with electron-dense precipitates; N, nucleus; PM, plasma membrane; RER, rough endoplasmic reticulum. (From Elder et al., 1982 [9].)

545

Harvey [27] discusses the wide range of methods that have been applied to biological tissue.

21.4.5 Low Temperature Embedding

Although not strictly a low temperature technique, inasmuch as the samples are not initially rapidly cooled, low temperature embedding procedures are assuming greater importance in the preservation of ultrastructure while at the same time retaining enzymatic activity and immunogenicity. Conventional (ambient temperature–wet chemical) preparative techniques impose limitations on the information that may be seen using electron beam instruments. Such techniques disrupt molecular complexes and invariably result in structural deformation of proteins. This is due to organic solvent denaturation during dehydration and local heat denaturation during the polymerization of the embedding resins. High resolution details are further obscured by the deposits of heavy metals used as fixatives and stains. There is now sufficient knowledge about the chemistry of cell components to show that in order to preserve the true ultrastructure of living tissue it is necessary to optimize the dielectric constant, polarity, and pH of the solvent used during preparation. As Fig. 7 shows, the macromolecular order will only be preserved if complete dehydration is avoided, mild inter- and intramolecular cross-links of proteins is achieved and a polar environment is maintained during preparation. In addition, low temperatures favor these processes. There have been a number of different approaches to this problem, and the papers by Sjostrand [29] and Armbruster et al. [28] should be consulted for detailed application of these procedures.

It would appear that as far as molecular integrity is concerned, the low temperature embedding methods are a useful compromise between the problems of phase separation invariably associated with standard cryofixation and the disruptive procedures associated with conventional wet chemical fixation.

21.4.6 Freeze-drying

Freeze-drying is a process of ice sublimation under vacuum and may be used to remove or etch water from frozen specimens. It is by no means the ideal technique and it is necessary to balance the problems associated with the inevitable formation and growth of ice crystals with the advantage of being able to avoid the tissue coming into contact with any chemicals during the preparative process.

Freeze-drying for microscopy and analysis is a somewhat empirical

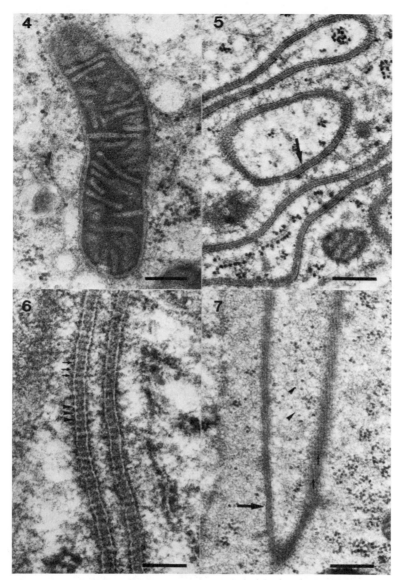

Figure 7. A, Mitochondrion from an isolated hepatocyte. Note contrast reversal of christae. Bar marker = 0.25 μm. B, Septate junction between two *Drosophila melanogaster*
epithelial cells (arrow indicates cross section). Bar marker = 0.25 μm. C, Higher magnification of part 5. Periodicities extend through the junction complex (arrowheads). Arrows
indicate septae with a bilobal substructure. Bar markers = 0.1 μm. D, A variant of a *D.
melanogaster* continuous junction (arrow indicates cross section). Striations are found in
tangential section of the membrane (small arrows). Cytoplasmic fibrils are marked by arrowheads. Bar markers = 0.25 μm. (From Armbruster et al. [28].)

547

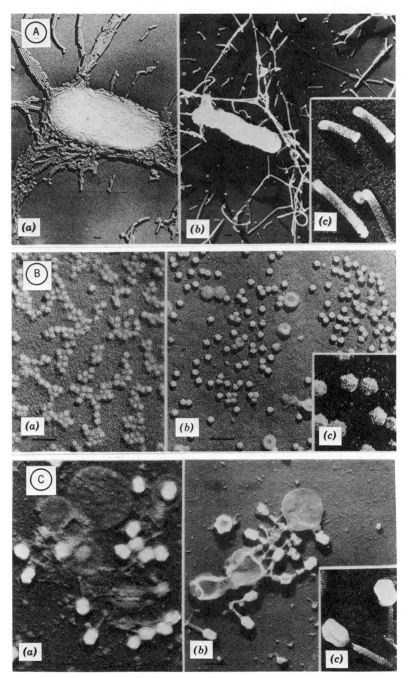

Figure 8. A, Tobacco rattle virus (TRV-Campinas strain) purified preparation mixed with bacteria: (*a*) air-dried, (*b*) freeze-dried, (*c*) higher magnification of freeze-dried particles standing on end. (*a*) and (*b*) 30,000×; (*c*) 170,000×. B, Lucerne transient streak virus

process and it is impossible to set out a protocol that will work for all specimens. Although much has been written about freeze-drying (it is an important process in the industrial preparation of drugs and foodstuffs), the recent paper by Franks [30] provides a good theoretical basis of the method.

The drying rate depends on a number of factors, including specimen temperature, size, and shape; the relative amounts of bound and free water in the sample; and to some extent, the vacuum pressure and partial pressure of water vapor in the drying chamber. The greatest impediment to fast drying is the object itself and in particular the resistance of the progressively increasing drying shell. Specimens dry from the outside inward and water molecules subliming inside the specimen must pass through the dried areas once occupied by ice crystals in order to be removed by the vacuum system. This dry organic shell increases in thickness as the freeze-drying proceeds through the specimen and offers progressively more resistance to the water molecules, which may undergo many collisions before reaching the surface. The drying rates for crystalline ice are slower than those measured in the metastable vitreous or amorphous ice. The drying rates for crystalline ice embedded in an organic matrix are even slower. The temperature at which drying should take place is a question of much debate. It is a compromise between structural preservation and sublimation at low specimen temperatures and a reasonable drying time (Fig. 8).

Samples should be as small as practicable, and as much as possible of the surface water should be removed. Prefixation with mild fixatives may only be used in connection with morphological studies. Penetrating cryofixatives should not be used, but nonpenetrating cryofixatives can be used, as they more readily give up their water during freeze-drying.

21.5 MICROSCOPY

Up to the present point we have been considering only the preparation of specimens using low temperature techniques. It is now necessary to consider some of the problems associated with the transfer and examination of the specimens by different forms of microscopy.

(LTSV), purified preparation: (a) air-dried, (b) freeze-dried, (c) individual particles showing faceted structure. (a) and (b) 90,000×; (c) 280,000×. C, T-even bacteriophage particles from crude plaque eluates: (a) air-dried, (b) freeze-dried, (c) selected particles showing helical structure on the tails. (a) and (b) 65,000×; (c) 130,000×. Bar ≡ 100 nm. (From Roberts et al., 1981 [31].)

21.5.1 Light Microscopy

The images obtained by light microscopy are usually evaluated qualitatively by observing and recording changes in morphology and texture together with differences in the phases of the liquid and solid components, the various manifestations of ice crystal formation, and cell injury (Fig. 9). Light microscopy allow us to study the dynamic phases of these processes and can give a real-time assessment of variations in cooling rates, shrinkage, volume changes, and crystallization. Image contrast is usually not a problem and considerable advances have been made in linking video recording systems to cryomicroscopes followed by digital computer processing of the images. The gross effects of ice crystal damage are readily observed, but the recognition, interpretation, and categorization of many of the consequences of sample freezing are difficult and frequently inaccurate.

21.5.2 Transmission Electron Microscopy

The main problem with visualizing frozen sections in the transmission electron microscope is the inherent lack of specimen contrast. This problem can be fairly easily resolved by introducing high atomic number elements into the sample—so-called staining. This may be achieved in the liquid or vapor phase and may result in positive or negative contrast. Alternatively, the image contrast of unstained sections can be increased by using dark field microscopy and the so-called Z contrast available in STEM instrument, as discussed by Carlemalm and Kellenberger [33].

The problem is even more complicated in unfixed frozen-hydrated sections where the presence of ice in the specimen gives rise to multiple electron scattering with a consequent degradation of image quality. The same effect is seen in images of frozen cell suspensions maintained in a microcrystalline ice matrix. This low contrast in frozen-hydrated specimens is of course due to the small mass differences between the aqueous and nonaqueous phases. There is considerable improvement in contrast when a frozen-hydrated specimen is frozen-dried and the low contrast of frozen-hydrated sections is accepted as being a good indicater that the samples contain ice.

12.5.3 Scanning Electron Microscopy

Fracture faces of frozen-hydrated bulk material are readily examined in the SEM (Echlin, [34]). The image quality is improved if a surface layer of

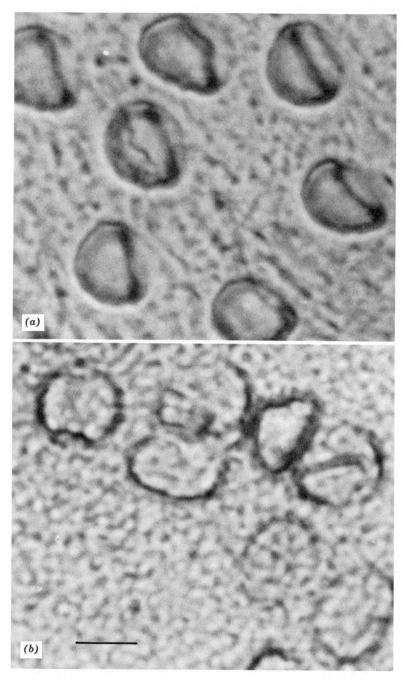

Figure 9. (*a*) Extracellular freezing of human erythrocytes. Scale bar ≡ 5 μm. (*b*) Intracellular freezing of human erythrocytes. Scale bar ≡ 5 μm. (From Diller [32].)

water is removed by sublimation, but this is now not considered to be absolutely necessary. Echlin et al. [35] have shown that good secondary electron images may be obtained from frozen-hydrated fracture faces which have been coated with gold. Acceptable images can also be obtained from carbon coated frozen-hydrated fracture faces (Echlin et al. [20]).

When examining new material it is probably useful to slightly etch the frozen hydrated fracture face in order to increase the quality of the image. However, with experience, it is possible to recognize the main features of cells in unetched samples. Contrast enhancement can be achieved by mixing the secondary and backscattered signals (Fig. 10).

It is much easier to obtain images of frozen samples using the scanning electron microscope in one or more of its reflective modes of operation. In contrast to sections, fracture faces are much rougher, and it is usually necessary to take stereopair photographs to interpret the surface properly. Many of the artifacts associated with sectioning are not to be seen on fractured faces of frozen samples. Provided the fracture is a real fracture and the knife is not allowed to scrape across the surface, knife marks are not apparent.

21.6   ANALYSIS

Most of the analytical studies that have been carried out on a frozen specimen have made use of x-ray microanalysis. The advantages and limitations of this technique are discussed elsewhere in this book and the processes and problems of specimen preparation are identical to those associated with morphological studies.

The analysis may be carried out on frozen sections or frozen-fractured bulk material. There are advantages and disadvantages to both approaches and both methods can be complementary; neither is mutually exclusive.

21.6.1   Thin Sections

The inherent difficulties of the thin section method center on the problems of tissue preparation and subsequent cell identification in the fully hydrated state, the ease with which the specimens can rapidly become dehydrated, and the relatively low mass of elements in the thin sections that are used. These disadvantages are, in many instances, outweighed by the improved spatial resolution (25–100 nm) compared to that which may be

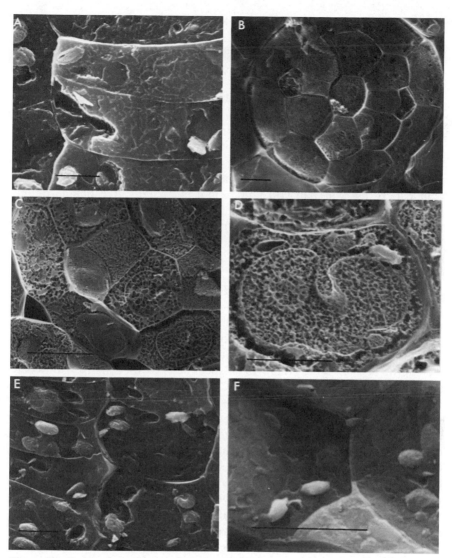

Figure 10. Frozen hydrated fracture faces of *Lemna minor* root tip. (A) Cortical region fractured in L.S. showing nucleus, chloroplasts, and starch grains. (B) Stelar region, slightly etched to reveal more of cell contents. (C) Stelar region, deep-etched, showing ice-crystal distribution. (D) Cortical region, frozen-dried; although there is extensive ice crystal formation, details of a mitochondrion can be seen (lower light). (E) Cortical region fractured in L.S. showing starch grains. (F) Stelar region, showing chloroplasts, and small vesicles. Marker = 10 μm. All samples fully hydrated, gold-coated, 15 kV, 100 K.

obtained from bulk specimens and the much simpler quantitative interpretation of the x-ray spectra.

21.6.2 Bulk Samples

The main advantages of using bulk frozen material are that samples are much easier to prepare and maintain in a frozen hydrated state and within the limits imposed by the reduced spatial resolution (2–5 μm), the morphological identification of tissue components is comparatively easy. Recent papers by Marshall [19] and Echlin et al. [20] give details of the progress that has been made with the analysis of bulk frozen samples, and the papers by Saubermann et al. [36], Bulger et al. [37], and Zierold [38] cover the recent advances in the analysis of thin sections.

21.7 SPECIMEN DAMAGE

An unfortunate consequence of irradiating an organic sample with a high energy beam is that the process of information transfer damages the specimen. This fact has been known for some time, but it has only been realized in the last few years that frozen specimens or specimens maintained at low temperature appear to suffer less damage than specimens maintained at ambient temperature. This effect is apparent irrespective of whether the specimen was prepared using the low temperature methods discussed earlier in this chapter or using ambient temperature-wet chemical methods. This apparent amelioration of the effect of beam damage, however, should not be taken as a signal that one can disregard these effects when working at low temperatures. Much of the energy transferred by the beam to the sample will eventually end up as heat. A temperature rise of 50–60 K could signal the onset of volatilization or organic material, phase transformation of crystallites, and a redistribution of solutes in the sample. Although the thermal conductivity of ice (5–7 W/m/K) is low compared with most metals (400–500 W/m/K), it is significantly higher than most organic materials (0.1 W/m/K). Talmon [39] has calculated that the thermal conductivity of frozen hydrated tissue (70% water, 30% organic matter) and finds a value of 2.4 W/m/K. All the evidence would suggest that in the transmission electron microscope, the STEM and the SEM, the thermal conductivity of the frozen specimens would limit temperature rises to between 15 and 20 K. The primary interactions of an electron beam with a specimen are excitation, ionization, and displacement of the constituent atoms. Because organic material is generally made up of light elements, the displacement of atoms is a relatively rare event, especially as most studies are carried out in the 15–100 KeV range. The excitation and ionization of atoms give rise to some of the signals we use in imaging and analysis (i.e., x-ray microanalysis, cathodoluminescence) and to secondary effects that can damage the specimen. Such secondary effects include heating, charging, phase transformations, bond rupture, cross-linking, and mass loss of both organic and inorganic constituents. By working at low temperatures, in some cases as low as 4 K, many of these deleterious effects are lessened.

21.8 CONCLUSION

There can be little doubt that low temperature specimen preparation, examination, and analysis can diminish the amount of damage brought about in organic samples by interaction with an electron beam. Such techniques avoid contact with deleterious chemical procedures and can be used in conjunction with all the analytical procedures discussed elsewhere in this book.

REFERENCES

1. Franks, F. (1982). In *Water: A Comprehensive Treatise.* Vol. 7, *Water and Aqueous Solutions at Sub-zero Temperatures,* Plenum Press, New York, 1982, chap. 3.
2. Fletcher, N. H. (1971). Rep. Prog. Phys. *34,* 913–993.
3. Hobbs, P. V. (1974). *Ice Physics,* Clarendon Press, Oxford, 1974.
4. Falls, A. H., S. T. Wellinghoff, Y. Talmon, and E. L. Thomas (1983). J. Mater. Sci. *18,* 2752.
5. Franks, F., and H. Skaer (1976). Nature *262,* 323.
6. Franks, F., M. H. Asquith, C. C. Hammond, H. Skaer, and P. Echlin (1977). J. Microsc. *110,* 223–238.
7. Echlin, P., H. Skaer, B. O. C. Gardiner, F. Franks, and M. H. Asquith (1977). J. Microsc. *110,* 239–255.
8. Skaer, H., F. Franks, M. H. Asquith, and P. Echlin (1977). J. Microsc. *110,* 257–270.
9. Elder, H. Y., C. C. Gray, A. G. Jardine, J. N. Chapman, and W. H. Biddlecombe (1982). J. Microsc. *126,* 45–61.
10. Knoll, G., G. Oebel, and H. Plattner (1982). Protoplasma *111,* 161–178.
11. Van Harreveld, A., and J. Crowell (1964). Anat. Rec. *149,* 381–386.
12. Heuser, J. E., T. S. Reese, M. J. Dennis, Y. Jan, L. Jan, and L. Evans (1979). J. Cell Biol. *81,* 275–300.

13. Escaig, J. (1982). J. Microsc. *126*, 221–229.

14. Bald, W. B. (1983). J. Microsc. *131*, 11.

15. Hemming F. J., Mesguich P., Morel G., and Dubois P. M. (1983) Cryoultramicrotomy versus plastic embedding. J. Micros. *131*, 25–34.

15a. Sitte, H. (1982). Proc. Eur. Cong. Hamburg, Vol. 19 Deutsche Gesellschaft fur Elektronenmikroskopie e.v D-6000 Frankfurt/Main 90.

16. Beeuwkes, R., A. Saubermann, P. Echlin, and S. Churchill (1982). *Proc. 40th Meeting EMSA*, Washington, D.C., pp. 754–757. San Francisco Press, San Francisco, California.

17. Biddlecombe, W. H., D. McEvan-Jenkinson, S. A. McWilliams, W. A. P. Nicholson, H. Y. Elder, and D. W. Demster (1982). J. Microsc. *126*, 63–67.

18. Rick R., Dorge A., and Thurau K. (1982). J. Micros. *125*, 239.

19. Marshall, A. T. (1982). Scanning Electron Microsc.—1982, *1*, 243–260.

20. Echlin, P., C. E. Lai, and T. L. Hayes (1982). J. Microsc. *126*, 285–306.

21. Beckett, A., and R. Porter (1982). Protoplasma *111*, 28–37.

22. Carr, K. E., T. L. Hayes, M. McKoon, and M. Sprague (1983). J. Microsc. *132*, 209.

23. Schmidt, D. G., and A. C. M. vun Hooydonk (1980). Scanning Electron Microsc.—1980, *3*, 653–658.

24. Pesheck, P. S., L. E. Scriven, and H. T. Davis (1981). Scanning Electron Microsc.—1981, *1*, 515–524.

25. Dalen H., Schere P., Myklebust R., and Saetersdal T. (1983) J. Microsc. *131*, 35.

26. Niedermeyer W. (1982). J. Microsc. *125*, 299.

27. Harvey, D. M. R. (1982). J. Microsc. *127*, 209–221.

28. Armbruster, B. L., E. Carlemalm, R. Chiovetti, R. M. Garavito, J. A. Hobot, E. Kellenberger, and W. Villiger (1982). J. Microsc. *126*, 77–85.

29. Sjostrand, F. (1982). J. Microsc., *128*, 279–286.

30. Franks, F. (1980). Scanning Electron Microsc.—1980 *2*, 349–360.

31. Roberts I. M., and Duncan G. H. (1981) J. Microsc. *124*, 295.

32. Diller, K. R. (1982). J. Microsc. *126*, 9.

33. Carlemalm, E., and E. Kellenberger (1982). EMBO J. *1*, 63–71.

34. Echlin, P. (1978). J. Microsc. *112*, 47–61.

35. Echlin, P., J. B. Pawley and T. L. Hayes (1979). Scanning Electron Microsc.—1979, *3*, 69–76.

36. Saubermann A. J., Echlin P., Peters P. D., and Beeuwkes R. (1981). Application of scanning electron microscopy 10 x-ray analysis of frozen-hydrated sections J. Cell. Biol. *88*, 257–267.

37. Bulger, R. E., R. Beeuwkes, and A. J. Saubermann (1981). J. Cell Biol. *88*, 274–280.

38. Zierold, K. (1982). J. Microsc. *125*, 149–156.
39. Talmon, Y. (1982). J. Microsc. *125*, 227–237.
40. Bachmann, L., and W. W. Schmitt (1982). J. Microsc. *126*, 45–61.

ELECTRON ENERGY LOSS SPECTROMETRY

DALE E. JOHNSON

Center for Bioengineering
University of Washington
Seattle, Washington

22.1 INTRODUCTION

Beginning as early as the 1940s [1,2] and intensifying in the last decade [3–9], theoretical discussions and experimental results have pointed to the considerable information obtainable from transmitted electron energy loss spectrometry (ELS). With electron energy loss spectrometers now readily available for use with transmission, scanning and scanning transmission electron microscopes, the full potential of this technique in analytical electron microscopy is being exploited. From these investigations the potential of ELS for biological analysis has remained clear, while the stringent requirements on experimental technique necessary to realize this potential have become much better appreciated.

It is the purpose of this chapter to describe the technique of energy loss spectrometry, the advantages of this technique for biological microanalysis, and the limitations of the technique that determine the operating conditions necessary for optimum results. Electron energy loss spectrometry (ELS) is an analytical technique that may be applied both in specialized, nonimaging instrumentation [10] and in the electron microscope. This chapter is limited only to the discussion of ELS as a component of analytical electron microscopy.

22.2 THE ANALYTICAL TECHNIQUE

22.2.1 Energy Loss Interactions

The use of electron energy loss spectrometry is based on the experimental fact that fast electrons passing near an atom will, with some probability, interact with the atomic electrons and lose energy in the process [11]. To the extent that the excitation energy of the atomic electrons is discrete and characteristic of the atom (molecule or system) involved, a characteristic feature will be produced in the spectrum of energy losses undergone by the incident electron. A typical energy loss spectrum is indicated qualitatively in Fig. 1.

22.2.2 Instrumentation

By far the most common instrumental approach to the observation and measurement of electron energy loss spectra involves the use of magnetic and/or electrostatic fields to separate spatially (i.e., to disperse) electrons of different amounts of energy loss. The dispersing fields are typically also imaging in nature, with large, and in this case, desirable chromatic aberration.

The spectrum of energy loss electrons that is produced at the image plane of a spectrometer can be measured in two basic ways. In serial collection, an adjustable slit is used to define a narrow energy window, and this window is scanned sequentially across the spectrum. Parallel detection systems record the entire spectrum at once by using multiple

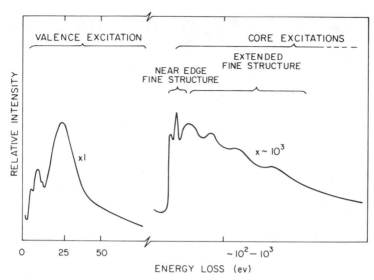

Figure 1. Features of a typical energy loss spectrum. Note the large intensity difference between valence and core excitations.

detectors. Parallel collection is inherently more efficient than serial collection and is particularly advantageous for analysis of radiation-sensitive specimens.

Electron detectors for serial collection typically involve scintillation materials coupled to a photomultiplier. At moderate count rates, the output pulses may be counted directly [12], and at high count rates, the photomultiplier output can be digitized before storage [13]. Detectors for parallel collection fall into two categories. Those that are exposed to a light image produced at an intermediate electron-to-photon conversion plate include television cameras [14] and photodiode arrays [15]. Image intensifiers may be used with light imaging systems. Alternatively, detectors such as photodiode or CCD arrays may be used directly to detect energy loss electrons [16]. Schematic diagrams of the two approaches to ELS detection are shown in Figs. 2 and 3.

22.2.3 Types of ELS Information

Valence Excitations

Dielectric Constants. The transmitted energy loss spectrum in the region from 0 to ~50 eV reflects the energy levels and transitions of valence shell electrons of the atoms and molecules condensed into the

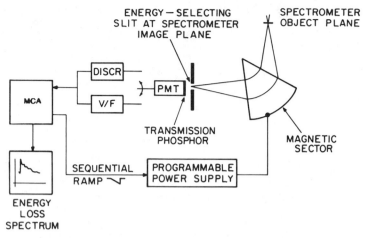

Figure 2. Schematic of a typical serial energy loss detection system.

solid. The response of this system to an applied electromagnetic field (e.g., passage of a fast electron) can be described in the dielectric theory by a complex dielectric constant $\varepsilon = \varepsilon_1 + i\varepsilon_2$. Electron energy loss spectrometry can be used very effectively to determine these dielectric constants of organic materials. The only alternative method over this energy range is the use of synchrotron radiation.

The optical constants of cytosine as derived from energy loss data are shown in Fig. 4. In addition to providing fundamental information about

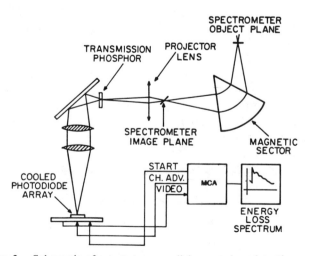

Figure 3. Schematic of a prototype parallel energy loss detection system.

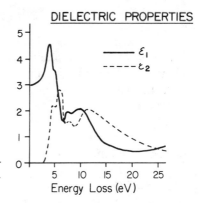

Figure 4. The optical constants of cytosine as calculated from electron energy loss data. (From Ref. 17.)

the energy level structure of these molecules, the dielectric constants obtained can be useful in determining the nature of particular energy loss events. This type of analysis has been carried out, for example, for the 20-eV energy loss peak in thin films of cytosine and has indicated the extensive single electron excitation nature of the peak, modified somewhat by collective effects [17].

Molecular Analysis. The energy loss spectrum of organic compounds in the region from 0 to 10 eV reflects the same transitions (modified slightly by collective effects) as ultraviolet and vacuum ultraviolet absorption spectra and is equally as characteristic. For example, the energy loss spectra of DNA, hemoglobin, and lecithin in this energy region are distinctly different and could be used to distinguish between regions containing different molecules.

An example of the characteristic energy loss spectrum in this energy region for a thin film of cytosine is included in Fig. 5.

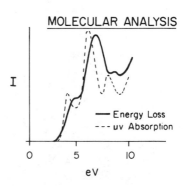

Figure 5. The low-lying energy loss spectrum of cytosine compared to UV absorption spectra.

Core Excitations

Elemental Analysis. The energy loss spectrum of organic materials in the region from a few hundred to several thousand electron volts reveals characteristic peaks due to ionization of inner shell electrons of the atoms present. The use of electron energy loss spectrometry for elemental microanalysis, because of its high geometrical collection efficiency and lack of dependence on the fluorescence yield [5], can result in a substantial increase in sensitivity compared to x-ray analysis for low atomic number biological elements ($Z < 20$) (see Section 22.5.1).

Near Edge Fine Structure. At the leading edge of energy absorption due to core electron excitations of biological materials, fine structure exists that can reflect the chemical bonding states of the atoms present. This fine structure has been explored, for example, in a group of nucleic acid base molecules and reveals carbon K energy levels that vary over a range of ~5 eV between the different carbon atoms in these aromatic hydrocarbons. An example of this type of spectral information is shown in Fig. 6.

One valuable use of this type of energy loss information might be in conjunction with elemental analysis for the determination of the relative concentrations of bound and free ions (e.g., Na^+).

Extended Energy Loss Fine Structure (EXELFS). Modulations in the energy loss spectrum past an ionization edge have been observed since at least the early 1970s [12,18]. The first limited attempt to analyze this energy loss structure was by Leapman and Cosslett [19] in studies of the Al K edge in Al and AL_2O_5. Their preliminary analysis was based on the successful explanation by Sayers, Stern, and Lytle [20] of the closely

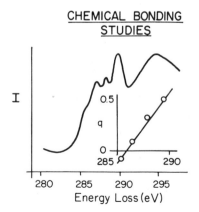

Figure 6. The near edge fine structure at the carbon K edge of the energy loss spectrum for cytosine. The insert plots the energy loss of the peaks versus the calculated charge on each carbon atom in the molecule. (From Ref. 17.)

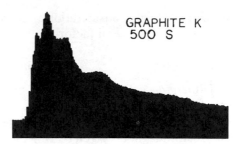

GRAPHITE K
500 S

Figure 7. Both near edge and extended fine structure at the carbon K edge in the energy loss spectrum of graphite. Full scale: 66K counts, 150 eV.

related phenomenon, the modulations beyond x-ray absorption edges. These modulations are called EXAFS (extended x-ray absorption fine structure). It was shown by Sayers et al. that the modulations in EXAFS are due to an interference between outgoing excited inner shell electron waves and waves scattered back from surrounding atoms. The modulations found in the differential inelastic electron scattering cross section, extended energy loss fine structure (EXELFS), is due to the same kind of interference. As in the x-ray absorption case, these modulations can be analyzed to provide information on the type, number, and spacing of neighboring atoms [21]. An example of this type of ELS information is shown in Fig. 7.

The extent to which this technique can be useful in the microanalysis of radiation-sensitive specimens with low concentrations of the elements of interest (the typical biological specimen) is much less clear. Since radiation damage is essentially a limit on spatial resolution (see Section 22.4.2) and the minimum useful concentration is dependent on the product of beam current and data gathering time, the question then is: Under practical microanalysis conditions, for what type of specimens and at what spatial resolution level can EXELFS provide useful information beyond simple elemental composition? These questions remain to be answered.

22.3 GATHERING OF ELS DATA

22.3.1 Factors Determining Spectrometer Performance

Several interrelated factors combine to determine the energy resolution and angular acceptance of any given energy loss spectrometer system. Ultimately the energy resolution will be determined by the detector spatial resolution (e.g., slit width or detector size in a parallel array) in relation to the dispersion of the spectrometer image in the dispersion plane (see Figs. 2 and 3). While dispersion and detector resolution are primarily design considerations, the size of the spectrometer image is determined by several factors.

Spectrometer Factors

Dispersion. Dispersion is the basic parameter determined by spectrometer design and simply indicates the spatial separation of images that differ in energy by δE. As an example the dispersion for a straight edge 90°, sector magnet is given by $D = 2R/E(\text{cm/eV})$ (R = radius of curvature, E = beam energy) [22].

To increase dispersion, the beam energy may be decreased before entering the spectrometer and then increased for detection following the spectrometer. It is also possible to increase dispersion by increasing physical size (e.g., increasing the radius of curvature of a magnetic sector spectrometer). Finally, dispersion can simply be increased by magnification of the dispersion plane with a magnetic lens.

Aberrations. Since most spectrometers in use are noncylindrically symmetric devices, they differ from ideal imaging in second and higher order terms. If all other contributions to the energy resolution are negligible, then the ultimate energy resolution is determined by second order aberrations. For example, in the case of straight edge magnetic sector spectrometers, this limit is given approximately by $\delta E/E \simeq \alpha^2$ (α = acceptance angle). The result is that high angular collection efficiency and high energy resolution in an uncorrected spectrometer are generally incompatible. This situation can be improved by correction of spectrometer aberrations [12,22,23] or by the use of cylindrically symmetric fields without second order aberrations (e.g., magnetic lenses) to reduce the angular divergence of the spectrometer object, thus reducing the effect of spectrometer aberrations [24–26].

Instrument Factors

The most fundamentally limiting instrumental contribution to the energy resolution is from the energy spread of the electron source itself ($\simeq 2$ eV/ hot filament, $\simeq 1$ eV/LaB_6, $\simeq 0.2$ eV/field emission). The specific energy resolution required for an application may in fact dictate the type of source to be used. For example, studies of "near edge" fine structure may require the use of field emission or at least LaB_6 sources.

In addition, stray ac fields and voltage and current instabilities throughout the instrument will contribute to the size of the image at the detector plane and thus degrade the energy resolution.

Post Specimen Optics

In the energy loss spectrometry of electron microscope specimens, the source of energy loss electrons can be characterized in terms of the spatial dimensions of the analyzed area and the maximum scattering angle accepted for analysis. If the area of the specimen selected for analysis is used directly as the spectrometer object, then the first order image of this area produced by the spectrometer will combine with the effect of spectrometer aberrations for the scattering angles selected, to produce the spectrometer image.

Ideally, it should be possible to select the specimen area and maximum angle of scattering based on the nature of the specimen and the type of information to be gathered and then to modify the size and angular divergence of the spectrometer object to optimize the performance of a given spectrometer. With certain limitations this flexibility is possible through the use of electron lenses that gather energy loss electrons, leaving the specimen, and focus them into a spectrometer object that differs, both in size and angular distribution, from the area and scattering angles selected at the specimen. A schematic drawing of the arrangement is shown in Fig. 8.

Choice of the Diameter and Angular Divergence of the Specimen Area Analyzed. The diameter of the area analyzed ($\equiv D$) will generally be determined by a combination of factors, including the spatial resolution

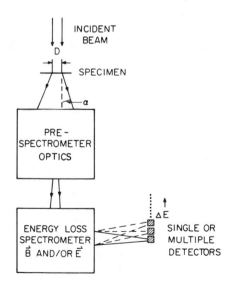

Figure 8. A schematic diagram of the role that prespectrometer optics play in matching a chosen D and α to the properties of the energy loss spectrometer.

desired, the beam current required, and specimen degradation due to radiation damage and contamination.

The angular acceptance of the system ($\equiv \alpha$), at least for elemental microanalysis, should be chosen to provide the highest peak-to-background ratio for the energy loss peak of interest. This "optimum aperture angle" is determined by the angular distributions of both the characteristic and background energy loss electrons. The angular distribution of characteristic energy loss electrons is generally proportional to $(\theta^2 + \theta_E^2)^{-1}$ with θ = scattering angle and $\theta_E = E/2E_0$ (E = energy loss, E_0 = incident energy).

This angular distribution is valid near an ionization edge where little kinetic energy is transferred to the ejected electrons. The range of scattering angles results from the fact that energy is conserved, but momentum need not be conserved between the incident electron and the scattered and ejected electrons, since momentum may be transferred to the atom.

The background energy loss electrons, however, typically result from collisions in which the ejected electrons receive kinetic energy large compared to their binding energy and thus the collisions resemble free electron collisions. This means that energy and momentum will be approximately conserved between incident and ejected electrons resulting in the distribution of background electrons being peaked at the scattering angle where this condition holds.

Because of these differing angular distributions, typical distributions for the peak-to-background ratio of an ionization edge as a function of aperture angle show a maximum at what is defined to be the optimum aperture angle ($\equiv \alpha_{opt}$) [27].

With D and α chosen as described earlier, postspecimen optics are then used to provide the best energy resolution for the given D and α. This is the subject of the next section.

Choice of Spectrometer Object—Operating Models. The principal choice of operating mode is between:

1. The use of an image of a *selected area* of the specimen as the spectrometer object ($\equiv$ IMAGE MODE) and,
2. The use of the diffraction pattern produced by a *selected angular cone* of scattering as the spectrometer object ($\equiv$ DIFFRACTION MODE).

These two operating modes have basically different characteristics that affect significantly the way in which the post-specimen lenses and the spectrometer combine to determine the final performance of the system.

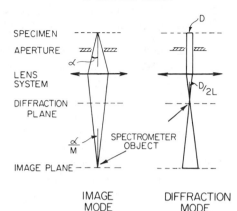

Figure 9. A schematic diagram of the two basic modes of object selection for an energy loss spectrometer.

The two operating modes are indicated schematically in Fig. 9. In the figure on the left, a small area of the specimen (diameter $= D$) is imaged by the lens system, and this image (diameter $= MD$, with $M =$ magnification) is used as the spectrometer object. The maximum scattering angle accepted α is determined by an aperture and results in an angular divergence of the spectrometer object $= \alpha/M$. In the figure on the right, electrons from a larger area (diameter $= D$) of the specimen but with a small angular divergence produce a diffraction pattern at the diffraction plane of the lens system, which is then used as the spectrometer object. The diameter of this object is $= 2\alpha L$ ($L =$ camera length) and the angular divergence is $= D/2L$. Since the spectrometer object plane is typically fixed, the excitation of the lens system is varied to place either the image or the diffraction pattern at the spectrometer object plane.

Ignoring for the moment any effect of lens aberrations, we see that in the image mode the first order contribution to the spectrometer image increases with the magnification ($= MD$) while the effect of spectrometer aberrations on the spectrometer image decreases with the magnification ($\propto M^{-2}$, since we assume second order aberrations in the spectrometer). Similarly, in the diffraction mode, the first-order contribution to the spectrometer image increases with the camera length ($= 2\alpha L$) while the effect of spectrometer aberrations decreases as L^{-2}.

In both the image mode and the diffraction mode, then, optimum values of M and L exist, respectively, that minimize the size of the spectrometer image and thus optimize the energy resolution for a given diameter of the specimen area analyzed, angular cone of scattering accepted, and particular energy loss spectrometer.

22.3.2   Specimen Preparation for ELS

There are two main points to be made regarding the preparation of specimens for energy loss spectrometry. These are:

1.   Because of the limitations of multiple scattering, energy loss specimens must be thin. The definition of *thin* in this context is open to some discussion, but it is certainly true that if a specimen is too thick for high resolution conventional transmission EM, it is too thick for ELS. In more quantitative terms, the optimum specimen thickness for ELS can be defined as that projected mass thickness ρt (ρ = density, t = thickness) that, at a given accelerating voltage, maximizes the probability that an energy loss event of interest occurs *and* no other scattering event also occurs. This is approximately equal to the mean free path for all scattering events. For example, this optimum thickness in carbon at 80 KeV is approximately 350 Å.

2.   The specimen preparation procedures must be adequate to preserve the detailed information that ELS is capable of providing. This is an obvious but very important point. For example, to use EXELFS analysis in biological specimens, rapidly frozen hydrated sections may be necessary to preserve the structural detail to be studied.

22.4   SENSITIVITY OF ELS

22.4.1   Spatial Resolution

In principle, and for sufficiently thin specimens, the spatial resolution of ELS can be reduced as the diameter of the incident electron beam is reduced. This continues until the long-range nature of the inelastic interaction provides a fundamental limitation to spatial resolution [28]. In practice, however, two other factors limit the spatial resolution of ELS in biological applications. These factors are counting statistics and radiation damage.

Counting Statistics

As with any component of analytical electron microscopy, one determinant of the spatial resolution possible will always be the limit imposed by the requirement of adequate beam current. The beam current necessary will be a function of the analysis time available, the concentration of the analyzed unit (atoms, molecules, etc.), the probability of the energy loss event of interest, and the signal-to-noise ratio required.

Once these factors have determined the minimum beam current necessary, the minimum size of the area analyzed will be determined principally by the electron source available and the electron optical system. For the necessary beam currents of $\lesssim 10^{-9}$ A, field emission sources should provide the best spatial resolutions of $\lesssim 5$ nm, while for larger required beam currents, hot filament sources should provide the best spatial resolution ranging up to ~ 1 μm resolution for $\sim 10^{-6}$ A beam current.

Radiation Damage

Depending on the radiation sensitivity of the specimen under study, in many cases it may not be possible to reduce the incident beam diameter to that determined by counting statistics. This is simply because for smaller beam diameters the specimen will not provide adequate ELS information before it is damaged to an extent that increased irradiation results only in reduced signal to noise for the energy loss peak of interest.

22.5 COMPARISON WITH OTHER TECHNIQUES

22.5.1 Comparison with EDS

The only comparable high spatial resolution technique involves the use of electron-beam-induced x-ray analysis. Such analysis has been used increasingly in recent years for low Z, elemental analysis of thin specimens, particularly of materials of biological interest. Although this technique has produced useful results, the sensitivity is limited by two main factors.

First, the fluorescent yield ω_K defined as the number of x-ray quanta emitted per K-shell excitation (similar definitions hold for other shells) decreases rapidly with atomic number Z. For example, using the value of ω_K for the carbon K-shell, only one out of 400 K-shell ionizations results in a carbon characteristic K-shell x-ray [29]. Even for a higher Z material, such as sodium ($Z = 11$), approximately only one out of every 40 K-shell ionizations results in a characteristic emitted x ray.

Second, this poor x-ray yield for low atomic number elements is coupled with relatively poor collection and detection efficiencies of most microprobe x-ray detectors. In thin specimens of the type used in transmission electron microscopy, the x rays are emitted uniformly of 4π steradians. But microprobe x-ray detectors generally subtend only very small solid angles at the specimen, and so the efficiency of collection of those emitted x rays can be quite small. For instance, the collection efficiencies of most wavelength dispersive detectors is roughly 10^{-3}–10^{-4}, and that of most energy dispersive (semiconductor) detectors is about 10^{-2} or less (with the additional qualification of less energy resolution and lower peak to background ratios) [30].

The detection of the characteristic energy loss of transmitted electrons holds particular advantage for low Z materials because of two factors. First, for each inner shell ionization, there exists an electron that has been transmitted through the specimen and lost a characteristic amount of energy in producing that ionization, regardless of the fluorescent yield, ω. That is, the yield of energy loss electrons to inner shell excitations and ionizations is unity. Second, the electrons that have lost energy in the event are scattered through relatively small angles, particularly for low atomic number elements (and correspondingly low energy inner shell levels). For instance, half of all electrons (in the 10–100 KeV energy range) that have been inelastically scattered and lost an amount of energy E in the process are scattered into angles smaller than about $2E/E_0$, where E_0 is the incident electron energy. Therefore, an electron spectrometer with a limited acceptance angle can still collect an appreciable fraction of the transmitted energy loss electrons.

This increase in collection efficiency over x-ray techniques can then be translated directly into an increase in elemental sensitivity or, for detection of a given concentration, into an increase in spatial resolution. The improvement in spatial resolution is possible because with improved collection efficiency less incident current (and thus smaller probe sizes) can produce the same count rate. If this increased collection efficiency is then combined with the ability to maintain high currents in small electron probes (e.g., field emission electron sources), elemental analysis at high resolution (~20 Å) appears promising. Of course, the ultimate spatial resolution obtainable in a microanalysis technique depends to a large extent on the damage incurred by the specimen due to the action of the incident electron beam.

22.5.2   Comparison with ESCA

The detection of x rays in the analysis of low atomic number materials is also limited, since the energy resolution for semiconductor detectors is just barely sufficient to separate the K-shell lines of carbon, nitrogen, and oxygen. Although diffractive detectors have higher energy resolution than semiconductor detectors, they are two or three orders of magnitude less efficient.

However, using transmitted energy loss electrons to detect inner shell excitations, the energy resolution can be quite good. This high energy resolution is not only of interest in the analysis of elements that have inner shell levels close to one another but can also provide information on the chemical binding states of the elements present.

In this regard, the energy loss technique becomes complementary to the well-known ESCA technique. In ESCA, one illuminates a sample with monoenergetic x rays and detects the photoelectrons emitted from the sample. The emitted electrons are characteristic of the elements present, and one obtains sharp peaks that correspond to the binding energies of the electrons in their respective shells in the constituent atoms of the sample. Shifts in the chemical state of the atoms change the binding energy and thus shift the emitted characteristic photoelectron peaks [31].

In the transmitted energy loss electron technique where the energy loss of the incident fast electron is detected, one does not have the requirement (as in ESCA) that the excited electron be emitted from the sample. That is, with this technique, one can detect electrons excited from some inner shell to a bound excited state. In fact, the characteristic energy loss spectra for inner shell excitation consist of sharp peaks that occur at the onset of the classical inner shell ionization edge and are due to such excitations. This preionization fine structure is then just the convolution of the initial and final density of states of the material and thus gives slightly different information about the chemical state of the constituent atoms from that which one obtains with the ESCA technique.

22.5.3   Comparison with EXAFS

a. EXELFS permits the extraction of fine structure information from low Z elements, which is still a difficult problem for synchrotron radiation EXAFS.

b. One can focus the electron beam to very small areas, which provides spatial resolution to study inhomogeneous samples.

c. Using the electron microscope one can image the irradiated area and also obtain the electron diffraction pattern of the sample, both of which can help to characterize the sample fully.

d. The data-gathering time is comparable to that of synchrotron sources.

e. One can study the momentum transfer dependence of the inelastic electron scattering cross section.

f. The instrumentation is more accessible and less expensive than synchrotron sources.

The subject of EXELFS in the analytical electron microscope has recently been reviewed [32].

22.6 POTENTIAL ARTIFACTS

22.6.1 Effects of Multiple Scattering

Multiple scattering events affect ELS spectra in two main ways: (1) multiple *inelastic* events degrade the energy loss spectrum in that they remove intensity from the event of interest and transfer it to higher energy loss regions of the spectrum. This transfer of intensity may reduce the visibility of an elemental edge by reducing the intensity of preionization peaks and may also complicate greatly the analysis of both near edge and extended energy loss fine structure by introducing structure mainly characteristic of multiple scattering. (2) The occurrence of any *elastic* scattering event in addition to the inelastic event of interest will reduce the effective beam current by scattering a large fraction of these electrons outside the angular acceptance angle of the spectrometer.

Because of this second factor, the energy loss signal for any given excitation reaches a maximum and then decreases again as specimen mass thickness increases. This produces particular problems in energy loss mapping. A simple example is shown in Fig. 10, in which a specimen feature is assumed to have the projected mass density profile shown. Any energy loss signal ($I_{\Delta E}$) of constant probability over this feature will, however, because of multiple scattering, map this feature into a distorted profile, as shown.

Varying specimen mass thickness can affect both the magnitude and shape of the background while multiple elastic scattering can affect the magnitude of both the background and the energy loss peak by removing intensity from the beam. In the energy loss analysis of a single region, the first effect is accounted for by fitting the background and extrapolating under the peak, and the second effect, by using a ratio of the peak area to, for example, a region of the background: This ratio is proportional to the concentration and independent of the amount of elastic scattering. Although rarely done in practice, these same two procedures are necessary for each point of an energy loss map. As pointed out by Jeanguillaume et al. [33], this may be approximated by obtaining energy loss maps using at least two energy loss intervals prior to the peak in addition to an interval containing the peak. This allows determination of both A and r in an assumed background dependence of the form $I_B(E) = AE^{-r}$.

Recent efforts to map both specific elements labeling molecules and naturally occurring elements using energy loss spectrometry have stimulated great interest in this technique. It is appreciated that mass thickness variations, as well as elemental variations, can produce contrast in energy loss maps, but the experimental techniques for recognizing and eliminat-

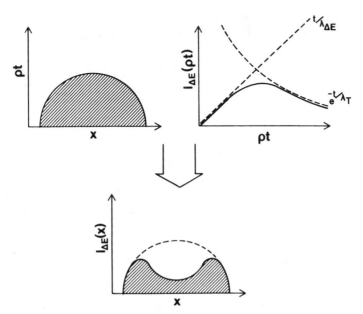

Figure 10. An illustration of the effect of multiple scattering on an energy loss signal. An object is assumed to have a projected mass density distribution as shown in the upper left, with uniform probability of energy loss, ΔE. The net effect of increasing mass thickness (ρt) on an energy loss signal is shown in the upper right, with the exponential decreasing term due to multiple scattering. (λ_Δ = mean free path for the energy loss region of interest, λ_T = mean free path for all scattering events. $I_{\Delta E}$ is the distorted energy loss signal.)

ing artifacts due to mass thickness variations are still underdeveloped and untested.

Although it is clear that ELS signals *can* produce useful elemental maps, with significantly higher signal levels than EDS, substantial questions remain regarding the validity of these maps for typical experimental techniques and specimen mass thickness variations.

22.6.2 Effects of Spatial Dispersion

One aspect of energy resolution through spatial dispersion is that any source of spatially varying intensity other than dispersion can mimic energy loss. These sources of spurious energy loss structure include variations in specimen transmission within the area analyzed if the spectrometer images this area at the dispersion plane, and variations in the angular scattering distribution of the area analyzed if the spectrometer images the diffraction pattern of this area at the dispersion plane.

The effects of the latter are most pronounced for specimens that diffract strongly, as the diffracted beams can appear as peaks in the energy loss spectrum. The presence of such spurious peaks is most easily detected on the energy gain side of the zero loss beam, but the absence of such "energy gain" peaks is no guarantee of their absence also on the energy loss side.

22.6.3   Effects of Radiation Damage

The effects of radiation damage are clearly not unique to ELS as a component of analytical electron microscopy. Only to the extent that the high energy resolution of ELS is used to ask more detailed questions than simple elemental composition (e.g., using near edge fine structure to determine chemical bonding states), is radiation damage a more severe limitation than in, for example, EDS.

The situation is one in which the same events that carry information, the inelastic scattering events, are also with some probability capable of destroying the source of the information. The signal-to-noise ratio ($P/B^{1/2}$) of an energy loss peak will initially increase with dose and then decrease again as the number of undamaged molecules decreases. The optimum dose will depend on the relative probabilities for the energy loss event of interest and for damage events that eliminate the characteristic energy loss events.

The maximum value of $P/B^{1/2}$ will depend on the magnitude of $(I\tau)^{1/2}$ (I = beam current, τ = irradiation time) at the optimum dose and thus will increase as $A^{1/2}$ (A = beam diameter), since dose = $I\tau/A$. In other words, for a given beam current, the maximum $p/B^{1/2}$ will be proportional to the beam diameter. In this sense, radiation damage can be considered as a limit to the spatial resolution of analysis: the more radiation sensitive the energy loss event, the less spatial resolution possible. This limitation will be especially important in the use of near edge or extended energy loss fine structure.

22.7   EXAMPLES OF THE APPLICATION OF ELS TO ORGANIC AND BIOLOGICAL SAMPLES

No attempt will be made here to present an exhaustive survey of all ELS applications to organic and biological materials; rather, only a few specific examples will be given to reflect the various kinds of information that have been obtained.

22.7.1 High Energy Resolution Studies of Valence and Core Electron Excitations

A detailed study of the valence shell excitations (including calculation of the optical constants) and core excitations (including near edge fine structure) of several organic molecules have been reported by Isaacson [33] and by Johnson [17]. These studies were carried out at an energy resolution of ~0.5 eV using a field emission source and using energy loss spectra and diffraction information to measure radiation damage [34]. The basic excitation information was gathered in the limit of zero damage by using very low spatial resolution.

22.7.2 Quantitative Elemental Analysis

Shuman [35] has measured the Ca content of terminal cisternae of frog skeletal muscle using the Ca L energy loss edge and compared these measurements directly with EDS measurements of the CaK x rays. The ELS measurements were made using an intensified TV-based, parallel detection system.

22.7.3 Elemental Mapping Using ELS

The area of application of ELS that probably has generated the most interest in the past several years is the use of ELS for elemental mapping. Several approaches have been pursued. Ottensmeyer [36], for example, has used an energy filter in a conventional transmission electron microscope to produce images interpreted as reflecting the phosphorus content of biological membranes. Costa et al. [37] have reported the use of a magnetic sector analyzer to form scanning transmission images with the contrast correlated with the fluorine content of labeled serotonin in the dense bodies of platelets. Recently, Shuman et al. [38] have used a magnetic sector analyzer to produce energy filter images from a conventional transmission microscope and have obtained micrographs with the contrast related to the Ca content of terminal cisternae in frog skeletal muscle.

22.8 SUMMARY

The overall characteristics of ELS are summarized in Table 1. It is clear that energy loss spectrometry, along with its high energy resolution and

TABLE 1   Energy Loss Spectrometry Summary of Characteristics

Characteristic	Implications
Advantages	
1. No dependence on fluorescent yield	High sensitivity
2. High angular collection efficiency	Microanalysis
3. High energy resolution	Energy level fine Structure information
Limitations	
1. High backgrounds peaked at large scattering angles	An optimum acceptance angle for maximum peak to backgrounds
2. Degrading effect of multiple scattering	An optimum specimen thickness
3. Degrading effect of radiation damage	An optimum dose (principally a limit to spatial resolution)
4. Aberrations of spectrometer and prespectrometer lenses	Optimum operation of prespectrometer optics in terms of magnification or camera length
5. Energy resolution through spatial dispersion	Multiple detector arrays necessary for high collection efficiency

high sensitivity, has a number of limitations that prescribe fairly tightly the operating conditions necessary for optimum results. An awareness of the limitations of ELS is crucial not only to obtain the full sensitivity of the technique but of equal importance, to avoid the production of spurious results.

With a growing awareness of both the potential and the limitations of ELS and with continuing instrumentation development, particularly parallel detection systems, ELS is now beginning to take its place as a valuable component of analytical electron microscopy in biological research.

REFERENCES

1. J. Hillier and R. F. Baker, J. Appl. Phys. *15*, 663 (1944).
2. H. Ruthemann, Ann. Phys. *2*, 135 (1948).
3. D. B. Wittry, R. P. Ferrier, and V. E. Cosslett, Brit. J. Appl. Phys. *2*, 1967 (1969).
4. A. V. Crewe, J. Wall, and L. M. Welter, Rev. Sci. Instrum. *39* 5861 (1968).

5. M. Isaacson and D. Johnson, Ultramicroscopy *1*, 33 (1975).

6. R. F. Egerton, Philos. Mag. *31*, 199 (1975).

7. C. Colliex, V. E. Cosslett, R. D. Leapman, and P. Treebia, Ultramicroscopy *1* 301 (1976).

8. D. E. Johnson, in J. Hren, D. C. Joy, and J. I. Goldstein, Eds., *Analytical Electron Microscopy,* North-Holland, Amsterdam, 1979, chap. 8.

9. D. C. Joy and D. M. Maher, J. Microsc. *114*, 117 (1978).

10. P. C. Gibbons, J. J. Ritsko, and S. E. Schatterly, Rev. Sci. Instrum. *46* 1546 (1975).

11. M. Inotkuti, Rev. Mod. Phys. *43*, 297 (1971).

12. A. V. Crewe, M. S. Isaacson, and D. E. Johnson, *Rev. Sci. Instrum. 41,* 411 (1971).

13. D. C. Joy and D. M. Maher, Ultramicroscopy *5*, 333 (1980).

14. H. Shuman, Ultramicroscopy *6* 163 (1981).

15. D. E. Johnson, K. L. Monson, S. Csillag, and E. A. Stern, in *Analytical Electron Microscopy,* R. H. Geiss, Ed., San Francisco Press, San Francisco, pp. 205–209.

16. R. F. Egerton and S. C. Cheng, J. Microsc. *127*, RP3 (1982).

17. D. E. Johnson, Rad. Res. *49*, 63 (1972).

18. C. Colliex and B. Jouffrey, Philos. Mag. *25* (1972).

19. R. D. Leapman and V. E. Cosslett, J. Phys. *D9*, L29 (1976).

20. D. E. Sayers, E. A. Stern, and F. W. Lytle, Phys. Rev. Lett. *27* 1204 (1971).

21. E. A. Stern, Contemp. Phys. *19*, 289 (1978).

22. H. H. Enge, in *Focusing of Charged Particles,* Vol. II, A. Septier, Ed., Academic Press, New York, 1967, p. 203.

23. H. Shuman, Ultramicroscopy *5*, 45 (1980).

24. A. V. Crewe, Optik, *47* 299 (1977).

25. D. E. Johnson, Scanning Electron Microsc. *1*, 33 (1980).

26. R. F. Egerton, Scanning Electron Microsc. *1*, 41 (1980).

27. D. C. Joy and D. M. Maher, Ultramicroscopy *3*, 69 (1978).

28. H. Rose, Optik *39*, 416 (1974).

29. W. Bambynck, Rev. Mod. Phys. *44*, 716 (1972).

30. T. A. Hall, in *Physical Techniques in Biological Research,* Vol. IA, 2nd ed., G. Oster, Ed., Academic Press, New York, 1971, chap. 3.

31. K. Siegbahn et al., ESCA, Almqvist and Wiksells Boktryckeri Ab., Uppsala (1967).

32. D. E. Johnson, S. Csillag, and E. A. Stern, *Scanning Electron Microsc. 1,* 105 (1981).

33. M. Isaacson, J. Chem. Phys. *56*, 1803 (1972).

34. M. Isaacson, D. Johnson, and A. V. Crewe, Rad. Res. *55*, 205 (1973).

35. H. Shuman, A. V. Somlyo, and A. P. Somlyo, in *Microprobe Analysis of Biological Systems,* T. E. Hutchinson and A. P. Somlyo, Eds., Academic Press, San Francisco (1981).

36. F. Ottensmeyer, *Proc. 40th Ann. Meeting EMSA,* Washington, D.C., 420. Claitor's Publishing, Baton Rouge, LA (1982).

37. J. L. Costa, D. C. Joy, D. M. Maher, K. Kirk, and S. Hui, Science *200,* 537 (1978).

38. H. Shuman, *Proc. 40th Ann. Meeting EMSA,* Washington, D.C., 416. Claitor's Publishing, Baton Rouge, LA (1982).

ELECTRON- AND PHOTON-STIMULATED DESORPTION

MICHAEL L. KNOTEK

Sandia National Laboratories
Albuquerque, New Mexico

23.1 INTRODUCTION

A primary concern of surface science is the understanding of the surface chemical bond. A wide array of spectroscopic tools is being employed in this pursuit, including several electron spectroscopes that attempt to detect chemical effects on either the valence or core levels of surface atoms. These highly successful techniques form the foundation of modern surface science studies. They are based on the energy and angular analysis of electrons leaving the surface after excitation by radiation in the form of electrons, photons, or ions. The makeup of a surface or the environment of an atom are deduced from the energy and angular spectrum of the emitted electrons. These techniques derive their inherent surface sensitivity from the fact that electrons in the energy range of 1–1000 eV have a very short mean free path, of the order a few lattice spacings.

A completely different perspective can be obtained by analysis of the atomic or molecular species leaving the surface as ions or neutrals. Techniques based on detection of ions include secondary ion mass spectroscopy (SIMS), electron-stimulated desorption (ESD), and photon-stimulated desorption (PSD). Electron- and photon-stimulated desorption are conceptually very simple techniques. The surface of a material is bombarded with low energy electrons or photons, which results in the desorption of either adsorbed species or elements of the surface itself. These techniques require high or ultrahigh vacuum techniques. Typically, the mass, kinetic energy, angular distribution, and threshold for desorption for desorbed ions and neutrals are measured. SIMS is based on the recoil scattering ions from a surface by an ion beam of intermediate energy (of the order 10^2–10^4 eV) and the physical processes that are responsible for the desorption of the ions have both kinematic and electronic components [1]. To the extent that SIMS is based on an electronic excitation of the surface, the discussion of the electron and photon-stimulated desorption presented here is applicable to it.

Although little ESD/PSD work has been done on organic surfaces per se, many examples exist that have a direct application to such systems due to the similarity of chemistry involved. These include studies of hydrogen, CO, NH_3, and various organic molecules on surfaces. In this paper we emphasize these examples to provide a picture of the general applicability of these techniques to organic surfaces.

ESD and PSD have long been recognized as proceeding via an electronic excitation of the surface [2–4]. Typically, the energy of an electron must exceed 3×10^5 eV before atoms in a solid can be displaced by an elastic or kinematic mechanism (billiard ball effect). In the experiments described here we will be concerned with ionizing radiation ranging in energy from ~10 to 10^4 eV. In this energy range we find that both valence and core levels can be efficiently ionized and that both kinds of excitation can lead to bond breakage or the desorption of ions or neutrals.

The history of ESD/PSD goes back to the days of electron tubes, which then formed the basis of the electronics industry [5]. It was felt that the electron beam caused desorption of positive ions from anode surfaces, which caused the oxide coatings of the cathode surfaces in the electron tube to degrade with time. Consequently, the early experiments were concerned with simulating the environment of an electron tube. These experiments concerned the desorption of various contaminant species from the metals or oxides commonly employed as anodes.

It was not until the next generation of experiments that an attempt was made to use ESD as an analytical tool to study the adsorption of species on clean surfaces [6–8]. A number of experimental observations form the basis for the eventual understanding of both the mechanisms of desorption of species from surfaces and the analytical power of desorption techniques.

Desorption cross sections for species from surfaces are first order in beam current and range from 10^{-17} to 10^{-23} cm^2 [6], considerably smaller than those for comparable gas phase dissociation, which lie in the range of 10^{-16} cm^2. Ion desorption cross sections are a factor of 10^2 or more smaller than neutral cross sections. The presence of the surface dramatically reduces the excited state lifetime, by absorbing energy from excited species, so that processes observed in molecules in the gas phase are quenched on the surface. In gas phase molecules, energy is trivially localized and is randomized in the vibrational modes of the molecule by "internal conversion" [9]. Desorption cross sections from surfaces are reduced relative to molecular dissociation by a large factor $e^{-\beta t_c}$, where β is the inverse lifetime of the excited state and t_c is the critical time necessary for the evolution to the desorptive state to occur.

The desorption yield is extremely sensitive to the details of the bonding to the surface; for example:

1. When adsorbing oxygen on Mo or W, desorption from maximal valency MoO$_3$-like units on the surface is orders of magnitude higher than for the lower oxide, for example, MO$_2$, phases [3,11–13]. An additional important observation is a strong preferential desorption of the electronegative atoms (e.g., O in an oxide), as opposed to the metal atoms, from compound surfaces [14].
2. The desorption yield for CO$^+$ from the β-precursor (bridge bonded) state on tungsten is much smaller than for the linearly top-bonded α-CO phase, while the O$^+$ yield in the β-presursor is higher than the α-CO [15,16].
3. The H$^+$ yield from H$_2$/W(100) goes through a distinct peak in yield with increasing coverage. The yield for the high coverage β_1-state is much smaller than that for the low coverage β_2-state [7,18–20]. The strength of H$^+$ ESD correlates with the C(2 × 2) LEED intensity, suggesting that the high ESD yielding sites are twofold reconstructed and the low yield sites twofold relaxed [20].

Ion energy distributions (IEDs) display peaked structures with peak energies ranging from 1 to 8 eV. Energy distributions extend as high as 15 eV. IEDs are in some cases distinctly different for different chemical states of the adsorbate, allowing this measurement to be used as an indicator of bonding configuration. For example, in the desorption of H$^+$ from W after exposure to H$_2$ and H$_2$O, the H$^+$ IED for the former peaks at 1 eV, but at 4.1 eV for the latter [21]. The former is due to an H–W bond, the latter is H in OH. Hence, a simple IED measurement can readily reveal the relative amounts of each. Figure 1 shows a recent example of a family of ion energy distribution curves (IEDs) for H$^+$ from Ta as a function of the energy of the exciting electron. The two distinct contributions in the IEDs are presumably from H in hydride and hydroxide states. The variation in their relative sizes is due to the fact that the different chemical states have different desorption thresholds.

Desorption yields show distinct thresholds with the energy of the exciting particle. In the desorption of neutrals and some ions the thresholds lie in the energy range of single one-electron excitations of the valence levels

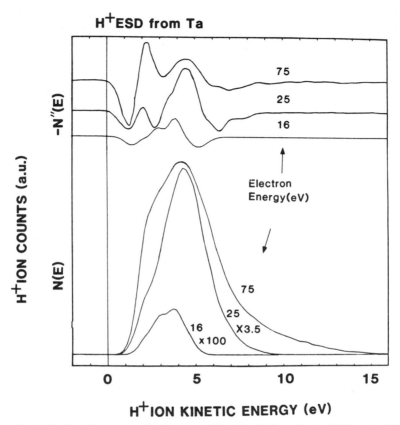

Figure 1. A family of ion energy distributions (IEDs) for H$^+$ from Ta, in $N(E)$ and $-N''(E)$, as a function of the energy of the exciting electron showing at least two different peaks in IEDs due to two different chemical states. Variations in relative strength are due to differences in desorption yield spectral dependence. (From M. M. Traum, unpublished, by permission.)

[22,23]. Most positive ion desorption thresholds lie in energy ranges corresponding to complex multielectron excitations, or ionizations of core levels [24–26]. For example, Fig. 2 shows typical ESD desorption thresholds for O$^+$ and H$^+$ from TiO$_2$ compared to a low energy electron loss (LEELs) spectrum from the same surface. The desorption thresholds correlate with the O(2s) and Ti(3p) core excitation thresholds, while no desorption occurs at the valence excitations. Desorption spectra show complex structures both in the region near threshold and well above threshold in the extended fine structure region [27,28]. As we show below, desorption spectral measurements can provide great insight into the electronic and structural features of the surface bond.

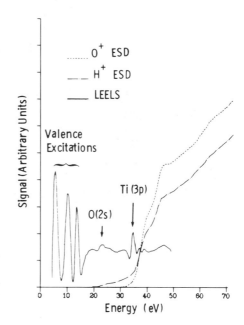

Figure 2. ESD desorption spectra for H⁺ and O⁺ from TiO₂ and the low energy electron loss spectrum (LEELS) from the same surface. Desorption occurs by excitation of O(2s) and Ti(3p) core levels. Spectra indicate H bonded to both O and Ti. O⁺ desorbed when its bonding site Ti is excited. (From Ref. 24.)

When the PSD or ESD ion angular distributions (PSDIAD or ESDIAD) of desorbed ions are measured, distinct lobes of emission are observed [16,29–31]. The azimuthal directions of these lobes or cones are correlated with the symmetry of the substrate and the angle relative to the surface normal reflects the bonding angle on the surface. For example, in studies of H₂ and O₂ on the fourfold W(100) and O₂ and CO on the threefold W(111) surfaces, lobes are observed due to both normal emission and off-normal emission with the same symmetry as the substrate. The observed patterns are generally superpositions arising from the adsorbate being bound to the surface in different bonding configurations. The angular widths of the lobes are sensitive to the degree of excitation of the bending and stretching modes of the bond, becoming sharper at lower temperatures [29]. An important feature of ESDIAD is that the appearance of distinct patterns does not require long-range order in the adsorbed layer, but does require azimuthal and planar alignment of equivalent bonding sites. A striking example of the use of ESDIAD to determine surface structure is shown in Fig. 3 [32]. On a clean Ni(111) surface, NH₃ is bonded through the nitrogen and is randomly oriented azimuthally. When oxygen is predosed on Ni(111), the NH₃ molecules orient by hydrogen bonding to the oxygen yielding a lobed ESDIAD pattern. The data allow a derivative of the bonding structure shown. Similar effects have been observed on Al(111) [33].

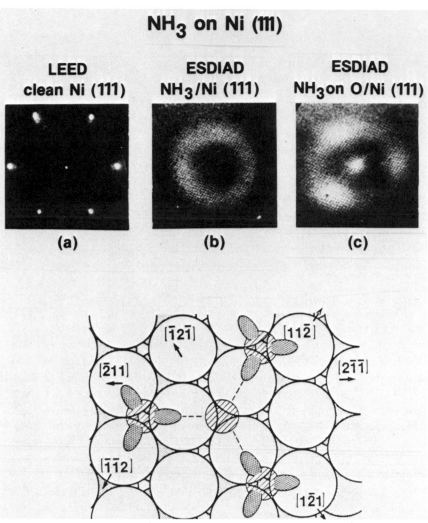

Figure 3. LEED and ESDIAD images for the Ni(111) surface when NH$_3$ is adsorbed on clean surface and one predosed with oxygen. On the clean surface the NH$_3$ is azimuthally ramdom yielding a halo in ESDIAD while presence of oxygen on surfaces causes aorientation of NH$_3$ by hydrogen bonding. Deduced structure is shown. (From Ref. 73.)

23.2 MECHANISMS OF DESORPTION

Much emphasis has been placed on the establishment of detailed mechanisms for desorption from surfaces in recent years. To understand the desorption process, we must describe first the physics of the electronic excitation of the surfaces and then the properties of the excited state that facilitate desorption. The most widely treated model is that of Menzel and Gomer [2] and Redhead [3], (the MGR model) which is quite general in its qualitative features. The MGR model is a one-dimensional picture that envisions desorption as a two-step process involving electronic excitation followed by complex nuclear and electronic motions. These can lead to desorption as ions or neutrals, or recapture to the surface as shown in Fig. 4. The ionizing particle excites the electronic structure of a general bonding site–adsorbate pair from the ground state to any of several excited states. In the excited state, the nuclei are not in their equilibrium position so nuclear motion occurs along some "reaction coordinate" either toward a new potential minimum or completely away from the site if a purely repulsive state is accessed. The great reduction in ion yield for desorption versus gas phase dissociation results from the absorption of energy from the excited species by the surface. This is dramatically demonstrated in isotope experiments where heavier isotopes are more effectively recaptured or reneutralized than lighter isotopes due to the longer residence time of the heavier species [34,35].

Although it is not generally assumed that the excitation is directly related to the dissociative state, several recent studies of neutral desorp-

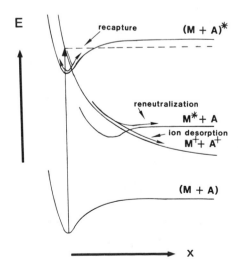

Figure 4. An energy level scheme typical of those used to discuss the Menzel Gomer, Redhead model of desorption. Curve crossings after nuclear motion can be diabatic or adiabatic leading to recapture, reneutralization, or desorption.

tion have suggested a correspondence between excitation thresholds and specific excitations between known energy levels. Such excitations typically lie in the energy range below 20 eV. Specifically, Rubio et al. [22] has observed a desorption threshold of ~5 eV for CO neutrals from W(110), with a peak in desorption at 8–9 eV, which is attributed to the $5\sigma \rightarrow 2\pi$ transition in CO. Feulner et al. [23] measure the neutral desorption of N_2 and CO from Ru(001) and conclude that the most likely candidate for the observed thresholds (~5 eV for N_2, ~10 eV for CO) is the charge transfer excitation in the $d_\pi \rightarrow 2\pi$ backbond. Both of these excitations are of levels strongly involved in bonding to the surface.

There is a second general mechanism for desorption that involves the excitation of core rather than valence levels. Although it has since been generalized to a wide range of chemical systems, the original idea was developed to explain several features of desorption from ionic systems [14,25]. The first of these is a general observation that it is most often the anion species that is desorbed from the surface. Furthermore, it is desorbed as a positive ion. Thus, while oxygen is bonded in a nominally O^{2-} state and fluorine in an F^- state, both are found to desorb as *positive* ions, which implies that there is a large charge transfer in the desorption process, up to three electrons in the case of the $O^{2-} \rightarrow O^+$ transformation. The second observation is that the desorption of positive ions is observed to have a threshold at the core level ionization potentials of either the desorbed species or its bonding site atom as illustrated in Fig. 2.

Consider a model maximal valency ionic compound MA. Maximal valency means that the cation is nominally ionized down to the noble gas configuration (e.g., K^+, Ca^{2+}, Sc^{3+}, Ti^{4+}, V^{5+}, etc., in eq. K_2O, CaO, Sc_2O_3, TiO_2, V_2O_5, etc.) and that the highest occupied level of the cation is its highest core level, of binding energy ≥ 20 eV. At the same time, suppose that the anions are negatively ionized to the noble gas configuration (e.g., F^-, O^{2-}, Cl^-, etc.). TiO_2 is the prototypical example of this electronic structure. The highest occupied level of the Ti^{4+} ion is the Ti(3p) at ~34 eV below the conduction band minimum. The valence band is almost exclusively O(2p) in character.

Figure 5 shows a simplified schematic of the process leading to desorption of O^+ when we begin with a ground state $Ti^{4+}O^{2-}$ surface configuration. If ionizing radiation removes the electron from the Ti atoms shallowest core, the predominant core hole decay will be an interatomic Auger process, due to the absence of higher energy electrons on the Ti atom. Thus a valence electron from the O^{2-} falls into the core hole and one (or two) electrons will be emitted from the anion to release the energy of the decay. The important point is that through the decay electrons are removed from an orbital that is centered on the oxygen. The loss of the

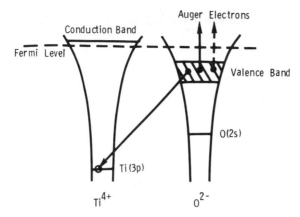

Figure 5. Schematic of desorption induced by "interatomic" Auger decay of a Ti(3p) core hole in TiO_2. In such systems electrons are always removed from anion species in Auger events resulting in highly repulsive final states and subsequent desorption.

Auger electrons transforms the oxygen to an O^+. The potential in which the O^{2-} sits before the excitation/Auger sequence is the sum of an attractive Madelung term and repulsive core overlap contribution. However, if the charge of the anion changes sign, the Madelung term becomes repulsive so that the O^+ is now in a totally repulsive potential and is driven to desorb.

There is an inherent valence sensitivity to this kind of desorption from ionic materials. When the metal (cation) is at less than maximal valency, there are valence electrons on the cation species (e.g., Ca^+, Sc^{2+}, Ti^{3+}, V^{4+}, etc.). These electrons on the cation will have two effects. First of all, the Auger decay of the metal atom core hole will have as a first step the decay of the metal atom valence electron into the metal core hole rather than from the anion. In addition, when an anion has been stripped of electrons, the electrons on first-neighbor cations provide a ready source of charge for a rapid reneutralization of the A^+, further stabilizing the surface. This simple prediction of the relative stability based on metal atom valency has been verified in three cases. First the class of oxides represented by NiO, Cr_2O_3, etc. (i.e., submaximal valency) shows no desorption, while the class represented by TiO_2, V_2O_5, WO_3, Nb_2O_5, Ta_2O_5, etc., the maximal valent compounds, shows exceedingly high desorption [25]. In the oxidation of metals such as W and Mo it is found that the initial oxidation states of the metal yield no stimulated desorption while the fully oxidized surface has a high yield [3,11,13]. Finally, in materials like TiO_2, WO_3, or Nb_2O_5 it is found that exposure to an e beam causes a reduction to a lower oxide such as Ti_2O_3, WO_2 or NbO_2 (i.e., a submaximal valent compound) after which the surface is stable [14,36,37].

We have singled out the excitation of the cation core hole because of the unique interatomic charge transfer process that occurs, but excitation of the anion core level is fully as efficient in causing desorption. Thus in TiO_2 we can excite the Ti M_{23}, M_1, L_{23}, L_1, etc., or the O L_1 or K and will observe desorption of O^+ of varying strengths, depending on the energy available from the core hole decay, which becomes important when the core hole is shallow enough that not enough energy is released by the decay to cause desorption.

The original formulation of the Auger decomposition model was to explain the desorption of ions from ionically bonded surfaces. It was soon demonstrated that ions were desorbed from covalently bonded surface complexes by essentially the same mechanism [43]. In the simplest picture the product of an Auger decay of a core hole in a covalent system is a two- or three-valence-hole final state, leaving the bond very highly ionized. The positively charged cores of the bonded atoms then see each other directly and we can envision a "Coulomb explosion" [38], as had been observed in gas phase molecules [39], where the repulsive reaction between the unscreened nuclei produce ion fragments. The important question in proceeding from gas phase molecules to molecules on a surface centers around the mechanism whereby the highly repulsive multiply ionized state is localized to the bond long enough for nuclear motion to occur. In the gas phase molecule the multiply ionized state cannot be relieved except by the fragmentation of the molecule. On a surface or in a solid, however, the solid can provide charge that neutralizes the highly ionized configuration. This neutralization must be inhibited for the desorption process to occur. A simple picture that helps us visualize how this occurs is the following.

When two "holes" exist on an atom or in a bond, there is an effective hole–hole Coulomb repulsion energy we will call U. When these two holes are created in a highly localized region the most obvious way to relieve the large repulsive energy is for the holes to move away before nuclear motion can occur. A simple effect we will now discuss often slows or completely stops the motion of the holes away from the highly repulsive configuration. This effect was first pointed out in explaining the existence of atomiclike Auger spectra in narrow d-band metals such as Ni and Cu [40][41], in insulators such as oxides and alkali halides [42], and in gas phase molecular spectra [43]. This effect makes the Auger-induced desorption process effective in covalent materials [43–45].

In any chemical system we can define a quantity, which we will call W, which is proportional to the strength of the electronic interaction between the atoms of the system. In solids, this quantity is the valence "bandwidth." In molecules it is more like the β-parameter in Hückel theory. The basic result is that if the repulsive term U is smaller than W, then the holes can separate on a time frame that is of the order of normal valence electron motion ($<10^{-15}$ s). If U is much greater than W, however, than it takes much longer for the holes to separate and the energy may stay in the bond up to $\sim 10^{-13}$ s, which is roughly the time necessary for nuclear motion to occur. Hence we have a simple criterion that can help us understand desorption in covalent systems. The important point is that we can qualitatively understand the properties of a bond that govern desorption by understanding their effect on U or W. Factors that increase U or decrease W will enhance desorption and vice versa.

U contains an intrasite Coulomb repulsion term that can be reduced by screening from the rest of the system. The intrasite (or intrabond) repulsion term is roughly approximated by the difference between the second and first ionization potentials and is inversely proportional to the ionic radius. The screening term is enhanced by higher coordination and the presence of unsaturated bonding, which allows very efficient charge transfer screening to occur. Thus we can deduce that low coordination and saturated bonding are factors that increase the likelihood of desorption. Increased coordination also increases W, which will further decrease localization. In general, W decreases with increasing ionicity [45] leading to greater desorption in ionic systems. Feibelman [46] points out in addition that due to the substantial reduction in intrasite screening when two electrons are removed from a single atom, the lowest unfilled orbitals have a considerably reduced spatial extent. Thus the overlap with wavefunctions on neighboring atoms is exponentially reduced, resulting in a like reduction in W. An additional factor in chemisorption systems is the energy difference between one-electron levels in surface and bulk atoms that can further enhance lifetimes [44].

Organic solids in general have not been widely studied using surface techniques. However, condensed multilayers and chemisorbed monolayers of similar general chemistry have been studied using ESD and PSD both because of the intrinsic interest in the physical properties of these molecules and the fact that they provide prototypical physical models for some of the ideas concerning desorption in covalent materials. Three examples provide a good picture of the factors influencing desorption in such materials. These include (1) CO chemisorbed on metals, (2) CH_3OH both chemisorbed and condensed in multilayers, and (3) condensed alkanes. We now discuss these three examples.

The adsorption of CO on metals has been one of the most widely studied problems in surface science [47,48]. This simple chemical system displays a complexity that makes study using a wide variety of techniques fruitful. For example, at temperatures below 150 K CO adsorbs on W(110) in an upright position with the carbon atom near the surface, bonded in an atom site to a single W atom. In this state, CO^+ with a smaller amount of O^+ [49] is seen to desorb in ESD. As the adsorbed layer is heated, the CO^+ yield falls and the O^+ yield increases as the system moves to the so-called β-precursor where the carbon in the CO is bridge bonded between two W atoms and the C–O bond order is reduced from ~3 to ~2.

As shown in Fig. 6, in the virgin state, increasing coverage up to ~0.5 monolayers where the CO–CO interaction becomes strong, the compression stage [50], causes the O^+ yield due to low energy excitation to go through a peak and then decrease while the CO^+ yield due to low energy excitation and the O^+ yield due to excitation of the deeper O(K) level increase linearly with coverage up to saturation [49]. Thus desorption of O^+ by low energy excitations is sensitive to the chemical state of the CO while desorption of CO^+ and O^+ due to deep core level excitation is not. PSD measurements of a similar CO on Ru(001) surface shown in Fig. 7 suggest at low energies a 3σ excitation on the CO with enhancement due to shake-up processes (the peak in PSD at ~41 eV) and 3σ-σ resonance

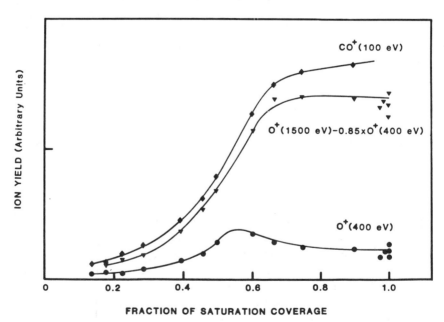

Figure 6.   The ion yields of O^+ and CO^+ at electron energies below 500 eV and the O^+ yield due to excitation of the O(K) as a function of coverage of CO. O^+ (400 eV) begins to fall off when layer enters compression stage at $\theta \sim 0.55$. (From Ref. 49.)

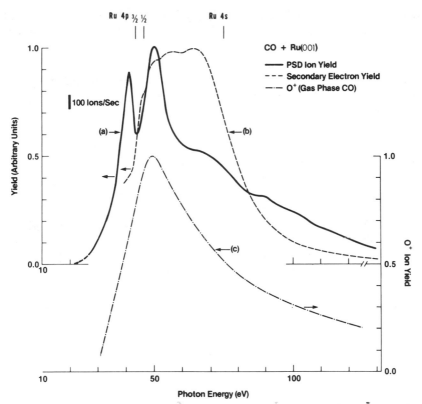

Figure 7. Comparison of PSD ion yield for CO on Ru(001) with secondary electron yield and gas phase photofragmentation. The PSD shows a peak, possibly due to 3σ + shake-up, not present in the CO (gas phase) curve. No correlation of Ru excitations is evident. (From Ref. 26.)

excitations (the PSD peak at ~50 eV) initiates desorption [51]. It is felt that an Auger decay from this initial excited state leads to the repulsive final state [49]. Also shown in Fig. 7 are the photoelectron yield from the Ru(001) surface, demonstrating that the excitation is intramolecular on the CO, not in the substrate. The gas phase CO dissociation to O^+ exhibits a spectral peak matching the 50 eV peak of O^+ PSD. Similar effects are seen on other surfaces [16,52–54]. The quenching, in the compressed layers, of the O^+ desorption by low energy excitation is presumably due to the combined effect of decreasing U and increasing W as the CO–CO interaction is increased.

Although it is known that core level excitation on the CO gives desorption [38], recent results suggest a complicated behavior. Gas phase CO displays a U of ~15 eV [55] and, in the gas phase, Auger decay of a C(K) or O(K) core hole is known to produce C^+ and O^+ efficiently [39]. Koel et al.'s Auger spectrum of CO on Ni(100), however, suggests a U of ~0 eV, indicating that the screening of the surface has totally relieved the repulsive energy of the Auger final state [56]. This may explain the observation by Jaeger et al. [57] that their threshold for desorbing O^+ at the O(K) edge for both CO and NO on Ni(100) occurs at an energy considerably higher than a simple core excitation followed by an Auger decay would predict, as seen in Fig. 8. This they attribute to a more complicated multielectron excitation. This more complicated excitation is necessary because the surface has completely screened out the simpler process. It is important to emphasize that, in detail, the desorption event in such systems can be quite complex and as yet is not fully understood.

In ESD and PSD studies of CH_3OH, both chemisorbed and condensed in multilayer on the Ti(001) surface, several important points have been demonstrated. The predominant yield from condensed CH_3OH is H^+ with heavier fragments <1% of the H^+ yield [58,59]. In isotope experiments using CD_3OH and CH_3OD it was determined that the ion yield was almost exclusively from the CH_3 moiety on the molecule. It is noted that even though there is often a striking "isotope effect" in the D^+ versus H^+ yields [35,60], with D^+ yields being much lower than H^+ due to its heavier mass, the D^+ yield from CD_3OH is dominant. This quenching of H^+ from

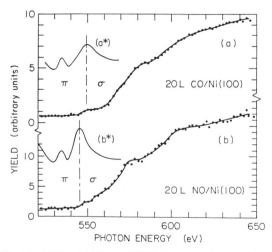

Figure 8. O^+ PSD yield of CO and NO on Ni(100) in comparison to the oxygen KLL Auger yield at the O(K) edge. The delayed onset of PSD relative to x-ray absorption (represented by the Auger yield) suggests a multielectron excitation (e.g., $1\sigma + 3\sigma$) leads to desorption. (From Ref. 74.)

the OH seems most likely due to hydrogen bonding involving the H in the OH, but may also involve the closer coupling of the oxygen to the lattice since the CH_3OH bonds to the surface through the oxygen. The hydrogen bonding may also explain the absence of larger fragments, since large fragments are observed in the condensed cyclic hydrocarbons C_6H_2 and C_8H_{16}, which do not have hydrogen bonding [61]. However, proton transfer reactions may also contribute to the quenching of larger fragments [59]. When CH_3OH is chemisorbed on Ti(001) at ~90 K, dissociation to methoxy (CH_3O-) and H occurs. No H^+ or other desorption fragments were seen to desorb from the methoxy for reasons not understood [59]. In comparative measurements of the H^+ PSD from the condensed (CH_3OH) on Ti(001) and the secondary electron yield from the Ti substrate, it was found that the excitation leading to desorption was exclusively intramolecular on the CH_3OH and did not involve the substrate Ti atoms. The desorption threshold for the H^+ from the CH_3 group in condensed CH_3OH is near the threshold for ionization of the $4a'$ orbital in methanol. This orbital is largely localized on the methyl group. A comparison of the ion yield with partial photoionization yields of the high-lying valence orbitals shows that one-electron excitation from these orbitals does not contribute to desorption of ions. These results demonstrate that slight modification in the environment of molecular species, for example, gas to condensed, or subgroups of them, as typified by the comparison of condensed methanol to methoxy, can have a profound effect on desorption processes.

The study of condensed organic species offers the experimenter a rich variation in the choice of the chemical environment of desorbed fragments. A recent ESD study of a series of condensed branched alkanes provides an example [68]. Neopentane ($C(CH_3)_4$) is the first molecule of the series. In neopentane all the carbons are in a tetrahedral environment with the central carbon having 4 carbon neighbors and the methyl carbons having 3 hydrogen and 1 carbon. Auger spectroscopy studies suggest that in neopentane the methyl carbons have a more localized state than the central carbons, the methyl carbons being effectively lower coordinated than the central carbons and, hence, less efficiently screened [43]. This suggests that excitations on the central carbon will be less efficient at causing desorption than will excitations on the methyl group. Figure 9 shows a series of ESD spectra for four condensed molecules, $C(CH_3)_4$, $Si(CH_3)_4$, $HC(CH_3)_2(CH_2D)$, and $DC(CH_3)_3$. From these molecules H^+, H_2^+, and CH_3^+ (or their isotopic analogues) are observed in desorption. For all the molecules in Fig. 9 the thresholds for H^+ and CH_3^+ desorption are equal, indicating that the same excitation can produce either fragment. Changing the central atom from C to Si results in only a small shift in threshold and minor changes in spectral shape, indicating that the

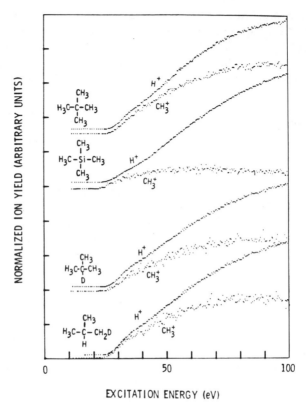

EXCITATION ENERGY (eV)

Figure 9. H⁺ and CH₃⁺ ESD yields from condensed C(CH₃)₄, Si(CH₃)₄, DC(CH₃)₃ and HC(CH₃)₂(CH₂D) as a function of excitation energy. On each molecule H⁺ and CH₃⁺ thresholds are equal. Excitations leading to desorption are of the methyl group, not central to the atom. (From Ref. 62.)

central atom is little involved in the desorption. This may be tested by removing one methyl from the central atom and replacing it with a D forming the DC(CH₃)₃. It is found that the D is not efficiently desorbed, supporting the hypothesis that the central carbon is not involved in ionic fragment production. If the D on the central carbon is exchanged with an H on a CH₃, then HC(CH₃)₂(CH₂D) is formed. From this molecule, D⁺ is observed with a threshold typical of a methyl group, but the yield is lower than an equivalent H due to the "isotope effect" mentioned above.

A comparison of the desorption thresholds of Fig. 9 to equivalent gas phase dissociation of methane [64,65] suggests that the excitation that initiates desorption is of the $3a_1$ (carbon 2s) orbital on the methyl group, which then decays by autoionization to give a localized multihole final state on the methyl group. These results can be generalized to suggest that in organic solids the highest desorption yield will be observed from terminal groups of low coordination to atoms of the solid. The desorption spectra reflect the electronic excitation of this terminal group, not the substrate. Hence the analysis of the spectra can yield information specific to the terminal group in question.

23.3   DESORPTION SPECTROSCOPY

We have examined a number of desorption spectra from the viewpoint of extracting mechanistic and/or chemical information from threshold energies and relative yields. An important point to note is that desorption processes, especially those involving Auger events or other multiple-hole final state phenomena, are intrinsically very highly localized. Hence if a given ion is observed in desorption, we have strong reason to believe, on mechanistic grounds just discussed, that either the desorbed ion or an atom in its bonding site was excited. We do not expect an appreciable yield due to second or further nearest neighbor excitations, at least for positive ion desorption. As we have discussed, the events leading to desorption consist of first an excitation, followed by electronic and/or nuclear relaxation, and then desorption. The ion desorption probability for an ionizing particle energy E, $P(E)$, then depends on these factors

$$P(E) \propto A(1 - f)\mu(E)$$

where A is the probability that an Auger or other event will lead to a repulsive final state, f is the probability of reneutralizing the ion before it escapes, and μ is the excitation probability as a function of the energy of the exciting particle. Since, to a first approximation, only μ depends on energy, the observed structure in the desorption spectrum is in most cases due exclusively to the excitation process, so the structure reflects the x-ray absorption of the excited atom. Thus desorption spectoscopy offers an atom specific, adsorbate specific, and surface specific probe of the structure of surface-bonded complexes. We now discuss the information contained in these spectra and its potential usefulness.

In the excitation of the electronic levels of atoms in molecules or solids there are spectral features intrinsic to the atom as well as those due to its environment. It is important to understand each contribution and also to contrast electron and photon excitation. The fundamental difference between the electron and photon excitations is that the electron cross section contains a Coulomb matrix element whereas the photon cross section contains a dipole matrix element [66]. A seemingly trivial but important fact is that the same excitations are involved in ESD and PSD. The different spectroscopic results reflect different excitation physics. Consider first the electron case. Following Henrich et al. a general rule can be obtained by looking at the electron excitation of He [67]. Transitions that are allowed by dipole selection rules have excitation functions that rise from zero below the threshold energy for the excitation, to a maximum at three to four times the threshold energy, and then fall off slowly with electron energy E_e as

$$f(E_e)_{\text{allowed}} \; \alpha \; \frac{\ln(E_e)}{E_e}$$

For optically forbidden transitions (nondipole), the cross section rises more rapidly to a sharp peak at less than twice the threshold energy and then falls off with increasing energy as

$$f(E_e)_{\text{forbidden}} \; \alpha \; \frac{1}{E_e}$$

Although these are only approximate functional forms, this general behavior is found to hold over a wide range of atomic and molecular species and is due to intrinsic atomic properties only. Figure 9 illustrates this general behavior. The ESD spectra have a rather abrupt threshold followed by a second more gradual increase above threshold. The first structure is the nondipole component and the second the dipole component.

The photon-excited desorption processes we consider will involve low energy photons, that is, of sufficiently low energy that the photon wavelength is much larger than the atomic structures being excited so that we need not consider photon momentum or Compton scattering (i.e., $h\nu \leq 50$ keV). Photoionization can be described in essence by a hydrogen-like model in which the electron is influenced primarily by the Coulomb field of the nucleus, with screening terms due to the other electrons [66]. In photoionization, the absorption cross section is seen to rise abruptly with increasing photon energy at the threshold energy for the excitation and then fall monotonically with photon energy $h\nu$ as

$$f(E_{h\nu}) \; \alpha \; \frac{1}{E_{h\nu}^n} \, , \; 2 < n < 3$$

Absorption edges are typically complicated by fine structure due to final state and many-body effects in both the atomic and solid state. The hydrogen-like model is seen to break down at lower photon energies where excitations occur in the outer levels of the atom.

There are several simple physical phenomena that govern the general shape of both electron and photon excitation edges. In the most inner reaches of an atom, the attractive Coulomb force is given by Ze^2/r^2 and at large r it is e^2/r^2. In the intermediate region the functional form is more of the order e^2/r^3. There is another force, termed the repulsive centrifugal force, which has the form:

$$F = \frac{l(l + 1)h^2}{mr^3}$$

for a state of angular momentum l.

The total potential due to these two contributions is

$$\phi(r) = V_{\text{Coulomb}}(r) - \frac{l(l + 1)h^2}{2mr^2}$$

which can be seen to have both attractive and repulsive regions of r and can result in a maximum in the total potential within 1 to 2 atomic units of the nucleus. This is the so-called centrifugal or angular momentum barrier, which can result in the absorption edge being effectively shifted to higher energies (the so-called delayed onset) for final states with $l \geq 2$ [66]. An example is presented in Fig. 10 where PSD of H^+ and F^+ from SiO_2 at the Si L_{23} edge and H^+ from BeO at the Be K edge are compared [62]. The overall envelope of the Si PSD, which is dominated by excitations from $p(l = 1)$ to $d(l' = 2)$ states, reaches a maximum at ~30 eV above the primary edge at 105 eV while the envelope of the Be PSD, which is dominated by $s(l = 0)$ to $p(l' = 1)$ excitation, rises abruptly at the edge and falls off monotonically with energy. The nature of the fine structure peaks will be discussed in more detail below.

In absolute magnitude, the electron and photon ionization cross section for a given subshell, where only atomic processes are considered, peak at approximately [68,69],

$$\sigma_{\text{electron,max}} \doteq \frac{1}{2} \frac{n}{E_I^2} \times 10^{-16} \text{ cm}^2 \text{ (Rydberg)}^2$$

and

$$\sigma_{\text{photon,max}} \doteq \frac{8n}{E_I} \times 10^{-18} \text{ cm}^2 \text{ (Rydberg)}$$

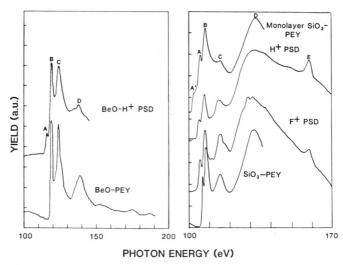

PHOTON ENERGY (eV)

Figure 10. Comparison of PSD and photoelectron yield (PEY) for BeO and SiO_2. The Si(L) edge structure shows an overall envelope typical of a delayed onset for p → d excitations while the Be(K) edge shows an abrupt edge and envelope typical of s → p. (From Ref. 62.)

where n is the number of electrons in the subshell and E_I is the ionization energy of the level in Rydbergs. The preceding forms provide only an order of magnitude estimate, but are useful in determining the relative strengths of processes.

There is a further difference between electron and photon excitation processes that occurs near threshold. Figure 11 shows an energy schematic for electron excitation of a surface. Typically hot cathode electron sources are used in ESD so that the energy of the electron is referenced to within kT of the vacuum level of the cathode. Hence, the energy of the electron relative to the Fermi level of the sample is

$$E_e = eV_{app} + e\phi_c + kT$$

If in the deexcitation process the incident electron has its final state at the Fermi level as on a metal (Fig. 11a), then the entire E_e is given up in the process and is available for exciting the surface atom. In most cases, however, there is a gap between the Fermi level and the lowest empty state in the solid, for example, in a semiconductor or insulator. Additional complications can arise in ESD when chemical modification of the surfaces results in band bending, which introduces an unknown chemical shift into the threshold energy [70].

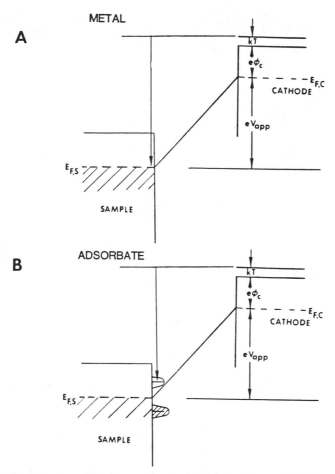

Figure 11. Energy schematic to be considered when relating observed ESD thresholds and the actual energy of the relevant excitation. ESD threshold excitation energies are complicated by uncertainties in ϕ_c and the energy of the final state of the incident electron relative to the sample Fermi level.

In some cases the final state of the electron may not be in the solid, but may reside in an adsorbed species that is weakly coupled to the solid as in Fig. 11 b. Here again the threshold for desorption for any excitation with an energy equal to E_{th} is shifted to higher energy by the energy Δ between the final state and the Fermi level so that the threshold value of the electron energy E_e is given by

$$(E_e)_{th} = E_{th} + \Delta$$

Photon-stimulated desorption has a distinct advantage in this regard in that all the photon's energy is given up in an inelastic event so that $h\nu$ at threshold is identically equal to E_{th}

$$(h\nu)_{th} = E_{th}$$

Although this uncertainty in determining the primary electrons final state can frustrate the accurate determination of threshold energies in an ESD experiment, the joint measurement of ESD and PSD thresholds can be used to define the energy of the electrons final state relative to the Fermi level in the ESD experiment, that is,

$$\Delta = (E_e)_{th} - (h\nu)_{th}$$

The fact that the final states of both the primary and excited electron lie in the same energy region leads to one further complication of ESD over PSD. In PSD or any x-ray absorption process the excitation probability $P_{h\nu}(E)$ is seen to reflect the density of final state $N(E)$:

$$P_{h\nu}(E) \; \alpha \; N(E)$$

after matrix element and symmetry effects are included. The fact that the primary electron also has a final state in the same empty state distribution mean that spectral structure above the edge reflects not $N(E)$ but a convolution of the density of empty states, that is,

$$P_e(E_0)\alpha \int_0^{E_0} N(E)N(E_0 - E) \, dE$$

where it has been assumed that the oscillator strength is constant across the empty state distribution. To the extent that this approximation is correct, the ESD spectral data above threshold should reflect the self-convolution of the density of empty final states.

Since ESD spectra do not provide a clear spectral picture of the environment of the excited atom, we discuss here the spectral features observed in PSD by deep core excitations. The historical usefulness of x-ray absorption spectroscopy has led to an extensive science in the interpretation of such spectra.

In ionic materials typified by BeO and SiO_2 the PSD and PEY spectra are marked by strong core–exciton peaks below the K and L edges which originate from excitations to a $(Z + 1)$ impurity-like level slightly below the threshold for excitations to the conduction-band continuum (the peaks marked *A* in Fig. 10). Slightly above threshold the spectra show "inner-well" resonances superimposed on continuum excitations. The volume between the central atom nucleus and the anion cage is called the "inner well." Resonance absorption in the inner well is caused by excited electrons scattering from nearest neighbor atoms and leads to a strong peak in the spectra. Although these are very strong in materials like BeO and SiO_2 they are less pronounced in more covalent systems such as in Fig. 12 where the H^+ PSD and PEY from diamond are shown [71]. In metallic systems contributions from far neighbors can be important in the x-ray absorption spectrum. There is an advantage in materials like BeO and SiO_2 because the spectrum is dominated by the first coordination shell, so that unknown PSD spectra can be modeled by bulk compounds of known

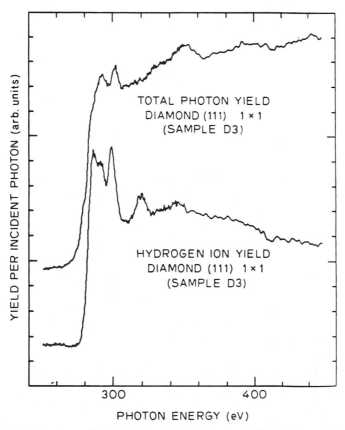

Figure 12.   A comparison of the H^+ PSD and photoelectron yield from diamond (111) 1 × 1. These spectra show a typical abrupt K-edge jump but do not show the strong "inner-well" resonances of Fig. 10. (From Ref. 71.)

environment and the unknown surface structure deduced [68]. In Fig. 9, for example, we find that, although H and F are both bonded at tetrahedral Si sites, the positions of the peaks in energy suggest that the site resembles that of a monolayer of SiO_2 on Si. In general, these near edge spectra are very difficult to analyze in detail since multiple scattering effects can be important and hence modeling is an attractive alternative.

In the energy region further above threshold, the so-called x-ray absorption extended fine structure (or EXAFS) region, the excited electron's interaction with the environment is simpler to analyze [28]. When the photoemitted electron leaves the excited atom, it is scattered by the cores of the atoms in its environment such that the wavefunction at the excited core contains a factor

$$1 + \sum_R A_{k,R} \sin[2kR + \phi(k)]$$

where k is the wavevector of the excited electron, R is the radial distance of a specific coordination shield, and $\phi(k)$ is the scattering phase shift. For a given coordination shell

$$\sum_{shell} A_{k,R} = \frac{N}{kR^2} f(k,\pi) e^{-2\sigma^2 k^2 - 2R/\lambda(k)}$$

where the first term arises from the spherical wave nature of the outgoing electron, f is a scattering amplitude, the σ^2 term in the exponential is the Debye–Waller factor, and λ is the inelastic mean free path. From these equations we see that the oscillatory structures in the EXAFS region of the spectrum can be related to the Fourier transform of the radial distribution function of the excited atom. Figure 13 shows the first example of the use of PSD to obtain EXAFS data which can be analyzed to obtain an RDF [28]. This data shows the O^+ PSD yield and the e-yield for an oxygen exposed Mo(100) surface. Analysis of this data showed: (1) The Mo–Mo separation for the excited atom is essentially identical to the bulk Mo–Mo separation. (2) Using the general form

$$N = \text{effective coordination number} = \sum_{\substack{atoms \\ in\ shell}} ((\tfrac{1}{3}) + \boldsymbol{\varepsilon} \cdot \boldsymbol{r}_l)$$

where $\boldsymbol{\varepsilon}$ is the polarization vector and $\boldsymbol{r}_l$ is the unit vector of the lth atom, the data show that the coordination of the excited Mo atom is one-half that of the bulk. (3) From the position of the threshold of this and other

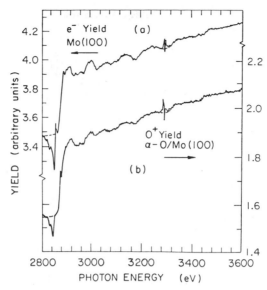

Figure 13. A comparison of O^+ and e^- yield from oxygen on Mo(100). Both spectra display EXAFS in the region from 50 to 750 eV above threshold. Analysis shows that the O-bonded Mo have half the Mo–Mo coordination of bulk Mo and that the Mo–Mo bond length is unaltered.

core edges [72] it was determined that the excited Mo was in an "oxidic" state.

23.4 CONCLUSION

The idea behind this chapter was to acquaint the reader with some of the basic concepts that have emerged recently in the stimulated desorption field. We have shown that in understanding the basic physics of the desorption process we derive a good deal of insight into surface chemistry as well as the factors governing radiation-induced damage. Stimulated desorption studies offer the experimenter a variety of measurable parameters that, due to the highly localized nature of the desorption process, provide an unprecedented specificity in deriving electronic and structural properties of the surface bond. Although these techniques were developed in pursuit of problems of historical interest to the classical surface scientist, the underlying general nature of the concepts and techniques described here promises an expansion of their use to disciplines such as the study of biological surfaces.

REFERENCES

1. P. Williams, Surf. Sci. *90*, 588 (1979).
2. D. Menzel and R. Gomer, J. Chem. Phys. *41*, 3311 (1964).
3. P. A. Redhead, Can. J. Phys. *42*, 886 (1964).
4. Y. Ishikawa, Rev. Phy. Chem. Japan. *16*, 117 (1942).
5. J. R. Young, J. Appl. Phys. *31*, 921 (1960).
6. T. E. Madey and J. T. Yates, Jr., J. Vac. Sci. Technol. 8, 525 (1971).
7. D. Menzel, Surf. Sci. 47, 370 (1975).
8. T. E. Madey, J. J. Czyzewski, and J. T. Yates, Jr., Surf. Sci. 63, 203 (1977).
9. C. Lifshitz, Adv. Mass Spectrosc. *73* (1977).
10. T. E. Madey and J. T. Yates, Jr., Surface Sci. *11*, 327 (1968).
11. D. A. King, I. E. Madey, and J. T. Yates, Jr., J. Chem. Soc., Faraday Trans. I *68*, 1347 (1972).
12. V. N. Ageev, A. I. Gubanov, S. T. Dzhalilov, and L. F. Ivantsov, Sov. Phys. Tech. Phys. *21*, 1535 (1976).
13. P. J. Feibelman and M. L. Knotek, Phys. Rev. *B18*, 6531 (1978).
14. M. L. Knotek and P. J. Feibelman, Surf. Sci. *90*, 78 (1979).
15. R. Gomer, Japan. J. Appl. Phys. Suppl. 2, Pt. 2 (1974) 213.
16. R. Jaeger and D. Menzel, Surf. Sci. *93*, 71 (1980).
17. T. E. Madey, Surface Sci. 36 (1973) 281.
18. A. Bennighoven, E. Loebach, N. Treitz, J. Vac. Sci. Technol. *9*, 600 (1972).
19. H. Jeland and D. Menzel, Surf. Sci. *40*, 295 (1973).
20. R. Jaeger and D. Menzel, Surf. Sci. *100*, 561 (1980).
21. M. Nishijima and F. M. Propst, Phys. Rev. *B2*, 2368 (1970).
22. J. Rubio, J. M. Lopez-Sancho, and M. P. Lopez-Sancho, J. Vac. Sci. Technol. *20*, 217 (1982).
23. P. Feulner, R. Treichler, and D. Menzel, Phys. Rev. *B24*, 7427 (1981).
24. M. L. Knotek, Surf. Sci. *91*, L17 (1980).
25. M. L. Knotek and P. J. Feibelman, Phys. Rev. Lett. *40*, 964 (1979).
26. T. E. Madey, R. Stockbauer, S. A. Flodström, J. F. van der Veen, F. J. Himpsel, and D. E. Eastman, Phys. Rev. *B23*, 6847 (1981).
27. M. L. Knotek, V. O. Jones, and V. Rehn, Surf. Sci. *102*, 566 (1981).
28. R. Jaeger, J. Feldhaus, J. Haase, J. Stöhr, Z. Hussain, D. Menzel, and D. Norman, Phys. Rev. Lett. *45*, 1870 (1980).
29. T. E. Madey, J. J. Czyzewski, and J. T. Yates, Jr., Surf. Sci. 63, 203 (1977).
30. T. E. Madey, J. J. Czyzewski, and J. T. Yates, Jr., Surf. Sci. 57 (1976) 580.
31. H. Niehus, Proc. of the 7th Int. Vacuum Congress and the 3rd Int. Conf. on Solid Surfaces, 1977, p. 2051.

32. F. P. Netzer and T. E. Madey, Interaction of NH_3 with oxygen-predosed Ni(111), Surf. Sci. (to be published).

33. R. Stockbauer, D. M. Hanson, S. Anders Flodström, and T. E. Madey, PSD and ultraviolet photoemission spectroscopic study of the interaction of H_2O with a Ti(001) surface, Phys. Rev. (to be published).

34. T. E. Madey, J. T. Yates, Jr., D. A. King, and C. J. Uhlaner, J. Chem. Phys. *52*, 5215 (1970).

35. T. E. Madey, Surf. Sci. *36*, 281 (1973).

36. T. T. Lin and D. Lichtman, J. Appl. Phys. *50*(3), 1298 (1979).

37. T. T. Lin and D. Lichtman, J. Materials Sci. *14*, 455 (1979).

38. R. Franchy and D. Menzel, Phys. Rev. Lett. *43*, 865 (1979).

39. R. B. Kay, Ph. E. Van der Leeuw and M. J. Van der Weil, J. Phys. *B10*, 2521 (1977).

40. M. Cini, Solid State Commun. *20*, 605 (1976).

41. G. A. Sawatsky, Phys. Rev. Lett. *39*, 504 (1977).

42. D. E. Ramaker, Phys. Rev. *B21*, 4608 (1980).

43. D. R. Jennison, J. A. Kelber, and R. R. Rye, Phys. Rev. (in press).

44. D. E. Ramaker, C. T. White, and J. S. Murday Phys. Lett. (to be published).

45. E. R. Johnson, *Radiation-Induced Decomposition of Inorganic Molecular Ions*, Gordon and Breach, New York, 1970.

46. P. J. Feibelman, Surf. Sci. *102*, L51 (1981).

47. R. Gomer, Japan. J. Appl. Phys. Suppl. 2, Pt. 2, 213 (1974).

48. L. D. Schmidt, *Topics in Applied Physics, Vol. 4: Interactions on Metal Surfaces*, R. Gomer, Ed., Springer, 1975, p. 63.

49. J. E. Houston and T. E. Madey, Core-level processes in the electron stimulated desorption of CO from the W(110) Surface, Surf. Sci. (to be published).

50. C. Steinbruchel and R. Gomer, Surf. Sci. *67*, 21 (1977).

51. T. E. Madey, R. Stockbauer, S. A. Flodström, J. F. van der Veen, F. J. Himpsel, and D. E. Eastman, Phys. Rev. *B23*, 6847 (1981).

52. P. Feulner, H. A. Engelhardt, and D. Menzel, Appl. Phys. *15*, 355 (1978).

53. T. E. Madey, Surf. Sci. *79*, 575 (1979).

54. T. E. Madey, J. T. Yates, A. M. Bradshaw, and F. M. Hoffman, Surf. Sci. *89*, 370 (1979).

55. J. A. Kelber, D. R. Jennison, and R. R. Rye, J. Chem. Phys. *75*, 682 (1981).

56. B. E. Koel, J. M. White, G. M. Loubriel (to be published).

57. R. Jaeger, J. Stöhr, R. Treichler, and K. Baberschke, Phys. Rev. Lett. *47*, 1300 (1981).

58. R. Stockbauer, E. Bertel, and T. E. Madey, J. Chem. Phys. (in press).

59. D. M. Hanson, R. Stockbauer, and T. E. Madey, J. Chem. Phys. (to be published).

60. T. E. Madey, J. T. Yates, Jr., D. A. King, and C. J. Uhlaner, J. Chem. Phys. *52*, 5215, (1970).

61. T. E. Madey and J. T. Yates, J. Surf. Sci. *76*, 397 (1978).

62. M. L. Knotek, R. H. Stulen, G. M. Loubriel, V. Rehn, R. A. Rosenberg, and C. C. Parks, Surf. Sci. (submitted).

63. D. R. Jennison, J. A. Kelber, and R. R. Rye, Phys. Rev. (in press).

64. C. G. Pantano and T. E. Madey, Surf. Sci. *7*, 115 (1981).

65. R. Locht, J. L. Olivier, and J. Momiguy. Chem. Phys. *43*, 425 (1979).

66. U. Fano and J. W. Cooper, Rev. Mod. Phys. *40*, 141 (1968).

67. V. E. Henrich, G. Dresselhaus, and H. J. Zeiger, Phys. Rev. *B22*, 4764 (1980).

68. E. J. McGuire, Phys. Rev. *A16*, 73 (1977).

69. E. J. McGuire, *The Photoionization Cross-sections of the Elements*, Vol. III: Magnesium to Potassium, Report SC-TM-67-2955, Sandia National Laboratories, Albuquerque, N.M., 1967.

70. M. L. Knotek, Surf. Sci. *101*, 334 (1980).

71. B. B. Pate, M. H. Hecht, C. Binns, I. Lindau, and W. E. Spicer, J. Vac. Sci. Technol. (Aug. 1982).

72. R. Jaeger, J. Stöhr, J. Feldhaus, S. Brennan, and D. Menzel, Phys. Rev. *B23*, 2102 (1981).

73. F. P. Netzer and T. E. Madey, Interaction of NH_3 with oxygen-predosed Ni(111), (to be published). Surf. Sci.

74. R. Jaeger, R. Treichler and J. Stöhr, Surf. Sci. *111*, 533 (1982).

ORGANIC AND BIOLOGICAL SURFACES: FLUORESCENCE MICROSCOPY

J. S. PLOEM

Department of Histochemistry and Cytochemistry
University of Leiden
Leiden, the Netherlands

24.1 INTRODUCTION

Fluorescence microscopy provides a powerful methodology for the study of biological surfaces. Immunofluorescence methods have resulted in numerous cytochemical techniques (Coons et al. [1], Nairn [2], Wick et al. [3]) for the demonstration of a large variety of biomolecules. Flow cytometry as a specialized form of fluorescence microscopy (Herzenberg et al. [4], Melamed et al. [5]) permits the examination of biological surfaces when cells pass a beam of excitation light obtained from a laser. A large number of cells can be analyzed in a relatively short period of time by this technology.

The cell surface has been studied extensively by immunological fluorescence methods as is, for example, demonstrated in the papers by Killander et al. [6], Greaves [7], Seligmann [8], Raff [9], Gonda et al. [10], Wick et al. [3].

24.2 FLUORESCENCE

The emission of light following on absorption of light is called photoluminescence and includes the processes of fluorescence and phosphorescence. Fluorescence is characterized by the fact that the emission occurs immediately after the absorption of light and lasts only 0.001–0.1 ns (Garland and Moore [11]. Phosphorescence is a much slower process.

Compounds exhibiting fluorescence are called fluorophores or fluorochromes. When a fluorophore absorbs light, the energy is taken up for excitation of electrons to higher energy states. The process of absorption is rapid and is immediately followed by return to lower energy states, which can be accompanied by emission of light. The spectral characteristics of a fluorochrome are related to the special electronic configurations of a molecule. Absorption and emission of light take place at different regions of the light spectrum (Fig. 1). According to Stokes law the wavelength of emission is almost always longer than the wavelength of excitation. This difference in wavelength enables the observation of light emitted by the fluorophore.

24.3 FLUORESCENCE MICROSCOPY

The fluorescence microscope must fulfil three main functions:

1. It must concentrate a maximum of light suitable for the excitation of fluorescence in the microscopic specimen.
2. It must collect as much as possible of the fluorescence emitted by the specimen in order to make the structures visible in the microscope.
3. It must excite with light of a wavelength close to the excitation peak of the fluorescent compound and select the fluorescence light of a wavelength close to the fluorescence peak of the compound.

All three factors have to receive full attention in the design of a fluorescence microscope. The first two factors contribute to the possibility of

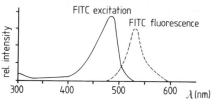

Figure 1. Excitation (absorption) and fluorescence (emission) spectra of fluorescein isothiocyanate (FITC).

visualizing relatively weak fluorescing objects. The third function enhances the contrast in the fluorescence image between the selected fluorescence of the fluorochrome to be studied and the unwanted fluorescence of other fluorescent substances in the same microscopic field.

Because modern filter technology can solve a substantial part of the special spectral requirements for a specific excitation of most fluorochromes, a considerable variety of optical components for fluorescence microscopy has been made available.

The image contrast in fluorescence microscopy depends mainly on the quality of the optical filters used. The image contrast is determined by the ratio between the fluorescence emission from the substances studied and the light observed in the background. The latter is composed of unwanted excitation light, the autofluorescence of optical parts and mounting medium. Unwanted excitation light is caused by imperfections of the filter systems used to separate excitation from fluorescence light. Filters should, however, not only have good spectral properties but also a sufficient transmission. Efforts in fluorescence microscopy aim at obtaining an optimal image contrast while maintaining a sufficiently bright image.

Reviews of fluorescence microscopy have been given by, for example, Young [12], Price [13], Trapp [14], Walter [15], Rost [16], Ploem [17] and lately Siegel [18].

24.4 THE FLUORESCENCE MICROSCOPE

The main components of a fluorescence microscope are depicted in Fig. 2. The excitation light is obtained from a light source with an arc or a filament of high intrinsic brilliancy. Light not wanted for the excitation of the fluorochrome must be removed from the excitation light path by a primary filter. This filter, also called the excitation filter, will reflect or absorb the unwanted excitation light. Concentration of the excitation light on the preparation is obtained with an illumination system such as a condenser (transmitted illumination) or an objective (epi-illumination). After the excitation light has reached the preparation a small part of this light, refracted or reflected by the specimen, will still enter the objective in the direction of the eyepieces. The unwanted excitation light must be absorbed by a secondary or barrier filter. The barrier filter should, in addition, selectively transmit the fluorescence light emitted by the fluorochrome.

If, as in some applications, more than one fluorescent compound is present in the preparation (Cornelisse and Ploem [19]), a fluorescence selection filter should be inserted as a secondary filter to isolate the fluorescence peak of the selected fluorescent substance from the total fluorescence emitted by the specimen.

24.5 LIGHT SOURCES

The choice of a light source is determined by the excitation spectrum of the fluorochrome and its quantum efficiency (ratio of energy emitted to energy absorbed). Weakly fluorescent fluorochromes (low quantum efficiency) will need excitation light of higher intensity than strongly fluorescent fluorochromes in order to be visible in the microscope. Some light sources emit more or less a continuum of wavelengths (halogen, tungsten, and xenon lamps), others have more distinct emission peaks (e.g., mercury lamps).

If emission peaks of equal height are present in both the short and long wavelength range of the emission spectrum of the light source, the shorter wavelengths—having a higher energy—will excite a stronger fluorescence than the longer wavelengths. The combination of a strong emission line in the emission spectrum of the light source and a moderate absorption in the excitation spectrum of the fluorochrome is often equally efficient as the combination of a low emission intensity of the lamp with a high absorption of the fluorochrome in the same wavelength range. This has been the reason that mercury lamps have been so successful in the excitation of the fluorochrome FITC with long wavelength blue light (480–490 nm). FITC has its excitation peak at 490 nm, far outside the location of the strong mercury lines (365, 406, and 435 nm). At 480 nm the mercury arc lamps emit a sufficiently strong continuum for the excitation of FITC.

Excitation with long-wavelength blue light has become possible by replacing the conventional ultraviolet and violet-blue transmitting colored glass filters with wide-bell-shaped transmittance curves (e.g., Schott ultraviolet UG 1 or blue BG 12) by narrow-band interference filters, transmitting only long-wavelength blue light (see Fig. 4). The relatively high transmission of this type of filter for long-wavelength blue light and the sharp cutoff of the transmittance curve toward the longer wavelength

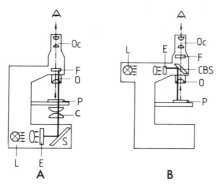

Figure 2. Schematic diagram of a microscope for fluorescence microscopy with (*A*) transmitted illumination and (*B*) incident illumination. The excitation light is depicted by solid lines and the fluorescent light by dashed lines. L, light source; E, excitation filter; S, mirror; C, condenser; P, preparation; O, objective; F, barrier filter; Oc, ocular; CBS, chromatic beam splitter.

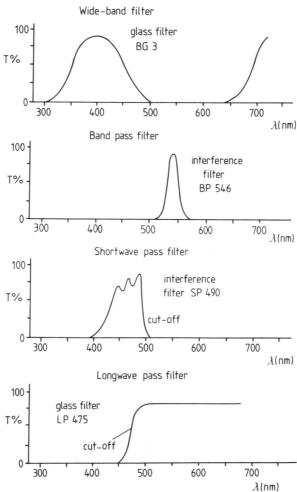

Figure 3. Transmission characteristics of filters used in FM. The X-axis gives the wavelength while the Y-axis shows the percentage transmission for each filter type.

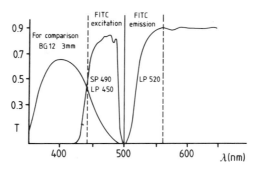

Figure 4. Transmission curves of excitation and barrier filters for FITC. The efficiency of BG 12 (colored glass filter) and SP 490 + LP 450 (interference filters) as excitation filters is compared.

ranges has enabled the use of light sources as the halogen quartz lamp, which emits mainly visible light and relatively little blue-violet light. This example shows that lamp performance must always be evaluated against the available types of excitation filters.

A *tungsten-halogen* lamp is a suitable and inexpensive light source for routine investigations (Tomlinson [20], provided that the specimen is stained with a fluorochrome with a relatively high quantum efficiency (e.g., FITC). These lamps can be used for both transmitted and incident light illumination. They are however, unsuited for irradiation with long ultraviolet and violet light, since halogen lamps do not emit sufficient light in this wavelength region (300–450 nm).

High pressure *mercury* lamps emit strong ultraviolet, violet, and green lines (Thomson and Hageage [21]). They also emit a sufficiently strong continuum in the blue region of the spectrum to enable an effective excitation of FITC. Some mercury lamps have smaller arcs than others. The arc sizes of the 50- and 100-W lamps permit optimal use of the high intrinsic brilliancy of these arcs in incident illumination since the light bundle can then be well collected by the objective. The relatively large arc of the 200-W lamp is, on the other hand, well suited to fill the entrance pupil of a substage condenser in transmitted illumination. Disadvantages of mercury lamps are their price and their limited lifetime (about 200 hr). The high pressure mercury arcs are operated on AC power supply. A 100-W high pressure mercury arc is available for operation on a stabilized DC power supply for microscope fluorometry.

High pressure *xenon* lamps emit a continuum without strong lines (Thomson and Hageage [21]). These lamps concentrate enough long blue excitation light on the preparation for an effective excitation of FITC. In

epi-illumination, the 75-W lamp with its small but brilliant arc is particularly effective. This lamp has a relatively long lifetime (400 hr). The 75-W high pressure lamp is operated on direct current, which requires a relatively expensive power supply. A high luminous flux and density in a stable arc is then obtained.

Lasers have been mostly used in experimental studies, such as those involving flow cytometry instrumentation. They offer monochromatic radiation of very high intensity (Bergquist and Nilsson [22], Wick et al. [23]). Recently a computer-controlled laser-scanning microscope fluorometer system has been described (Wilke [24]). The short lifetime and high cost of lasers limit their use in routine fluorescence microscopy.

24.6 LAMP HOUSINGS

Lamp housings should permit an easy alignment of the light source, a correct positioning of the collecting lens of the lamp housing, and a possibility to house heat-absorbing filters. To render maximal energy of the light source in the specimen plane, some lamp houses are provided with a three-lens condenser with a relatively large opening. A short illumination tube length in the exciting light path avoids energy losses. Some lamp housings have the option of changing the collecting lens when a lamp with different size or arc is used.

In fluorescence microscopy careful alignment of light source, collecting lens, condenser, and objective is essential.

If transmitted illumination with a dark-field condenser is used, a piece of transparent paper should be held on the entrance pupil of the condenser to visualize the image of the filament or arc in the center of the entrance pupil.

For alignment in epi-illumination, one of the objectives should be removed and an image of the light source projected on a piece of paper lying on the microscope stage. By centering the light source in the lamp housing, the image of the arc or filament should be brought exactly above a light spot projected by the substage condenser, which is previously aligned with the objective to be used. Thereafter the microscope stage is raised about 3 cm above its normal position. It then occupies the place of the pupil of most objectives. By adjustment of the collector lens, a sharp image of the arc or filament can now be obtained on the piece of paper. The objective to be used can then be turned to its working position and its pupil will be filled with an image of the light source (Koehler illumination). Some vertical illuminators are provided with a ground glass screen to check lamp alignment.

24.7   EXCITATION FILTERS

The three main types of excitation filters most frequently used in fluorescence microscopy are colored-glass filters, narrow-band interference filters, and wide-band interference filters (Fig. 3). Filters will be considered here of the wide-band type if their half-width (HW = the width of the main transmission band at 50% of the peak transmission) is more than 25 nm and as narrow-band filters if their half-width is less than 25 nm. According to this definition, the conventional colored-glass filters are wide-band filters.

Conventional filters such as colored-glass filters approximately transmit light of their own colour and absorb complementary colors. They have a wide transmittance curve that is bell-shaped. These filters generally do not have a steep cutoff toward the longer wavelength side. The transmittance of a barrier filter to be used in combination with such an excitation filter may only start far above the peak transmittance of the exciter filter.

With the introduction of fluorochromes such as fluorescein isothiocyanate (FITC) and tetramethyl rhodamine isothiocyanate (TRITC), which have excitation and emission spectra that are only separated by about 30 nm (Porro et al. [25], excitation filters were required that would allow the use of barrier filters with a transmittance starting much nearer to the exciter filter peak transmittance. This led to development of very specialized interference filters with a particular steep edge toward the longer wavelength side (Rygaard and Olson [26], Kraft [27], Ploem [28], Herzog et al. [29], Lea and Ward [30]). These filters were originally described as short-wave pass interference filters (Figs. 3, 4 and 5). A modern filter for

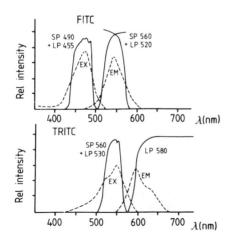

Figure 5.   Examples of filter combinations for FITC and TRITC fluorescence.

FITC excitation can have a transmittance of more than 80% at 490 nm with residual transmittance of less than 0.1% at 525 nm (the fluorescence peak of FITC). A barrier filter can then be used that already transmits maximally in this wavelength range. In most applications these especially designed interference filters for FITC excitation also meet the estimated permitted low levels of unwanted (green) excitation light for dark-field transmitted illumination with objectives of moderate numerical aperture and a halogen quartz lamp as a light source.

With objectives of higher numerical aperture (NA > 1.00) the permissible transmittance of unwanted excitation light by the excitation filter is much lower, since such objectives collect the unwanted excitation light more effectively than objectives with low NA. In epi-illumination with objectives of maximal numerical aperture (NA 1.30–1.40) and light sources of extreme brilliancy (e.g., 100-W mercury or 75-W xenon arcs), these levels should be even less. This would require highest quality interference filters or two interference filters in series in the excitation light path.

Some short-wave pass interference filters for FITC excitation are wide-band filters and may transmit a considerable amount of ultraviolet light (Fig. 4). This can excite autofluorescence of optical parts, immersion oil, and fluorescent substances in the specimen. To eliminate short-wavelength light a barrier-type filter to absorb ultraviolet, violet, and short-wavelength blue light (e.g., 1-mm GG 475) should be inserted in the excitation light path.

If unwanted excitation light is still observed in the background (epi-illumination) or as a rim around microscopic structures (dark-field illumination), it should be realized that factors other than the transmittance characteristics of the excitation filter may play a role in the problem of unwanted excitation light: (1) the luminance of the light source and the presence of strong lines in the emission spectrum of the lamp in the wavelength region transmitted by the secondary filters; (2) the number of light-scattering or reflecting structures in the preparation; (3) the numerical aperture and focal distance of the illuminating condenser or objective; (4) the level of fluorescence of the stained specimen to be detected; (5) the numerical aperture and magnification of the observing objective; (6) the use of dry or immersion objectives in epi-illumination.

An important problem in employing red fluorescent fluorochromes is the necessity to remove unwanted infrared and red exciting radiation from the excitation beam. This unwanted red excitation light must be removed with infrared and red absorbing glass filters (e.g., a Calflex infrared reflecting interference filter, a KG1 infrared absorbing glass filter, and a 4–800 nm BG 38 red absorbing filter) in the excitation light path. In addition, a filter reflecting red light above 600 nm (e.g., KP 560 or KP 555) can be inserted in the excitation light path. Xenon and halogen quartz lamps having a strong red continuum require more infrared and red filtering than the high pressure mercury arcs to remove the unwanted excitation light from the irradiating beam.

24.8 ILLUMINATING SYSTEMS

The two illuminating systems that are at present considered most effective for fluorescence microscopy are dark-field illumination with transmitted light (Young [12], Nairn [2]) and bright-field epi-illumination (Brumberg and Krylova [31], Ploem [32,47], Kraft [33], Nairn and Ploem [34]. The choice depends on several factors such as price, availability of high quality excitation filters, numerical aperture, and magnification of available objectives and eyepieces, required total magnification of the specimen, and eventual necessity to combine fluorescence microscopy with simultaneous phase or interference contrast microscopy. In general, dark-field transmitted illumination gives good results for a low total magnification of the specimen (e.g., <250×), while a high total microscope magnification (e.g., >250×) often requires epi-illumination for optimal brilliancy of the fluorescent images. Furthermore, dark-field illumination permits the use of (cheap) glass filters, whereas epi-illumination requires the use of (expensive) interference filters for optimal results. Objects that should be observed in phase contrast can be simultaneously observed with fluorescence epi-illumination.

If a simple routine microscope must be converted to a fluorescence microscope this can often be achieved at acceptable cost by adding a dark-field condenser, an effective interference filter, and a low voltage tungsten or halogen lamp as a light source. A sufficient level of fluorescence can then be obtained in many applications.

24.8.1 Dark-Field Illumination

In fluorescence microscopy with dark-field illumination, the cone of light illuminating the specimen should not enter the microscope objective directly. When no specimen is present, the light passing the condenser will not be scattered and the microscope field remains dark. The light path in a normal and Tiyoda dark-field condenser is schematically presented in Fig. 6. The numerical aperture of the objective should be limited since otherwise a part of the outer cone of the transmitted dark-field illumination would still be collected by the objective. Especially the high power objec-

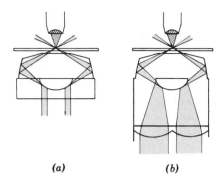

Figure 6. Light path in dark-field transmitted illumination (shaded area) with (*A*) normal dark-field condenser and (*B*) a Tiyoda dark-field condenser.

(*a*)   (*b*)

tives with maximal numerical apertures (NA > 1.00) cannot be used with their apertures fully opened.

To achieve optimal results, the dark-field condenser must be immersed to the object slide. Otherwise the very oblique cone of light rays will not enter the glass slide. The thickness of the slide is determined by the focal length of the condenser. Since one dark-field condenser has often to be used with objectives of different power, the field illuminated in the specimen must be large enough for the lower power objectives. The eventual use of a second dark-field condenser with a short focus, which would concentrate all the exciting radiation in a relatively small field, would only be useful for higher power objectives, which would themselves only visualize a small field in the specimen.

For specimens stained with relatively strong fluorescent dyes and examined at low total microscope magnifications (e.g., <200×), dark-field condensers are very satisfactory. The difficulties in dark-field illumination include careful centering and focusing of the condenser. Dark-field condensers with highest NA are of the short focus type and have to be immersed to the slide. In that case smearing of oil or glycerine and air bubbles in the immersion fluid may become a problem. Focusing is very critical for obtaining an optimal image.

24.8.2 Epi-Illumination (Incident Light Illumination)

In fluorescence microscopy with incident illumination, vertical illuminators are used to direct the exciting light onto the microscopic specimen. Bright-field incident light illumination is the illumination of choice in epi-illumination, since dark-field incident illumination only allows the use of objectives at limited numerical apertures.

Vertical illuminators are equipped with a dichromatic beam splitter placed at an angle of 45 degrees to the optical axis of the microscope. The dichromatic beam splitter is an interference mirror that reflects a high percentage of the light striking the mirror. It has a high reflectance for the excitation light below a certain wavelength while longer wavelengths are almost entirely transmitted. The reflected light is directed into the objective, which then concentrates the exciting light onto the microscopic specimen. The excitation beam is automatically focused and centered as soon as a sharp image of the specimen is obtained by the objective, and seen in the eyepieces (Ploem [47]).

In incident illumination the area of the specimen that is illuminated by the objective is exactly the same size as the area of which the fluorescence is observed. High power objectives used in incident light excitation therefore concentrate intense excitation light onto a much smaller area than low power objectives. The latter objectives, however, will spread the exitation light more thinly over a much larger area of the microscopic specimen while permitting examination of fluorescence in the same illuminated large area.

The dichromatic beam splitter in a vertical illuminator will in the first place deflect the incident excitation light onto the specimen, thus acting as a primary filter that will isolate those wavelengths from the exciting beam that are suited for the excitation of a specific fluorochrome in the specimen (Fig. 7). In the example of Fig. 7 it reflects green light. All other excitation light emitted by the light source and not wanted for exciting the fluorochrome in the microscopic specimen will for a large part be removed from the exciting light path because this light is transmitted by the

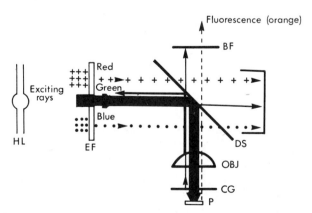

Figure 7. Light path in incident illumination. The vertical illuminator equipped with a chromatic beam splitter has a high reflectance for green excitation light and a high transmittance for orange fluorescent light. HL, light source; EF, excitation filter; DS, chromatic beam splitter; OBJ, objective; CG, cover glass; P, preparation; BF, barrier filter.

dichromatic beam splitter. The spectral characteristics of the dichromatic beam splitter in the vertical illuminator are therefore chosen in such a way that the other wavelengths are transmitted by the dichromatic beam splitter. In the example given in Fig. 7, blue and red light are transmitted by the dichromatic mirror. This unwanted light is absorbed into a dark space and thus removed from the excitation light path.

The dichromatic beam splitter also has the function of a secondary filter (barrier filter) by reflecting almost all excitation light that might be scattered by the glass surfaces of the objective lenses, the cover glass, and the microscopic specimen. An extra barrier filter has been inserted in the vertical illuminator to absorb this scattered unwanted excitation light, which due to imperfections in the spectral performance of the beam splitter, may still find its way toward the eyepieces.

Dichromatic beam splitters are designed to separate excitation and fluorescence light, which have peaks that often differ less than 30 nm in wavelength. As a consequence special dichromatic beam splitters are needed for the excitation of a specific fluorochrome. Excitation and emission spectra of fluorochromes may occur over the whole range of the spectrum. With four dichromatic beam splitters for excitation with respectively ultraviolet, violet, blue, and green light most fluorochromes can, however, be excited effectively.

Present vertical illuminators permit excitation with a larger part of the light spectrum by incorporating several interchangeable dichromatic beam splitters, each coated for the reflectance of a limited wavelength range. These vertical illuminators also make use of the short-wave pass interference filters as excitation filters. Consequently, light from a halogen or high pressure lamp, preferably with a small arc, can then be used. Advantages of epi-illumination are its brightly fluorescent images resulting in the possibility that small weakly fluorescing structures can be examined; its ability to concentrate excitation light on a relatively small spot of the specimen; the possibility that it can be easily combined with transmitted illumination, and the option to employ easily interchangeable complete filter systems.

24.9 SECONDARY (BARRIER) FILTERS

To obtain optimal brightness and image contrast, appropriate barrier filters have to be selected in combination with each excitation filter. Barrier types of filters transmit light of wavelengths longer than the cutoff value of the filter. It is inadvisable to apply a barrier filter that transmits only part of the fluorescence spectrum. In practice this means that for the observation of green (525 nm) FITC fluorescence a barrier filter with a cutoff value (50% transmittance and absorbance) at 530 nm, as has been used in combination with wide-band blue glass excitation, is undesirable. On the other hand, the cutoff value of the barrier filter should not be chosen at too short a wavelength to prevent the transmittance of un-wanted blue excitation light. Excitation and barrier filters should be care-fully matched; the effect of an expensive short-wave pass interference filter for FITC can be ruined by the wrong barrier filter.

Narrow-band filters that only transmit the peak fluorescence of a cer-tain fluorochrome can also be used as a secondary filter. The latter type of filters will be referred to as fluorescence selection filters. Interference filters without colored glasses are satisfactory for this purpose. They can be made less than 3 mm thick and may be inserted in the slides normally used to insert barrier filters. The use of fluorescence selection filters is only permitted if measures are taken to diminish the autofluorescence of tissue components to a sufficiently low level. When fluorochromes are used with an excitation peak outside the ultraviolet region, this can often be achieved with narrow-band excitation of a wavelength close to the excitation peak of the fluorochrome. Autofluorescence will then generally be low.

24.10   OBJECTIVES AND EYEPIECES

In dark-field fluorescence microscopy, the illumination is performed with the condenser and the objective only collects the fluorescence light. The numerical aperture of the condenser is of importance, since its light con-centration is proportional to the square power of its NA.

In epi-illumination the illumination of the specimen and the collection of the fluorescence light are both performed by the objective. The light collection of the specimen is also proportional to the square power of the NA of the objective. Thus the efficiency in epi-illumination depends on the fourth power of the numerical aperture of the objective. Consequently it has been worthwhile to develop special objectives for epi-illumination with moderate magnification and maximal NA. Several such objectives are now available (e.g., 22× water immersion, NA 0.65; 50× water im-mersion, NA 1.00; and 40× oil immersion, NA 1.30).

Immersion objectives are to be preferred to dry objectives, since the amount of incident light reflected on glass to air interfaces is much higher than on glass to oil, glycerine, or water interfaces.

Achromates or fluorites having fewer lenses than apochromates are more transparent and therefore well suited for fluorescence microscopy.

Since the fluorescence image is often monochromatic, the color correction of an objective is of less importance.

Autofluorescence of objectives has been a widely discussed subject. With the increased use of long-wavelength blue excitation light this problem has lost most of its significance. If a fluorochrome has to be used, which has its excitation peak in the ultraviolet region, one should realize that some highly corrected apochromates of maximal NA do not transmit much ultraviolet light.

The use of low power eyepieces (e.g., 4–6.3×) is always indicated, since the brightness per unit of area of the microscope preparation decreases with the square power of the magnification of the eyepiece.

24.11 FADING

Fading is a reduction in fluorescence intensity and is caused by photochemical reactions that cause decomposition of fluorescing molecules (Patzelt [35]).

Fading of fluorescence should generally be kept minimal. Since fading occurs more rapidly at higher intensities of excitation light than with lower intensities, the excitation light should not be more intense than necessary to visualize the fluorescent specimen with low power eyepieces. Efforts should be directed to collect as much as possible of the fluorescence light by using objectives with maximal NA and moderate magnification, and by the use of barrier filters having cutoff wavelengths as far below the fluorescence spectrum of the tracer as the excitation filter permits by eliminating unwanted excitation light.

Another cause for reduction in fluorescence intensity consists of quenching. Quenching occurs in the presence of other fluorophores, oxidizing agents, or other compounds that influence the electronic configuration of the fluorescing fluorophore. For these reasons preparations to be studied in fluorescence microscopy are stored at 4°C in the dark.

According to Gill [36] fading can be inhibited by adding dithionite to the mounting medium. Giloh and Sedat [37] propose the addition of *n*-propyl gallate to diminish fading.

24.12 AUTOFLUORESCENCE

Organic or biological material may have components that present a considerable amount of autofluorescence when excited with ultraviolet or violet light. The primary (excitation) filter should be chosen in such a way that no autofluorescence occurs, if such autofluorescence cannot be recognized by, for example, its blue color from the fluorescence of the tracer. For instance, if a barrier filter transmitting only green fluorescence is used, very little autofluorescence should be tolerated. The green wavelength range of the broad blue-green emission spectrum of the autofluorescence of tissue components will then be transmitted by the barrier filter and be indistinguishable from the green fluorescence of the tracer (e.g., FITC). If a barrier filter only absoring ultraviolet light is used, the blue part of the autofluorescence spectrum is easily distinguished from the green fluorescence of the tracer (e.g., FITC).

Excitation filters with a transmission between 460 and 490 nm only will excite so little blue-green autofluorescence in tissue sections that they can be used in combination with a yellow barrier filter, transmitting no blue, but only green light. With this barrier filter, care has to be taken that the primary filter eliminates the green (520-nm) excitation light to less than 10^{-5} of the original intensity of the wavelength.

On the other hand, autofluorescence can be used to study plant structures. Examples of such studies are given in the book by O'Brien and McCullen [38]. Audran and Willemse [39] studied wall development in pollen by means of autofluorescence and Teichmueller and Wolf [40] in coal petrology.

24.13 DOUBLE STAINING

In the past, double staining in fluorescence microscopy has been hampered by the lack of filter systems that would enable the use of the full range of excitation light from ultraviolet to green wavelengths, which is required for the efficient excitation of two fluorochromes with different excitation spectra. The development of efficient filter systems for excitation with blue (470–490 nm) and green (540–560 nm) light has stimulated the use of two fluorochromes in one preparation (e.g., Klein et al. [41], Cornelisse and Ploem [19]). The use of double staining in turn has led to the development of vertical illuminators provided with an internal turret containing four chromatic beam splitters and barrier filters and a revolver for several excitation filters. This makes the procedure of exchanging filters and beam splitters for excitation of each fluorochrome considerably more practical. For the visualization of double-stained cells, special illuminators are available, which permit an exchange of a complete combination of an excitation filter, a chromatic beam splitter, a barrier filter, and a fluorescence selection filter in one movement.

24.14   FLUORESCENCE COMBINED WITH OTHER
MICROSCOPE TECHNIQUES

Incident fluorescence microscopy can, for example, be combined with one or two substage illumination light paths for phase contrast images, absorption images or fluorescence images of double-stained preparations (Ploem [42]). In such stands, change of illumination is performed by electronically operated shutters. Incident illumination can also be realized in inverted microscopes, which are used in the visualization of cells in tissue culture tubes and immunological test trays (Jongkind et al. [43]).

A new instrument (Ploem [44], Ploem and Thaer [45], Wouters and Koerten [46]) permits the combined use of scanning electron microscopy (SEM) and light microscopy (LM). The normal SEM stage has been replaced by a special stage incorporating light microscopic optics, without significantly restricting the SEM possibilities (Figs. 8 and 9). The optical parts for LM are permanently present in the vacuum chamber of the SEM. This avoids the difficulty of alternatively applying LM and SEM to the same specimen by moving it from the LM into the vacuum chamber of the SEM. The cells or thin sections are placed on a quartz cover glass that is inclined by 45 degrees relative to the vertical electron beam. The specimen facing the electron beam is treated in a conventional way for SEM. A high power immersion oil objective with a numerical aperture of 1.4 is placed under the cover glass. A vacuum-resistant fluorescence-free oil is used. A modified medium power objective with long free working distance is mounted above the cover glass. The second objective serves as a condenser for transmitted light illumination. For undisturbed collecting

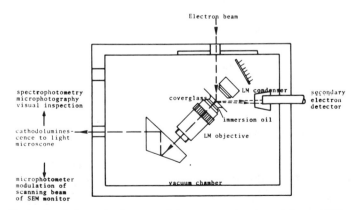

Figure 8.   Schematic diagram of the vacuum chamber of the LM/SEM.

Figure 9. Internal view of the special specimen stage, showing the LM optics: objective, condenser, and specimen holder.

efficiency of the electron and x-ray detectors this objective can be swung out. The vacuum chamber of the SEM is connected via an optical bridge with an LM permitting ordinary light microscopy as well as fluorescence microscopy. The oil immersion objective below the specimen collects the images. Cathodoluminescence of the specimen excited by the scanning electron beam is also collected by the objective below the cover glass. Cathodoluminescence can be recorded either by direct photomicrography through the LM or by modulating the scanning beam of the SEM monitor.

An LM fluorescence image can be obtained by using cytochemical staining techniques. The interaction of the excitation processes (by light and electrons) can be observed and perhaps used for studying the binding conditions of fluorochromes.

REFERENCES

1. Coons, A. H., H. J. Creech, and R. N. Jones (1941). Proc. Soc. Exp. Biol. Med. *47*, 200–202.
2. Nairn, R. C. (1976). *Fluorescent Protein Tracing,* Livingstone, Edinburgh.
3. Wick, G., K. N. Traill, and K. Schauenstein, Eds. (1982). *Immunofluorescence Technology: Selected Theoretical and Clinical Aspects,* Elsevier Biomedical Press, Amsterdam.
4. Herzenberg, L. A., R. G. Sweet, and L. A. Herzenberg (1976). Sci. Am. *234* (3), 108–117.
5. Melamed, M. R., P. F. Mullaney, and M. L. Mendelsohn, Eds. (1979). *Flow Cytometry and Sorting,* Wiley, New York.

6. Killander, D., U. Hellstrom, N. Johansson, E. Klein, A. Levin, and P. Perlmann (1973). In *Fluorescence Techniques in Cell Biology*, A. A. Thaer and M. Sernetz, Eds., Springer Verlag, Berlin, pp. 163–171.

7. Greaves, M. F. (1975). In *Immune Recognition*, A. S. Rosenthal, Ed., Academic Press, New York, 1975, pp. 3–19.

8. Seligmann, M. (1975). Brit. J. Haematol. *31*, suppl., 1–4.

9. Raff, M. C. (1976). Sci. Am. *234*(5), 30–40.

10. Gonda, M. A., R. V. Gilden, and K. C. Hsu (1979). J. Histochem. Cytochem. *27*, 1445–1454.

11. Garland, P. B., and C. H. Moore (1979). Biochem. J. *183*, 561–572.

12. Young, M. R. (1961). Quart. J. Microsc. Sci. *102*(4), 419–449.

13. Price, Z. H. (1965). Am. J. Med. Technol. *31*, 45–60.

14. Trapp, L. (1965). Acta Histochem., Suppl. VII.

15. Walter, F. (1970). Leitz-Mitt Wiss. Tech. V(2), 33–40.

16. Rost, F. W. (1972). In *Histochemistry: Theoretical and Applied*, Vol. 2, A. G. Everson Pearse, Ed., Churchill Livingstone, Edinburgh, pp. 1171–1206.

17. Ploem, J. S. (1973). In *Immunopathology of the skin: Labeled antibody studies*, E. H. Beutner, T. P. Chorzelski, S. F. Bean, and R. E. Jordon, Eds., Dowden, Hutchinson & Ross, Stroudsburg, Pa., pp. 248–270.

18. Siegel, J. I. (1982). Int. Lab. *12*, 46–51.

19. Cornelisse, C. J., and J. S. Ploem (1976). J. Histochem. Cytochem. *24*, 72–81.

20. Tomlinson, A. H. (1971). Proc. Royal Microsc. Soc. *7*, 27–37.

21. Thomson, L. A., and G. J. Hageage (1975). Appl. Microbiol. *30*, 616–624.

22. Bergquist, N. R., and P. Nilsson (1975). Ann. N.Y. Acad. Sci. *254*, 157–162.

23. Wick, G., K. Schauenstein, F. Herzog, and A. Steinbatz (1975). Ann. N.Y. Acad. Sci. *254*, 172–174.

24. Wilke, V. (1982). Laser scanning in microscopy, Proc. Royal Micr. Soc. *17*(4), 21.

25. Porro, T. J., S. P. Dadik, M. Green, and H. T. Morse (1963). Stain Technol. *38*, 37–48.

26. Rygaard, J., and W. Olson (1969). Acta Path. Microbiol. Scand. *76*, 146–148.

27. Kraft, W. (1970). Leitz-Mitt. Wiss. Techn. V(2), 41–44.

28. Ploem, J. S. (1971). Ann. N.Y. Acad. Sci. *177*, 414–429.

29. Herzog, F., B. Albini, and G. Wick (1973). J. Immunol. Meth. *3*, 211–220.

30. Lea, D. J., and D. J. Ward (1974). J. Immunol. Meth. *5*, 213–215.

31. Brumberg, E. M., and T. N. Krylova (1953). Zh, Obshch. Biol. *14*, 461–464.

32. Ploem, J. S. (1965). Acta Histochem., Suppl. VII, 339–343.

33. Kraft, W. (1973). Leitz Tech. Inform. *2*, 97–109.

34. Nairn, R. C., and J. S. Ploem (1974). Leitz-Mitt. Wiss. Tech. VI(3), 91–95.

35. Patzelt, W. (1972). Letiz-Mitt. Wiss. Techn. V(7), 226–228.

36. Gill, D. (1979). Experientia *35*, 400–401.

37. Giloh, H., and J. W. Sedat (1982). Science *217*, 1252–1255.

38. O'Brien, T. P., and M. E. McCully (1981). *The Study of Plant Structure: Principles and Selected Methods,* Termarcarphi Pty. Ltd., Melbourne.

39. Audran, J. C., and M. T. M. Willemse (1982). Protoplasma *110*, 106–111.

40. Teichmueller, M., and M. Wolf (1977). J. Microsc. *109*, 49–73.

41. Klein, G., L. Gergely, and G. Goldstein (1971). Clin. Exp. Immunol. *8*, 593–602.

42. Ploem, J. S. (1982). In *Immunofluorescence Technology.* G. Wick et al., Eds., Elsevier Biomedical Press, Amsterdam, pp. 73–94.

43. Jongkind, J. F., J. S. Ploem, H. Galjaard, and A. J. J. Reuser (1974). Histochemistry *40*, 221–230.

44. Ploem, J. S. (1981). Cytometry *2*, 121.

45. Ploem, J. S., and A. Thaer (1981). Proc. Royal Microscopical Society *16*, 253.

46. Wouters, C. H., and H. Koerten (1982). Cell Biol. Int. Rep. *6*, 955–959.

47. Ploem, J. S. (1967). Z. Wiss. Mikrosk. Mikrosk. Tech. *68*, 129–142.

INDEX